Amphetamine and Its Analogs

Amphetamine and Its Analogs
Psychopharmacology, Toxicology, and Abuse

Edited by

Arthur K. Cho
Department of Pharmacology
School of Medicine
University of California, Los Angeles
Los Angeles, California

David S. Segal
Department of Psychiatry
School of Medicine
University of California, San Diego
La Jolla, California

ACADEMIC PRESS
A Division of Harcourt Brace & Company
San Diego   New York   Boston
London   Sydney   Tokyo   Toronto

This book is printed on acid-free paper. ∞

Copyright © 1994 by ACADEMIC PRESS, INC.
All Rights Reserved.
No part of this publication may be reproduced or transmitted in any form or by any means, electronic or mechanical, including photocopy, recording, or any information storage and retrieval system, without permission in writing from the publisher.

Academic Press, Inc.
525 B Street, Suite 1900, San Diego, California 92101-4495

United Kingdom Edition published by
Academic Press Limited
24–28 Oval Road, London NW1 7DX

Library of Congress Cataloging-in-Publication Data

Amphetamine and its analogs : psychopharmacology, toxicology, and
  abuse / edited by Arthur K. Cho, David S. Segal.
      p.   cm.
    Includes bibliographical references and index.
    ISBN 0-12-173375-0
    1. Amphetamines. 2. Amphetamines--Toxicology. 3. Amphetamine
  abuse.   I. Cho, A. K. (Arhtur K.), Date.   II. Segal, David S.
    [DNLM: 1. Amphetamines--pharmacology. 2. Substance Abuse. QV
  102 A5255 1994]
  RM666.A493A463   1994
  615'785--dc20
  DNLM/DLC
  for Library of Congress   93-38062
                                                                  CIP

PRINTED IN THE UNITED STATES OF AMERICA
94 95 96 97 98 99   BB   9 8 7 6 5 4 3 2 1

Contents

Contributors xv
Preface xix

I BIOCHEMISTRY AND CHEMISTRY OF AMPHETAMINES

1 Medicinal Chemistry and Structure–Activity Relationships
David E. Nichols

I. Introduction 3
II. Derivatives That Release Neuronal Catecholamines 4
    A. N Substitution 6
    B. Analogs Oxidized at the Benzylic Carbon 6

C. Modification of the Side Chain: Rigid Analogs   7
        D. Mechanism of Action   9
        E. Summary of Structure–Activity Relationships   9
  III. Amphetamine Derivatives That Release Neuronal Serotonin   9
        A. Effects of N Substitution   14
        B. Side Chain Modifications   15
        C. Aromatic Ring Substituents   17
        D. Mechanism of Action   21
        E. Summary of Structure–Activity Relationships   21
  IV. Hallucinogenic Amphetamine Derivatives   22
        A. N Substition   23
        B. Aromatic Ring Substitution   24
        C. Side Chain Modifications   29
        D. Mechanism of Action   33
        E. Summary of Structure–Activity Relationships   33
     References   34

2 Metabolism of Amphetamine and Other Arylisopropylamines
Arthur K. Cho and Yoshito Kumagai

  I. General Overview   43
        A. Enzymology   44
        B. Reactions   45
  II. Specific Compounds   49
        A. Amphetamine   49
        B. Methamphetamine   53
        C. Phentermine   56
        D. Ephedrine   57
        E. Methylenedioxyamphetamines   59
        F. *p*-Chloroamphetamine   62
        G. Fenfluramine   63
        H. Deprenyl   65
        I. Cathinone   67
  III. Conclusions   68
     References   70

II NEUROPSYCHOPHARMACOLOGY

3 Neurochemistry of Amphetamine
Ronald Kuczenski and David S. Segal

  I. Introduction   81
  II. Mechanisms of Interaction of Amphetamine with Catecholamine Neurons   82

             A. Uptake Carrier   83
             B. Vesicle   86
             C. Monoamine Oxidase   87
    III. Postmortem Studies of Synthesis and Metabolism   88
             A. Dopamine Synthesis and Turnover   88
             B. Dopamine Metabolites   89
    IV. Extracellular Fluid Sampling in Awake Animals   90
             A. Extracellular Dopamine   91
             B. Extracellular 3-Methoxytryamine   93
             C. Extracellular Dopamine Acid Metabolites   93
             D. Exchange–Diffusion Model   94
    V. Regional Characteristics of the Extracellular Dopamine Response   97
    VI. Effects of Amphetamine on Extracellular Concentrations of Other Neurotransmitters   99
             A. Norepinephrine   100
             B. Serotonin   101
             C. Acetylcholine   101
             D. Ascorbic Acid   102
    VII. Conclusions and Speculations   103
          References   104

4 Behavioral Pharmacology of Amphetamine
David S. Segal and Ronald Kuczenski

    I. Introduction   115
    II. Factors That Influence the Emergent Behavioral Profiles   116
             A. Dosage Parameters   117
             B. Situational and Experiential Factors   121
             C. Individual Differences   122
    III. Amphetamine-Induced Behavior–Neurotransmitter Relationships   127
             A. Dopamine   127
             B. Other Neurotransmitter Systems   137
    IV. Conclusions and Speculations   140
         References   141

5 Neurochemistry of Ring-Substituted Amphetamine Analogs
Christopher J. Schmidt

    I. Introduction   151
    II. Methoxylated Amphetamines (Hallucinogens)   152
    III. Methylenedioxyamphetamines   153
             A. *In Vitro* Neurochemistry   154

B. *In Vivo* Neurochemistry   158
C. Electrophysiology   164
IV. Ring-Alkylated Amphetamines (Fenfluramine)   166
A. *In Vitro* Neurochemistry   166
B. *In Vivo* Neurochemistry   167
V. Future Directions   170
References   171

6 Behavioral Pharmacology of Ring-Substituted Amphetamine Analogs
Mark A. Geyer and Clifton W. Calloway

I. Introduction   177
II. Behavioral Effects after Acute Administration   179
  A. Effects in Unconditioned Behavior Paradigms   179
  B. Effects of Substituted Amphetamines on Startle Responding   191
  C. Effects of Substituted Amphetamines on Antinociception   195
III. Behavioral Effects after Repeated Administrations   195
  A. Effects in Conditioned Behavior Paradigms   195
IV. Conclusions and Future Directions   197
References   201

7 Neurochemistry of Halogenated Amphetamines
Ray W. Fuller and Mark G. Henderson

I. Introduction and Background   209
II. Acute Neurochemical Effects of p-Chloroamphetamine   210
III. Long-Term Depletion of Brain Serotonin by p-Chloroamphetamine   210
  A. Characteristics of Long-Term Effects   212
  B. Use of p-Chloroamphetamine for Chemical Lesioning of Serotonin Pathways   213
  C. Comparison of p-Chloroamphetamine to Hydroxylated Indoleamine Neurotoxins   214
IV. Structure–Activity Relationships   217
  A. Ring-Substituted Amphetamines   218
  B. Side-Chain-Substituted Amphetamines   219
V. Are p-Chloroamphetamine and Its Analogs Substrates for the Serotonin Transporter?   221
VI. Is a Metabolite of p-Chloroamphetamine Involved in Its Neurochemical Effects?   221

VII. Possible Mechanisms in Acute and Long-Term Serotonin Depletion 222
   A. Prevention of *p*-Chloroamphetamine Effects by Uptake Inhibitors 222
   B. Other Protective Measures 225
   C. Role for Dopamine in *p*-Chloroamphetamine Neurotoxicity 229
VIII. Functional Effects Associated with Long-Term Neurochemical Deficits Induced by *p*-Chloroamphetamine 229
IX. Summary and Conclusions 230
References 232

8 Mechanisms of Abuse
Alison McGregor and David C. S. Roberts

I. Introduction 243
II. Dopamine 244
   A. Dopamine Antagonists and Amphetamine Self-Administration 244
   B. Dopamine Projection Areas and Amphetamine Self-Administration 246
   C. Microdialysis and Amphetamine Self-Administration 249
   D. Conditioned Place Preference 250
III. Serotonin 252
IV. Noradrenaline 254
V. Discussion 255
   A. Drug-Induced Perseveration 255
   B. Multiple Paradigms to Measure Reinforcement 258
   C. Neural Mechanisms of Reinforcement 259
   D. Mechanisms of Abuse 260
References 261

III TOXICOLOGY

9 Neurochemical Mechanisms of Toxicity
James W. Gibb, Glen R. Hanson, and Michel Johnson

I. History of Neurotoxicity by Amphetamine and Its Analogs 269
II. Response of Dopaminergic System to Methamphetamine 270
III. Serotonergic Response to Methamphetamine 270
IV. Neurochemical Alterations Persist for Extended Periods of Time 272

    V. Does Methamphetamine Cause Generalized Neurotoxicity? 273
   VI. Neurochemical Alterations by Methylenedioxy Congeners of
       Amphetamine 274
  VII. Role of Dopamine in Methamphetamine-Induced Chemical
       Changes in Dopaminergic Systems 275
 VIII. Role of Dopamine in Methamphetamine-Induced Alterations in the
       Serotonergic System 277
   IX. Role of Toxic Metabolites in Neurotoxicity 279
    X. Role of Oxidative Processes in the Neurotoxic Response 280
   XI. Role of Glutaminergic and GABA-ergic Systems in Neurotoxicity
       Induced by Amphetamine Congeners 283
       A. Effect of Amphetamine and Methamphetamine on
          Glutamate Release 286
       B. Role of γ-Aminobutyric Acid in Amphetamine
          Toxicity 287
  XII. Development of Tolerance to Methamphetamine-Induced
       Toxicity 287
 XIII. Conclusion 288
       References 289

10 Functional Consequences of Neurotoxic Amphetamine Exposure
George A. Ricaurte, K. E. Sabol, and Lewis Seiden

    I. Introduction 297
   II. Functional Role of Brain Dopamine and Serotonin 298
       A. Dopamine 298
       B. Serotonin 299
  III. Neurotoxic Amphetamine Exposure in Animals 299
       A. Positive Findings 300
       B. Negative Findings 304
   IV. Considerations 307
    V. Summary and Conclusions 308
       References 309

11 Structural Features of Amphetamine Neurotoxicity in the Brain
Karen J. Axt, Laura A. Mamounas, and Mark E. Molliver

    I. Introduction 315
       A. Neurotoxicity: Level of Analysis 316
       B. Neuroanatomic Methods: Parameters of Neurotoxicity 317
   II. Basic Anatomy of Ascending Serotonergic and Dopaminergic
       Projections 318

        A. Serotonin   318
        B. Dopamine   321
III. Toxic Effects of Amphetamine Derivatives: Evidence for Persistent Axon Loss   323
        A. Regional Specificity of Axon Loss   324
        B. Loss of Monoamine Uptake Sites   335
        C. Loss of Axonal Transport   337
IV. Morphological Evidence of Axonal Damage and Degeneration   340
        A. Evidence of Structural Changes in Axons   341
        B. Silver Impregnation Studies   344
        C. Reactions of Glial Cells   346
V. Reinnervation: Serotonergic Projections   348
VI. Summary   352
        A. Phases of Neurotoxicity   352
        B. Future Directions   357
References   358

IV USE AND ABUSE

12 Use and Abuse of Ring-Substituted Amphetamines
Una D. McCann and George A. Ricaurte

I. Introduction   371
II. Methylenedioxy Methamphetamine   372
        A. Behavioral Effects   372
        B. Biochemical Effects   373
        C. Use Patterns   374
        D. Adverse Consequences   375
III. Other Ring-Substituted Amphetamines Currently Used Clinically or Recreationally   377
        A. 2,5-Dimethoxy-4-methylamphetamine   377
        B. 2,5-Dimethoxy-4-bromoamphetamine   377
        C. Mescaline   378
        D. 3,4-Methylenedioxyamphetamine   378
        E. 3,4-Methylenedioxy-N-ethylamphetamine   379
        F. N-Methyl-1-(1,3-benzodixol-5-yl)-2-butanamine   379
        G. p-Methoxymethamphetamine   379
        H. p-Methoxyamphetamine   379
        I. p-Chloroamphetamine   380
        J. Fenfluramine   380
IV. Summary and Implications   380
References   381

13 Amphetamine Psychosis: Clinical Variations of the Syndrome
Burt Angrist

I. Introduction   387
II. Data Sources   388
III. Individual Variations in Response to Central Nervous System Stimulants   389
IV. Possible Substrates for Variable Sensitivity to Central Nervous Systems Stimulants   390
    A. Genetic Variables   390
    B. Tolerance   391
    C. Sensitization   392
    D. Premorbid Psychiatric Status   395
V. Clinical Variations and Unresolved Issues in Amphetamine Psychosis   396
    A. Why Was Amphetamine Psychosis Originally Considered a Rare Condition and Later Seen Frequently?   396
    B. Is Amphetamine Psychosis an Idiosyncratic Reaction to the Drug or a Manifestation of Latent Psychosis?   397
    C. Is a Minimum Dose Required for Amphetamine Psychosis?   397
    D. Is a Minimal Duration of Use Required for Amphetamine Psychosis?   398
    E. Is the Symptomatology of Amphetamine Psychosis Similar or Different in Relatively Naive and Chronic Abusers?   399
    F. Does Amphetamine Abuse Lead to Spontaneous or Stress-Induced Psychosis?   400
    G. Can Amphetamine Cause Persistent Psychosis?   402
    H. Does Pre-existing Schizophrenia Affect the Quality of Symptoms That Occur after Amphetamine Abuse?   404
    I. Varations and Range of Acute Clinical Symptoms of Amphetamine Psychosis   405
    J. What Does Amphetamine Psychosis "Model"?   409
VI. Summary   410
    References   411

14 Use and Abuse of Amphetamines in Japan
Kyohei Konuma

I. Introduction   415
II. Historical Background   416
    A. Advent of Amphetamines   416
    B. Amphetamine Use in the First Epidemic   416

C. Development of Methamphetamine Psychosis   417
D. Use of Amphetamines after the First Epidemic   420
E. Spread of Methamphetamine Use during the Second Epidemic   422
F. Patterns of Methamphetamine Use during the Second Epidemic   422
III. Types of Methamphetamine-Induced Mental Disorders   425
A. Methamphetamine Acute Intoxication   425
B. Methamphetamine Dependence Syndrome   426
C. Methamphetamine Psychosis   428
D. Sequelae of Methamphetamine-Induced Mental Disorders   430
IV. Differences between Japan and the West in the Conception of Methamphetamine Psychosis   433
V. Conclusion   434
References   434

V EPIDEMIOLOGY

15 Epidemiology of Amphetamine Use in the United States
Marissa A. Miller and Arthur L. Hughes

I. Introduction   439
A. Epidemiologic Triangle: Agent, Host, Environment   439
B. Factors Influencing Amphetamine Epidemics   440
II. History of Epidemic Use   441
A. Drug Development and Early Medical Use   441
B. Inoculation of a Large Population   441
C. Amphetamine Use and Abuse during and after World War II   442
D. Escalation of Amphetamine Abuse during the 1960s and Development of a Drug Subculture   443
E. Oversupply of Amphetamines   443
III. Current Patterns and Trends of Abuse   445
A. Introduction and Overview   445
B. National Household Survey on Drug Abuse   446
C. Drug Abuse Warning Network   447
D. Monitoring the Future Study   450
E. Ethnographic and Field Study Data   452
IV. Conclusion   454
References   454

16 Epidemiology of Amphetamine Abuse in Japan and Its Social Implications
Susumu Fukui, Kiyoshi Wada, and Masaomi Iyo

I. Introduction   459
II. Period Prior to 1945   460
III. First Period of Methamphetamine Abuse: 1945–1957   461
    A. Pre-epidemic Stage: 1945–1951   461
    B. Epidemic Stage: 1951–1954   462
    C. Ebbing Stage: 1954–1957   463
IV. Latent Period: 1957–1969   463
V. Second Period of Methamphetamine Abuse: 1970–Present   464
    A. History   464
    B. Social Background   467
    C. Methamphetamine: The Constant Supply to Society   467
    D. Demographic Features of Methamphetamine Abusers   468
    E. Long-Term Abuse and Methamphetamine Psychosis   474
VI. Current Problems in Methamphetamine Abuse   476
    A. Social Environment   476
    B. Methamphetamine as a Source of Income for Organized Gangs   476
    C. Demographics of Methamphetamine Abusers   477
VII. Conclusion   477
References   478

17 Prospects in Amphetamine Research
Charles R. Schuster and Christine R. Hartel

I. Multiple Roles of Amphetamine in Neuropharmacology   480
II. Future of Amphetamine Research   482
    A. Individual Differences in Response to Amphetamine   482
    B. Effects on Monoamines, Hormones, and the Immune System   485
    C. Amphetamine and Age-Related Changes in Brain   486
    D. Neuroscience and Behavior   486
III. Summary   488
References   489

Index   493

Contributors

Numbers in parentheses indicate the pages on which the authors' contributions begin.

Burt Angrist (387), Department of Psychiatry, New York University School of Medicine, and Psychiatry Service, New York Veterans Affairs Medical Center, New York, New York 10010

Karen J. Axt (315), Department of Neuroscience, School of Medicine, The Johns Hopkins University, Baltimore, Maryland 21206

Clifton W. Calloway (177), Department of Psychiatry, School of Medicine, University of California, San Diego, La Jolla, California 92093

Arthur K. Cho (43), Department of Pharmacology, School of Medicine, University of California, Los Angeles, Los Angeles, California 90024

Susumu Fukui (459), Division of Drug Dependence and Psychotropic Drug Clinical Research, National Institute of Mental Health, National Center of Neurology and Psychiatry, Chiba 272, Japan

Ray W. Fuller (209), Lilly Research Laboratories, Eli Lilly and Company, Indianapolis, Indiana 46285

Mark A. Geyer (177), Department of Psychiatry, School of Medicine, University of California, San Diego, La Jolla, California 92093

James W. Gibb (269), Department of Pharmacology and Toxicology, University of Utah, Salt Lake City, Utah 84112

Glen R. Hanson (269), Department of Pharmacology and Toxicology, University of Utah, Salt Lake City, Utah 84112

Christine R. Hartel (479), Division of Basic Research, National Institute on Drug Abuse, National Institutes of Health, Rockville, Maryland 20857

Mark G. Henderson (209), Merck & Co., Inc., Charlotte, North Carolina 28209

Arthur L. Hughes (439), Division of Epidemiology and Prevention Research, National Institute on Drug Abuse, Rockville, Maryland 20857

Masaomi Iyo (459), Division of Drug Dependence and Psychotropic Drug Clinical Research, National Institute of Mental Health, National Center of Neurology and Psychiatry, Chiba 272, Japan

Michel Johnson (269), Department of Pharmacology and Toxicology, University of Utah, Salt Lake City, Utah 84112

Kyohei Konuma (415), National Psychiatric Institute of Shimohusa, Chiba-Ken 266, Japan

Ronald Kuczenski (81, 115), Department of Psychiatry, School of Medicine, University of California, San Diego, La Jolla, California 92093

Yoshito Kumagai (43), Research Team for Health Effects of Air Pollutants, The National Institute for Environmental Sciences, Tsukuba, Japan

Laura A. Mamounas (315), Laboratory of Biochemical Genetics, NIMH Neuroscience Center at St. Elizabeth's, Washington D.C. 20032

Una D. McCann (371), Section of Anxiety and Affective Disorders, Biological Psychiatry Branch, National Institutes of Mental Health, National Institutes of Health, Bethesda, Maryland 20892

Alison McGregor (243), Department of Psychology, Carleton University, Ottawa, Canada K1S 5B6

Marissa A. Miller (439), National Vaccine Program Office, Washington, D.C. 20201

Mark E. Molliver (315), Department of Neuroscience and Neurology, The Johns Hopkins University, School of Medicine, Baltimore, Maryland 21205

David E. Nichols (3), Department of Medicinal Chemsitry and Pharmacognosy, School of Pharmacy and Pharmacal Sciences, Purdue University, West Lafayette, Indiana 47907

George A. Ricaurte (297, 371), Department of Neurology, The Johns Hopkins University School of Medicine, Francis Scott Key Medical Center, Baltimore, Maryland 21224

David C. S. Roberts (243), Department of Psychology, Carleton University, Ottawa, Canada K1S 5B6

K. E. Sabol (297), Department of Pharmacology and Physiological Sciences, University of Chicago, Chicago, Illinois 60637

Christopher J. Schmidt (151), CNS Research, Marion Merrell Dow Research Institute, Cincinnati, Ohio 45212

Charles R. Schuster (479), Addiction Research Center, National Institute of Drug Abuse, Baltimore, Maryland 21224

David S. Segal (81, 115), Department of Psychiatry, School of Medicine, University of California, San Diego, La Jolla, California 92093

Lewis Seiden (297), Department of Pharmacology and Physiological Sciences, University of Chicago, Chicago, Illinois 60637

Kiyoshi Wada (459), Division of Drug Dependence and Psychotropic Drug Clinical Research, National Institute of Mental Health, National Center of Neurology and Psychiatry, Chiba 272, Japan

Preface

Amphetamine, phenylisopropylamine, contains the same basic structure as a number of centrally active compounds that range in activity from general stimulants to hallucinogens. As a group, the amphetamines have a checkered history. Amphetamine has been used therapeutically in the treatment of catalepsy, attention deficit disorders, and obesity. However, its central stimulant actions have led to extensive abuse with attendent sociological and psychiatric problems. Other members of this class of centrally active compounds have also been used in recreational settings and are classified as controlled substances.

Amphetamine itself has been abused in the United States since the 1930s (Spotts and Spotts, 1980) when it was introduced as an over-the-counter nasal decongestant, formulated as an inhalant (Griffith, 1977). The stimulant properties of the compound were quickly recognized and, in some instances of abuse, the drug was extracted from its formulation and ingested directly. The problem was further exacerbated when the drug, primarily in tablet form, was used widely to suppress appetite and to enhance performance at repetitive and tedious tasks. In this century, several epidemics of amphetamine abuse have occurred in the United States, Sweden, and Japan (Blum, 1984). Of particular concern recently is the availability of a pure preparation of (+)-methamphetamine hydrochloride, called "ice," which constitutes a major drug problem in Hawaii (Cho, 1990). Ring-substituted amphetamines have been abused for their hallucinogenic properties. These compounds are related to the peyote alkaloid mescaline (3,4,5-trimethoxyphenylethylamine); several derivatives achieved notoriety in the United States during the 1960s as hallucinogens. In the 1980s and 1990s, methylenedioxymethamphetamine (MDMA) has been widely abused in the United States (Peroutka, 1987) and in the United Kingdom (Henry et al., 1992).

The synthesis of N-methylamphetamine (methamphetamine) occurs in clandestine laboratories most commonly located in the western United States (Beaty, 1989). In 1987, the California State Attorney General's Office estimated that 25% of all methamphetamine laboratories seized in 1985 were in San Diego County; 67% of all laboratories were found in California (Los Angeles Times, 1987). The synthesis of this and other phenylisopropylamines is a relatively simple chemical task. In part for this reason, a family of central nervous system (CNS) agents with varied actions of abuse potential is available as the product of clandestine synthesis. More recently, "designer drugs" have been synthesized that retain the pharmacological activity of many of these substances in a novel chemical entity that is not included in the Controlled Substance Act. Since the act identifies specific compounds, substitution of a methyl or an ethyl for a hydrogen (e.g., MDMA and MDEA, the N-methyl and N-ethyl analogs of methylenedioxyamphetamine, respectively) results in structures not covered by the Controlled Substance Act. Similar circumstances exist for the narcotic analgesics including, for example, fentanyl and meperidine analogs (U.S. Department of Justice, 1986). These activities have resulted in new legislation in the form of the Analog Drug Act, which controls not only existing compounds but also derivatives likely to exhibit similar pharmacology.

Much of the research involving amphetamine and its derivatives has been directed at elucidating the mechanisms underlying their abuse liability. Interest in these substances is further motivated by the observation that they appear to induce or precipitate various forms of psychopathology; therefore, additional understanding of their mechanisms of actions might provide insight into the pathophysiology of these disorders. As these investigations continue, periodic reviews are desirable to summarize and reevaluate the past and to consider possible changes in the focus of new research. In this volume we have assembled reviews by established investigators addressing different aspects of the chemistry, pharmacology, behavior, and toxicology of this large group of compounds with the hope that the diverse perspectives and varied information will facilitate progress in these and related areas of study.

Arthur K. Cho
David S. Segal

REFERENCES

Baum, R. M. (1985). New variety of street drugs poses growing problem. *Chem. Eng. News*, Sept. 9, 7–16.
Beaty, J. (1989). *Time*, 24 April, 1013.
Blum, K. (1984). "Handbook of Abusable Drugs." Gardner, New York.
Cho, A. K. (1990). ICE: A new dosage form of an old drug. *Science* **249**, 631–634.
Griffith, J. D. (1977). *In* "Drug Addiction II: Amphetamine, Psychotogen and Marihuana Dependence" (W. R. Martin, ed.), pp. 277–296. Springer-Verlag, Berlin.
Henry, J. A., Jeffreys, K. J., and Dawling, S. (1992). Toxicity and deaths from 3,4-methylenedioxymethamphetamine ("ecstasy"). *Lancet* **340**, 384–387.
Los Angeles Times (1987). April 22, 1987, Part II, p. 6.
Peroutka, S. J. (1987). Incidence of recreational abuse of 3,4-methylenedioxymethamphetamine (MDMA, "Ecstasy") on an undergraduate campus. *N. Engl. J. Med.* **317**, 1542–1543.
Spotts, J. V., and Spotts, C. A. (1980). "Use and Abuse of Amphetamine and Its Substitutes." National Institute on Drug Abuse Research Issues Number 25. U.S. Department of Health, Education, and Welfare, Washington, D.C.
U.S. Department of Justice, Drug Enforcement Administration, Office of Diversion Control (1986). "Controlled Substances Analogs." U.S. Government Printing Office, Washington, D.C.

I
BIOCHEMISTRY AND CHEMISTRY OF AMPHETAMINES

David E. Nichols

1
Medicinal Chemistry and Structure–Activity Relationships

I. INTRODUCTION

> Amphetamine remains the chemist's "Cinderella" molecule. No other compound has displayed such a plethora of pharmacological, biochemical, and physiological effects, . . . nor have many other molecules served as so versatile a starting base for the synthetic elaboration of a host of novel therapeutic agents.
>
> *(Biel and Bopp, 1978)*

This description of amphetamine remains as true today as it was in 1978. As a simple molecule of only nine carbon atoms with one chiral center, amphetamine is dwarfed by the structural complexity of most other classes of drugs. Nevertheless, the pharmacophoric template, an aromatic ring separated by two carbon atoms from a basic nitrogen, is a motif that occurs repeatedly in many classes of biologically active molecules. Perhaps a remarkable characterisitic of amphetamine is that relatively minor structural changes can alter the pharmacology of the molecule completely. Thus, for example, simple changes in the aromatic substitution pattern lead to compounds that may have potent actions on presynaptic neuronal amine carriers and stores of dopamine, norepinephrine, or serotonin. Other substitution patterns give potent agonist actions at postsynaptic serotonin receptor subtypes. Central actions of substituted amphetamines in humans range from sedation to stimulation to hallucinogenic effects.

The reader should be aware, however, that in this chapter no attempt is being made to provide an encyclopedic compendium of all possible substituted amphetamine derivatives, with a companion commentary on their particular pharmacologies. Such an endeavor would occupy at least one large volume itself, and would not be a manageable chapter in this book. Further, such compendia fail to "distill" the essence of structure–activity relationships, leaving the reader full of facts but lacking a coherent framework within which to understand how molecular structure influences biological activity. Therefore, after reading this chapter and examining the structure of a substituted amphetamine, the author hopes that the reader will be able to predict with reasonable certainty the type of pharmacology that molecule would likely possess. The author has attempted to provide a representative and reasonably comprehensive set of examples, with sufficient discussion to categorize these compounds in the context of structure–activity relationships.

In general, one may recognize that substituted amphetamines interact with receptors and biological targets that have evolved to accommodate the monoamine neurotransmitters norepinephrine, epinephrine, dopamine, and serotonin. Since all these chemical molecules ultimately are derived from the aromatic amino acids tyrosine or tryptophan which, after further biosynthetic elaboration, lead to molecules containing an aromatic system spaced two carbons away from a basic amine, this action of substituted amphetamines is not surprising.

One can gain an appreciation for the pharmacology of substituted amphetamines by understanding the normal physiology of the neurotransmitter systems and receptors for the monoamine transmitters. With a knowledge of how a particular amphetamine derivative affects those processes, the pharmacology of that molecule can be understood. For example, the pressor effects of amphetamine can be largely understood from the knowledge that amphetamine causes the release of norepinephrine from peripheral sympathetic neurons, leading to increased vascular tone.

II. DERIVATIVES THAT RELEASE NEURONAL CATECHOLAMINES

Amphetamine (**1**) is an acronym for alpha-**m**ethyl**ph**en**et**hyl**amine**, also known as 1-phenyl-2-aminopropane or phenylisopropylamine. The homolog lacking the α-methyl group, 2-phenethylamine, lacks central effects after systemic administration because of its rapid degradation by monoamine oxidase (MAO). However, the simple addition of a methyl group to the α carbon yields amphetamine, which is a rather poor substrate for MAO and thus survives in the body and penetrates the central nervous system (CNS) to exert its effects. The presence of the methyl group also introduces chirality into the side chain; the isomer with the S-(+) configuration, as shown, is now well known to be more potent. A 4- to 10-fold stereoselectivity typically is observed for the enantiomers, depending on the assay used.

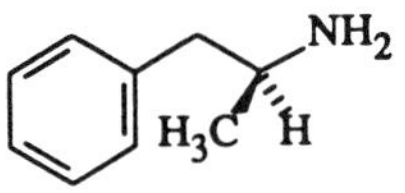

1

Amphetamine itself might be called the prototype psychostimulant. The more prominent actions of this molecule include CNS stimulation, production of euphoria, increased motor activity, and appetite suppression. These effects all are believed to be the result of the release of central stores of the endogenous catecholamines dopamine and norepinephrine. The focus of most recent studies of amphetamine has been on its dopaminergic effects; the role of norepinephrine is less well understood.

Perhaps also unique about amphetamine is the surprisingly small number of structural modifications that can be made to the structure to lead to compounds that retain psychostimulant activity. For example, Higgs and Glennon (1990) compared the three isomeric ring methyl-substituted amphetamines **2, 3,** and **4** in the two-lever drug discrimination paradigm using rats trained to discriminate 1 mg/kg (+)-amphetamine sulfate from saline. Whereas the ED_{50} for substitution with amphetamine itself was 0.42 mg/kg, only the *ortho*-methyl compound gave full substitution with an ED_{50} of 4.1 mg/kg, approximately one order of magnitude less potent than amphetamine itself. The *meta* and *para* substituted compounds **3** and **4** produced disruption only at higher doses.

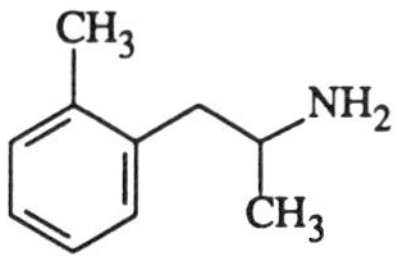

2 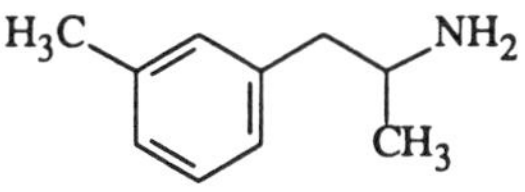   **3** 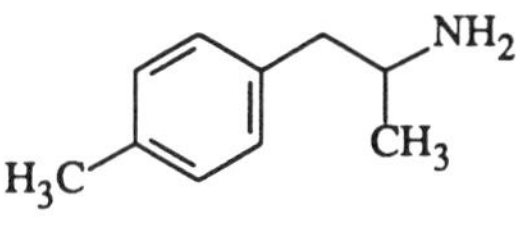   **4**

Indeed, simple monosubstitution of amphetamine at the *para* position with a halogen such as chlorine or iodine gives compounds that seem to possess, as their predominant pharmacology, the ability to release neuronal serotonin rather than dopamine (Johnson *et al.*, 1990a; Fuller, 1992), although this activity does not preclude many of these compounds from having significant effects on catecholamine systems. Indeed, p-chloroamphetamine, p-methoxyamphetamine, 3,4-methylenedioxyamphetamine, and 3,4-methylenedioxymethamphetamine all have the ability to induce the release of neuronal catecholamines. Although all these compounds possess significant potency in this respect, they are discussed in the section on serotonin-releasing compounds since this activity appears to be a more prominent feature of their mechanisms of action.

Apparently, then, no group or moiety can be placed on the aromatic ring and allow retention of the simple catecholamine-releasing activity of amphetamine. Structures substituted at the 4 or 3,4 positions that do retain an amphetamine-like psychostimulant component of action typically also seem to possess significant potency as serotonin-releasing agents, complicating interpretation of their pharmacology. However, a number of side-chain modifications of amphetamine have been studied; several have psychostimulant properties similar to those of amphetamine, although most are generally less potent.

A. N Substitution

The types of substituents that can be added to the amino group of amphetamine are relatively restricted. N-Methyl substitution of amphetamine gives methamphetamine, which has nearly twice the *in vivo* potency of amphetamine. However, the N-ethyl and N-n-propyl derivatives have about half the activity of amphetamine (Van der Schoot *et al.*, 1961). In drug discrimination tests in rats, both N-ethyl and N-hydroxy amphetamine gave complete substitution in (+)-amphetamine-trained rats (Glennon *et al.*, 1988). The N,N-dimethyl compound has only about one-fifth the potency of amphetamine. Note that tertiary amine congeners may be N-dealkylated *in vivo* to generate secondary amines, which are inherently more potent. Except for these few modifications, catecholamine-releasing activity seems to be lost. This result might be anticipated since the monoamine transport systems have evolved to accommodate either a primary or an N-methyl-substituted amine.

B. Analogs Oxidized at the Benzylic Carbon

Analogs oxidized at the benzylic position also retain activity. For example, cathinone (5) has stimulant effects similar to those of amphetamine. Using the drug discrimination paradigm in rats trained to discriminate amphetamine from saline, racemic cathinone was essentially equipotent to (+)-amphetamine (Glennon, 1986). Schechter *et al.* (1984) trained rats to discriminate 0.6 mg/kg racemic cathinone from saline. In these rats, (+)-amphetamine gave full substitution with a dose–response curve parallel to that of cathinone, the training drug. The ED_{50} values for racemic cathinone and (+)-amphetamine were identical. Glennon *et al.* (1984a) established that, in (+)-amphetamine-trained rats, S-(−)-cathinone has about twice the potency of the racemic mixture. In cathinone-trained rats, S-(−)-cathinone is equipotent to the racemate, whereas the R-(+) enantiomer is only about one-third as potent. Thus, the more potent enantiomers of both amphetamine and cathinone possess the S configuration at the α side-chain carbon. N,N-Diethyl substitution on the nitrogen gives diethylpropion, a related compound with similar but quantitatively lower (about half) activity.

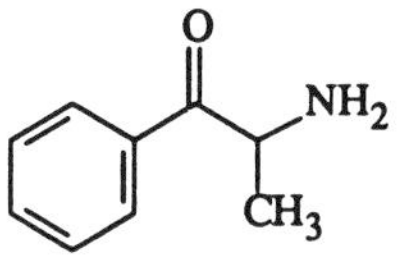

5

Reduction of the benzylic ketone of cathinone and incorporation of the oxygen and nitrogen into a morpholine ring gives compounds such as phenmetrazine and phendimetrazine. Although these molecules are also potent CNS stimulants, strictly speaking they are not amphetamines. Such structures will not be addressed in this discussion. A general overview of psychostimulants and anorectics related to amphetamine can be consulted for a comprehensive discussion (e.g., see Biel and Bopp, 1978).

C. Modification of the Side Chain: Rigid Analogs

Extension of the side-chain α-methyl to an ethyl group dramatically attenuates amphetamine-like activity. For example, the (+)-α-ethyl homolog **6** does not produce full substitution in rats trained to discriminate saline from 1 mg/kg (+)-amphetamine sulfate. Using the same paradigm, the racemic α-ethyl homolog of N-methylamphetamine (**7**) produced full substitution but had only about one-tenth the potency of amphetamine itself.

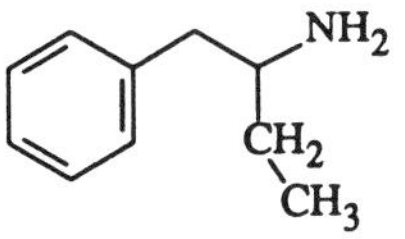

6 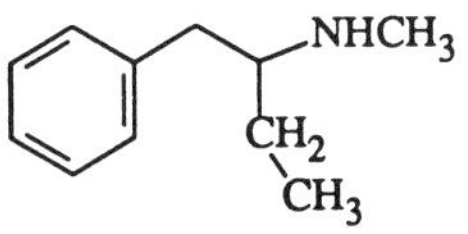   **7**

Interesting effects occur when the side chain of amphetamine is incorporated into carbocyclic rings. For example, 2-aminoindan (**8**) can be viewed as an amphetamine analog in which the α-methyl has been "attached" to the aromatic ring. Using the drug discrimination paradigm in rats trained to discriminate 1.0 mg/kg (+)-amphetamine sulfate, **8** fully substituted with an ED_{50} of 2.1 mg/kg whereas, for comparison, (+)-amphetamine sulfate had an ED_{50} of 0.42 mg/kg (Glennon et al., 1984a). Using a different training schedule but the same training drug and dose, a more recent report failed to observe substitution with **8** (Oberlender and Nichols, 1991). In this latter report, **8** gave a maximum of 75% amphetamine-appropriate responding at nearly three times the training drug dose.

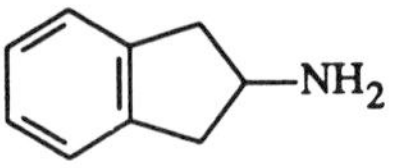

8 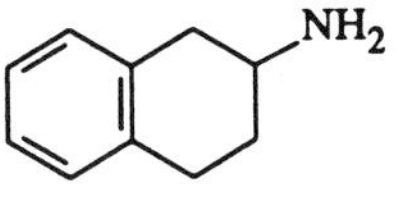   **9**

Expansion of the carbocyclic ring leads to 2-aminotetralin (**9**). Glennon et al. (1984a) also report complete substitution with this analog in (+)-amphetamine sulfate-trained animals. An ED_{50} of 1.2 mg/kg places the tetralin at about one-third the potency of (+)-amphetamine. Similarly, Oberlender and Nichols (1991) observed complete substitution for this compound with an ED_{50} of 11.9 μmol/kg, whereas the ED_{50} for (+)-amphetamine sulfate was 1.57 μmol/kg (0.29 mg/kg). Thus, **9** had about one-eighth the potency of amphetamine in this study.

Although the two groups did not report consistent results for 2-aminoindan (**8**), both studies were in agreement that 2-aminotetralin (**9**) was more potent than 2-aminoindan (**8**) as an amphetamine-like agent. Further expansion of the carbocyclic ring to seven carbon atoms resulted in a compound that no longer gave (+)-amphetamine-appropriate responding (Glennon et al., 1984a).

Curiously, introduction of a double bond at the 3,4 position of 2-aminotetralin (**9**) gave compound **10** which, in the drug discrimination paradigm, completely substituted in (+)-amphetamine-trained rats, with very little behavioral disruption. The S-(−)-isomer of **10** had nearly half the activity of (+)-amphetamine, whereas the R-(+)-enantiomer failed to possess amphetamine-like properties. These results are in agreement with those of an earlier study of the effects of the enantiomers of **10** on spontaneous locomotor activity in mice (Hathaway et al., 1982).

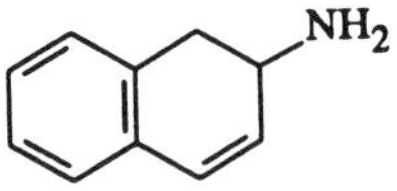

10

D. Mechanism of Action

Although the mechanism by which amphetamine causes catecholamine release is not well understood, several reports (Sulzer *et al.*, 1992, 1993) have suggested that amphetamine and other weak bases reduce intracellular pH gradients of synaptic vesicles. Once the buffering capacity of the vesicle has been exceeded, the decreased proton gradient reduces the driving force for transmitter uptake. Deprotonated catecholamine then may diffuse from the vesicle, following its concentration gradient. This elevated cytosolic monoamine then may be released from the cell by reversal of the uptake carrier.

E. Summary of Structure–Activity Relationships

One may conclude that the amphetamine molecule tolerates very little structural variation without great attenuation, or even complete loss, of amphetamine-like effects. If one assumes, as most research results suggest, that the effects of amphetamine are produced by a drug-induced efflux of presynaptic neuronal catecholamines (principally dopamine), one can infer that the structural requirements for this process are very rigid. The optimal compound may be amphetamine itself, lacking any aromatic ring substituents.

III. AMPHETAMINE DERIVATIVES THAT RELEASE NEURONAL SEROTONIN

The class of substituted amphetamines that release serotonin had relatively few members until recently. Even now, this group comprises a relatively small number of compounds. Although amphetamine itself can cause release of neuronal serotonin, this action is fairly weak (e.g., see Steele *et al.*, 1987). However, the introduction of a single substituent in the *para* position can increase this effect dramatically. The most well-known example is *p*-chloroamphetamine (PCA; **11**). Although PCA itself seems not to have received clinical study, its N-methyl derivative was evaluated briefly as an antidepressant agent (Verster and Van Praag, 1970). PCA potently releases serotonin from neuron terminals, but in relatively small doses also leads to profound and long-lasting depletion of central serotonin (Fuller and Snoddy, 1974; Harvey *et al.*, 1975; for an overview see Fuller, 1992). This depletion is accompanied by loss of serotonin (5-HT) neuronal markers such as tryptophan hydroxylase, a decrease in the B_{max} for the serotonin uptake site (Battaglia *et al.*, 1987; Huang *et al.*, 1992), and a loss of serotonin immunoreactivity (Mamounas and Molliver, 1988; O'Hearn *et al.*, 1988). Therefore, PCA generally is considered a serotonergic neurotoxin. Although this neurotoxic action has been the subject of intense scrutiny for nearly two decades, the mechanism still remains elusive. This topic is discussed at greater length in other chapters of this volume.

11 *p*-chloroamphetamine structure

12 fenfluramine structure

13 *p*-methoxyamphetamine structure

Another compound that interacts potently with serotonin neurons and shares certain toxicological similarities with PCA is fenfluramine (**12**; N-ethyl-*meta*-trifluoromethylamphetamine). Marketed many years ago as an appetite suppressant under the trade name Pondimin, this drug is no longer available in the United States but is prescribed widely in Europe as the *dextro* isomer dexfenfluramine. Fenfluramine also has been studied for efficacy in infantile autistism (Du Verglas *et al.*, 1988). Fenfluramine appears to cause the same types of neuronal serotonin deficits produced by PCA (e.g., see Johnson and Nichols, 1990, and references therein). However, some researchers argue that fenfluramine is not neurotoxic and only depletes serotonin levels acutely (e.g., see Kalia, 1991).

A third compound with potent effects on serotonin neurons is *p*-methoxyamphetamine (PMA; **13**). Although for legal purposes PMA has been classified as a "hallucinogenic" amphetamine, the only description of the clinical effects of PMA suggests that it bears little resemblance to a hallucinogenic amphetamine, but may act more like a stimulant (Shulgin and Shulgin, 1991). Further, in the chronic spinal dog, Martin *et al.* (1978) were able to distinsuigh PMA from hallucinogenic amphetamines as more "amphetamine-like." Although PMA has been known for a long time to have significant indirect adrenergic and peripheral cardiovascular effects (e.g., see Cheng *et al.*, 1974a,b; Nichols *et al.*, 1975, and references therein), Tseng *et al.* (1976, 1978) and Loh and Tseng (1978) first provided evidence for the potent serotonin-releasing action of PMA. Nichols *et al.* (1982) later compared the ability of the PMA enantiomers to induce the release of [^{3}H]5-HT from prelabeled rat brain synaptosomes; the isomers had similar potency. Nevertheless, although researchers generally recognize the "hallucinogenic" and adrenergic effects of PMA, researchers seem to have had less awareness of its potent indirect serotonergic actions.

Until recently, these three compounds—PCA, fenfluramine, and PMA —were the only well-known examples in this class of drug. With the exceptions of the use of PCA as an experimental serotonergic neurotoxin and of numerous studies focused on the pharmacological effects of fenfluramine, serotonin-releasing agents seemed to generate little scientific interest until the mid-1980s.

However, in 1984, 3,4-methylenedioxymethamphetamine (**14**; MDMA, "Ecstasy") began to gain popularity as a new recreational drug, used primarily among college students and young professionals. Beginning in 1985 and for the subsequent 4–5 years, the popular literature contained numerous articles about "Ecstasy" (for a review, see Peroutka, 1990). The drug is still generating controversy, primarily in the context of its use in connection with raves—all-night dance parties with elaborate sound and light systems.

Since MDMA was the catalyst that focused attention on serotonin-releasing agents and returned attention to the issue of serotonin neuron toxicity, a portion of the discussion in this section is appropriately devoted to this compound. However, information also must be presented on 3,4-methylenedioxyamphetamine (MDA; **15**). Not only does MDA serve as the chemical progenitor to MDMA, but certain features of the pharmacology of MDA also are retained in MDMA. A good review of the earlier literature on the effects of MDA was presented by Thiessen and Cook (1973). The clinical effects of MDA first described by Alles (1959) seemed to suggest that MDA had mild hallucinogenic actions. Clearly, racemic MDA does have effects similar to those of other hallucinogenic amphetamine derivatives (Shulgin, 1978). However, in contrast to the effects of virtually all other hallucinogenic phenethylamine derivatives, MDA was reported to produce in its users a need to be with and talk to other people (Jackson and Reed, 1970). In addition, this unusual effect was associated with a feeling of enhanced emotional closeness that earned MDA a reputation as the "love drug" (Weil, 1976). Studies in chronic spinal dogs by Martin and his colleagues (1978) reported MDA to have an action that was both "amphetamine-like" and "LSD-like," whereas other substituted amphetamines simply gave an LSD-like effect. However, additional clarification of the mechanism of action of MDA required studies of its enantiomers.

14   **15**

Marquardt *et al.* (1978) were the first to report studies with the optical isomers of MDA. Behavioral scoring in mice showed that the LSD-like effects of racemic MDA were completely attributable to the *R*-(−) isomer of MDA, where the *S*-(+)-enantiomer possessed an amphetamine-like profile. In cats, the *S* enantiomer also produced a marked pressor response. Pretreatment with reserpine to deplete endogenous norepinephrine stores, or with the α-adrenergic antagonist phenoxybenzamine, blocked the pressor response to MDA, suggesting its action to be one of releasing endogenous neuronal norepinephrine. Indeed, the potent adrenergic effect of MDA has been described earlier (Fujimori and Himwich, 1969; Nichols et al., 1975), but Marquardt et al. (1978) first showed that the S isomer of MDA was primarily responsible for these effects.

Nichols et al. (1982) subsequently demonstrated that MDA was also a potent releaser of serotonin from synaptosomes prelabeled with [^{3}H]5-HT. Weak stereoselectivity for the S isomer was observed in these experiments. MDA then also was shown to release [^{3}H]5-HT from prelabeled slices of rat hippocampus (Johnson et al., 1986); no selectivity was observed for its enantiomers. MDA was a moderately potent inhibitor of [^{3}H]5-HT uptake into rat brain hippocampal synaptosomes, with IC_{50} values for the S-(+) and the R-(−) enantiomers that indicated an approximately 3-fold stereoselectivity (Steele et al., 1987). This study also showed that MDA was a potent inhibitor of [^{3}H]norepinephrine uptake into rat hypothalamic synaptosomes and had modest inhibitory effects against [^{3}H]dopamine uptake into rat striatal synaptosomes. Johnson et al. (1986) also showed that MDA induced the release of [^{3}H]dopamine from rat caudate slices; the S enantiomer had approximately twice the potency of the R antipode.

Animal studies, particularly using the two-lever drug discrimination paradigm, have shown clearly that substitution of MDA in animals trained to discriminate saline from LSD or the hallucinogenic amphetamine 2,5-dimethoxy-4-methylamphetamine (DOM) is the result of the effects of the R enantiomer (Glennon et al., 1982b; Nichols et al., 1986a; Oberlender and Nichols, 1988). The rabbit hyperthermia model also shows that the R enantiomer of MDA is a more potent hallucinogen (Anderson et al., 1978).

Thus, although racemic MDA has behavioral effects characteristic of both hallucinogens and psychostimulants, the preceding discussion demonstrates that this activity can be attributed to "hallucinogenic" actions of the R enantiomer and indirect adrenergic and serotonin-releasing actions of the S enantiomer. Additional experiments using the drug discrimination paradigm support this hypothesis (Shannon, 1980; Glennon et al., 1981, 1982; Glennon and Young, 1984a,b; Nichols et al., 1986a; Nichols and Oberlender, 1990a,b). Whereas the studies of Glennon and his colleagues suggested that S-MDA had an amphetamine-like action, the lack of symmetrical substitution and generalization of serotonergic training cues to the S isomer in other studies (Nichols and Oberlender, 1990a,b) suggest that the primary discriminative cue of S-(+)-MDA is serotonergic in nature. Studies by Shannon (1980) also indicated that racemic MDA was not amphetamine-like.

With this background information for MDA, we can proceed to a discussion of the N-methyl derivative of MDA, MDMA [N-methyl-1-(3,4-methylenedioxyphenyl)-2-aminopropane; 3,4-methylenedioxymethamphetamine]. Although MDMA first was synthesized in 1912 (Shulgin, 1990), this compound did not become widely popular as a recreational drug until the mid- to late 1980s. Although the United States Army carried out toxicological tests on MDMA in the early 1950s, these results remained classified until 1969. The first public report of the effect of MDMA in humans was in 1978 (Shulgin and Nichols, 1978). Use of MDMA by a number of psychiatrists as an adjunct to psychotherapy apparently began at about this time; use gradually escalated into the population at large over the next decade. As for MDA, the popularity of MDMA could be attributed to its ability to induce a sense of emotional "closeness" with other individuals, while lacking the sensory disrupting effects characteristic of hallucinogens (Peroutka et al., 1988). Thus, whereas MDA had been referred to as the "love drug," MDMA became known as the "hug drug."

Although amphetamine itself could be classified as a psychostimulant and other substituted amphetamines could be classified as hallucinogens, a pharmacological classification for MDA or MDMA was less certain. Indeed, these drugs have been proposed to belong to a new pharmacological category with unique psychoactive effects. The name "entactogen" has been suggested for this drug class (Nichols, 1986). Further, because of the unique psychopharmacology of these compounds—particularly their powerful actions on affect, empathy, and emotion—the neuronal substrates of their mechanisms of action were of great interest.

As discussed elsewhere in this volume, both MDA and MDMA induce the selective loss of brain serotonin axons in a manner that appears similar to the toxic actions of PCA and fenfluramine. Thus, additional interest was raised in the mechanisms of action of MDA and MDMA from the perspective of the mechanisms underlying this neurotoxic effect. On the other hand, despite apparent similarities in the pharmacology of these agents, no evidence suggests that the human psychopharmacology of PCA or fenfluramine in any way resembles that of MDA or MDMA.

What is the predominant pharmacology of MDMA? The *in vitro* pharmacologies of MDMA and MDA are very similar, with the exception of the hallucinogen-like effects seen with MDA. Researchers generally believe that the animal behavioral effects of molecules known to be hallucinogenic in humans, such as LSD and DOM, are mediated primarily by an interaction with postsynaptic serotonin 5-HT$_2$ or 5-HT$_{1C}$ receptors. Further, the R enantiomer of hallucinogenic amphetamines is the more potent hallucinogen (for a review, see Nichols et al., 1991b). As might be anticipated, therefore, the R enantiomer of MDMA also has higher affinity than the S isomer for the 5-HT$_2$ receptor (Lyon et al., 1986). However, the S isomer is more potent *in vivo*. Further, the discriminative cue of MDMA is not blocked by the 5-HT$_2$ antagonist ketanserin (Nichols and Oberlender, 1990a,b), nor is the disruption of operant responding in mice caused by either enantiomer of MDMA blocked by the 5-HT$_2$ antagonist pirenperone (Rosecrans and Glennon, 1987). Also, N-methylation of hallucinogenic amphetamines is known to decrease hallucinogenic activity by about one order of magnitude (Shulgin, 1978). Thus, in drug discrimination studies, R-(−)-

MDA substitutes for hallucinogenic training drugs (Glennon et al., 1981b, 1982; Nichols et al., 1986a) whereas R-(−)-MDMA does not (Glennon et al., 1982b; Nichols et al., 1986).

Thus, the R enantiomer of MDA has hallucinogen-like pharmacology, presumably by acting directly at postsynaptic 5-HT_2 receptors, and the S enantiomer induces the release of endogenous stores of serotonin and catecholamines. N-Methylation of MDA to yield MDMA seems to abolish the activity of the R enantiomer but has little effect on the activity of the S enantiomer. Therefore, although racemic MDA has a complex pharmacological action resulting from the combined effects of both isomers, racemic MDMA has a pharmacology derived predominantly from its S enantiomer. All studies carried out to date with the enantiomers of MDA and MDMA seem consistent with this concept.

A. Effects of N Substitution

Clearly the preceding discussion shows that the addition of an N-methyl group to MDA to yield MDMA has little effect on the ability of the compound to release endogenous neuronal stores of serotonin. PCA and N-methyl PCA also have similar neurochemical effects (Fuller, 1992). Although N-methyl PCA is metabolized rapidly to PCA in rats (Fuller and Baker, 1977) and MDA is a metabolic N-demethylation product of MDMA in rats, numerous *in vitro* studies, some already discussed, have established clearly that the N-methyl group does not have a significant deleterious effect on the ability of substituted amphetamines to induce the release of endogenous neuronal monoamines.

However, the range of substituent modification that is tolerated on the nitrogen atom seems fairly limited, although comprehensive pharmacology has not been carried out on many other N-alkylated derivatives. The most well studied other N-substituted compound in this family is N-ethyl MDA (16; MDE, MDEA). MDE apears to have subjective effects in humans that are somewhat similar to those produced by MDMA (Hermle et al., 1993). In addition, the animal behavioral effects (Boja and Schechter, 1987) and the *in vitro* actions of MDE on the uptake and release of serotonin, dopamine, and norepinephrine appear similar to those of MDMA (Johnson et al., 1986; Boja and Schechter, 1991; McKenna et al., 1991a; Paulus and Geyer, 1992). However, the dopaminergic effects of MDE are weaker than those of MDMA (Schmidt, 1987; McKenna et al., 1991a; Nash and Nichols, 1991).

In a study of the analgesic and psychopharmacological effects of a series of N-alkylated MDA derivatives, Braun et al. (1980) reported that only the N-methyl, N-ethyl, and N-hydroxy compounds were active. However, evidence suggests that the N-hydroxy compound may be reduced metabolically to MDA, a process known to occur with the N-hydroxy derivative of PCA (Fuller et al., 1974). MDE also produces long-term serotonin deficits similar to those induced by MDMA, but MDE is somewhat less potent in this regard (Ricaurte et al., 1987; Schmidt, 1987; Stone et al., 1987).

Some mention should be made of fenfluramine, which has an N-ethyl group. Both fenfluramine and norfenfluramine are potent serotonin-releasing agents and inhibitors of serotonin uptake (e.g., see Boja and Schechter, 1988; McKenna et al., 1991a; Berger et al., 1992). Similar to PCA and MDMA, both fenfluramine and norfenfluramine induce persistent deficits in rat brain serotonin markers after a single high dose (e.g., Johnson and Nichols, 1990, and references therein). Racemic fenfluramine and norfenfluramine are nearly equipotent to PCA, (+)-MDA, and (+)-MDMA in inhibition of serotonin uptake into rat brain synaptosomes (McKenna et al., 1991a). However, these compounds are somewhat less potent than the other drugs in inhibiting synaptosomal dopamine uptake.

Only one other N substituent is known to produce a compound in this class with pharmacology similar to that of the parent: N-cyclopropyl-*p*-chloroamphetamine. However, Fuller et al. (1987) have shown that this activity can be attributed to rapid metabolic N-dealkylation, leading to PCA itself.

N,N Dialkylation appears to lead to compounds that lack *in vivo* activity. For example, N,N-dimethyl MDA lacked effects in humans (Shulgin and Shulgin, 1991). However, no *in vitro* pharmacology studies have been carried out on this compound to ascertain whether or not it retains effects on serotonin neurons.

B. Side-Chain Modifications

The two-carbon side chain appears optimal for serotonin release, as might be anticipated since it inherently represents a key feature of the arylethylamine pharmacophoric motif. In a series of PCA analogs, changing the length of the side chain gave compounds that were inactive or much less active than PCA (Fuller et al., 1972). Addition of a second α-methyl to the side chain of MDA or MDMA gave compounds that were clinically inactive (Shulgin and Shulgin, 1991) and also lacked significant potency in inducing the release of [^{3}H]labeled serotonin from prelabeled rat brain synaptosomes (Nichols et al., 1982).

The α-methyl substituent, however, can be extended to an α-ethyl group in serotonin-releasing agents. This modification was demonstrated first with the α-ethyl homolog of MDMA (MBDB; **17**, e.g., Nichols, 1986; Nichols *et al.*, 1986a; Oberlender and Nichols, 1988). *In vitro* studies of monoamine uptake and release from rat brain synaptosomes and brain slices show a similar pharmacological profile for MDMA and MBDB (Johnson et al., 1986; Steele et al., 1987). However, whereas the actions of MDMA and MBDB at serotonin neurons are of comparable potency, MBDB is considerably weaker at dopamine neurons (Johnson et al., 1986; Steele et al., 1987; Nash and Nichols, 1991). This result is consistent with the structure–activity relationships of catecholamine-releasing amphetamine analogs discussed earlier, in which an α-ethyl group dramatically attenuates the ability of the compounds to interact with the dopamine carrier. Although MBDB does induce long-term serotonin deficits after repeated dosing, the magnitude of the effect is less than with MDMA (Johnson and Nichols, 1989) and does not occur after a single dose (Nash and Nichols, 1991). Although complete pharmacological characterization of compounds with an α-alkyl group longer than ethyl has not been carried out, none of those compounds with longer α-alkyl groups tested clinically possessed activity (Shulgin and Shulgin, 1991).

17 **18**

Another compound in this series is the α-ethyl homolog of PCA reported by Johnson et al. (1990a). This compound, 1-(4-chlorophenyl)-2-aminobutane (**18**, CAB), was compared with PCA in drug discrimination assays in animals trained to discriminate MDMA or (+)-MBDB from saline. Using microdialysis, CAB also was compared with PCA in its ability to induce *in vivo* dopamine release and to increase dihydroxy phenylacetic acid (DOPAC) concentrations in striatum. CAB and PCA also were compared for their ability to inhibit the uptake of [^{3}H]5-HT and [^{3}H]dopamine in rat whole brain synaptosomes. Finally, the relative potencies of CAB and PCA to induce persistent deficits in serotonin markers were examined. In this study, CAB was only 2-fold less potent at inhibiting serotonin uptake but 5-fold less potent at inhibiting dopamine uptake. In drug discrimination assays, CAB had about one-third the potency of PCA. A single 10 mg/kg dose of PCA caused a very large increase in extracellular dopamine (2000% of basal!), whereas twice the molar dose of CAB led only to an approximate doubling of extracellular dopamine. These studies indicate that the α-ethyl analog **18** retains the ability to affect serotonergic transmission, as noted from the drug discrimination and monoamine uptake studies, but has a marked loss of dopaminergic action. Further, whereas 10 mg/kg PCA led to approximately 80% reductions in brain serotonin markers at sacrifice 1 wk after drug treatment, twice the molar dose of CAB (22 mg/kg) gave only 30–50% reductions.

Thus, two clear examples—MBDB and CAB—illustrate how the extension of the α-methyl of the amphetamine derivative to an α-ethyl gives compounds that largely retain their ability to induce release of neuronal serotonin stores. The marked attenuation of the ability of these compounds to release dopamine, however, compared with the analogous amphetamine parent, indicates that this structural modification strategy is effective in increasing the selectivity of serotonin-releasing amphetamines when dopaminergic effects complicate the pharmacology.

C. Aromatic Ring Substituents

Although our knowledge is limited in the area of structure–activity relationships of aromatic ring substituents, amphetamines with a single substituent at the 3 position (or at least the *meta*-trifluoromethyl group, as in fenfluramine), at the 4 position (PCA or PMA), or at the 3,4 positions (MDMA or MBDB) provide potent serotonin-releasing agents. 3,4-Methylenedioxy-5-methoxyamphetamine (MMDA; **19**) also is a moderately potent inhibitor of serotonin uptake (McKenna *et al.*, 1991).

However, when the amphetamine contains an *ortho*-methoxy substituent, the compound appears to lose its ability to interact with monoamine carriers (Steele *et al.*, 1987; Johnson *et al.*, 1991b; McKenna *et al.*, 1991a). For example, 3-methoxy-4-methylamphetamine (**20**) is a potent indirect-acting serotonin-releasing agent (Johnson *et al,*. 1991a,b). However, when an *ortho*-methoxy group is added to the molecule to give 2,5-dimethoxy-4-methylamphetamine (the potent hallucinogenic amphetamine DOM, discussed subsequently), all effects on serotonin, dopamine, and norepinephrine transporters are lost (Steele *et al.*, 1987; Johnson *et al.*, 1991b; McKenna *et al.*, 1991a). In addition, 2-methoxy-4,5-methylenedioxyamphetamine, an *ortho*-methoxy derivative of MDA, and a 2,5-dimethoxyamphetamine in which the 3,4 positions are bridged to a cyclopentane ring (a benzobicycloheptane derivative; see **37**) both lack affinity for the serotonin and dopamine uptake carriers (McKenna *et al.*, 1991a). Thus, although an *ortho*-methoxy group gives optimal activity in hallucinogenic amphetamines, as discussed later in this chapter, it prevents the amphetamines from interacting with the monoamine uptake carriers.

1. Other 4-Substituted Derivatives

In addition to PCA, the other halogens also give compounds that are potent serotonin-releasing agents. Early studies compared the *p*-fluoro, *p*-chloro, *p*-bromo, and *p*-iodoamphetamines (Fuller *et al.*, 1975, 1980). The member of the series that appeared most similar to PCA was *p*-iodoamphetamine (Fuller *et al.*, 1980). A more recent study also compared PCA with *p*-iodoamphetamine (Nichols *et al.*, 1991a). In this report, *p*-iodoamphetamine was somewhat more potent than PCA in inhibiting the uptake of [^{3}H]5-HT into rat brain synaptosomes, but had only about one-fourth the potency of PCA in producing substitution in MDMA-trained rats in the drug discrimination assay. Although a single dose of *p*-iodoamphetamine was able to induce long-term serotonin deficits in rats, PCA was much more potent in this regard.

Another selective serotonin-releasing agent is *p*-methylthioamphetamine (MTA; **21**; Huang *et al.*, 1992). In drug discriminationn assays in rats trained to discriminate either racemic MDMA or (+)-MBDB from saline, MTA was equipotent to PCA. MTA was twice as potent as PCA in inhibiting synaptosomal serotonin uptake but, whereas PCA was not selective for inhibition of serotonin over norepinephrine uptake and had only about 2-fold selectivity for serotonin over dopamine, MTA had 40-fold and 30-fold selectivity, respectively, for uptake inhibition of serotonin over dopamine and norepinephrine. In experiments with superfused rat frontal cortex slices prelabeled with [^{3}H]5-HT, MTA was equipotent to PCA in inducing tritium overflow. Further, a single large dose of MTA in rats had no effect on serotonin markers in rat brain on sacrifice 1 wk after treatment, whereas half the molar dose of PCA produced over 90% depletion. Thus, MTA appears to be potent, selective, and nontoxic to neurons that release serotonin.

21

Earlier, we noted that PMA is also a potent serotonin-releasing agent. Although some uncertainty exists regarding its exact pharmacological classification, PMA appears to possess complex components of action, including not only an ability to induce the release of neuronal serotonin but also hallucinogen-like effects by an agonist action at 5-HT$_2$ receptors as well as a potent tyramine-like releasing action on catecholamine stores. By analogy to MDA, one might anticipate that N methylation of PMA would attenuate the hallucinogen-like actions of PMA while allowing the molecule to retain its serotonin-releasing action. Consistent with this idea, Glennon and Higgs (1992) showed that the N-methyl derivative of PMA (PMMA) fully substituted in rats trained to discriminate MDMA hydrochloride (1.5 mg/kg) from saline. Based on ED_{50} values, PMMA had approximately three times the potency of MDMA, the training drug. Thus, as for MDA, the N methylation of PMA allows the molecule to retain its serotonin-releasing action. The structurally related 3,4-dimethoxyamphetamine failed to substitute in these animals, suggesting the lack of MDMA-like behavioral effects. Although these investigators characterized the MDMA cue in these experiments as "an additional non-amphetamine-like effect," little doubt exists that the cue is mediated by endogenous serotonin release.

The similarities between MDMA and PMMA are also evident in the finding that PMMA can induce persistent serotonin deficits in rat brain (Steele *et al.*, 1992). However, both PMA and PMMA require doses in rats that are four times higher than those of MDMA to induce serotonin deficits that were comparable to those induced by MDMA. Despite these apparent similarities between PMMA and MDMA, the human psychopharmacology seems to differ (Shulgin and Shulgin, 1991). Shulgin and Shulgin reported that an acute 100-mg dose of PMMA hydrochloride failed to produce MDMA-like psychopharmacology in humans.

2. Other 3,4-Substituted Derivatives

The most extensively studied 3,4-disubstituted compounds are those with a 3,4-methylenedioxy function (more correctly called a 1,3-dioxole ring): MDA, MDMA, MDE, and MBDB. The dioxole ring has been modified in several ways. First, one or two methyl groups have been added to the central methylene atom to afford ethylidenedioxyamphetamine (EDA; **22**) and isopropylidenedioxyamphetamine (IDA; **23**). EDA and IDA were tested for their ability to substitute in rats trained to discriminate saline from MDMA, and were compared with MDA. Compound **22** had half the potency of MDMA, whereas **23** was only one-fifth as potent (Nichols *et al.*, 1989). In the same report, these compounds also were compared for their ability to induce the release of [^{3}H]5-HT or [^{3}H]dopamine from superfused slices of rat hippocampus or caudate, respectively. MDA and EDA were nearly equipotent in inducing the release of serotonin, whereas IDA only had about one-tenth the potency. The drug-induced efflux of dopamine from caudate slices was most pronounced with MDA; EDA was somewhat less potent and IDA again was about one-thenth as potent as MDA. Thus, the progressive methylation of the methylenedioxy function of MDA decreases potency with each additional methyl group added. The monoamine uptake carriers evidently do not tolerate the addition of steric bulk to the dioxole ring.

The oxygen atom at the 4 position of a dioxole ring, in a compound that also possesses an *ortho*-methoxy group, has been replaced with a sulfur atom. This compound has slightly greater potency than its oxygen isostere MMDA-2 in inducing the release of [^{3}H]5-HT from rat brain synaptosomes (McKenna *et al.*, 1991a). Although the presence of the *ortho*-methoxy group confounds interpretation of the results, one might infer that the 4-thio isostere of MDA or MDMA retains effects on serotonin neurons.

Expansion of the dioxole ring of MDMA by an additional methylene gives ethylenedioxymethamphetamine (EDMA; 24). This compound retains serotonin- and dopamine-releasing potency comparable to that of racemic fenfluramine (McKenna *et al.*, 1991a). Nevertheless, the one report of clinical experiments with this compound revealed it to be inactive at a dose that was more than twice the active dose for MDA (Shulgin and Shulgin, 1991).

Very recently, the oxygen atoms in the dioxole ring of MDA were replaced individually with methylene units to give compounds 25, 26, and 27. In addition, the ring-expanded compound 28 was prepared for comparison.

Studies of inhibition of uptake of [^{3}H]5-HT into rat whole brain synaptosomes revealed that the presence of the oxygen atoms in 25 and 26 was not critical for inhibiting serotonin uptake. Indeed, the carbocyclic analog 27 proved to be the most active in this series, although all the compounds were relatively potent, with only 3- to 4-fold variation in activity (Monte *et al.,* 1993). These compounds were much less potent in inhibiting catecholamine uptake; 25 and 27 were nearly equipotent and about 3-fold more active than 26 or 28.

Finally, a compound mentioned earlier, 3-methoxy-4-methylamphetamine (MMA; 20), has shown the highest selectivity for serotonin over catecholamines for inducing the release and inhibiting the uptake of monoamines (Johnson *et al.,* 1991a,b). The ratios of the IC_{50}s for inhibition of serotonin versus dopamine and norepinephrine were 120 and 33, respectively. This particular amphetamine appears to be the most serotonin selective of all the substituted amphetamines studied to date. Based on the earlier discussion, one could anticipate that extension of the α-methyl group in this structure would lead to a compound with even greater serotonin selectivity.

D. Mechanism of Action

As for the catecholamine-releasing agents, the exact mechanism by which these drugs induce the release of neuronal serotonin is unclear. These compounds are not simply uptake inhibitors; studies of many of these compounds in superfused preparations have shown that they actively cause the release of endogenous neuronal serotonin. The mechanism may be similar to that proposed for the amphetamine-type catecholamine-releasing agents. Berger *et al.* (1992) provided evidence that MDMA, PCA, and fenfluramine all release serotonin by a common mechanism. Rudnick and Wall (1992a,b) showed that both PCA and MDMA release serotonin by an action that must involve the serotonin transporter protein. The results of their studies suggest that direct interactions with the amine transporter, for example, in serotonin–MDMA exchange, may be at least partially responsible for serotonin efflux from vesicles. Also, passive diffusion into synaptic vesicles may increase the pH, leading to efflux of uncharged serotonin from the storage vesicle. This serotonin, in turn, may be transported out of the neuron by a reverse action of the uptake pump across the plasma membrane.

E. Summary of Structure–Activity Relationships

One can draw some general conclusions regarding structure–activity relationships for serotonin-releasing agents. First, monosubstitution at the 4 position leads to compounds that have a significant serotonergic component of action. The halogens or a methoxy group at this position also have appreciable adrenergic properties, attributable to a neuronal catecholamine-releasing action. Second, a 3-trifluoromethyl substituent, as in fenfluramine and norfenfluramine, gives a relatively selective serotonin-releasing agent. A 3,4-disubstitution pattern also leads to compounds with selective serotonergic effects. If this substitution takes the form of a dioxole ring, as in

MDA or MDMA, the compounds also have appreciable catecholaminergic effects. However, other substituents may give a very selective serotonin effect. Note that the addition of an *ortho* methoxy group to any of these structures appears to decrease markedly their ability to release neuronal serotonin.

Generally, primarily amines are most potent, but their effects are attenuated only slightly by an N-methyl or N-ethyl substituent. However, the latter substituent dramatically reduces the ability of the compound to interact with catecholamine carriers and improves serotonin selectivity of the compound (e.g., MDE). All compounds studied to date that have potent releasing actions on both serotonin and catecholamine neurons, when given in high or repeated doses, lead to long-term serotonin deficits and apparent degeneration of brain serotonin terminals and axons.

Finally, extension of the α-methyl to an α-ethyl group allows retention of serotonin effects and improves serotonin selectivity by attenuating effects of the compound on catecholaminergic neurons. Based on the limited data currently available, α-alkyl groups longer than ethyl are inactive.

However, all these conclusions should be considered subject to further refinement. In actuality, the number of substituted amphetamines that have been tested for a selective serotonin-releasing action is very small. Consequently, the database used to derive these structure–activity relationship conclusions is incomplete. With further effort in this field, undoubtedly other substitution patterns and structural modifications will be uncovered that will lead to further refinement of these structure–activity relationship features.

IV. HALLUCINOGENIC AMPHETAMINE DERIVATIVES

The hallucinogens are a fascinating class of molecules. These drugs have been called psychotomimetic, hallucinogenic, or psychedelic at various times by various groups of people. For many years, the term "psychotomimetic," which implies that these drugs produce a psychosis-like effect, was used most frequently in the scientific literature as the descriptor for these compounds. Gradually, "psychotomimetic" has been replaced by the more neutral term "hallucinogen." Although it has a less negative connotation, this descriptor nevertheless implies that these drugs produce hallucinations, which is not strictly correct. Although some compounds, generally at high dosages, may produce hallucinations in some individuals, this action of this drug class is not reliable and reproducible. The term "psychedelic" has been used widely in the lay press but was never favored in the scientific literature because of the positive connotation of "mind manifesting," meaning that these drugs bring out desirable qualities of the mind.

However, Jaffe (1990) argued for the use of "psychedelic" as the appropriate designation for this drug class. He noted that "the feature that distinguishes the psychedelic agents from other classes of drugs is their capacity reliably to induce states of altered perception, thought, and feeling that are not experienced otherwise except in dreams or at times of religious exaltation." This definition illustrates nicely the profound effects these drugs may exert on the human psyche. The chemical types that are included in this definition are substituted phenethylamines and tryptamines, of which the potent compound (+)-lysergic acid diethylaminde (LSD) can be considered representative of the latter.

As the most structurally simple class of molecules possessing hallucinogenic effects, a large number of substituted amphetamines now has been studied. Of particular note are the extensive studies by Shulgin and collaborators, the results of which have been compiled into a book (Shulgin and Shulgin, 1991). Most of the human data discussed in this section (and in this chapter) are from this reference.

The starting compound for this series was mescaline (**29**). Addition of an α-methyl group led to the "substituted amphetamine" 3,4,5-trimethoxyamphetamine (TMA; **30**; Peretz *et al.*, 1955) and began the journey that ultimately led to the synthesis and pharmacological evaluation of nearly 200 potentially hallucinogenic substituted amphetamines. Medicinal chemists readily envision several parts of the molecule that are modified easily to develop structure–activity relationships for the series, including (1) the basic nitrogen atom, (2) aromatic ring substituents, and (3) the side chain. The following discussion focuses on structural modifications of these three portions of the molecule, in that order.

29 **30**

A. *N* Substitution

In contrast to the catecholamine- or serotonin-releasing amphetamines discussed earlier in this chapter in which an *N*-methyl group or, in some cases, an *N*-ethyl group may be tolerated, *N* alkylation of hallucinogenic amphetamines dramatically attenuates or abolishes hallucinogenic activity (Shulgin, 1981; Shulgin and Shulgin, 1991). Not only does receptor affinity decrease (Shannon *et al.*, 1984), but *in vivo* activity also is diminished. Whereas an *N*-methyl group seems to reduce activity by about one order of magnitude, substituting the nitrogen with *N*,*N*-dialkyl groups, even as small as methyls, completely abolishes hallucinogenic activity (Shulgin, 1978). *N*

Substitution with a single alkyl larger than a methyl also seems to abolish hallucinogenic activity. Incorporation of the basic nitrogen into a heterocyclic ring leads to inactive compounds as well (Wolters *et al.*, 1974).

B. Aromatic Ring Substitution

1. 2,4,5- and 3,4,5-Substituent Orientation

Although the 3,4,5-trimethoxy pattern of mescaline was the example of aromatic ring substitution, this configuration also seems to afford the lowest potency of all the substitutions studied to date. However, moving the 3-methoxy group to the 2 position and/or replacing the 4-methoxy group with a more hydrophobic group leads to highly active compounds, that is, compounds with a 2,4,5-substituent orientation in which the 2 and 5 substituents are methoxy groups possess the highest potency. Some represen-

TABLE I Representative 4-Substituted Hallucinogenic Amphetamines Based on 2,5-Dimethoxyamphetamine

R	Trivial name	Approximate dose (mg)[a]
H	2,5-DMA	80–160
OCH_3	TMA-2	20–40
OCH_2CH_3	MEM	20–50
SCH_3	*p*-DOT	5–10
NO_2	DON	3–4.5
CH_3	DOM	3–10
CH_2CH_3	DOEt	2–6
$CH_2CH_2CH_3$	DOPr	2.5–5
Br	DOB	1–3
I	DOI	1.5–3
CF_3	DOTFM	—[b]

[a] Human data are for the hydrochloride salts, and are from Shulgin and Shulgin (1991).
[b] Although DOTFM has not been tested in humans, it has somewhat higher affinity for the rat brain 5-HT_2 receptor labeled with [^{125}I]-DOI than does DOI itself (D. E. Nichols, unpublished data).

tative examples are given in Table I, in approximate order of increasing potency.

Of the 2,5-dimethoxy-substituted analogs, a wide variety of 4 substituents has been studied. The compounds in Table I exemplify the range of atoms and groups chosen. Further, potency does not seem to depend on electronic character, since high activity is observed not only with alkyl substituents but also with highly electronegative groups such as nitro and trifluoromethyl. In general, highest activity seems to occur in congeners in which the *para* substituent is relatively hydrophobic and fairly resistant to oxidative metabolism.

Over the years, much attention has focused on the reasons for the importance of the 4 substituent in this series. Researchers have suggested that this moiety may force the 5-methoxy group to adopt an "anti" orientation (Nichols *et al.*, 1986b). For example, the 2,3-dihydrobenzofuran-6-yl compounds **31** and **32** have beeen shown to lack LSD-like activity, whereas the dihydrobenzofuran-4-yl compound **33** is as potent as its flexible 5-methoxy analog DOB (Table I; Nichols *et al.*, 1991c). These studies clearly show that the preferred orientation of the 5-methoxy group is anti with respect to the 4 substituent. A hydrogen-bond donor in the receptor binding site may require this orientation of the oxygen unshared electron pairs, or perhaps this orientation of the methoxy group is required for the aromatic system to appear electronically similar to the indole that serves as the nucleus of serotonin.

31

32

33 X = Br

34 X = H

The calculated free energies of binding for **33** and **34** at the rat brain serotonin 5-HT$_2$ receptor indicate a contribution by the bromine that appears to be too large simply to represent a hydrophobic interaction (Nichols *et al.*, 1991c). This result suggests the probability that the 4 substituent might interact with a specific complementary residue in the receptor binding site. Studies with compounds containing isopropyl, *tert*-butyl (Shulgin and Shulgin, 1991), or isomeric 2-butyl substituents (Oberlender *et al.*, 1984; Oberlender, 1989) indicate that the receptor is also not tolerant of branching at this position. However, branching more distal from the aromatic ring, for example, when the substituent is an isobutyl group, gives active compounds (Oberlender *et al.*, 1984).

Although resistance to metabolism might contribute to *in vivo* potency, the *para* substituent also seems to be the major determinant of *in vitro* serotonin 5-HT$_2$ receptor affinity in this series. Based on recently proposed three-dimensional models of serotonin receptor geometries (e.g., see Trumpp-Kallmeyer *et al.*, 1992) the *para* substituent seems most likely to interact with specific amino acid residues or perhaps to lie in a hydrophobic pocket or groove in the binding site of the receptor.

In the 2,5-dimethoxy-4-*n*-alkyl homolog series, optimum activity seems to reside in the *n*-propyl group. The *n*-butyl substituent retains activity, but the *n*-pentyl is much less active. The human data generally parallel the ability of the compounds to elicit a contraction in isolated sheep umbilical artery strips (Shulgin and Dyer, 1975). However, the descriptions of the clinical effects of the *n*-propyl, *n*-butyl, and *n*-pentyl homologs suggest that they differ significantly from the lower alkyl homologs in the nature of their human psychopharmacology (Shulgin and Shulgin, 1991). Glennon and co-workers (Seggel *et al.*, 1990) provided evidence that, as the alkyl group is lengthened in this series from methyl to *n*-octyl, the action of the compounds may change from agonist to antagonist. In this same study, radioligand binding data for the ketanserin-labeled serotonin 5-HT$_2$ site were presented for several analogs possessing polar 4 substituents such as OH, NH$_2$, and carboxylic acid esters. The receptor affinities for these compounds were reduced markedly relative to those of compounds with hydrophobic 4 substituents, indicating again the apparent nonpolar nature of the complementary portion of the binding site in the receptor.

Some evidence suggests that the 4 substituent may adopt a conformation that is out of the aromatic ring plane. For example, in 3,4,5-substituted compounds, the buttressing of the adjacent methoxy groups is known to force the 4-methoxy group into a conformation that lies in a plane nearly perpendicular to the aromatic ring plane (Ernst and Cagle, 1973). Further, homologous compounds with a 4-ethoxy or even a 4-isopropyloxy substituent are quite potent (Braun *et al.*, 1978). These more bulky substituents clearly will be forced out of the aromatic ring plane (see Fig. 1 in Nichols and Glennon, 1984). This conformation is depicted here:

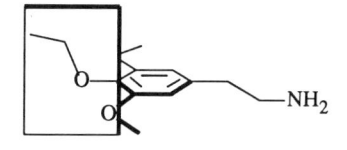

In contrast, in the 2,5-dimethoxy-substituted series, a 4-alkoxy group larger than methoxy does *not* increase potency. Indeed, 2,5-dimethoxy-4-ethoxyamphetamine has somewhat lower potency in humans than does the 4-methoxy congener TMA-2 (Shulgin and Shulgin, 1991) and has about half the affinity of TMA-2 at the ketanserin-labeled serotonin 5-HT$_2$ receptor (Seggel *et al.*, 1990). The different effect on potency of a large 4-alkoxy substituent in the 3,4,5 compared with the 2,4,5 series could be explained as a consequence of the need for the 4 substituent to lie in a plane perpendicular to the aromatic ring plane, as just illustrated. On the other hand, in the 2,4,5 series, the 4-alkoxy group will tend to lie in a conformation that maximizes the overlap of the oxygen unshared electron pairs with the π system of the aromatic ring (Anderson *et al.*, 1979; Makriyannis and Knittel, 1979; Knittel and Makriyannis, 1981). This overlap will force the alkyl attached to this oxygen to lie in the aromatic ring plane, which must be postulated to be an unfavorable conformation.

If a 2,5-dimethoxy compound is substituted with a sulfur atom in the *para* position, large and bulky alkyls can be attached to the sulfur to give compounds that possess considerable potency. For example, hydrochloride salts of the 4-ethylthio and 4-*n*-propylthio compounds are active in humans in the 4–8 mg range, whereas the 4-isopropylthio compound is only slightly less active in a dose range of 7–12 mg (Shulgin and Shulgin, 1991). Qualitatively, the effects, particularly of the 4-*n*-propylthio compound, seem to deviate from classical psychedelic action, tending to produce a state of emotional detachment and anhedonia. Nevertheless, if these compounds are assumed to bind to the same receptor population as their oxygen isosteres, receptor complementarity still can be rationalized by hypothesizing that the alkyl substituent is directed into an out-of-plane conformation. Because sulfur is a larger atom than oxygen, its orbital structure differs from that of oxygen; when the unshared electrons of sulfur overlap with the aromatic π cloud, the alkyl group is not forced to lie in the ring plane. Sulfur is also considerably more lipophilic than oxygen. Thus, in 4-alkylthio compounds, a hydrophobic substituent may be envisioned where the alkyl group projects above the aromatic ring, as shown here:

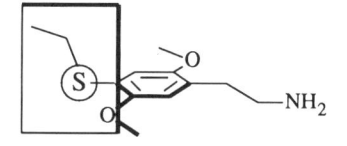

Biological activity is low in compounds in which the oxygen atom of either the 2- or the 5-methoxy group has been replaced with a sulfur, illustrating the difficulty in developing bioisosteres of the 2,5-dimethoxy-substituted aromatic nucleus. However, if relative importance were assigned to the two methoxy groups, the 2-methoxy group would appear to be more critical for optimal activity (Jacob *et al.*, 1977). For example, referring to Table I, when the 2-methoxy group of DOEt is replaced with a methylthio group, *in vivo* activity is reduced by more than one order of magnitude (Jacob and Shulgin, 1983; Shulgin and Shulgin, 1991). However, the replacement of the 5-methoxy oxygen with a sulfur reduces activity only 4- to

6-fold. Similarly, when the 2-methoxy group of DOM is replaced with a methylthio group, activity drops by a factor of 10–20, whereas similar replacement of the 5-methoxy only reduces activity 5 to 10 fold (Jacob et al., 1977; Shulgin and Shulgin, 1991).

2. Other Substituent Orientations

Although the majority of compounds and the most extensive pharmacological studies have been carried out on molecules with 2,4,5 or 3,4,5 substitution patterns, several other orientations have been examined and some afford psychoactive compounds. In disubstituted compounds, the hydrochlorides of 2,4- and 2,5-dimethoxyamphetamine are reported to be psychoactive in humans at oral doses of 60 and 50 mg, respectively (Shulgin, 1978). However, the published descriptions of the effects of these substances leave some doubt about their exact potencies and whether they truly may be categorized as hallucinogenic amphetamines (Shulgin and Shulgin, 1991). The biological activity of 3,4-dimethoxyamphetamine, which would perhaps bear greatest similarlity to the catecholamines, is controversial. No clear evidence exists that it is psychoactive in humans, but this compound appears to be active in the range of a few hundred milligrams (Shulgin, 1978).

Perhaps some of the more interesting, but certainly less well studied, derivatives are the bicyclic derivatives such as 35, 36, and 37. These compounds all appear to be quite potent in humans, with oral dosages of the hydrochlorides in the 12–32 mg range, but the action is very long, typically exceeding 12 hr. The benzonorbornane derivative 37 has a good deal of steric bulk in the *para* region and, based on the known structure–activity relationships, might have been predicted to be inactive. However, no pharmacological studies exist to indicate whether the mechanism of action is simlar to that of other 2,5-dimethoxy-substituted amphetamines. This compound does not, however, have any effect on uptake or release of serotonin or dopamine from rat brain synaptosomes (McKenna et al., 1991a).

35   **36**   **37**

The 2,4,6 trisubstitution pattern has received very little attention, but appears quite interesting. The 2,4,6-trimethoxyamphetamine 38 appears to be active in humans in the 30–40 mg range, not too far removed from the potency of 2,4,5-trimethoxyamphetamine (Shulgin and Shulgin, 1991). Further, 2,6-dimethoxy-4-methylamphetamine (**39**), a positional isomer of DOM, has been reported to be active in humans in the 15–25 mg range (Shulgin and Shulgin, 1991). Based on the known structure–activity relationships in the 2,4,5-substituted series, one might anticipate that more hydrophobic 4 substituents in this series would lead to quite active compounds. However, no additional members of the series have been reported, nor have any animal or biochemical pharmacological studies been carried out to indicate whether the mechanism of action of the 2,4,6-substituted series is similar to that of compounds with the other substituent orientations.

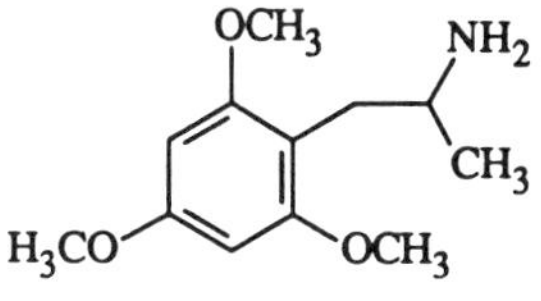

38

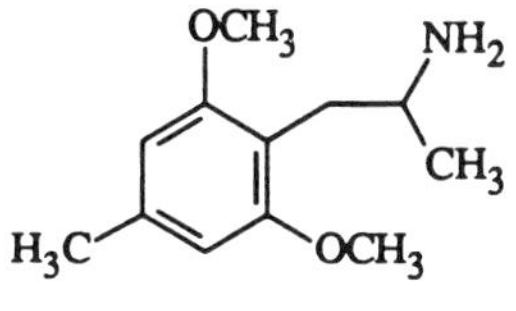

39

Finally, 2,3,4,5-tetramethoxyamphetamine was reported earlier to be active in humans (Shulgin, 1978). However, the more recent description of its effects leaves some doubt about this conclusion (Shulgin and Shulgin, 1991).

C. Side Chain Modifications

The topic of this volume is "amphetamines." Strictly speaking, amphetamine is α-methyl-substituted 2-phenethylamine. However, much has been learned about structure–activity relationships of "amphetamine" from examination of side-chain-substituted hallucinogenic phenethylamine derivatives. The simplest modification is to remove the α-methyl group completely, since mescaline lacks an α-methyl group and is active. On the other hand, 2,4,5-trimethoxyphenethylamine is completely inactive whereas its α-methylated analog 2,4,5 trimethoxyamphetamine (TMA-2; Table I) is quite potent (Shulgin, 1978). Many of the non-α-methylated analogs of hallucinogenic amphetamines retain potency within about one order of magnitude of their amphetamine congeners (e.g., Shulgin and Carter, 1975). Although a decrease of this magnitude may seem dramatic from the perspective of structure–activity relationships, these compounds still remain active in humans with relatively small acute oral dosages. For example, 2,5-dimethoxy-4-bromophenethylamine (2C-B) and 2,5-dimethoxy-4-iodophenethylamine (2C-I) possess only about one-tenth the potency of their amphetamine counterparts DOB and DOI, respectively. Nevertheless, DOB and

DOI are two of the most potent hallucinogenic amphetamines known. Therefore, oral human dosages of 2C-B and 2C-I are in the 5–20 mg range.

The presence or absence of an α-methyl group has a much less dramatic effect if the 4- substituent is an alkylthio group. Although an approximately 10-fold difference in potency is seen if the substituent is a methylithio group, if the alkyl portion is larger—for example, ethyl, *n*-propyl, or isopropyl—the difference in potency is at most 2- to 3-fold (Shulgin and Shulgin, 1991).

α-Methylation also seems to have less of an effect on potency in 3,4,5-substituted compounds, with perhaps a 2-fold increase of activity from mescaline to its amphetamine counterpart. Fewer examples are available in this substitution series, but the α-methyl congener of escaline (3,5-dimethoxy-4-ethoxyphenethylamine) is virtually equipotent to escaline (Shulgin and Shulgin, 1991). Although Clare (1990) was not able to demonstrate a statistically significant correlation between potency and the presence or absence of an α-methyl group in the two substitution patterns, only limited data are available. To find a differential effect on metabolism that could affect *in vivo* potency would not be surprising, that is, one might speculate that 3,4,5-substituted phenethylamines are generally less susceptible to side-chain deamination than are 2,4,5-substituted phenethylamines. Thus, an α-methyl group would have more of an effect on *in vivo* potency in the latter series. As discussed later, an α-methyl group does not enhance receptor affinity, so any increase in potency as a consequence of adding this structural element would appear to be related to pharmacokinetic effects. To date, however, no one has tested this hypothesis.

The *R* and *S* side-chain enantiomers of a large number of substituted hallucinogenic amphetamines have been studied. Early on Shulgin (1973) discovered that the clinical effects of racemic DOM were reproduced by its *R*-(−)-enantiomer (**40**), which was more potent than the *S*-(+) antipode. The development of an asymmetric synthesis for these compounds (Nichols *et al.*, 1973) was followed by several studies that clearly showed that, in all cases, hallucinogen-like activity both *in vivo* and *in vitro* was higher for the *R* enantiomers (Benington *et al.*, 1973; Dyer *et al.*, 1973; Cheng *et al.*, 1974b; Snyder *et al.*, 1974; Anderson *et al.*, 1978; Glennon *et al.*, 1982a,b; Johnson *et al.*, 1987). However, this difference is not stereospecific. Rather, the difference in the activity of the enantiomers is stereoselective, typically being 3- to 6-fold in most assays, whether *in vivo* or *in vitro*.

40

Receptor binding studies have shown that the α-methyl group does not enhance affinity for the serotonin 5-HT$_2$ receptor and actually may reduce affinity for the 5-HT$_{1A}$ receptor. Indeed, the basis for the steroselectivity of the R over the S enantiomer of hallucinogenic amphetamines now can be understood based on receptor binding studies. Studies of the rat cortical 5-HT$_2$ receptor, labeled with the agonist ligand 2,5-dimethoxy-4-[^{125}I]iodophenethylamine (2C-[^{125}I]I), have shown that the R isomer of DOI (**41**) has the same affinity as the non-α-methylated compound 2C-I (**42**). On the other hand, the affinity of the S enantiomer of DOI is reduced, that is, in the more active R enantiomer the α-methyl group has no effect on receptor affinity whereas in the less active S enantiomer the group has a deleterious effect. As discussed earlier, the higher *in vivo* activity of hallucinogenic amphetamines relative to their nonmethylated phenethylamine homologs is explained most easily if the α-methyl group provides protection from side-chain deamination. Note that the increased hydrophobicity contributed by an α-methyl group also will increase CNS penetration for compounds with hydrophobicities below the optimum. Using human data, Barfknecht *et al.* (1975) calculated this optimum octanol–water log P to be 3.1.

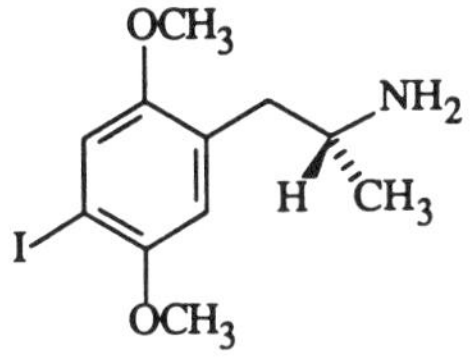

41 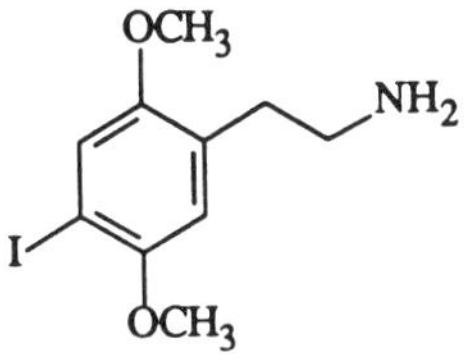   **42**

Addition of a second α-methyl group to the side chain to give α,α-dimethyl compounds abolishes activity (Barfknecht *et al.*, 1978). "Linking" these two methyl groups to form a cyclopropyl ring, as in **43**, restores activity. An explanation for these differences in activity may lie in the inability of the α,α-dimethyl compound to adopt an antiperiplanar conformation, as a consequence of nonbonded steric interactions between the side chain and the aromatic ring, whereas conformational mobility is restored in the less bulky cyclopropyl compound (Weintraub *et al.*, 1980).

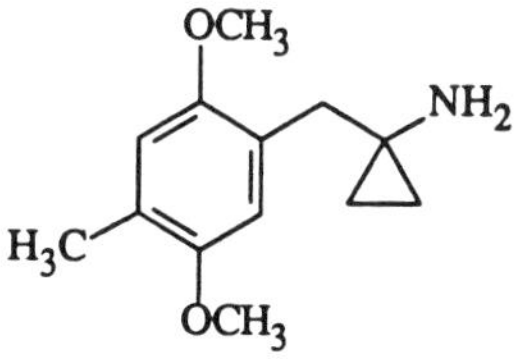

43 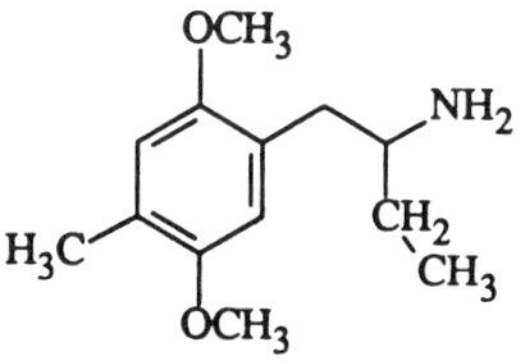   **44**

Extensions of the α-methyl group to longer alkyls abolishes activity completely. For example, the α-ethyl analogs of mescaline (Shulgin, 1963; Shulgin and Shulgin, 1991) and DOM (44; Standridge et al., 1976) are completely inert. Solution NMR and molecular mechanics studies have failed to provide any explanation for this loss of activity based on conformational preferences. A steric effect at the receptor seems most likely to be the reason. In fact, Johnson et al. (1990b) showed that the α-ethyl homolog of DOM (44) had only about 1/100th the affinity of DOM for the agonist-labeled 5-HT$_2$ receptor in rat cortex.

However, the α-methyl group of the amphetamines can be incorporated into a cyclopropane ring with retention of activity. Cooper and Walters (1972) first compared the *cis* and *trans* cyclopropylamine analogs of mescaline for behavioral activity in rats. Only the *trans* compound (45) was active. Aldous et al. (1974) subsequently examined several *trans* cyclopropylamine analogs of hallucinogenic amphetamines. This group reported that these congeners had activity and potency similar to those of their amphetamine counterparts. The cyclopropylamine analog of DOM, DMCPA (46), first reported by this group subsequently was resolved into its enantiomers, which were tested by Nichols et al. (1979). The 1R,2S-(−) enantiomer (46, as shown) proved to be most active. This result is, perhaps, not surprising since the stereochemistry at the α carbon of the cyclopropyl ring is identical to that of the R isomer of the amphetamines. Further, although the difference in affinity for the 5-HT$_2$ receptor between the R and S enantiomers of the amphetamines was small, Johnson et al. (1990b) reported that the enantiomers of DMCPA have a ~30-fold difference in affinity.

45

46

Steric effects operative in the amphetamines still apply to these more rigid congeners. Addition of a methyl group to the cyclopropane ring at C3 abolishes activity (Jacob and Nichols, 1982), as does expansion of the cyclopropane ring to an cyclobutane (Nichols et al., 1984). Both these systems could be envisioned as conformationally constrained α-ethyl analogs. Rigid analogs in which the α-methyl group is incorporated into a carbocyclic 5- or 6-membered ring also lack hallucinogen-like activity in animal models (Nichols et al., 1974), although the tetralin analog is a potent agonist in dog vascular smooth muscle (Cheng et al., 1974c). The steric demands of the receptor(s) involved seem to be very stringent.

β-Methoxy-substituted phenethylamines also have been examined briefly. Lemaire *et al.* (1985) noted that a series of four β-methoxy phenethylamines, substituted with 3,4,5-trimethoxy, 3,4-methylenedioxy, 2,5-dimethoxy-4-methyl, and 2,5-dimethoxy-4-bromo substituents, were slightly more potent than the homologs lacking the β-methoxy group but were less active than the corresponding amphetamines.

D. Mechanism of Action

Although the intent of this discussion is not to delve into the mechanism of action for hallucinogenic agents, note that hallucinogenic amphetamine derivatives seem to have pharmacological properties similar to those of the other structural types of hallucinogens, including tryptamines such as psilocin or ergolines such as LSD. Thus, the following discussion of the structure–activity relationships of hallucinogenic amphetamines generally assumes that the mechanisms of action being discussed are the same. However, the reader should be aware that the measurement of biological activity, at least *in vivo* or in clinical studies, is not entirely precise.

The previous sections of this chapter have focused on amphetamine derivatives that act at monoamine uptake carriers, leading to "indirect" pharmacological effects as a consequence of the release of endogenous neuronal monoamines. In contrast, the hallucinogenic amphetamines apparently have a direct postsynaptic agonist action (e.g., see Nichols *et al.*, 1991b, for a review). Further, the receptor that seems to be most important is the serotonin 5-HT_2 subtype. All hallucinogenic agents have high affinity for the agonist state of this receptor (Glennon *et al.*, 1984a; Titeler *et al.*, 1988). However, all hallucinogenic agents also have high affinity for the 5-HT_{1C} receptor subtype. The possibility that this site is also important for the actions of hallucinogens cannot be ruled out (Burris *et al.*, 1991; Sanders-Bush and Breeding, 1991). In addition, although the hallucinogenic amphetamines have relatively selective affinity for these two serotonin receptor subtypes, the tryptamines and LSD also have high affinity for the serotonin 5-HT_{1A} receptor (McKenna *et al.*, 1991b). Thus, the hallucinogenic amphetamines may have a fairly simple pharmacological mechanism, whereas other structural types may owe some of their effects to actions at various other receptors. Despite this relatively recent knowledge concerning the receptors that may be important for the mechanism of action of these chemicals, we really are very far away from any satisfactory explanation of how effects at receptors produce changes in consciousness.

E. Summary of Structure–Activity Relationships

Hallucinogenic activity generally is found in phenethylamines with a primary amino group, containing 3,5- or 2,5-dimethoxy substituents and a hydrophobic group at the 4 position. Although the 4 substituent has not been studied as extensively in the 3,4,5 orientation, a wide variety of groups has been examined in 2,4,5-trisubstituted compounds. In general, the most active compounds contain, at the 1 position, an unbranched alkyl no longer than three carbons, a halogen larger than fluorine, or an alkylthio group. Some indication exists that 2,6-dimethoxy-4-substituted compounds also may possess high potency.

When an α-methyl group is present, these molecules commonly are called "hallucinogenic amphetamines" and usually are 2–10 times more potent than the nonmethylated phenethylamine. The stereochemistry at the α carbon of the more active enantiomer has the R absolute configuration, and the compounds are levorotatory. The α-methyl group can be incorporated into a cyclopropane ring, but no other structural modification is known that leads to compounds that clearly retain hallucinogenic activity. Although certain other derivatives appear to have psychoactive properties (e.g., a side-chain β-methoxy group), whether their mechanism of action can be considered to be the same as that of the "classical" hallucinogens remains unknown.

ACKNOWLEDGMENTS

The author is most grateful for grants DA02189 and DA04758 from the National Institute on Drug Abuse, which were the major source of funding for most of the studies carried out in his laboratory.

REFERENCES

Aldous, F. A. B., Barrass, B. C., Brewster, K., Buxton, D. A., Green, D. M., Pinder, R. M., Rich, P., Skeels, M., and Tutt, K. J. (1974). Structure–activity relationships in psychotomimetic phenylalkylamines. *J. Med. Chem.* **17**, 1100–1111.

Alles, G. (1959). Some relations between chemical structure and physiological action of mescaline and related compounds. *In* "Neuropharmacology: Transactions of the 4th Conference" (H. Abramson, ed.), pp. 181–268. Madison Printing, Madison, New Jersey.

Anderson, G. M., III, Braun, G., Braun, U., Nichols, D. E., and Shulgin, A. T. (1978). Absolute configuration and psychotomimetic activity. *In* "Quantitative Structure–Activity Relationships of Analgesics, Narcotic Antagonists, and Hallucinogens" (G. Barnett, M. Trsic, and R. E. Willette, eds.), pp. 27–32. National Institute on Drug Abuse Research Monograph 22. DHEW Pub. No. (ADM) 78-729. U.S Government Printing Office, Washington, D.C.

Anderson, G. M., III, Kollman, P. A., Domelsmith, L. N., and Houk, K. N. (1979). Methoxy group nonplanarity in *o*-dimethoxybenzenes. Simple predictive models for conformations and rotational barriers in alkoxyaromatics. *J. Am. Chem. Soc.* **101**, 2344–2352.

Barfknecht, C. F., Nichols, D. E., and Dunn, W. J., III (1975). Correlation of psychotomimetic activity of phenethylamines and amphetamines with 1-octanol-water partition coefficients. *J. Med. Chem.* **18**, 208–210.

Barfknecht, C. F., Caputo, J. F., Tobin, M. B., Dyer, D. C., Standridge, R. T., Howell, H. G.,

Goodwin, W. R., Partyka, R. A., Gylys, J. A., and Cavanagh, R. L. (1978). Congeners of DOM: Effect of distribution on the evaluation of pharamcologic data. *In* "Quantitative Structure–Activity Relationships of Analgesics, Narcotic Antagonists, and Hallucinogens" (G. Barnett, M. Trsic, and R. E. Willette, eds.), pp. 16–26. National Institute on Drug Abuse Research Monograph 22. DHEW Pub. No. (ADM) 78-729. U.S. Government Printing Office, Washington, D.C.

Battaglia, G., Yeh, S. Y., O'Hearn, E., Molliver, M. E., Kuhar, M. J., and De Souza, E. B. (1987). 3,4-Methylenedioxymethamphetamine and 3,4-methylenedioxyamphetamine destroy serotonin terminals in rat brain: Quantification of neuro degeneration by measurement of [^{3}H]-paroxetine-labeled serotonin uptake sites. *J. Pharmacol. Exp. Ther.* **242**, 911–916.

Benington, F., Morin, R. D., Beaton, J., Smythies, J. R., and Bradley, R. J. (1973). Comparative effects of stereoisomers of hallucinogenic amphetamines. *Nature New Biol.* **242**, 185–186.

Berger, U. V., Gu, X. F., and Azmitia, E. C. (1992). The substituted amphetamines 3,4-methylenedioxymethamphetamine, methamphetamine, p-chloroamphetamine and fenfluramine induce 5-hydroxytryptamine release via a common mechanism blocked by fluoxetine and cocaine. *Eur. J. Pharmacol.* **215**, 153–160.

Biel, J. H., and Bopp, B. A. (1978). Amphetamines: Structure–activity relationships. *In* "Handbook of Psychopharmacology" (L. L. Iversen, S. D. Iversen, and S. H. Snyder, eds.), Vol. 11, pp. 1–39. Plenum Press, New York.

Boja, J. W., and Schechter, M. D. (1987). Behavioral effects of N-ethyl-3,4-methylenedioxyamphetamine (MDE, "Eve"). *Pharmacol. Biochem. Behav.* **28**, 153–156.

Boja, J. W., and Schechter, M. D. (1988). Norfenfluramine, the fenfluramine metabolite, provides stimulus control: Evidence for serotonergic mediation. *Pharmacol. Biochem. Behav.* **31**, 305–311.

Boja, J. W., and Schechter, M. D. (1991). Possible serotonergic and dopaminergic mediation of the N-ethyl-3,4-methylenedioxyamphetamine discriminative stimulus. *Eur. J. Pharmacol.* **202**, 347–353.

Braun, U., Braun, G., Jacob, P., Nichols, D. E., and Shulgin, A. T. (1978). Mescaline analogs: Substitutions at the 4-position. *In* "Quantitative Structure–Activity Relationships of Analgesics, Narcotic Antagonists, and Hallucinogens" (G. Barnett, M. Trsic, and R. E. Willette, eds.), pp. 27–32. National Institute on Drug Abuse Research Monograph 22. DHEW Pub. No. (ADM) 78-729. U.S. Government Printing Office, Washington, D.C.

Braun, U., Shulgin, A. T., and Braun, G. (1980). Centrally active N-substituted analogs of 3,4-methylenedioxyphenylisopropylamine (3,4-methylenedioxyamphetamine). *J. Pharm. Sci.* **69**, 192–195.

Burris, K. D., Breeding, M., and Sanders-Bush, E. (1991). (+)Lysergic acid diethylamide, but not its nonhallucinogenic congeners, is a potent serotonin 5HT$_{1C}$ receptor agonist. *J. Pharmacol. Exp. Ther.* **258**, 891–896.

Cheng, H. C., Long, J. P., Nichols, D. E., and Barfknecht, C. F. (1974a). Effects of para-methoxyamphetamine on the cardiovascular system of the dog. *Arch. Int. Pharmacodyn. Ther.* **212**, 83–88.

Cheng, H. C., Long, J. P., Nichols, D. E., and Barfknecht, C. F. (1974b). Effects of psychotomimetics on vascular strips: Studies of methoxylated amphetamines and optical isomers of 2,5-dimethoxy-4-methylamphetamine (DOM) and 2,5-dimethoxy-4-bromoamphetamine (DOB). *J. Pharmacol. Exp. Ther.* **188**, 114–123.

Cheng, H. C., Long, J. P., Nichols, D. E., Barfknecht, C. F., and Rusterholz, D. B. (1974c). Effects of rigid amphetamine analogs on vascular strips: Studies of 2-aminotetrahydronaphthalene and 2-aminoindane derivatives. *Arch. Int. Pharmacodyn. Ther.* **208**, 264–273.

Clare, B. W. (1990). Structure-activity correlations for psychotomimetics. 1. Phenylalkylamines: Electronic, volume, and hydrophobicity parameters. *J. Med. Chem.* **33**, 687–702.

Cooper, P. D., and Walters, G. C. (1972). Stereochemical requirements of the mescaline receptor. *Nature (London)* **238**, 96–98.
Du Vergler, G., Banks, S. R., and Gruver, K. C. (1988). Clinical effects of fenfluramine on children with autism: A review of the research. *J. Autism Dev. Dis.* **18**, 297–308.
Dyer, D. C., Nichols, D. E., Rusterholz, D. B., and Barfknecht, C. F. (1973). Comparative effects of stereoisomers of psychotomimetic phenylisopropylamines. *Life Sci.* **13**, 885–896.
Ernst, S. R., and Cagle, F. W., Jr. (1973). Mescaline hydrobromide. *Acta Crystallogr.* **B29**, 1543–1546.
Fujimori, M., and Himwich, H. E. (1969). Electroencephalographic analysis of amphetamine and its methoxy derivatives with reference to their sites of EEG alerting in the rabbit brain. *Int. J. Neuropharmacol.* **8**, 601–613.
Fuller, R. W. (1992). Effects of *p*-chloroamphetamine on brain serotonin neurons. *Neurochem. Res.* **17**, 449–456.
Fuller, R. W., and Baker, J. C. (1977). The role of metabolic N-dealkylation in the action of *p*-chloroamphetamine and related drugs on brain 5-hydroxytryptamine. *J. Pharm. Pharmacol.* **29**, 561–562.
Fuller, R. W., and Snoddy, H. D. (1974). Long-term effects of 4-chloroamphetamine on brain 5-hydroxyindole metabolism in rats. *Neuropharmacology* **13**, 85–90.
Fuller, R. W., Schaffer, R. J., Roush, B. W., and Molloy, B. B. (1972). Drug disposition as a factor in the lowering of brain serotonin by chloroamphetamines in the rat. *Biochem. Pharmacol.* **21**, 1413–1417.
Fuller, R. W., Perry, K. W., Baker, J. C., Parli, C. J., Lee, N., Day, W. A., and Molloy, B. B. (1974). Comparison of the oxime and the hydroxylamine derivatives of 4-chloroamphetamines as depletors of brain 5-hydroxyindoles. *Biochem. Pharmacol.* **23**, 3267–3272.
Fuller, R. W., Baker, J. C., Perry, K. W., and Molloy, B. B. (1975). Comparison of 4-chloro-, 4-bromo-, and 4-fluoroamphetamine in rats: Drug levels in brain and effects on brain serotonin metabolism. *Neuropharmacology* **14**, 483–488.
Fuller, R. W., Snoddy, H. D., Snoddy, A. M., Hemrick, S. K., Wong, D. T., and Molloy, B. B. (1980). *p*-Iodoamphetamine as a serotonin depletor in rats. *J. Pharamcol. Exp. Ther.* **212**, 115–119.
Fuller, R. W., Snoddy, H. D., and Perry, K. W. (1987). *p*-Chloroamphetamine formation responsible for long-term depletion of brain serotonin after N-cyclopropyl-*p*-chloroamphetamine injection in rats. *Life Sci.* **40**, 1921–1927.
Glennon, R. A. (1986). Discriminative stimulus properties of phenylisopropylamine derivatives. *Drug Alcohol Dep.* **17**, 119–134.
Glennon, R. A., and Higgs, R. (1992). Investigation of MDMA-related agents in rats trained to discriminate MDMA from saline. *Pharmacol. Biochem. Behav.* **43**, 759–763.
Glennon, R. A., and Young, R. (1984a). MDA: An agent that produces stimulus effects similar to those of 3,4-DMA, LSD, and cocaine. *Eur. J. Pharmacol.* **99**, 249–250.
Glennon, R. A., and Young, R. (1984b). MDA: A psychoactive agent with dual stimulus effects. *Life Sci.* **34**, 379–383.
Glennon, R. A., Rosecrans, J. A., and Young, R. (1981). Behavioral properties of psychoactive phenylisopropylamines in rats. *Eur. J. Pharmacol.* **76**, 353–360.
Glennon, R. A., Young, R., Benington, F., and Morin, R. D. (1982a). Behavioral and serotonin receptor properties of 4-substituted derivatives of the hallucinogen 1-(2,5-dimethoxyphenyl)-2-aminopropane. *J. Med. Chem.* **25**, 1163–1168.
Glennon, R. A., Young, R., Rosecrans, J. A., and Anderson, G. M. (1982b). Discriminative stimulus properties of MDA analogues. *Biol. Psychiatry* **17**, 807–814.
Glennon, R. A., Titeler, M., and McKenney, J. D. (1984a). Evidence for 5-HT$_2$ involvement in the mechanism of action of hallucinogenic agents. *Life Sci.* **35**, 2505–2511.
Glennon, R. A., Young R., Hauck, A. E., and McKenney, J. D. (1984b). Structure-activity

studies on amphetamine analogues using drug discrimination methodology. *Pharmacol. Biochem. Behav.* **21**, 895–901.

Glennon, R. A., Yousif, M., and Patrick, G. (1988). Stimulus properties of 1-(3,4-methylenedioxyphenyl)-2-aminopropane (MDA) analogs. *Pharmacol. Biochem. Behav.* **29**, 443–449.

Harvey, J. A., McMaster, S. E., and Yunger, L. M. (1975). p-Chloroamphetamine: Selective neurotoxic actions in brain. *Science* **187**, 841–843.

Hathaway, B. A., Nichols, D. E., Nichols, M. B., and Yim, G. K. W. (1982). A new, potent, conformationally restricted analogue of amphetamine 2-Amino-1,2-dihydronaphthalene. *J. Med. Chem.* **25**, 535–538.

Hermle, L., Spitzer, M., Borchardt, D., Kovar, K.-A., and Gouzoulis, E. (1993). Psychological effects of MDE in normal subjects. *Neuropsychopharmacology* **8**, 171–176.

Higgs, R. A., and Glennon, R. A. (1990). Stimulus properties of ring-methyl amphetamine analogs. *Pharmacol. Biochem. Behav.* **37**, 835–837.

Huang, X., Marona-Lewicka, D., and Nichols, D. E. (1992). p-Methylthioamphetamine is a potent new nonneurotoxic serotonin-releasing agent. *Eur. J. Pharmacol.* **229**, 31–38.

Jackson, B., and Reed, A. Jr. (1970). Another abusable amphetamine. *J. Am. Med. Assoc.* **211**, 830.

Jacob, J. N., and Nichols, D. E. (1982). Isomeric cyclopropyl ring-methylated homologues of trans-2-(2,5-dimethoxy-4-methylphenyl)cyclopropylamine, an hallucinogen analogue. *J. Med. Chem.* **25**, 526–530.

Jacob, P., III, and Shulgin, A. T. (1983). Sulfur analogues of psychotomimetic agents. 2. Analogues of (2,5-dimethoxy-4-methylphenyl)- and (2,5-dimethoxy-4-ethylphenyl)isopropylamine. *J. Med. Chem.* **26**, 746–752.

Jacob, P., III, Anderson, G., III, Meshul, C. K., Shulgin, A. T., and Castagnoli, N., Jr. (1977). Monomethylthio analogues of 1-(2,4,5-trimethoxyphenyl)-2-aminopropane. *J. Med. Chem.* **20**, 1235–1239.

Jaffe, J. H. (1990). Drug addiction and drug abuse. In "Goodman and Gilman's Pharmacological Basis of therapeutics" (A. G. Gilman, T. W. Rall, A. S. Nies, and P. Taylor, eds.), 8th Ed., pp. 521–573. Pergamon Press, New York.

Johnson, M. P., and Nichols, D. E. (1989). Neurotoxic effects of the alpha-ethyl homologue of MDMA following subacute administration. *Pharmacol. Biochem. Behav.* **33**, 105–108.

Johnson, M. P., and Nichols, D. E. (1990). Comparative serotonin neurotoxicity of the stereoisomers of fenfluramine and norfenfluramine. *Pharmacol. Biochem. Behav.* **36**, 105–109.

Johnson, M. P., Hoffman, A. J., and Nichols, D. E. (1986). Effects of the enantiomers of MDA, MDMA and related analogues on [^{3}H]serotonin and [^{3}H]dopamine release from superfused rat brain slices. *Eur. J. Pharmacol.* **132**, 269–276.

Johnson, M. P., Hoffman, A. J., Nichols, D. E., and Mathis, C. A. (1987). Binding to the serotonin 5-HT$_2$ receptor by the enantiomers of [^{125}I]-DOI. *Neuropharmacology* **26**, 1803–1806.

Johnson, M. P., Huang, X., Oberlender, R., Nash, J. F., and Nichols, D. E. (1990a). Behavioral, biochemical and neurotoxicological actions of the α-ethyl homologue of p-chloroamphetamine. *Eur. J. Pharmacol.* **191**, 1–10.

Johnson, M. P., Mathis, C. A., Shulgin, A. T., Hoffman, A. J., and Nichols, D. E. (1990b). [^{125}I]-2-(2,5-dimethoxy-4-iodophenyl)aminoethane ([^{125}I]-2C-I) as a label for the 5-HT$_2$ receptor in rat frontal cortex. *Pharmacol. Biochem. Behav.* **35**, 211–217.

Johnson, M. P., Conarty, P. F., and Nichols, D. E. (1991a). [^{3}H]Monoamine releasing and uptake inhibition properties of 3,4-methylenedioxymethamphetamine and p-chloroamphetamine analogues. *Eur. J. Pharmacol.* **200**, 9–16.

Johnson, M. P., Frescas, S. P., Oberlender, R., and Nichols, D. E. (1991b). Synthesis and pharmacological examination of 1-(3-methoxy-4-methylphenyl)-2-aminopropane and 5-methoxy-6-methyl-2-aminoindan: Similarities to 3,4-(methylenedioxy)methamphetamine (MDMA). *J. Med. Chem.* **34**, 1662–1668.

Kalia, M. (1991). Reversible, short-lasting, and dose-dependent effect of (+)-fenfluramine on neocortical serotonergic axons. *Brain Res.* **548,** 111–125.
Knittel, J. J. and Makriyannis, A. (1981). Studies on phenethylamine hallucinogens. 2. Conformations of arylmethoxyl groups using ^{13}C NMR. *J. Med. Chem.* **24,** 906–909.
Lemaire, D., Jacob, P., III, and Shulgin, A. T. (1985). Ring-substituted β-methoxyphenethylamines: A new class of psychotomimetic agents active in man. *J. Pharm. Pharmacol.* **37,** 575–577.
Loh, H. H., and Tseng, L.-F. (1978). Role of biogenic amines in the actions of monomethoxyamphetamines. *In* "The Psychopharmacology of Hallucinogens" (R. C. Stillman and R. E. Willette, eds.), pp. 13–22. Pergamon Press, New York.
Lyon, R. A., Glennon, R. A., and Titeler, M. (1986). 3,4-Methylenedioxymethamphetamine (MDMA): Stereoselective interactions at brain 5-HT$_1$ and 5-HT$_2$ receptors. *Psychopharmacology* **88,** 525–526.
Makriyannis, A., and Knittel, J. J. (1979). The conformational analysis of aromatic methoxyl groups from carbon-13 chemical shifts and spin-lattice relaxation times. *Tet. Lett.* 2753–2756.
Mamounas, L. A., and Molliver, M. E. (1988). Evidence for dual serotonergic projections to neocortex: Axons from the dorsal and median raphe nuclei are differentially vulnerable to the neurotoxin *p*-chloroamphetamine (PCA). *Exp. Neurol.* **102,** 23–36.
Marquardt, G. M., DiStefano, V., and Ling, L. L. (1978). Pharmacological effects of (±)-, (S)-, and (R)-MDA. *In* "The Psychopharmacology of Hallucinogens" (R. C. Stillman and R. E. Willette, eds.), pp. 85–104. Pergamon Press, New York.
Martin, W. R., Vaupel, D. B., Sloan, J. W., Bell, J. A., Nozaki, M., and Bright, L. D. (1978). The mode of action of LSD-like hallucinogens and their identification. *In* "The Psychopharmacology of Hallucinogens" (R. C. Stillman and R. E. Willette, eds.) pp. 118–125. Pergamon Press, New York.
McKenna, D. J., Guan, X.-M., and Shulgin, A. T. (1991a). 3,4-Methylenedioxyamphetamine (MDA) analogues exhibit differential effects on synaptosomal release of ^{3}H-dopamine and ^{3}H-5-hydroxytryptamine. *Pharmacol. Biochem. Behav.* **38,** 505–512.
McKenna, D. J., Repke, D. B., Lo, L., and Peroutka, S. J. (1991b). Differential interactions of indolealkylamines with 5-hydroxytryptamine receptor subtypes. *Neuropharmacology* **29,** 193–198.
Monte, A. P., Marona-Lewicka, D., Cozzi, N. V., and Nichols, D. E. (1993). Synthesis and pharmacological examination of benzofuran, indan, and tetralin analogues of 3,4-methylenedioxyamphetamine (MDA). *J. Med. Chem.,* in press.
Nash, J. F., and Nichols, D. E. (1991). Microdialysis studies on 3,4-methylenedioxyamphetamine and structurally related analogues. *Eur. J. Pharmacol.* **200,** 53–58.
Nichols, D. E. (1986). Differences between the mechanism of action of MDMA, MBDB, and the classical hallucinogens: Identification of a new therapeutic class: Entactogens. *J. Psychoact. Drugs* **18,** 305–313.
Nichols, D. E., and Glennon, R. A. (1984). Medicinal chemistry and structure-activity relationships of hallucinogens. *In* "Hallucinogens: Neurochemical, Behavioral, and Clinical Perspectives" (B. L. Jacobs, ed.), pp. 95–142. Raven Press, New York.
Nichols, D. E., and Oberlender, R. (1990a). Structure–activity relationships of MDMA and related compounds: A new class of psychoactive drugs? *Ann. N.Y. Acad. Sci.* **600,** 613–625.
Nichols, D. E., and Oberlender, R. (1990b). Structure–activity relationships of MDMA and related compounds: A new class of psychoactive agents. *In* "Ecstasy: The Clinical, Pharmacological and Neurotoxicological Effects of the Durg MDMA" (S. J. Peroutka, ed.), pp. 105–131. Kluwer, Boston.
Nichols, D. E., Barfknecht, C. F., Rusterholz, D. B., Benington, F., and Morin, R. D. (1973). Asymmetric synthesis of enantiomers of psychotomimetic amphetamines. *J. Med. Chem.* **16,** 480–483.

Nichols, D. E., Barfknecht, C. F., Long, J. P., Standridge, R. T., Howell, H. G., Partyka, R. A., and Dyer, D. C. (1974). Potential psychotomimetics. II. Rigid analogs of 2,5-dimethoxy-4-methylphenylisopropylamine (DOM, STP). *J. Med. Chem.* **17**, 161–166.

Nichols, D. E., Ilhan, M., and Long, J. P. (1975). Comparison of cardiovascular, hyperthermic and toxic effects of *para*-methoxyamphetamine (PMA) and 3,4-methylenedioxyamphetamine (MDA). *Arch. Int. Pharmacodyn. Ther.* **214**, 133–140.

Nichols, D. E., Woodard, R., Hathaway, B., Lowy, M. T., and Yim, G. K. W. (1979). Resolution and absolute configuration of trans-2-(2,5-dimethoxy-4-methylphenyl)cyclopropylamine, an hallucinogen analogue. *J. Med. Chem.* **22**, 458–460.

Nichols, D. E., Lloyd, D. H., Hoffman, A. J., Nichols, M. B., and Yim, G. K. W. (1982). Effects of certain hallucinogenic amphetamine analogues on the release of [^{3}H]serotonin from rat brain synaptosomes. *J. Med. Chem.* **25**, 530–535.

Nichols, D. E., Jadhav, K. P., Oberlender, R. A., Zabik, J. E., Bossart, J. F., Hamada, A., and Miller, D. D. (1984). Synthesis and evaluation of substituted 2-phenylcyclobutylamines as analogues of hallucinogenic phenethylamines: Lack of LSD-like biological activity. *J. Med. Chem.* **27**, 1108–1111.

Nichols, D. E., Hoffman, A. J., Oberlender, R. A., Jacob, P., III, and Shulgin, A. T. (1986a). Derivatives of 1-(1,3-benzodioxol-5-yl)-2-butanamine: Representatives of a novel therapeutic class. *J. Med. Chem.* **29**, 2009–2015.

Nichols, D. E., Hoffman, A. J., Oberlender, R. A., and Riggs, R. M. (1986b). Synthesis and evaluation of 2,3-dihydrobenzofuran analogues of the hallucinogen 1-(2,5-dimethoxy-4-methylphenyl)-2-aminopropane: Drug discrimination studies in rats. *J. Med. Chem.* **29**, 302–304.

Nichols, D. E., Oberlender, R., Burris, K., Hoffman, A. J., and Johnson, M. P. (1989). Studies of dioxole ring substituted 3,4-methylenedioxyamphetamine (MDA) analogues. *Pharmacol. Biochem. Behav.* **34**, 571–576.

Nichols, D. E., Johnson, M. P., and Oberlender, R. (1991a). 5-Iodo-2-aminoindan, a nonneurotoxic analogue of *p*-iodoamphetamine. *Pharmacol. Biochem. Behav.* **38**, 135–139.

Nichols, D. E., Oberlender, R., and McKenna, D. J. (1991b). Stereochemical aspects of hallucinogenesis. *In* "Biochemistry and Physiology of Substance Abuse" (R. R. Watson, ed.), Vol. III, pp. 1–39. CRC Press, Boca Raton, Florida.

Nichols, D. E., Snyder, S. E., Oberlender, R., Johnson, M. P., and Huang, X. (1991c). 2,3-Dihydrobenzofuran analogues of hallucinogenic phenethylamines. *J. Med. Chem.* **34**, 276–281.

Oberlender, R. (1989). "Stereoselective Aspects of Hallucinogenic Drug Action and Drug Discrimination Studies of Entactogens." Ph.D. Thesis. Purdue University, West Lafayette, Indiana.

Oberlender, R., and Nichols, D. E. (1988). Drug discrimination studies with MDMA and amphetamine. *Psychopharmacology* **95**, 71–76.

Oberlender, R., and Nichols, D. E. (1991). Structural variation and (+)-amphetamine-like discriminative stimulus properties. *Pharmacol. Biochem. Behav.* **38**, 581–586.

Oberlender, R. A., Kothari, P. J., Nichols, D. E., and Zabik, J. (1984). Substituent branching in phenethylamine-type hallucinogens: A comparison of 1-[2,5-dimethoxy-4-(2-butyl)-phenyl]-2-aminopropane and 1-[2,5-dimethoxy-4-(2-methylpropyl)phenyl]-2-aminopropane. *J. Med. Chem.* **27**, 788–792.

O'Hearn, E., Battaglia, G., De Souza, E. B., Kuhar, M. J., and Molliver, M. E. (1988). Methylenedioxyamphetamine (MDA) and methylenedioxymethamphetamine (MDMA) cause selective ablation of serotonergic axon terminals in forebrain: Immunocytochemical evidence for neurotoxicity. *J. Neurosci.* **8**, 2788–2803.

Paulus, M. P., and Geyer, M. A. (1992). The effects of MDMA and other methylenedioxy-substituted phenylalkylamines on the structure of rat locomotor activity. *Neuropsychopharmacology* **7**, 15–31.

Peretz, D. I., Smythies, J. R., and Gibson, W. C. (1955). New hallucinogen: 3,4,5-Trimethoxy-

β-aminopropane, with notes on stroboscopic phenomenon. *J. Ment. Sci.* **101**, 317–329.
Peroutka, S. J. (1990). "Ecstasy: The Clinical Pharmacological and Neurotoxicological Effects of the Drug MDMA." Kluwer Academic Publishers, Boston.
Peroutka, S. J., Newman, H., and Harris, H. (1988). Subjective effects of 3,4-methylenedioxymethamphetamine in recreational users. *Neuropsychopharmacology* **1**, 273–277.
Ricaurte, G. A., Finnegan, K. F., Nichols, D. E., DeLanney, L. E., Irwin, I., and Langston, J. W. (1987). 3,4-Methylenedioxyethylamphetamine (MDE), a novel analogue of MDMA, produces long-lasting depletion of serotonin in the rat brain. *Eur. J. Pharmacol.* **137**, 265–268.
Rosecrans, J. A., and Glennon, R. A. (1987). The effect of MDA and MDMA ("ecstasy") isomers in combination with pirenpirone on operant responding in mice. *Pharmacol. Biochem. Behav.* **28**, 39–42.
Rudnick, G., and Wall, S. C. (1992a). p-Chloroamphetamine induces serotonin release through serotonin transporters. *Biochemistry* **31**, 6710–6718.
Rudnick, G., and Wall, S. C. (1992b). The molecular mechanism of "ecstasy" (3,4-methylenedioxymethamphetamine (MDMA)]: Serotonin transporters are targets for MDMA-induced serotonin release. *Proc. Natl. Acad. Sci. U.S.A.* **89**, 1817–1821.
Sanders-Bush, E., and Breeding, M. (1991). Choroid plexus epithelial cells in primary culture: A model of $5HT_{1C}$ receptor activation by hallucinogenic drugs. *Psychopharmacology (Berlin)* **105**, 340–356.
Schechter, M. D., Rosecrans, J. A., and Glennon, R. A. (1984). Comparison of behavioral effects of cathinone, amphetamine and apomorphine. *Pharmacol. Biochem. Behav.* **20**, 181–184.
Schmidt, C. J. (1987). Acute administration of methylenedioxymethamphetamine: Comparison with the neurochemical effects of its N-desmethyl and N-ethyl analogs. *Eur. J. Pharmacol.* **136**, 81–88.
Seggel, M. R., Yousif, M. Y., Lyon, R. A., Titeler, M., Roth, B. L., Suba, E. A., and Glennon, R. A. (1990). A structure–affinity study of the binding of 4-substituted analogues of 1-(2,5-dimethoxyphenyl)-2-aminopropane at $5\text{-}HT_2$ serotonin receptors. *J. Med. Chem.* **33**, 1032–1036.
Shannon, H. E. (1980). MDA and DOM: Substituted amphetamines that do not produce amphetamine-like discriminative stimuli in the rat. *Psychopharmacology* **67**, 311–312.
Shannon, M., Battaglia, G., Glennon, R. A., and Titeler, M. (1984). $5\text{-}HT_1$ and $5\text{-}HT_2$ binding properties of derivatives of the hallucinogen 1-(2,5-dimethoxyphenyl)-2-aminopropane (2,5-DMA). *Eur. J. Pharmacol.* **102**, 23–29.
Shulgin, A. T. (1963). Psychotomimetic agents related to mescaline. *Experientia* **19**, 127–128.
Shulgin, A. T. (1973). Stereospecific requirements for hallucinogenesis. *J. Pharm. Pharmacol.* **25**, 271–272.
Shulgin, A. T. (1978). Psychotomimetic drugs: Structure–activity relationships. *In* "Handbook of Psychopharmacology" (L. L. Iversen, S. D. Iversen, and S. H. Snyder, eds.), Vol. 11, pp. 243–333. Plenum, New York.
Shulgin, A. T. (1981). Hallucinogens. *In* "Burger's Medicinal Chemistry" (M. E. Wolff, ed.), Part III, 4th Ed. pp. 1109–1137. Wiley, New York.
Shulgin, A. T. (1990). History of MDMA. *In* "Ecstasy: The Clinical, Pharmacological and Neurotoxicological Effects of the Drug MDMA" (S. J., Peroutka, ed.), pp. 1–20. Kluwer Academaic Publishers, Boston.
Shulgin, A. T., and Carter, M. F. (1975). Centrally active phenethylamines. *Psychopharmacol. Commun.* **1**, 93–98.
Shulgin, A. T., and Dyer, D. C. (1975). Psychotomimetic phenylisopropylamines. 5. 4-Alkyl-2,5-dimethoxyphenyl isopropylamines. *J. Med. Chem.* **18**, 1201–1204.
Shulgin, A. T., and Nichols, D. E. (1978). Characterization of three new psychotomimetics. *In* "The Psychopharmacology of Hallucinogens" (R. C. Stillman and R. E. Willette, eds.), pp. 74–83. Pergamon Press, New York.

Shulgin, A., and Shulgin, A. (1991). "PIHKAL: A Chemical Love Story." Transform Press, Berkeley, California.

Snyder, S. H., Unger, S., Blatchley, R., and Barfknecht, C. F. (1974). Stereospecific actions of DOET (2,5-dimethoxy-4-ethylamphetamine) in man. *Arch. Gen. Psychiatry* **31**, 103–106.

Standridge, R. T., Howell, H. G., Gylys, J. A., Partyka, R. A., and Shulgin, A. T. (1976). Phenylalkylamines with potential psychotherapeutic utility. I. 2-Amino-1-(2,5-dimethoxy-4-methylphenyl)butane. *J. Med. Chem.* **19**, 1400–1404.

Steele, T. D., Nichols, D. E., and Yim, G. K. W. (1987). Stereochemical effects of 3,4-methylenedioxymethamphetamine (MDMA) and related amphetamine derivatives on inhibition of uptake of [^{3}H]monoamines into synaptosomes from different regions of rat brain. *Biochem. Pharmacol.* **36**, 2297–2303.

Steele, T. D., Katz, J. L., and Ricaurte, G. A. (1992). Evaluation of the neurotoxicity of N-methyl-1-(4-methoxyphenyl)-2-aminopropane (*para*-methoxymethamphetamine, PMMA). *Brain Res.* **589**, 349–352.

Stone, D. M., Johnson, M., Hanson, G. R., and Gibb, J. W. (1987). A comparison of the neurotoxic potential of methylenedioxyamphetamine (MDA) and its N-methylated and N-ethylated derivatives. *Eur. J. Pharmacol.* **134**, 245–248.

Sulzer, D., Pothos, E., Sung, H. M., Maidment, N. T., Hoebel, B. G., and Rayport, S. (1992). Weak base model of amphetamine action. *Ann. N.Y. Acad. Sci.* **654**, 525–528.

Sulzer, D., Maidment, N. T., and Rayport, S. (1993). Amphetamine and other weak bases act to promote reverse transport of dopamine in ventral midbrain neurons. *J. Neurochem.* **60**, 527–535.

Thiessen, P. N., and Cook, D. A. (1973). The properties of 3,4-methylenedioxyamphetamine (MDA). 1. A review of the literature. *Clin. Toxicol.* **6**, 45–52.

Titeler, M., Lyon, R. A., and Glennon, R. A. (1988). Radioligand binding evidence implicates the brain 5-HT$_2$ receptor as a site of action for LSD and phenylisopropylamine hallucinogens. *Psychopharmacology* **94**, 213–216.

Trumpp-Kallmeyer, S., Hoflack, J., Bruinvels, A., and Hibert, M. (1992). Modeling of G-protein-coupled receptors: Application to dopamine, adrenaline, serotonin, acetylcholine, and mammalian opsin receptors. *J. Med. Chem.* **35**, 3448–3462.

Tseng, L.-F., Menon, M. K., and Loh, H. H. (1976). Comparative actions of monomethoxyamphetamines on the release and uptake of biogenic amines in brain tissue. *J. Pharmacol. Exp. Ther.* **197**, 263–271.

Tseng, L.-F., Harris, R. A., and Loh, H. H. (1978). Blockage of *para*-methoxyamphetamine induced serotonergic effects by chlorimipramine. *J. Pharmacol. Exp. Ther.* **204**, 27–38.

Van der Schoot, J. B., Ariens, E. J., Van Rossum, J. M., and Hurkmans, J. A. (1961). Phenylisopropylamine derivatives, structure and action. *Arzneim. Forsch.* **9**, 902–907.

Verster, J., and Van Praag, H. M. (1970). A comparative investigation of methylamphetamine and 4-chloro N-methylamphetamine in healthy test subjects. *Neuropsychopharmakologie* **3**, 239–248.

Weil, A. (1976). The love drug. *J. Psychoact. Drugs* **8**, 335–337.

Weintraub, H. J. R., Nichols, D. E., Makriyannis, A., and Fesik, S. W. (1980). Conformational energy differences between side chain-alkylated analogues of the hallucinogen, DOM. *J. Med. Chem.* **23**, 339–341.

Wolters, R. J., Bei, A. J., and Tanner, N. S. (1974). Conformationally constrained analogs of mescaline. *J. Pharm. Sci.* **63**, 1379–1382.

Arthur K. Cho
Yoshito Kumagai

2
Metabolism of Amphetamine and Other Arylisopropylamines

I. GENERAL OVERVIEW

This chapter reviews the pharmacokinetics and metabolism of amphetamines and other arylisopropylamines, emphasizing more recent findings, particularly for Phase I metabolism. An overview of the metabolism is followed by a discussion of specific compounds. One of the more notable advances in drug metabolism in the past 15 years or so is the regiochemical and molecular biological characterization of isozymes of cytochrome P450, the major drug-metabolizing enzyme system. This system is responsible for most of the Phase I transformations of the amphetamines, including those that lead to active metabolites. The system is also responsible for individual differences in metabolism and for certain pharmacokinetically based drug interactions. For these reasons, when available, information on the participation of specific cytochrome P450 isozymes in the metabolic transformations has been included.

AMPHETAMINE (1)

As a group, the amphetamines are relatively strong bases with pK_a values ranging from 9.5 to 10 (Vree, 1973). These compounds are lipid soluble as free bases but water soluble as the hydrochloride or sulfate salts, which are the common dosage forms. The compounds enter the brain rapidly and appear to localize in this tissue relative to plasma (Clay et al., 1971; Cho et al., 1973; Melega et al., 1992). The high pK_a values of these compounds probably account for their significant urinary elimination as unchanged drugs, since ion trapping of the cations in the slightly acidic (relative to plasma) urine reduces the passive reabsorption of the parent drug by the renal tubules. Consistent with this notion, a decrease of urine pH has the expected effect of increasing urinary excretion (Beckett and Rowland, 1965) and decreasing plasma half-life (Davis et al., 1971) of the drug. Another characteristic of these compounds is the chiral center at the α carbon. The optical antipodes of this center exhibit metabolic and pharmacological differences that are discussed in this chapter and throughout this volume.

A. Enzymology

Drug metabolism is organized into Phase I and Phase II reactions. Phase I metabolic transformations functionalize drugs, making them more polar and more susceptible to conjugative Phase II reactions. The Phase I reactions associated with amphetamine metabolism are catalyzed by two enzyme systems, the cytochrome P450 system and the flavin monooxygenase (FMO) system. The latter enzyme oxidizes secondary or tertiary amines (Ziegler, 1991) and can oxidize methamphetamine at the nitrogen atom (Yamada et al., 1984). Amphetamine is an effective inhibitor of monoamine oxidase (MAO) with K_i values in the micromolar range (Robinson, 1985). The dominant enzyme for Phase I reactions is the cytochrome P450 monooxygenase system (Ortiz de Montellano, 1986; Nebert et al., 1989). This enzyme system consists of a heme protein—cytochrome P450—and a flavoprotein—cytochrome P450 reductase—that are imbedded in the lipid matrix of the endoplasmic reticulum of different cells. A second cytochrome P450 system with an additional protein component is present in mitochondria, but its role in drug metabolism is less important. Because of its location in the lipid membrane of the endoplasmic reticulum, the cytochrome P450 system can "extract" lipid-soluble compounds present in the cytoplasm and oxygenate them at various sites to generate phenols, carbinolamines, and hydroxylamines. Cytochrome P450 is expressed as numerous isozymes that are regulated differentially by hormones and xenobiotics. Further, since isozymes within a specific subclass can differ in amino acid sequence, their regiochemistry may differ also. Therefore, identification of isozymes responsible for a particular reaction is necessary to understand the basis for species differences and drug-based changes in metabolism (Smith, 1991).

The important isozymes for metabolism of amphetamines are members of the CY2B, 2C, 2D, and 3A families. Members of the 2B family, 2B1 and 2B4, are the major isozymes induced by phenobarbital in rats and rabbits, respectively, as are members of the 3A family. These isozymes have a rather broad substrate specificity but differ in their regiochemistry. For example, whereas 2B4 efficiently catalyzes the N-oxidation and deamination reactions of amphetamine, 2B1 does so very poorly (see subsequent discussion). The 2D family and its human form 2D6 are especially important because of their participation in the metabolism of drugs such as imipramine and metroprolol (Otton *et al.*, 1988; Lennard, 1990) and because of their polymorphism. A significant percentage (10%) of the Caucasian population of the United Kingdom (Mahgoub *et al.*, 1977; Brosen and Gram, 1989) has a genetically based deficiency in CYP2D6. This group of individuals exhibits a greater sensitivity to certain drugs, presumably because of reduced metabolic capability. Evidence is accumulating that supports the notion that some of the amphetamines are substrates for the 2D6 isozyme (Kumagai *et al.*, 1992c).

The cytochrome P450 system is distributed extensively. Although the organs with the greatest activity are the liver, lungs, and kidneys, the enzyme is present in most organs examined to date, including the brain (Naslund *et al.*, 1988; Ravindranath and Anandatheerthavarada, 1989; Bergh and Strobel, 1992). Brain cytochrome P450 is important because it can convert lipid-soluble compounds to polar metabolites after brain entry. Further, since the distribution of cytochrome P450 isozymes within the brain is not uniform (Walther *et al.*, 1986), formation of area-specific metabolites is also possible. The pharmacokinetic implications of selective brain metabolism of drugs has been considered by Britto and Wedlund (1992).

B. Reactions

1. Phase I reactions

The common carbon skeleton and functional group in the amphetamines results in common metabolic transformations. Thus, as a group, the amphetamines undergo hydroxylation on the α and β carbon, the nitrogen, and the aromatic ring (1). These initial metabolites subsequently are oxidized or conjugated to the excreted metabolites. The following sections describe the reactions and the amphetamines for which the reactions have been investigated.

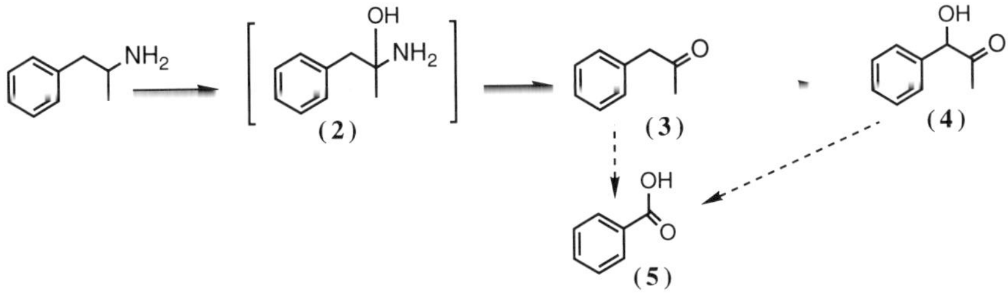

Scheme 1

a. Dealkylation   The dealkylation reaction is actually two reactions, a cytochrome P450-mediated oxidation to a carbinolamine (2) and decomposition of the latter to form phenylacetone (3) and ammonia (Scheme 1). The reaction was reported first by Axelrod (1955) and subsequently has been examined with amphetamine, methamphetamine, methylenedioxyamphetamine (MDA), methylenedioxymethamphetamine (MDMA), and fenfluramine as substrates. Phenylacetone (3), the metabolite of amphetamine, has no pharmacological activity and has been proposed to be the precursor to benzoic acid (5), a metabolite common to most species. Phenylacetone is hydroxylated at the β carbon to form the hydroxyketone (4); subsequent reduction of the ketone forms the 1,2-diol. The diol is a possible precursor to benzoic acid, but the details of the reaction and its enzymology have not been elucidated to date. N-Demethylation of N-methylamphetamines can occur by a reaction identical to that just described. However, N-demethylation also can be mediated by FMO by hydroxylation of the nitrogen to the hydroxylamine, which rearranges to a carbinolamine (Lindeke and Cho, 1982; Baba et al., 1987). *In vivo* studies of the metabolism of methamphetamine have shown demethylation to be a major pathway, but not always the dominant one (Caldwell et al., 1972; Yamamoto et al., 1984).

b. Aromatic Ring Hydroxylation   The importance of the aromatic ring hydroxylation reaction is its generation of pharmacologically active metabolites (Scheme 2). The secondary metabolite for amphetamine, *p*-hydroxynorephedrine (7), is the product of dopamine β-hydroxylase action on *p*-hydroxyamphetamine (6) and affects serotonin (5-HT) systems in the cau-

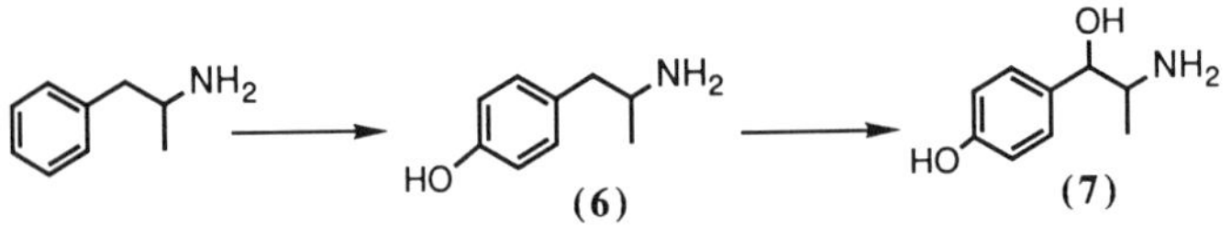

Scheme 2

date (Matsuda et al., 1989). p-Hydroxynorephedrine is also active peripherally as a false transmitter (Thoenen, 1966) and may be responsible for the tolerance that develops to some peripheral actions after amphetamine use (Brodie et al., 1969, 1970; Lewander, 1971a,b).

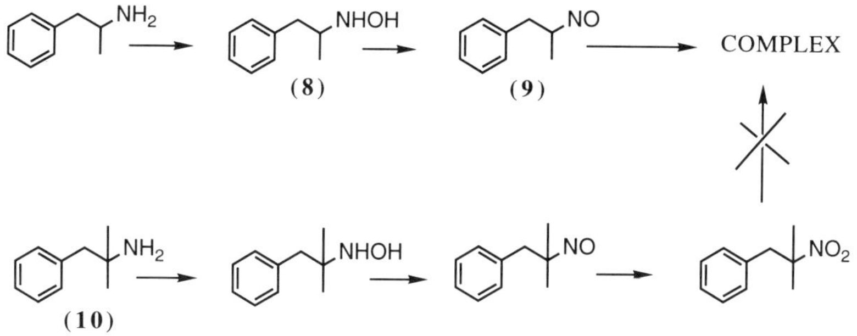

Scheme 3

c. N-Hydroxylation N-Hydroxylation (Scheme 3) does not appear to be of pharmacological relevance to the neurochemical actions of amphetamines, but may be involved in pharmacokinetic interactions with cytochrome P450. The initial N-hydroxylation product (8) is oxidized further to the nitroso state (9), a functionality that inhibits cytochrome P450 by forming a complex with the heme group (Mansuy et al., 1976; Franklin, 1977; Lindeke and Paulsen-Sorman, 1988). This inhibitory reaction sequence is especially effective in rat liver microsomes from phenobarbital-pretreated animals (Lindeke et al., 1982); although inhibition has been shown *in vitro* (Franklin, 1974), it has not been shown *in vivo*. The complex is detected by its absorption in the 450-nm range, is dependent on oxidation of hydroxylamines, and requires a free α hydrogen. The last requirement may be a steric effect, since phentermine (10), the α-methyl analog of amphetamine, is N oxidized but does not form the complex.

d. β Hydroxylation Dopamine β-hydroxylase, a norepinephrine biosynthesis enzyme, can hydroxylate phenethylamine derivatives such as amphetamine (1) and p-hydroxyamphetamine (6) (Creveling et al., 1962) to form norephedrine (11; Scheme 4) and p-hydroxynorephedrine (7; Scheme

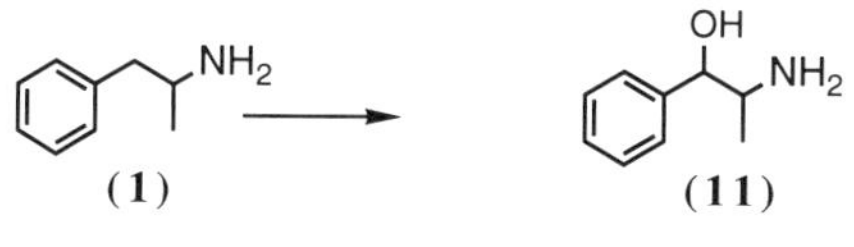

Scheme 4

4), respectively, which are pharmacologically active and therefore contribute to the overall actions of amphetamine. The corresponding metabolites of other phenylisopropylamines such as methamphetamine and fenfluramine have not been investigated thoroughly. Cathinone (55), α-keto amphetamine, is reduced to norephedrine and norpseudoephedrine (Brenneisen et al., 1985) but the enzymology has not been established.

2. Phase II Reactions

The Phase II reactions generate conjugates of the drug or its metabolite that are usually pharmacologically inactive and are eliminated rapidly.

a. Glucuronidation Ring-hydroxylated metabolites of the amphetamines are excreted as their glucuronides (12; Scheme 5). This pathway may be an important elimination reaction because the hydroxylated amphetamines are accumulated and retained in nerve terminals (Cho et al., 1975a; Dougan et al., 1986) whereas the gluconuronide is excreted rapidly through the kidneys.

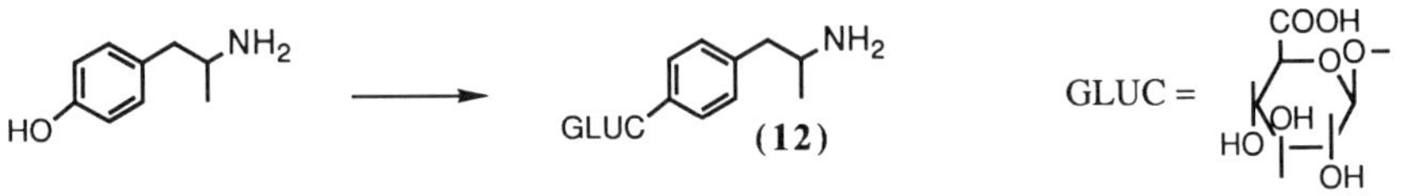

Scheme 5

b. Sulfation The sulfate ester (13) of an enol of phenylacetone (3; Scheme 6) has been reported (Smith and Dring, 1970) and was examined further by Dring et al. (1970), but the compound is unstable and attempts to synthesize it failed. Identification of the compound was based on indirect evidence and no subsequent reports of its presence have appeared.

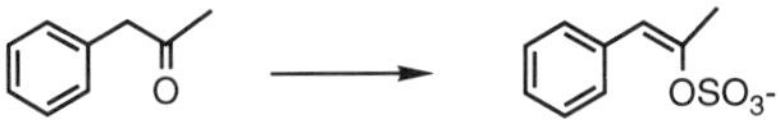

Scheme 6

c. Schiff Base Formation The formation of a Schiff base was reported in in vitro studies of N-hydroxyamphetamine and -methamphetamine metabolism by Baba and associates (Baba et al., 1987). These workers reported the formation (14) of the acetaldehyde adduct N-[(1-methyl-2-phenyl)-ethyl]ethanimine N-oxide (15) in incubates of liver homogenates (Scheme 7). Presumably, acetaldehyde or its equivalent is present in the homogenate and reacts rapidly with the hydroxylamine.

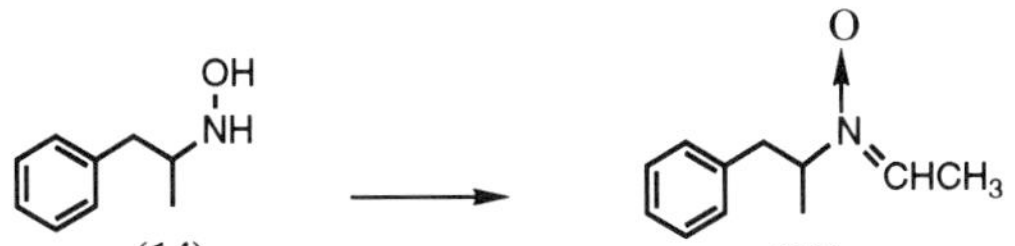

Scheme 7

II. SPECIFIC COMPOUNDS

A. Amphetamine

1. *In Vivo* Studies and Species Differences

The initial *in vivo* studies of amphetamine metabolism were reported by Axelrod (1954), who examined its disposition in dogs. Subsequently, Dring *et al.* (1970; Smith and Dring, 1970) compared its metabolism in different species by examining the urinary excretion of radioactivity after administration of [^{14}C]-labeled amphetamine. Considerable species variation was observed, but human, monkey, and dog appear to be quantitatively similar in their excretion patterns; benzoic acid was the major metabolite. Rats were the exception, since they excreted about 60% of the dose as the *p*-hydroxy metabolite, which was a minor product for most other species. On the other hand, benzoic acid was only a minor product in rats. The possibility that chronic amphetamine use may alter its metabolism was evaluated in humans. Examinations of the excretion patterns of amphetamine in two chronic abusers (Sever *et al.*, 1973) showed slight increases in the urine levels of norephedrines.

2. Pharmacokinetics

Several reports are available on the pharmacokinetics of amphetamine in humans (Rowland, 1969; Davis *et al.*, 1971; Vree, 1973). The half lives of amphetamine are 7 ± 1.2 and 11 ± 2.1 hr for the (+) and (−) enantiomers, respectively (Vree, 1973). About 30% of the dose of (±)-amphetamine is excreted unchanged; excretion is sensitive to urinary pH. When individuals were given ammonium chloride to acidify their urine (pH ~5.0), the urinary excretion half-life was ~4.6 hr (Rowland, 1969). Similarly, Davis *et al.* (1971) altered urinary pH with dietary changes and determined the plasma half-life of [^{3}H]amphetamine (presumably racemic) to range from 16 to 31 hr in alkaline (pH > 7.5) and from 8 to 10 hr in acid (pH 6.0) urine. Other parameters such as volume of distribution were not reported in this study, but Rowland (1969) reported values ranging from 3.5 to 4.6 liter/kg in human subjects.

In studies of amphetamine pharmacokinetics in domestic animals, Baggot and Davis (1973) reported half-lives ranging from 6.56 to 0.62 hr with volumes of distribution ranging from 1.81 to 3.08 liter/kg. The longest half-life was found in the cat and the shortest in the goat. The volume of distribution at the steady state in rats, the species most often used in experimental pharmacology studies, was 2.30 liter/kg (Cho et al., 1973). The elimination rate constant in this species was $0.012 \pm .001$ min^{-1}, corresponding to a half-life of about 1 hr. Plasma and brain levels were determined in several studies (Clay et al., 1971; Cho et al., 1973; Melega et al., 1992). After an intravenous dose of 5 mg/kg, initial brain levels of amphetamine were 10–12 times higher than plasma levels and gradually approached 7–8 times higher at steady state. Thus, amphetamine enters the brain rapidly and is accumulated in the tissue, possibly as the result of an intracellular pH that is lower than that of plasma. Studies of the distribution of amphetamine in different areas of the brain (cortex, striatum, cerebellum) do not show significant differences (Jori and Caccia, 1974; Melega et al., 1992). Thus, although amphetamine interacts specifically with catecholamine terminals, it does not appear to accumulate selectively in those areas rich in catecholamine terminals, such as the striatum, at the moderate doses used in these studies.

3. Metabolic Studies

a. Ring Hydroxylation Although ring hydroxylation (Scheme 2) is most important in rats, this reaction has been difficult to study *in vitro* with preparations of rat tissue because of very low activity, in part the result of an apparent substrate inhibition (Jonsson, 1974; Cho et al., 1975b; Billings et al., 1978). The reaction is mediated by cytochrome P450, as shown by its sensitivity to carbon monoxide and other typical inhibitors of this enzyme system. In rats, the reaction appears to involve a constitutive isozyme and exhibits some stereoselectivity with a maximal velocity for the S-(+) enantiomer that was 1.32 times that for the R-(−) form (Jonsson, 1974). Although most of the hydroxylation activity was found in microsomes, some evidence for mitochondrial activity exists (Rommelspacher et al., 1974). Studies comparing hydroxylation activity in different levels of tissue organization with rat preparations noted that microsomal activity was about 50% that of isolated hepatocytes, in contrast to other substrates for which microsomal activity usually exceeds hepatocyte activity (Billings et al., 1978). The cytochrome P450 isozymes responsible for the transformationn have not been identified to date.

Evidence exists for aromatic hydroxylation in the brain. The reaction has been demonstrated in brain slices (Kuhn et al., 1978), microsomes, and mitochondria (Liccione and Maines, 1989). Kuhn et al. (1978) also demonstrated that *p*-hydroxylation occurred after direct injection of amphetamine into the brain. The formation of the catecholamine α-methyldopamine also has been reported to occur in brain microsomal preparations in the presence of amphetamine (Hoffman et al., 1979). Although p-hydroxyamphetamine can enter the brain from the periphery (e.g., Matsuda et al., 1989), entrance of α-methyldopamine is much less likely. Generation of this compound in the brain indicates that such polar metabolites could contribute to the central nervous system (CNS) effects of amphetamine.

b. Deamination

i. Carbinolamine vs. Hydroxylamine–Oxime The reaction pathway leading to phenylacetone (3; Scheme 8) was thought to be a typical N-dealkylation reaction, proceeding through the carbinolamine (2), until a report appeared demonstrating the formation of phenylacetone oxime (16; Hucker et al., 1971) and its subsequent conversion to phenylacetone. The reaction has been studied *in vitro* with rabbit liver preparations (Hucker et al., 1971; Parli et al., 1971; Wright et al., 1977a; Wright et al., 1977b; Cho and Wright, 1978). The prevailing evidence suggests that, although N-oxidation is a pathway that could lead to phenylacetone, the dominant pathway is α-carbon hydroxylation in a typical cytochrome P450 reaction. An enzyme system also exists in microsomal preparations that oxidizes N-hydroxyamphetamine (8) to phenylacetone oxime (16), and does not appear to be cytochrome P450 (Matsumoto and Cho, 1982).

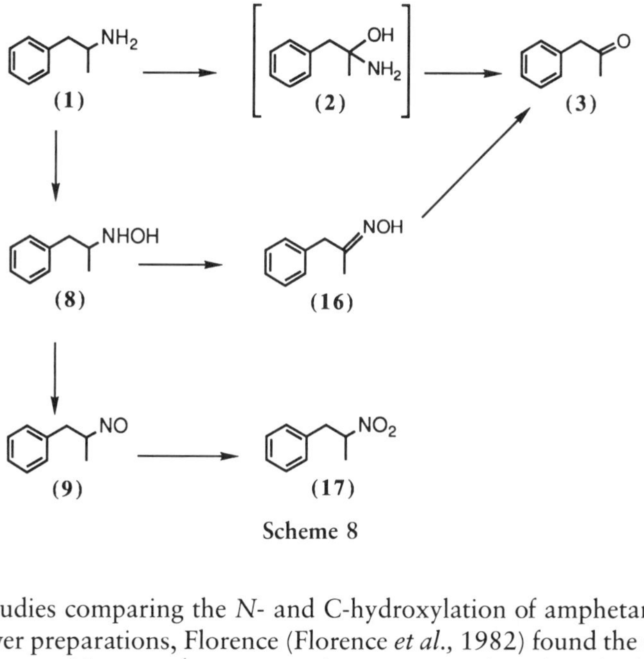

Scheme 8

In studies comparing the N- and C-hydroxylation of amphetamines by rabbit liver preparations, Florence (Florence et al., 1982) found the activities to be comparable in microsomes and to increase after phenobarbital pretreatment. In this study, N-hydroxyamphetamine (8) was shown to be oxidized further to the nitro compound (17) by superoxide generated in the incubation mixtures.

ii. Benzoic Acid and Deamination Deamination has been proposed as the initial step in the conversion of amphetamine to benzoic acid (5, Scheme 9) (Dring *et al.*, 1970) but the details of the pathway are not clear. The reaction requires scission of a carbon–carbon bond, which is relatively rare for metabolic reactions. A carbon–carbon bond cleavage, catalyzed by cytochrome P450, does occur in corticosterone biosynthesis. A key step is the conversion of cholesterol 21,22-diol to pregnenolone. In this conversion, the heme of cytochrome P450 is proposed to accept the σ-bond electrons (Van Lier and Rousseu, 1976; Walsh, 1979) and effect cleavage to the corresponding ketone and benzaldehyde. However, direct incubation of phenylacetone with rabbit liver 9000 g supernatant (a mixture of microsomes and the soluble fraction of liver homogenate) yielded only very small amounts of benzoic acid (Kammerer *et al.*, 1978). Studies by Yoshimura and associates (Yamada *et al.*, 1992) suggest that the reaction is considerably more complex. In experiments with guinea pig liver preparations, these investigators found benzoic acid but no evidence for benzaldehyde when amphetamine, N-hydroxyamphetamine, and phenylacetone were incubated with either microsomes or 9000 g supernatant from this species. Further, formation of benzoic acid from phenylacetone was 20 times greater than from amphetamine; formation from both was enhanced markedly by the addition of ATP. The enhancement by ATP was found to be general for energy-rich phosphates; phosphocreatine was found to be the most effective additive, increasing conversion from phenylacetone almost 10-fold. The reaction also involves a cytochrome P450-mediated step, as evidenced by carbon monoxide, SKF 525A, and metyrapone sensitivity. The biochemical details of this conversion remain to be elucidated.

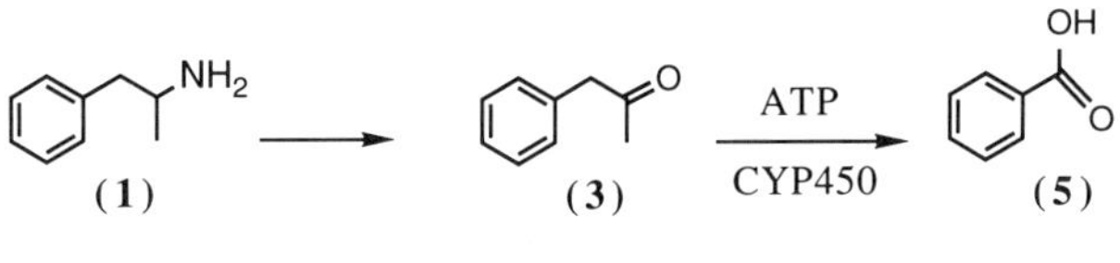

Scheme 9

iii. Cytochrome P450 Isozymes in Amphetamine Metabolism Much of the work identifying isozymes involved in specific reactions has been done using rabbit liver preparations. Rabbit liver microsome preparations have been used to demonstrate participation of cytochrome P450 in both C- and N-hydroxylation reactions (Florence *et al.*, 1982). Yamada *et al.* (1989) reported that a constitutive isozyme (CYP2C3) is capable of catalyzing deamination, but N-hydroxylation was not monitored. This finding would account for the minimal induction observed with phenobarbital (Florence *et* al., 1982). If deamination were the result of the actions of both a constitutive and an inducible isozyme, induction would increase overall microsomal activity only slightly. Since both N-hydroxylation and deamination are induced in the same manner, constitutive isozymes may catalyze N-hydroxylation as well. The p-hydroxylation of amphetamine is of interest because of its pharmacological implications. The inability to induce the reaction is consistent with the involvement of a constitutive isozyme. A study by Tyndale et al. (1991) with canine brain suggests that amphetamine has a high affinity for CYP2D1. Smith (1986) reported indirect evidence for the involvement of CYP2D6, the human isoform of this family, in the reaction.

iv. N-Oxidation

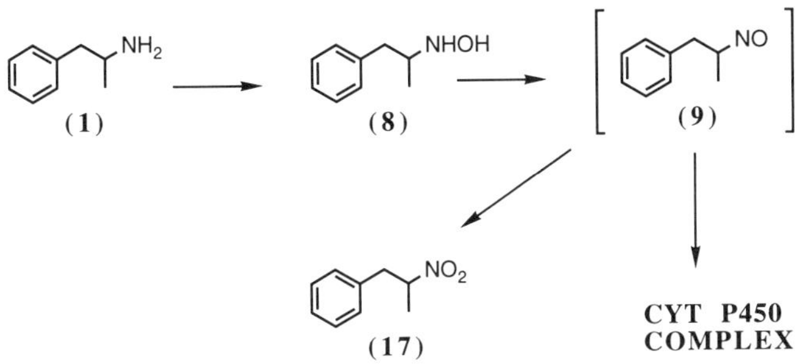

$$(1) \xrightarrow{NH_2} (8) \xrightarrow{NHOH} [(9) \xrightarrow{NO}]$$

$$\downarrow$$

$$(17) \; NO_2 \qquad \text{CYT P450 COMPLEX}$$

Scheme 10

N-Oxidation (Scheme 10) is a major reaction in rabbit liver microsomes (Florence et al., 1982) and occurs at rates comparable to that of α-carbon hydroxylation. N-Oxidation of amphetamine also leads to formation of an inhibitory complex with cytochrome P450 (Mansuy et al., 1976; Franklin, 1977; Jonsson and Lindeke, 1976). This complex, referred to as a metabolic intermediate complex, has been reviewed extensively (Lindeke and Paulsen-Sorman, 1988). The reaction sequence leading to complex formation (Scheme 10) includes N-hydroxylation but also requires a second NADPH-dependent oxidation of the initial product. The ligand for cytochrome P450 is thought to be the corresponding nitroso compound (9: Mansuy et al., 1976; Lindeke and Paulsen-Sorman, 1988). The effectiveness of inhibition depends on the stability of the complex, but its formation suggests that amphetamine could inhibit the metabolism of other drugs.

B. Methamphetamine

1. General

A comparative study of the urinary excretion patterns of methamphetamine (18) in human, guinea pig, and rat has been performed (Caldwell et al., 1972). Humans excrete a substantial portion (23%) of the dose as

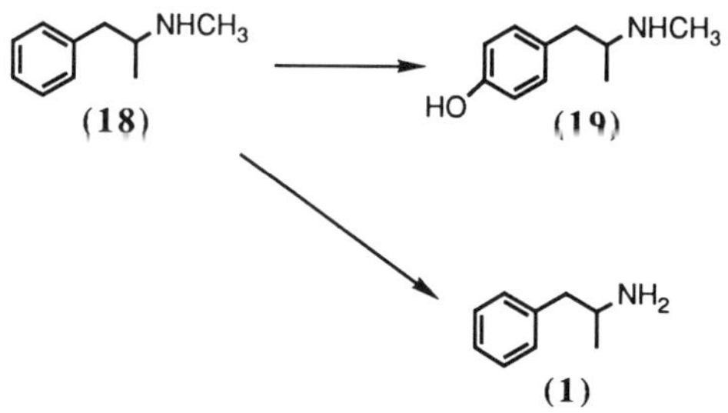

(Scheme 11: compound **18** (PhCH2CH(NHCH3)CH3) → **19** (HO-Ph-CH2CH(NHCH3)CH3) and → **1** (PhCH2CH(NH2)CH3))

Scheme 11

unchanged drug; 18% and 14% are excreted as ring-hydroxylated (**19**; Scheme 11) and demethylated metabolites, respectively. The rat excretes 53% of the dose as the ring-hydroxylated metabolite and 28% as amphetamine. The guinea pig excretes 64% as the demethylated metabolite and no ring-hydroxylated metabolite.

2. Pharmacokinetics

The pharmacokinetics of methamphetamine have been studied in human subjects whose urine was acidified with ammonium chloride (Beckett and Rowland, 1965). With a urine pH of ~5.0, the half-life, determined by urinary excretion, was ~4.9 hr. Under these conditions, 70% of the dose is eliminated as methamphetamine, in contrast to the 23% reported by Caldwell (Caldwell et al., 1972), so urine acidification has a substantial effect on the disposition of this drug. In studies relating plasma concentration to pharmacological actions, Melega et al. (1992) determined the pharmacokinetic parameters of methamphetamine in rats after intravenous doses. The terminal half-life was approximately 45 min and the volume of distribution was 4.1 liter/kg. Brain levels were 7–10 times aminesthe plasma concentration. In urinary excretion studies with rats, Yamada et al. (1986) observed a reduction in p-hydroxymethamphetamine levels as the subcutaneously administered dose increased from 5 to 40 mg/kg. This observation may reflect the substrate inhibition of ring hydroxylation of these compounds, as described earlier.

3. Biochemical Studies

a. *N*-Demethylation Demethylation of methamphetamine to amphetamine can occur by two pathways: *N*-oxidation and C-hydroxylation. The two pathways are mediated by different enzymes and therefore are subject to different inhibitors and inducers. The *N*-demethylation reaction mediated by cytochrome P450 in rats is mediated by both constitutive and phenobarbital-inducible isozymes (Baba et al., 1988). *N*-Demethylation also can occur through *N*-oxidation, catalyzed by FMO to form the nitrone (**21**) and

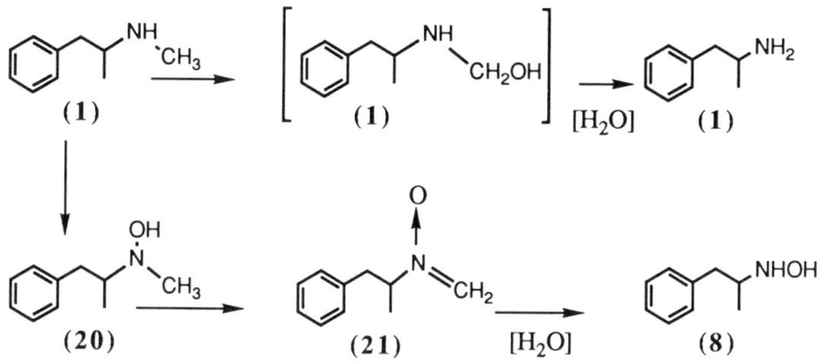

Scheme 12

the Schiff base of N-hydroxymethamphetamine (**22**), which then undergoes hydrolysis to the N-hydroyamphetamine and formaldehyde (Yamada et al., 1984; Baba et al., 1988). Lindeke et al. (1979) demonstrated that N-hydroxyamphetamine also could be oxidized to the nitroso compound (**9**), the metabolic intermediate complex ligand. The role of FMO in the N-hydroxylation-mediated pathway of N-demethylation was demonstrated by Yamada et al. (1984) with selective inhibitors of FMO and cytochrome P450. The results showed that in rat liver microsomes both enzymes participate, whereas in guinea pigs only FMO is involved. These observations were confirmed in reconstitution studies with purified P450 isozymes (Baba et al., 1987b, 1988). Evidence for the inhibition of cytochrome P450 action by methamphetamine, presumably through the metabolic intermediate complex, has been reported by Yamamoto et al. (1988). These workers showed that pretreatment of rats with methamphetamine resulted in reduced demethylation and ring-hydroxylation activities in liver microsomes.

b. Ring Hydroxylation Yamamoto et al. (1984) showed that pretreatment of rats with increasing doses of methamphetamine (5–20 mg/kg over 7 days) reduced ring-hydroxylation activity in liver microsomes. This inhibition may account for the reduction in the excretion of p-hydroxymethamphetamine (**19**) after high methamphetamine doses (Yamada et al., 1986). Based on the reversibility of the inhibition by dialysis and on direct in vitro experiments with N-hydroxymethamphetamine (**20**), these workers suggested that metabolic intermediate complex formation was responsible for the inhibition. However, no spectral evidence for complex formation could be obtained in microsomes from treated animals. Studies by Baba et al. (1988) showed that two isozymes, CYP2C11 and CYP2C13, present as constitutive proteins in the rat liver are responsible for the majority of ring-hydroxylation activity. The levels of these isozymes may be too low to generate measurable levels of the complex.

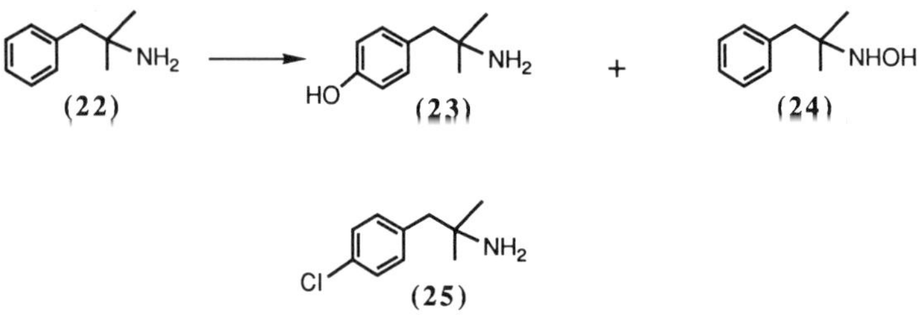

C. Phentermine

1. *In Vivo* Metabolism

The *in vivo* disposition of phentermine (**22**; Scheme 13) has been examined in rats (Cho *et al.*, 1973) and humans (Beckett and Belanger, 1974; Cho, 1974). In rats, 64% of the administered dose is excreted as the 4-hydroxy metabolite (**23**) and 10% as the unchanged drug. Humans excreted less of the 4-hydroxy metabolite but the fraction of the dose was not determined (Cho, 1974). N-Hydroxyphentermine (**24**) at low levels (<5% of the administered dose) and its subsequent metabolites have been found as urinary excretion products in humans. N-Hydroxyphentermine appears to be reduced rapidly in rats, however, since the plasma concentration of phentermine was higher after N-hydroxyphentermine administration than after administration of the amine itself (Cho *et al.*, 1974). In studies with 4-chlorophentermine (**25**) Caldwell *et al.* (1975) found that 25% of the administered dose was excreted as N-oxidation products by rats. These workers suggested that the high N-oxidation-based metabolism was the result of the inability of rats to hydroxylate the aromatic ring. Higher levels of N-oxidation products (10%) also were found in humans after 4-chlorophentermine administration (Beckett and Belanger, 1974; Caldwell *et al.*, 1975).

2. Pharmacokinetics

The pharmacokinetics of phenetermine have been examined in rats (Cho *et al.*, 1973) and humans (Hinsvark *et al.*, 1973). In humans, the elimination rate constant was .029 hr^{-1}, corresponding to a half-life of approximately 24 hr. This half-life is about twice that of amphetamine. The volume of distribution of the drug was estimated to be 3–5 times body volume. In rats, the compound has a half-life of 86 min and a volume of distribution of 2.9 liter/kg. Brain levels of the drug were approximately 20 times higher than the plasma concentration after an intravenous dose (Cho *et al.*, 1973).

3. Biochemistry

The ring-hydroxylation reaction of phentermine has been examined *in vitro* with rat liver microsomes (Cho *et al.*, 1975b). Like amphetamine, this

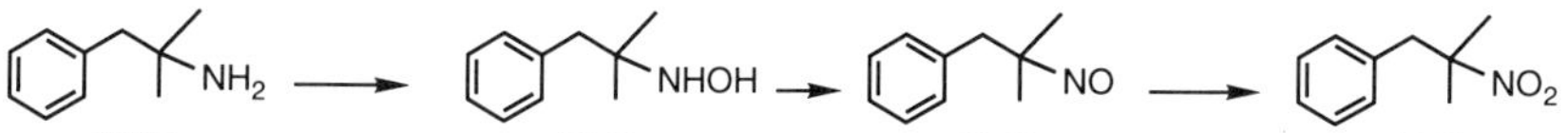

Scheme 14: (22) PhCH₂C(CH₃)₂NH₂ → (24) PhCH₂C(CH₃)₂NHOH → (26) PhCH₂C(CH₃)₂NO → (27) PhCH₂C(CH₃)₂NO₂

compound is *p*-hydroxylated at a very low rate *in vitro* and exhibits substrate inhibition. In contrast, the compound is metabolized actively by rabbit liver preparations (Beckett and Belanger, 1974; Sum and Cho, 1977), but only on the nitrogen. The *N*-oxidation of phentermine proceeds to the corresponding nitro derivative (27; Scheme 14), the product of a reaction sequence involving an initial cytochrome P450-mediated oxidation to form *N*-hydroxyphentermine, uncoupling of the enzyme system to generate superoxide, oxidation of the hydroxylamine by superoxide to the nitroso compound (26), and its autoxidation to the nitro compound (27; Cho et al., 1991). No ring hydroxylation was detected in incubatin mixtures of rabbit liver preparations.

D. Ephedrine

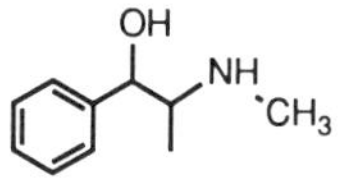

(28) PhCH(OH)CH(NHCH₃)CH₃   (29) PhCH(OH)CH(NH₂)CH₃

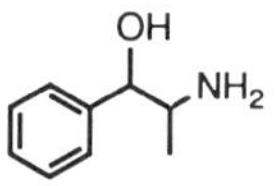

The ephedrines (28) are a group of compounds used as over the counter sympathomimetics and weight loss pills. They consist of two stereoisomers, ephedrine and pseudoephedrine, and their desmethyl analogs (11). Ephedrine is *erythro*-2-methylamino-1-hydroxy-1-phenylpropane and pseudoephedrine is its *threo* isomer. The compounds are related to the amphetamines by β hydroxylation. A racemic mixture of the norephedrines (29) is referred to as phenylpropanolamine (Merck Index).

1. Pharmacokinetics

The pharmacokinetics of ephedrine (Wilkinson and Beckett, 1968), norephedrine (Wilkinson and Beckett, 1968; Lönnerholm et al., 1984), and pseudoephedrine (Graves and Rotenberg, 1989) have been examined in humans. The half-life of (−)-ephedrine is 3.3 hr; that of norephedrine, 2.99 hr (Wilkinson and Beckett, 1968); and that of pseudoephedrine, 4–7 hr (Lönnerholm et al., 1984). The volume of distribution of phenylpropanolamine is 4.4 liter/kg (Lönnerholm et al., 1984).

2. Metabolism

As for other amphetamine derivatives, considerable species differences exist in metabolism of the ephedrines (Scheme 15). Humans excrete 90% of

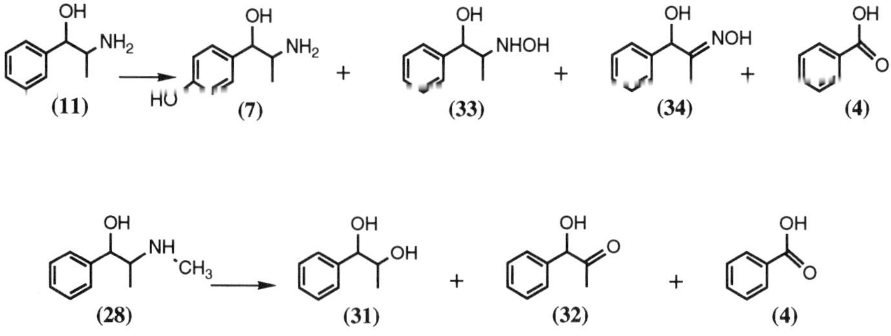

Scheme 15

administered norephedrine as the unchanged compound, whereas rat and guinea pig excrete 38 and 29% of the dose as *p*-hydroxynorephedrine (7) and benzoic acid (4), respectively (Sinsheimer *et al.*, 1973). Benzoic acid is a major metabolite of norephedrine for rabbits but not for humans or rats (Sinsheimer *et al.*, 1973). Baba and co-workers (Baba *et al.*, 1972) found that benzoic acid is also a major metabolite of ephedrine in rabbits (52%) and guinea pigs (26%). The propanediol metabolite (31) also is found in rabbit and guinea pig urine after ephedrine administration. Feller *et al.* (1973) and Feller and Malspeis (1977) have made similar observations with rabbits. These workers also found greater metabolism of the (−) isomer.

3. Biochemistry

A comparison of the metabolism of ephedrine enantiomers by rabbit liver preparations showed an enantiomeric perference to metabolize the (−) isomers more quickly (Feller *et al.*, 1973). Benzoic acid was formed from the (−) isomer at 3 times the rate of the (+) isomer. Products of the deamination reaction, phenylpropanediol (31) and 1-hydroxy-1-phenyl-2-phenylpropane (32), were much greater (~6-fold) after administration of the (−) isomer. These metabolites could be the precursors to benzoic acid, the major *in vitro* metabolite of norephedrine. Baba *et al.* (1969) found roughly equal amounts of norephedrine and benzoic acid with the (−) isomer. Feller *et al.* (1973) found a norephedrine to benzoate ratio of 4 to 1. When the stereoisomers were compared, the order for microsomal conversion to benzoic acid was: (+)-pseudoephedrine ≥ (−)-ephedrine > (−)-pseudoephedrine ≥ (+)-ephedrine (Feller *et al.*, 1973). However, in studies of norephedrine metabolism with similar rabbit liver preparations, Beckett *et al.* (1974) report that N-oxidized metabolites are important. Although no quantitative data were presented, these investigators stated that compounds such as the oxime (34) and N-hydroxynorephedrine (33) are major metabolites and suggested that N-oxidation was at least as important as α-carbon oxidation. These data are difficult to reconcile with the studies of Sinsheimer *et al.* (1973), who account for the majority (~90%) of administered drug in terms of C-oxidation products (propanediol, keto-alcohol, and benzoic acid). One possible explanation is that N-oxidation also may lead to benzoic acid formation.

E. Methylenedioxyamphetamines

(35) methylenedioxyamphetamine with NH$_2$ group; (36) methylenedioxymethamphetamine with NHCH$_3$ group.

Methylenedioxyamphetamine (MDA; 35) and its N-methyl analog methylenedioxymethamphetamine (MDMA; 36) are abused amphetamine derivatives. The pharmacology of the compounds is described elsewhere in this volume. The neurotoxocity of these compounds has been a subject of considerable investigation, in part because of the abuse of MDMA in recent years (Barnes, 1988; Henry et al., 1992). The neurotoxicity appears to be related to metabolism, so extensive studies of the metabolism and the potential toxicity of the metabolites have been done.

1. Pharmacokinetics

No information is available on the human pharmacokinetics of the compounds. Since these compounds are Drug Enforcement Agency (DEA) Schedule 1 agents and have no therapeutic indication, they have not been investigated clinically. In studies of the pharmacokinetics of MDMA enantiomers in rats (Cho et al., 1990b), differences in the levels of MDA formed (36; Scheme 16) were noted; the (+) enantiomer was greater. Conversion of MDMA to MDA is not a major pathway of metabolism, representing about 10% and 3% of the plasma decay for the (+) and (−) isomers, respectively. The half-life of the (+) isomer is 74 min and that of the (−) isomer is 100 min. The demethylation products (37, 38; Scheme 16) are major *in vitro* metabolites and may be the major products for elimination, but urinary excretion was not measured. Compared with the *in vivo* formation of MDA from MDMA, the enantioselectivity is reversed *in vitro*; in microsomal preparations, the enantiomers generate comparable levels of MDA, and the (−) isomer is slightly higher. The biochemical basis for this discrepancy is not clear. The notion that FMO participates in N-demethylation of MDMA was not supported by subsequent experiments.

2. Metabolism

In vivo and *in vitro* metabolism of these compounds was described first by Lim and Foltz (1988), who identified metabolites from urine of animals treated with MDMA and from incubation mixtures of liver preparations. The metabolites, shown in Scheme 16, were established. In a subsequent

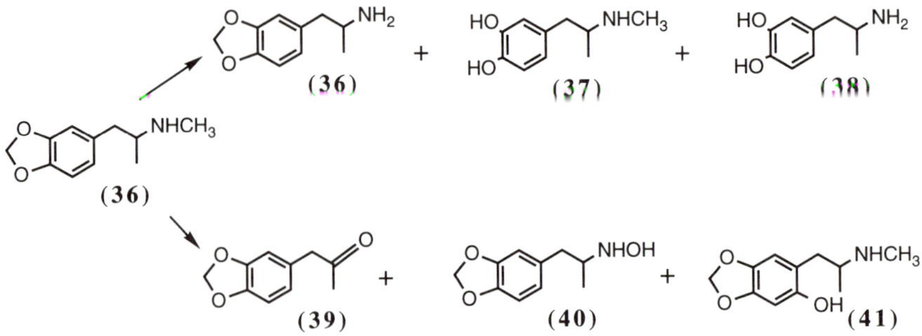

Scheme 16

report, these investigators established the presence of the same metabolites in human urine samples obtained from subjects that had taken MDMA. Subsequent experiments from our laboratories confirmed the pathways, but also demonstrated participation of the Fenton reaction, that is, a hydroxyl radical, in the demethylenation (Kumagai et al., 1991) and ring hydroxylation (T. Chu, unpublished observations). The ease of cleavage of the methylenedioxy ring by hydroxyl radical suggests complications in the analysis of MDMA metabolites, particularly in brain tissue, which contains substantial levels of ascorbate and iron. Phosphate salts used in buffers also contain significant amounts of iron and any reducing agent, in conjunction with hydrogen peroxide, can initiate the reaction. This reaction may have been a complicating factor in the demethylenation activity of brain tissue reported by Steele (Steele et al., 1991). These investigators reported demethylation activity in brain tissue comparable to that of liver, whereas our laboratory found about 1/1000th the activity. Using one batch of phosphate buffer, we also found much higher demethylenation activity that was suppressed by hydroxyl radical scavengers such as thiourea and by the iron chelator desferal (Lin et al., 1992).

3. Biochemistry

a. N-Demethylation The N-demethylation of MDMA requires NADPH and is sensitive to carbon monoxide and SKF-525A (Gollamudi et al., 1989). This reaction also was induced by phenobarbital pretreatment, but 3-methylcholanthrene pretreatment decreased activity. A significant enhancement (4- to 6-fold) of the V_{max}/K_m ratio for the reaction, as catalyzed by liver microsomes from phenobarbital-pretreated rats, suggested that the CYP2B subfamily of cytochrome P450 catalyzed the reaction. Consistent with this notion, the reaction was inhibited by more than 80% by antibody prepared against CYP2B1–CYP2B2. Yamada et al. (1984) reported that the N demethylation of methamphetamine was catalyzed by FMO as well as cytochrome P450, raising the possibility that MDMA may be a FMO substrate. However, denaturation of FMO by heating at 37°C for 60 min (Dyroff and Neal, 1983; Tynes and Hodgson, 1983) had no effect on microsomal N-demethylation.

b. Demethylenation The two consequences of the oxidation of the methylenedioxy group are cleavage to the catechol or oxidation to an intermediate, possibly a carbene (Mansuy et al., 1976b; Delaforge et al., 1985), that forms a complex with the heme of cytochrome P450. Experiments by our group (Hiramatsu et al., 1990) and others (Lim and Foltz, 1988) showed that MDMA is demethylenated to the catechol dihydroxymethamphetamine (DHMA) by cytochrome P450. DHMA is oxidized further by superoxide to the o-quinone, a highly electrophilic species that was identified as its glutathione adduct (Hiramatsu et al., 1990). The demethylenation reaction can be effected by hydroxyl radical as well; in subsequent studies, we noted (Lin et al., 1992) that DHMA is formed from MDMA in incubation mixtures of rat brain microsomes by both chemical and enzymatic processes.

Kinetic studies with rabbit liver enzyme preparations suggested that several cytochrome P450 isozymes participate in the demethylenation of different methylenedioxyphenyl derivatives. For example, P450 isozymes responsible for demethylenation of methylendioxybenzene are phenobarbital inducible, whereas MDMA and MDA oxidation are catalyzed predominantly by constitutive isozymes (Kumagai et al., 1991). Results of experiments using selective inhibitors, immunoinhibition, and reconstitution demonstrated that CYP2B4, but not CYP2B1, oxidizes methylendioxybenzene but that as-yet-unidentified constitutive isoforms are primarily responsible for MDA demethylenation (Kumagai et al., 1992a).

The cytochrome P450 isozymes catalyzing demethylenation of MDMA have been characterized in the rat (Kumagai et al., 1992b). The kinetics of MDMA demethylenation by liver microsomes from male Sprague–Dawley rats were obviously biphasic with low K_m values in the micromolar range and high K_m values in the millimolar range. The low K_m demethylenation was unaffected by pretreatment with typical inducers for CYP1A, 2B, 2E, or the 3A subfamily and was not inhibited by an antibody raised against CYP2C11 that cross-reacts with other members of the 2C family. In contrast, the high K_m demethylenation was induced marginally by phenobarbital or pregnenolone 16α-carbonitrile and was suppressed by ~60% by addition of anti-CYP2B1. The available evidence suggests that cytochrome P450 isozymes belonging to the constitutive CYP2D subfamily catalyze the low K_m demethylenation, whereas members of the CYP2B or other phenobarbital-inducible isoform families such as CYP3A1 catalyze the high K_m reaction. CYP2B and CYP2D isoforms have been found in the brain (Fonne-Pfister et al., 1987; Naslund et al., 1988; Strobel et al., 1989; Anandatheerthavarada et al., 1990; Niznik et al., 1990; Perrin et al., 1990; Tyndale et al., 1991). If these isozymes are catalytically active, they could generate these polar metabolites in concentrations that would not be achieved if formed peripherally.

c. Aromatic Hydroxlation Lim and Foltz (1991) and Kumagai *et al.* (1992c) have reported the formation of 2-hydroxy-4,5-methylenedioxymethamphetamine (2-OH MDMA) after incubation of MDMA with rat or rabbit liver preparations in the presence of an NADPH-generating system. 2-OH MDMA can undergo demethylation to 2,4,5-trihydroxymethamphetamine, a 6-hydroxydopamine analog with the analogous capability of generating hydroxyl radical (Sachs and Jonsson, 1975). The neurotoxicity of 2,4,5-trihydroxymethamphetamine has been evaluated and its possible role in MDMA neurotoxicity has been discussed (Zhiyang *et al.*, 1992). Lim and Foltz (1991) demonstrated that the demethylenation of 2-OH MDMA by the 9000 *g* supernatant of rat liver was inhibited drastically by quinine, an inhibitor of CYP2D1, indicating participation of this isozyme in the reaction. Thus, the demethylenation reaction catalyzed by members of the CYP2D family generate two potentially toxic metabolites—DHMA, a precursor to the electrophilic *o*-quinone, and trihydroxymethamphetamine, a 6-hydroxydopamine analog.

F. *p*-Chloroamphetamine

1. *In Vivo* Metabolism

The neurotoxic properties of *p*-chloroamphetamine (PCA; **42**) on 5-HT neurons have been the basis for its extensive use in experimental pharmacological studies (Fuller, 1978). However, very little is known of either the mechanism of this toxic action and or its pharmacokinetic properties. Early studies on the *in vivo* disposition of PCA were performed with analytical procedures that may have had sensitivity and specificity problems. PCA is accumulated extensively in the brain after peripheral administration (Fuller and Hines, 1967). The compound or its metabolites persist in that tissue. In direct comparisons, PCA has a much longer half-life (8.5 hr) in the brain than amphetamine (1 hr) (Fuller and Hines, 1967). The plasma half-life is about 10 hr when monitored over 24 hr after intraperitoneal dosage. No information on its volume of distribution has been published.

2. Biochemistry

One of the urinary metabolites of PCA (**42**; Scheme 17) is the 3-chloro-4-hydroxy compound (**43**), presumably generated by the NIH shift (Parli and Schmidt, 1975). However, this compound was not observed in brain after direct intraventricular administration of PCA (Gal and Sherman, 1978); *p*-chloronorephedrine (**44**) and 3,4-dimethoxyamphetamine (**45**) were found instead. Formation of the ephedrine derivative suggests dopamine β-hydroxylase involvement, but this possibility has not been assessed. The formation of dimethoxyamphetamine is of interest since its possible precursor dihydroxyamphetamine (**46**) has been reported as an *in vitro*

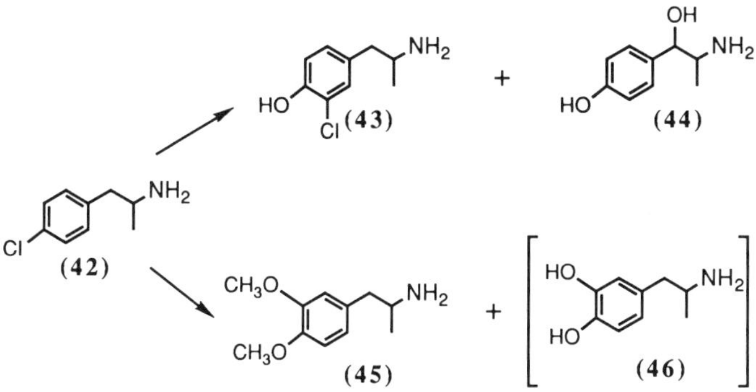

Scheme 17

metabolite of amphetamine (Hoffman *et al.*, 1979) and methylenedioxy-amphetamine (Lim and Foltz, 1988). However, the dimethoxy derivative has not been reported as a metabolite of either compound. Results of an *in vitro* study demonstrating the formation of covalent bonds to brain tissue by PCA have been reported (Miller *et al.*, 1986). Covalent bond formation is NADPH and oxygen dependent, but insensitive to carbon monoxide and SKF 525A. Thus, although reactive metabolites are formed by PCA, details of their generation and their disposition are not understood.

G. Fenfluramine

1. *In Vivo* Usage

Fenfluramine is used in the treatment of obesity, as a hypoglycemic agent, and in the treatment of autistic children. Also, some evidence suggests that the stereoisomers have different pharmacological properties; the S-$(+)$ enantiomer exhibits the anorectic actions and the R-$(-)$ enantiomer exhibits neuroleptic properties, so dispositional differences of the enantiomers is of therapeutic interest.

2. Pharmacokinetics

a. Humans The observation that fenfluramine (**47**; Scheme 18) enantiomers may exhibit different pharmacological properties has resulted in studies of their pharmacokinetics. The half-life of racemic fenfluramine is about 20 hr, with a range of 13.67–30 hr. As for other amphetamine derivatives, the half-life of this compound can be shortened to 11 hr with urine acidification (Pinder *et al.*, 1975; Caccia *et al.*, 1985). This relatively long half-life (vs. 11 hr for amphetamine) may be the result of the slow conversion of the compound to *m*-trifluoromethyl-benzoic acid or its conjugate, the corresponding hippuric acid. This metabolite represents 60–90% of the

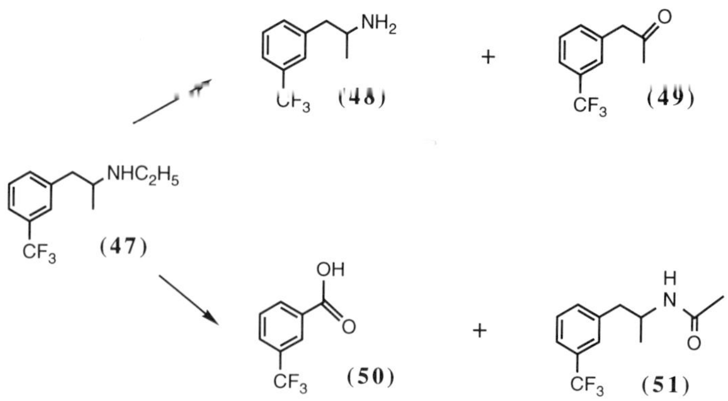

Scheme 18

urinary excretion products of the drug. In a chronic pharmacokinetic study of fenfluramine enantiomers, Caccia et al. (1985) reported steady state plasma levels of 43 ng/ml for the (+) enantiomer and 67 ng/ml for the (−) enantiomer. Corresponding levels of norfenfluramine were 34 and 50 ng/ml, respectively. The plasma half-lives for the two enantiomers under these steady state conditions were significantly different: 18 and 25 hr for the (+) and (−) isomers, respectively. The half-life of norfenfluramine was longer than that of the parent compound; that of the *d* isomer was shorter than that of the *l* isomer (32 vs 50 hr).

b. Animals The half-life differences of the fenfluramine enantiomers are reversed in rats. In pharmacokinetic studies after oral doses in this species, Caccia et al. (1981) reported that the brain half-lives varied with dose, but that the half-life for the (+) isomer was always longer than that for the (−) isomer. The volume of distribution after intravenous administration of the (+) isomer is 8.7 liter/kg (Fracasso et al., 1988); that for the (−) isomer is 6.3–8.1 liter/kg (Spinelli et al., 1988). The plasma half-lives are 2 hr for the (+) isomer (Fracasso et al., 1988) and 1–1.4 hr for the (−) isomer (Spinelli et al., 1988). The (+) and (−) isomers of norfenfluramine (**48**; Scheme 18) have much longer half-lives, about 13 hr (Fracasso et al., 1988; Spinelli et al., 1988). The conversion to norfenfluramine is also 4–5 times greater than for the (−) enantiomer over a dose range of 5–13 mg/kg. The major metabolic pathway in rats appears to be N-deethylation; norfenfluramine (**48**) accounts for 74% of the elimination product for the (+) isomer and 90% for the (−) isomer after an oral dose of 5 mg/kg. As the dose is increased from 5 to 40 mg/kg, however, the proportion of N-deethylation decreases, so at 40 mg/kg N-deethylation represents only 50% of the elimination products. The bioavailability exhibits a marked stereoselectivity in favor of the (+) isomer but, because of its greater metabolism, (−)-

fenfluramine provides a much greater bioavailability of norfenfluramine (Caccia et al., 1981). A comparison of intravenous and oral doses of (−)-fenfluramine showed a bioavailability of only 20%, reflecting a first-pass effect of 80%.

3. Disposition

The large volume of distribution of (+)-fenfluramine (8.7 liter/kg in rats after a dose of 1.25 mg/kg; Spinelli et al., 1988) suggests that the drug is distributed extensively and/or is accumulated. The compound is only 40% bound to plasma protein, so this factor is not important in its distribution. The brain-to-plasma ratios after an oral dose of 5 mg/kg were 10:1 for the (+) isomer and 6:1 for the (−) enantiomer. The brain levels of (+)-fenfluramine were also 4–5 times greater than those of (−)-fenfluramine, whereas the levels of (−)-norfenfluramine were about 2 times greater than those of (+)-norfenfluramine (Caccia et al., 1981).

4. Metabolism

In human studies, 90% of the administered radioactivity from [^{14}C]-labeled (+)-fenfluramine was recovered in urine (Richards et al., 1989) over a 144-hr interval, indicating that the majority of the dose is excreted through the kidneys. Similar observations were made with racemic fenfluramine (Brownsill et al., 1991), since some 45% of the dose was recovered within 24 hr. The urinary metabolites are summarized in Scheme 18 and include most of the common pathways except for aromatic ring hydroxylation (Brownsill et al., 1991). Cytochrome P450-mediated aromatic ring hydroxylation favors an electron-rich ring; the trifluoromethyl group may reduce the electron density of the ring sufficiently to reduce activity of this pathway of metabolism. The N-dealkylation reaction occurs in both directions yielding norfenfluramine (**48**) and *m*-trifluoromethyl phenylacetone (**49**). The subsequent metabolites are those from the phenylacetone derivative. *m*-Trifluorobenzoic acid (**50**) is the major excreted metabolite.

Norfenfluramine is also the major metabolite of fenfluramine in rats and other experimental animals, including the guinea pig (Fuller et al., 1988). Although most of the conversion takes place in the liver it also occurs in the brain, since direct intracerebral–ventricular injections of fenfluramine resulted in conversion to norfenfluramine and its N-acetyl conjugate (**51**).

H. Deprenyl

Deprenyl (**52**; Scheme 19), the N-propargyl derivative of methamphetamine (**18**), is an MAO-B inhibitor currently used to treat Parkinson's disease. The actions of this compound appear to include an immediate therapeutic action and a protective effect against the progression of the disease (Rinne et al., 1978; Karoum et al., 1982). The compound is a

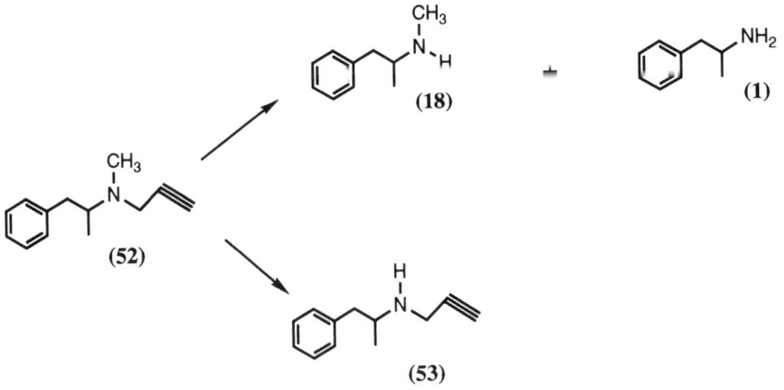

Scheme 19

mechanism-based inhibitor of MAO-B, that is, it is converted by the enzyme to a reactive intermediate that irreversibly inhibits the enzyme. The inhibition is based on the oxidation of the amine by MAO to an imine, followed by covalent addition of the propargyl group to the flavin in the enzyme (Maycock et al., 1976). In addition, however, the compound is converted rapidly and efficiently to methamphetamine by cytochrome P450 (Yoshida et al., 1986), leading to the suggestion that much of its pharmacology is the result of the generated methamphetamine (Karoum et al., 1982; Engberg et al., 1991). The R-(−) enantiomer is 25 times more effective as an MAO-B inhibitor than the S-(+) enantiomer (Robinson, 1985) and is converted to R-(−)-methamphetamine. The presence of a propargyl group markedly alters the base strength; deprenyl has a pK_a of 7.4, compared with 9.8 for methamphetamine and 9.6 for amphetamine (Robinson, 1985). Therefore, at physiological pH, deprenyl is present in much greater proportion as the neutral uncharged species and, when corrected for the proportion of neutral form, the K_i of deprenyl against MAO-B is approximately one-tenth that of other amphetamines (Robinson, 1985). The difference in MAO inhibition by deprenyl is its irreversible nature.

1. Disposition

Studies with human subjects have shown an apparent stereochemical difference in clearance since the (+) enantiomer is cleared rapidly and the (−) enantiomer is retained in the brain. The steady state concentrations of the drug in the plasma and cerebrospinal fluid (CSF) were very similar, indicating that the compound readily enters the brain (Heinonen et al., 1989).

2. Metabolism

Studies with rats liver microsomes have shown that deprenyl is converted to methamphetamine, amphetamine, and desmethyl deprenyl (53;

Scheme 19), with methamphetamine (20) as the major (60–70%) metabolite in liver microsomes. The microsomal metabolism exhibits sex differences. Preparations from male rats are approximately 2 times as active as preparations from females (Yoshida *et al.*, 1987). *In vivo* studies have shown that up to 25% of the administered dose is excreted in 24 hr as amphetamine derivatives, that is, compounds without the propargyl group. The proportion is even greater in humans, since 44–60% of the drug is eliminated as methamphetamine and 15–27% as amphetamine. Desmethyl deprenyl (53) represents only 1% of the administered dose in urine, so this pathway would seem to be very minor (Heinonen *et al.*, 1989). The depropargylation reaction appears to be dependent on cytochrome P450 in the liver because it is induced by phenobarbital (Weli *et al.*, 1985; Yoshida *et al.*, 1987) and inhibited by SKF 525A (Yoshida *et al.*, 1987). When the effects of cytochrome P450 induction and inhibition on MAO-B inhibition were examined, the liver was found to respond in a manner consistent with the notion that cytochrome P450-dependent metabolism removes the active inhibitor. However, inhibition of brain MAO-B after administration of deprenyl was not affected by any of these treatments (Yoshida *et al.*, 1986), suggesting that the brain disposes of the drug differently and that the peripheral metabolism of deprenyl does not affect the brain levels of the drug.

I. Cathinone

Cathinone (55) is an alkaloid of the khat plant, found in East Africa and the Arab Peninsula. The leaves of the khat plant are chewed for their ability to induce psychic excitement. Chemically, the compound is the β-keto derivative of *d*-amphetamine and has very similar central and peripheral actions (Brenneisen *et al.*, 1990). The β-keto structure makes the α carbon susceptible to racemization, which results in a reduction of activity. This process occurs as the leaf tissue decomposes, so users desire fresh leaves. One of the experimental findings of studies of the pure compound is its rapid action; cathinone response occurs within 15 min under conditions that require 30 min for amphetamine response (Schechter, 1989). Limited data have appeared reporting the pharmacokinetics and metabolism of cathinone. In a small human subject study, Brenneisen *et al.* (1990) found maximal plasma levels 1–2 hr after an oral dose and a half-life of about 3 hr. The major metabolite, norephedrine (30), rises to a plateau in approximately 1 hr and persists for at least 8 hr after cathinone dose. The natural form of the alkaloid, *S*-(−)-cathinone, is metabolized to norephedrine and norpseudoephedrine in a 9:1 ratio, whereas the *R*-(+) form is metabolized in favor of norpseudoephedrine (1:2). The reduction is stereospecific and only the *R* configuration is obtained (Brenneisen *et al.*, 1985), but the enzymology of this reaction does not appear to have been established. Note that the ephedrines and cathinone only account for about 40% of the administered compound, so additional pharmacologically active metabolites may be formed.

III. CONCLUSIONS

The functionalization of amphetamine derivatives takes place primarily at two centers, the basic nitrogen and the aromatic ring. The reactions are catalyzed by a number of cytochrome P450 isozymes, both constitutive and inducible. The flavin-dependent monooxygenase FMO also participates in the oxidation of the basic nitrogen. The aryl acetone formed by oxidation of the α-carbon is thought to be the precursor of benzoic acid, a common terminal metabolite. However, the details of formation of this compound from amphetamines appear to be highly complex and remain to be elucidated.

The oxidation of the aromatic ring is also a common metabolic route that generates pharmacologically active metabolites. The reaction has been demonstrated for unsubstituted and methylenedioxy- and chloro-substituted derivatives. These reactions also appear to be catalyzed by several P450 isozymes, including members of the CYP2D family. This family is of interest in human drug metabolism because it exhibits a polymorphism within the Caucasian population. Individuals who are deficient in this isozyme have limited ability to metabolize this and other CNS drugs and exhibit a greater sensitivity to their actions.

Although the biochemical details of Phase II metabolism are not as well characterized, urinary excretion studies of amphetamines (e.g., Ellison *et al.*, 1965; Dring *et al.*, 1970), methamphetamine (Caldwell *et al.*, 1972), ephedrine (Baba *et al.*, 1972; Feller and Malspeis, 1977), and fenfluramine (Brownsill *et al.*, 1991) all indicated that the glycine and glucuronic acid conjugates of the benzoic acids and the glucuronides of ring-hydroxylated metabolites are common excretion products.

Table I lists selected phrarmacokinetic parameter values from the literature. The values are listed for humans and for rats, the latter because of their extensive use in experimental behavioral and neurochemical studies of this group of drugs. The varied conditions of the pharmacokinetic studies may account for the descrepancies between values. The reader is advised to consult the primary literature to assess the differences and their relevance to the experiment being considered.

Overall, the metabolism of these compounds affects their pharmacology and duration of action through the formation of active and inactive metabolites. The notion that the more polar metabolites do not have access to the

TABLE I Pharmacokinetic Properites of Some Amphetamines

Compound[a]	Species	Vd[b] (1/Kg)	$t_{1/2}$	Reference
Amphetamine (+)	Human	NA[c]	7 ± 1.2 hr	Vree (1973)
Amphetamine (−)	Human	NA	11 ± 2.1 hr	Vree (1973)
Amphetamine (+)	Rat	2.3	57 ± 4.8 min	Cho et al. (1973)
Methamphetamine (+)	Rat	4.1	45 min	Melega et al. (1992)
Phentermine	Human	NA	24 hr	Hinsvark et al. (1973)
Phentermine	Rat	2.9	86 min	Cho et al. (1973)
Ephedrine (−)	Human	NA	3.3 hr	Wilkinson and Beckett (1968)
Norephedrine (−)	Human	NA	2.99 hr	Wilkinson and Beckett (1968)
Phenylpropanolamine	Human	4.4	4–7 hr	Wilkinson and Beckett (1968)
MDMA[d] (+)	Rat	1.65	74 min	Cho et al. (1990a)
MDMA (−)	Rat		100 min	Cho et al. (1990a)
p-Chloroamphetamine	Rat (?)	NA	8.5 hr	Fuller and Hines (1967)
Fenfluramine (±)	Human	NA	13.7–30 hr	Caccia et al. (1985)
Fenfluramine (+)	Human	NA	18 hr	Caccia et al. (1985)
Fenfluramine (−)	Human	NA	25 hr	Caccia et al. (1985)
Fenfluramine (+)	Rat	8.7	2 hr	Caccia et al. (1981)
Fenfluramine (−)	Rat	6.3–8.1	1–1.4 hr	Caccia et al. (1981)

[a] Enantiomer (+ or −) given in parentheses where appropriate.
[b] Vd, Volume of distribution.
[c] NA, Not available.
[d] MDMA, Methylenedioxymethamphetamine.

brain because of the blood–brain barrier may be questioned now in light of recent demonstrations (Warner et al., 1988; Tyndale et al., 1991c; Bergh and Strobel, 1992) of the presence of different cytochrome P450 isozymes in the brain. The data indicate, further, that different isozymes are present in different cell types and in different regions. If these isozymes are catalytically active, they would generate metabolites, including some polar compounds, in a highly localized manner. Therefore, future research on drug metabolism by the brain will impact our understanding of the pharmacology of the amphetamines as well as of other centrally active compounds.

ACKNOWLEDGEMENTS

Work from our laboratories described in this chapter was supported by the National Institute on Drug Abuse through USPHS grants DA-04206 and DA-02411.

REFERENCES

Anandatheerthavarada, H. K., Shankar, S. K., and Ravindranath, V. (1990). Rat brain cytochrome P-450: Catalytic, immunochemical properties and inducibility of multiple forms. *Brain Res.* 536, 339–343.
Axelrod, J. (1954). Studies on sympathomimetic amines. II. The biotransformation and physiological disposition of D-amphetamine, D-p-hydroxyamphetamine and D-methamphetamine. *J. Pharmacol. Exp. Ther.* 110, 315–320.
Axelrod, J. (1955). The enzymatic deamination of amphetamine–benzedrine. *J. Biochem.* 214, 753–763.
Baba, S., Matsuda, A., and Nagase, Y. (1969). Studies on the analysis of drugs by use of radioisotopes. V. Identification of metabolites of ephedrine in the rabbit liver. *Yakugaku Zasshi* 89, 833–836.
Baba, S., Enogaki, K., Matsuda, A., and Nagase, Y. (1972). Studies on the analysis of drugs by use of radioisotopes. VIII. Species differences in 1-ephedrine metabolism. *Yakugaku Zasshi* 92, 1270–1274.
Baba, T., Yamada, H., Oguri, K., and Yoshimura, H. (1987). A new metabolite of methamphetamine: Evidence for formation of N-[(1-methyl-2-phenyl)ethyl]ethanimine N-oxide. *Xenobiotica* 17, 1029–1038.
Baba, T., Yamada, H., Oguri, K., and Yoshimura, H. (1988). Participation of cytochrome P450 isozymes in N-demethylation, N-hydroxylation and aromatic hydroxylation of methamphetamine. *Xenobiotica* 18, 475–484.
Baggot, J. D., and Davis, L. E. (1973). A comparative study of pharmacokinetics of amphetamine. *Res. Vet. Sci.* 14, 207–214.
Barnes, D. M. (1988). New data intensify the agony over ectasy. *Science* 239, 864–866.
Beckett, A. H., and Rowland, M. (1965). Urinary excretion kinetics of methamphetamine in man. *J. Pharm. Pharmacol.* (Suppl.) 17, 109S–114S.
Beckett, A. H., and Belanger, P. M. (1974). Metabolism of chlorphentermine and phentermine in man to yield hydroxylamino, C-nitroso- and nitro-compounds. *J. Pharm. Pharmacol.* 26, 205–206.
Beckett, A. H., Jones, G. R., and Al-Sarraj, S. (1974). Metabolic N- and alpha-C-oxidation of norephedrine by rabbit liver microsomal fractions and synthesis of the metabolic products. *J. Pharm. Pharmacol.* 26, 945–951.
Bergh, A. F., and Strobel, H. W. (1992). Reconstitution of the brain mixed function oxidase system: Purification of NADPH–cytochrome P450 reductase and partial purification of cytochrome P450 from whole rat brain. *J. Neurochem.* 59, 575–581.
Billings, Ruth E., Murphy, Patrick J., McMahon, Robert E., and Ashmore, John (1978). Aromatic hydroxylation of amphetamine with rat liver microsomes, perfused liver, and isolated hepatocytes. *Biochem. Pharmacol.* 27, 2525–2529.
Brenneisen, R., Gesshusler, S., and Schorno, X. .(1985). Metabolism of cathinone to (−)-norephedrine and (−)-norpseudoephedrine. *J. Pharm. Pharmcol.* 38, 298–300 872.
Brenneisen, R., Fisch, H.-U., Koelbing, U., Geisshusler, S., and Kalix, P. (1990). Amphetamine-like effects in humans of khat alkaloid cathinone. *Br. Clin. Pharmacol.* 30, 825–828.
Britto, M. R., and Wedlund, P. J. (1992). Cytochrome P450 in the brain. Potential evolutionary and therapeutic relevance of localization of drug-metablizing enzymes. *Drug Metab. Disp.* 20, 446–450.
Brodie, B. B., Cho, A. K., Stefano, F. J. E., and Gessa, G. L. (1969). On mechanisms of norepinephrine release by amphetamine and tyramine and tolerance to their effects. *In* "Advances in Biochemical Psychopharmacology" (E. Costa and P. Greengard, eds.), Vol. 1, pp. 219–238. Raven Press, New York.
Brodie, B. B., Cho, A. K., and Gessa, G. L. (1970). Possible role of p-hydroxynorephedrine in the depletion of norepinephrine induced by d-amphetamine and in tolerance to this drug.

In "Amphetamines and Related Compounds" (Proceedings of the Mario Negri Institute for Pharmacological Research, Milan, Italy) (E. Costa and S. Garattini, eds.), pp. 217–230. Raven Press, New York.

Brosen, K., and Gram, L. F. (1989). Clinical significance of the sparteine/debrisoquine oxidation polymorphism. *Eur. J. Clin. Pharamcol.* **36**, 537–547.

Brownsill, R., Wallace, D., Taylor, A., and Campbell, B. (1991). Study of human urinary metabolism of fenfluramine using gas chromatography–mass spectrometry. *J. Chromatogr.* **562**, 267–277.

Caccia, S., Dagnino, G., Garattini, S., Guiso, G., Madonna, R., and Zanini, M. G. (1981). Kinetics of fenfluramine isomers in the rat. *Eur. J. Drug Metab. Pharmacokinet.* **6**, 297–301.

Caccia, S., Conforti, I., Duchier, J., and Garattini, S. (1985). Pharmacokinetics of fenfluramine and norfenfluramine in volunteers given D- and D,L-fenfluramine for 15 days. *Eur. J. Clin. Pharmacol.* **29**, 221–224.

Caldwell, J., Dring, L. G., and Williams, R. T. (1972). Metabolism of ^{14}C methamphetamine in man, the guinea and the rat. *Biochem. J.* **129**, 11–22.

Caldwell, J., Koster, U., Smith, R. L., and Williams, R. T. (1975). Species variations in the N-oxidation of chlorphentermine. *Biochem. Pharmacol.* **24**, 2225–2232.

Cho, A. K. (1974). The identification of *p*-hydroxyphentermine as an urinary metabolite of phentermine. *Res. Commun. Chem. Pathol. Pharmacol.* **7**, 67–78.

Cho, A. K., and Wright, J. (1978). Minireview: Pathways of metabolism of amphetamine and related compounds. *Life Sci.* **22**, 363–372.

Cho, A. K., Hodshon, B. J., Lindeke, B., and Miwa, G. (1973). Application of quantitative GC–mass spectrometry to study of pharmacokinetics of amphetamine and phentermine. *J. Pharm. Sci.* **62**, 1491–1494.

Cho, A. K., Lindeke, B., and Jenden, D. J. (1974). GC/MS in the distribution and metabolism of phentermine. *In* "Mass Spectrometry in Biochemistry and Medicine" (A. Frigerio and N. Castagnoli, Jr., eds.), pp. 83–90. Raven Press, New York.

Cho, A. K., Schaeffer, J. C., and Fischer, J. F. (1975a). The accumulation of 4-hydroxyamphetamine by rat striatal homogenates. *Biochem. Pharmacol.* **24**, 1540–1542.

Cho, A. K., Hodshon, B. J., Lindeke, B., and Jonsson, J. (1975b). The *p*-hydrodxylation of amphetamine and phentermine by rat liver microsomes. *Xenobiotica* **5**, 531–538.

Cho, A. K., Hiramatsu, M., DiStefano, E. W., Chang, A. S., and Jenden, D. J. (1990a). Stereochemical differences in the metabolism of 3,4-methylenedioxymethamphetamine *in vivo* and *in vitro*. A pharmacokinetic analysis. *Drug Metab. Disp.* **18**, 686–691.

Cho, A. K., Hiramatsu, M., Nabeshima, T., and Kameyama, T. (1990b). Pharmacokinetic and pharmacodynamic properties of phencyclidine analogs. *In* "NMDA Related Agents: Biochemistry, Pharmacology and Behavior," Proceedings of the Satellite Symposium of 17th Congress of Collegium Internationale Neuropsychopharmacologicum, Nagoya, Japan, pp. 361–370.

Cho, A. K., Duncan, J. D., and Fukuto, J. M. (1991). Studies on the N-oxidation of phentermine: Evidence for an indirect pathway of N-oxidation mediated by cytochrome P450. *In* "N-Oxidation of Drugs: Biochemistry, Pharmacology, Toxicology" (P. Hlavica, L. A. Damani, and J. W. Gorrod, eds.). pp. 207–216. Ellis Horwood, Chichester.

Clay, G. A., Cho, A. K., and Roberfroid, M. (1971). Effect of diethylaminoethyl diphenylpropylacetate hydrochloride (SKF-525A) on the norepinephrine-depleting actions of *d*-amphetamine. *Biochem. Pharmacol.* **20**, 1821–1831.

Creveling, C. R., Daly, J. W., Witkop, B., and Udenfriend, S. (1962). Substrates and inhibitor of dopamine-beta-oxidase. *Biochim. Biophys. Acta* **64**, 125–134.

Davis, J. M., Kopin, I. J., Lemberger, L., and Axelrod, J. (1971). Effects of urinary pH on amphetamine metabolism. *Ann. N.Y. Acad. Sci.* **179**, 493–501.

Delaforge, M., Ioannides, C., and Parke, D. V. (1985). Ligand–complex formation between

cytochromes P-450 and P-448 and methylene-dioxyphenyl compounds. *Xenobiotica* **15**, 333–342.
Dougan, D. F. H., Duffield, A. M., Duffield, P. H., and Wade, D. N. (1986). Stereoselective accumulation of hydroxylated metabolites of amphetamine in rat striatum and hypothalamus. *Br. J. Pharmacol.* **88**, 285–290.
Dring, L. G., Smith, R. L., and Williams, R. T. (1970). The metabolic fate of amphetamine in man and other species. *Biochem. J.* **116**, 425–435.
Dyroff, M. C., and Neal, R. A. (1983). Studies on the mechanism of metabolism of thioacetamide S-oxide by rat liver microsomes. *Mol. Pharmacol.* **23**, 219–227.
Ellison, T., Gutzait, L., and Van Loon, E. J. (1965). The comparative metabolism of d-amphetamine-C14 in the rat, dog, and monkey. *J. Pharmacol. Exp. Ther.* **152**, 383–387.
Engberg, G., Elebring, T., and Nissbrandt, H. (1991). Deprenyl (Selegiline), a selective MAO-B inhibitor with active metabolites; effects on locomotor activity, dopaminergic neurotransmission and firing rate of nigral dopamine neurons. *J. Pharmacol. Exp. Ther.* **259**, 841–847.
Feller, D. R., and Malspeis, L. (1977). Biotransformation of D(−)-ephedrine in the rabbit, in vivo and in vitro. *Drug Metab. Disp.* **5**, 37–46.
Feller, D. R., Basu, P., Mellon, W., Curott, J., and Malspeis, L. (1973). Metabolism of the ephedrine isomers in rabbit liver. *Arch Int. Pharmacodyn. Ther.* **203**, 187–199.
Florence, V. M., Di Stefano, E. W., Sum, C. Y., and Cho, A. K. (1982). The metabolism of R-(−)-amphetamine by rabbit liver microsomes: Initial products. *Drug Metab. Disp.* **10**, 312–315.
Fonne-Pfister, R., Bargetzi, M. J., and Meyer, U. A. (1987). MPTP, the neurotoxin inducing parkinson's disease, is a potent competitive inhibitor of human and rat cytochrome P450 isozymes (P450bufI, P450dbl) catalyzing debrisoquine 4-hydroxylation. *Biochem. Biophys. Res. Commun.* **148**, 1144–1150.
Fracasso, C., Guiso, G., Garratini, S., and Caccia, S. (1988). Disposition of D-fenfluramine in lean and obese rats. *Appetite* **10**, 45–55.
Franklin, M. R. (1974). Inhibition of the metabolism of N-substituted amphetamines by SKF 525-A and related compounds. *Xenobiotica* **4**, 143–150.
Franklin, M. R. (1977). Inhibition of mixed function oxidations by substrates forming reduced cytochrome P450 metabolic intermediate complexes. *Pharmacol. Ther. A* **2**, 227–245.
Fuller, R. (1978). Neurochemical effects of serotonin neurotoxins: An introduction. *Ann. N.Y. Acad. Sci.* **305**, 178–181.
Fuller, R. W., and Hines, C. W. (1967). Tissue levels of chloroamphetamines in rats and mice. *J. Pharm. Sci.* **56**, 302–303.
Fuller, R. W., Snoddy, H. D., and Perry, K. W. (1988). Metabolism of fenfluramine to norfenfluramine in guinea-pigs. *J. Pharm. Pharmacol.* **40**, 439–441.
Gal, E. M., and Sherman, A. D. (1978). Cerebral metabolism of some serotonin depletors. *Ann. N.Y. Acad. Sci.* **305**, 119–128.
Gollamudi, R., Ali, S. F., Lipe, G., Newport, G., Webb, P., Lopez, M., Leakey, J. E. A., Kolta, M., and Slikker, W. (1989). Influence of inducers and inhibitors on the metabolism in vitro and neurochemical effects in vivo of MDMA. *Neurotoxicology* **10**, 455–466.
Graves, D. A., and Rotenberg, K. S. (1989). Pseudoephedrine absorption from controlled release formulations: Absorption rate constant estimation methods. *Biopharmaceutics Drug Disp.* **10**, 127–136.
Heinonen, E. H., Myllyla, V., Sotaniemi, K., Lammintausta, R., Salonen, J. S., Anttila, M., Savijarvi, M., Kotila, M., and Rinne, U. K. (1989). Pharmacokinetics and metabolism of selegiline. *Acta Neurol. Scand.* **126**, 93–99.
Henry, J. A., Jeffreys, K. J., and Dawling, S. (1992). Toxicity and deaths from 3,4-methylenedioxymethamphetamine ("ecstasy"). *Lancet* **340**, 384–387.
Hinsvark, O. N., Truant, A. P., Jenden, D. J., and Steinborn, J. A. (1973). The oral bio-

availability and pharmacokinetics of soluble and resin-bound forms of amphetamine and phentermine in man. *J. Pharmacokinet. Biopharmaceut.* **1**, 319–328.

Hiramatsu, M., Kumagai, Y., Unger, S. E., and Cho, A. K. (1990). Metabolism of methylenedioxymethamphetamine (MDMA): Formation of dihydroxymethoamphetamine and quinone identified as its glutathione adduct. *J. Pharamcol. Exp. Ther.* **254**, 521–527.

Hoffman, A. R., Rama Sastry, B. V., and Axelrod, J. (1979). Formation of alpha-methyldopamine ("Catecholamphetamine") from p-hydroxyamphetamine by rat brain microsomes. *Pharmacology* **19**, 256–260.

Hucker, H. B., Michniewicz, B. M., and Rhodes, R. E. (1971). Phenylacetone oxime—An intermediate in the oxidative deamination of amphetamines. *Biochem. Pharmacol.* **20**, 2123–2128.

Jonsson, J. A. (1974). Hydroxylation of amphetamine to prahydroxyamphetamine by rat liver microsomes. *Biochem. Pharmacol.* **23**, 3191–3197.

Jonsson, J., and Lindeke, B. (1976). On the formation of cytochrome P-450 product complexes during the metabolism of phenylalkylamines. *Acta Pharm. Suec.* **13**, 313–320.

Jori, A., and Caccia, S. (1974). Distribution of amphetamine and its hydroxylated metabolites in various areas of the rat brain. *J. Pharm. Pharmacol.* **26**, 746–748.

Kammerer, R. C., Cho, A. K., and Jonsson, J. (1978). In vitro metabolism of pheylacetone, phenyl-2-butanone and 3-methyl-1-phenyl-2-butanone by rabbit liver preparations. *Drug Metab. Disp.* **6**, 396–402.

Karoum, F., Chuang, L.-W., Eisler, T., Calne, D. B., Liebowitz, M. R., Quitkin, F. M., Klein, D. F., and Wyatt, R. J. (1982). Metabolism of (−)deprenyl to amphetamine and methamphetamine may be responsible for deprenyl's therapeutic benefit: A biochemical assessment. *Neurology* **32**, 503–509.

Kuhn, C. M., Schanberg, S. M., and Breese, G. R. (1978). Metabolism of amphetamine by rat brain tissue. *Biochem. Pharmacol.* **27**, 343–351.

Kumagai, Y., Lin, L. Y., Schmitz, D. A., and Cho, A. K. (1991). Hydroxyl radical mediated demethylenation of (methylenedioxy)phenyl compounds. *Chem. Res. Toxicol.* **4**, 330–334.

Kumagai, Y., Lin, L. Y., Philpot, R. M., Yamada, H., Oguri, K., Yoshimura, H., and Cho, A. K. (1992a). Regiochemical differences in cytochrome P450 isozymes responsible for the oxidation of methylenedioxyphenyl groups by rabbit liver. *Mol. Pharmacol.* **42**, 695–702.

Kumagai, Y., Lin, L. Y., and Cho, A. K. (1992b). Cytochrome P450 isozymes responsible for the metabolic activation of methylenedioxymethamphetamine (MDMA) in rat. *FASEB J.* **6**, A2567.

Kumagai, Y., Schmitz, D. A., and Cho, A. K. (1992c). Aromatic hydroxylation of methylenedioxybenzene (MDB) and methylenedioxymethamphetamine (MDMA) by rabbit cytochrome P450. *Xenobiotica* **22**, 395–403.

Lennard, M. S. (1990). Genetic polymorphism of sparteine/debrisoquine oxidation: A reappraisal. *Pharmacol. Toxicol.* **67**, 273–283.

Lewander, T. (1971a). A mechanism for the development of tolerance to amphetamine in rats. *Psychopharmacologia* **21**, 17–31.

Lewander, T. (1971b). On the presence of p-hydroxynorephedrine in the rat brain and heart in relation to changes in catecholamine levels after administration of amphetamine. *Acta Pharmacol. Toxicol.* **29**, 33–48.

Liccione, J. J., and Maines, M. D. (1989). Manganese-mediated increase in the rat brain mitochondrial cytochrome P450 and drug metabolism activity: Susceptibility of the striatum. *J. Pharmacol. Exp. Ther.* **248**, 222–228.

Lim, H. K., and Foltz, R. L. (1988). In vivo and in vitro metabolism of 3,4-(methylenedioxy)methamphetamine in the rat: Identification of metabolites using an ion trap detector. *Chem. Res. Toxicol.* **1**, 370–378.

Lim, H. K., and Foltz, R. L. (1991). Ion trop tandem mass spectrometric evidence for the metabolism of 3,4 (methylenedioxy) methamphetamine to the potent neutrotoxins 2,4,5-

trihydroxymethamphetamine and 2,4,5-trihydroxyamphetamine. *Chem. Res. Toxicol.* **4**, 626–632.

Lin, L. Y., Kumagai, Y., and Cho, A. K. (1992). Enzymatic and chemical demethylenation of (methylenedioxy)amphetamine and (methylenedioxy)methamphetamine by rat brain microsomes. *Chem. Res. Toxicol.* **5**, 401–406.

Lindeke, B., and Cho, A. K. (1982). N-Dealkylation and Deamination. *In* "Metabolic Basis of Detoxication. Metabolism of Functional Groups" (W. B. Jakoby, J. R. Bend, and J. Caldwell, eds.). pp. 105–127. Academic Press, New York.

Lindeke, B., and Paulsen-Sorman, U. (1988). Nitrogenous compounds as ligands to hemoporphyrins—The concept of metabolic intermediary complexes. *In* "Biotransformation of Organic Nitrogen Compounds" (A. K. Cho and B. Lindeke, eds.), pp. 63–102. Karger, Basel.

Lindeke, B., Paulsen, U., and Anderson, E. (1979). Cytochrome P450 complex formation in the metabolism of phenylakylamines. IV. Spectral evidences for metabolic conversion of methamphetamine to N-hydroxyamphetamine. *Biochem. Pharmacol.* **28**, 3629–3635.

Lindeke, B., Paulsen-Sorman, U., Hallstrom, G., Khuthier, A.-H., Cho, A. K., and Kammerer, R. C. (1982). Cytochrome P-455-nm complex formation in the metabolism of phenylalkylamines. VI. Structure–activity relationships in metabolic intermediary complex formation with a series of a α-substituted 2-phenylethylamines and corresponding N-hydroxylamines. *Drug Metab. Disp.* **10**, 700–705.

Lönnerholm, G., Grahnén, G., and Lindström, B. (1984). Steady-state kinetics of sustained-release phenylpropanolamine. *Int. J. Clin. Pharmacol. Ther. Toxicol.* **22**, 39–41.

Mahgoub, A., Idle, J. R., Dring, L. G., Lancaster, R., and Smith, R. L. (1977). Polymorphic hydroxylation of debrisoquine in man. *Lancet* **2**, 584–586.

Mansuy, D., Beaune, P., Chottard, J. C., Bartoli, J. F., and Gans, P. (1976). The nature of "455-nm absorbing complex" formed during the cytochrome P450-dependent oxidative metabolism of amphetamine. *Biochem. Pharmacol.* **25**, 609–612.

Matsuda, L. A., Hanson, G. R., and Gibb, J. W. (1989). Neurochemical effects of amphetamine metabolites on central dopaminergic and serotonergic systems. *J. Pharmacol. Exp. Ther.* **251**, 901–908.

Matsumoto, R. M., and Cho, A. K. (1982). Conversion of N-hydroxyamphetamine to phenylacetone oxime by rat liver microsomes. *Biochem. Pharmacol.* **31**, 105–108.

Maycock, A. L., Abeles, R. H., Salach, J. I., and Singer, T. P. (1976). The structure of the covalent adduct formed by the interaction of 3-dimethylamino-1-propyne and the flavine of mitochondrial amine oxidase. *Biochemistry* **15**, 114–125.

Melega, W. P., Williams, A. E., Schmitz, D., Stefano, E. D., and Cho, A. K. (1992). Pharmacokinetic/pharmacodynamic analysis of D-amphetamine and D-methamphetamine. *Soc. Neurosci. Abstr.* **18**, 363.

Miller, K. J., Anderholm, D. C., and Ames, M. M. (1986). Metabolic activation of the serotonergic neurotoxin para-chloroamphetamine to chemically reactive intermediates by hepatic and brain microsomal preparations. *Biochem. Pharmacol.* **35**, 1737–1742.

Naslund, B. M. A., Glaumann, H., Warner, M., Gustafsson, J.-A., and Hansson, T. (1988). Cytochrome P-450 *b* and *c* in the rat brain and pituitary gland. *Mol. Pharmacol.* **33**, 31–37.

Nebert, D. W., Nelson, D. R., Adesnik, M., Coon, M. J., Estabrook, R. W., Gonzalez, F. J., Guengerich, F. P., Gunsalus, I. C., Johnson, E. F., Kemper, B., Levin, W., Phillips, I. R., Sato, R., and Waterman, M. R. (1989). The P450 superfamily: Updated listing of all genes and recommended nomenclature for the chromosomal loci. *DNA* **8**, 1–13.

Niznik, H. B., Tyndale, R. F., Sallee, F. R., Gonzalez, F. J., Hardwick, J. P., Inaba, T., and Kalow, W. (1990). The dopamine transporter and cytochrome P450IID1 (debrisoquine 4-hydroxylase) in brain: Resolution and identification of two distinct [^{3}H]GBR-12935 binding proteins. *Arch. Biochem. Biophys.* **276**, 424–432.

Ortiz de Montellano, P. R., ed. (1986). "Cytochrome P-450 Structure, Mechanism, and Biochemistry." Plenum Press, New York.

Otton, S. V., Crewe, H. K., Lennard, M. S., Tucker, G. T., and Woods, H. F. (1988). Use of quinidine inhibition to define the role of the sparteine/debrisoquine cytochrome P450 in metroprolol oxidation by human liver microsomes. *J. Pharmacol. Exp. Ther.* **247**, 242–247.

Parli, C. J., and Schmidt, B. (1975). Metabolism of 4-chloroamphetamine to 3-chloro-4-hydroxyamphetamine in rat: Evidence for an in vivo "NIH SHIFT" of chlorine. *Res. Commun. Chem. Pathol. Pharmacol.* **10**, 601–604.

Parli, C. J., Wang, N., and McMahon, R. E. (1971). The enzymatic "N-hydroxylation of an imine. A new cytochrome P450-dependent reaction catalyzed by hepatic microsomal monooxygenases. *J. Biol. Chem.* **246**, 6953–6955.

Perrin, R., Minn, A., Ghersi-Egea, J.-F., Grassiot, M.-C., and Siest, G. (1990). Distribution of cytochrome P450 activities towards alkoxyresorufin derivatives in rat brain regions, subcellular fractions and isolated cerebral microvessels. *Biochem. Pharmacol.* **40**, 2145–2151.

Pinder, R. M., Brogden, R. N., Sawyer, P. R., Speight, T. M., and Avery, G. S. (1975). Fenfluramine: A review of its pharmacological properties and therapeutic efficacy in obesity. *Drugs* **10**, 241–323.

Ravindranath, V., and Anandatheerthavarada, H. K. (1989). High activity of cytochrome P450-linked aminopyrine N-demethylase in mouse brain microsomes, and associated sex-related difference. *Biochem. J.* **261**, 769–773.

Richards, R. P., Gordon, B. H., Ings, R. M. J., Campbell, D. B., and King, L. J. (1989). The measurement of d-fenfluramine and its metabolite, d-norfenfluramine in plasma and urine with an application of the method to pharmacokinetic studies. *Xenobiotica* **19**, 547–553.

Rinne, V. K., Siirtola, T., and Sonninen, V. (1978). L-Deprenyl treatment of on-off phenomenon in parkinson's disease. *J. Neural Transm.* **43**, 279–286.

Robinson, J. B. (1985). Stereoselectivity and isoenzyme selectivity of monoamine oxidase inhibitors. Enantiomers of amphetamine, N-methylamphetamine and deprenyl. *Biochem. Pharmacol.* **34**, 4105–4108.

Rommelspacher, H., Honecker, H., Schulze, G., and Strauss, S. M. (1974). The hydroxylation of D-amphetamine by liver microsomes of the male rat. *Biochem. Pharmacol.* **23**, 1065–1071.

Rowland, M. (1969). Amphetamine blood and urine levels in man. *J. Pharm. Sci.* **58**, 508–509.

Sachs, C., and Jonsson, G. (1975). Mechanisms of action of 6-hydroxydopamine. *Biochem. Pharmacol.* **24**, 1–8.

Schechter, M. D. (1989). Temporal parameters of cathinone, amphetamine and cocaine. *Pharmacol. Biochem. Behav.* **34**, 289–292.

Sever, P. S., Caldwell, J., Dring, L. G., and Williams, R. T. (1973). The metabolism of amphetamine in dependent subjects. *Eur. J. Clin. Pharmacol.* **6**, 177–180.

Sinsheimer, J. E., Dring, L. G., and Williams, R. T. (1973). Species differences in the metabolism of norephedrine in man, rabbit, and rat. *Biochem. J.* **136**, 763–771.

Smith, D. A. (1991). Species differences in metabolism and pharmacokinetics: Are we close to an understanding? *Drug Metab. Rev.* **23**, 355–373.

Smith, R. L. (1986). Introduction. *Xenobiotica* **16**, 361–365.

Smith, R. L., and Dring, L. G. (1970). Patterns of metabolism of beta-phenylisopropylamines in man and other species. In "Amphetamines and Related Compounds" (Proceedings of the Mario Negri Institute for Pharmacological Research, Milan, Italy) (E. Costa and S. Garattini, eds.), pp. 121–139. Raven Press, New York.

Spinelli, R., Fracasso, C., Guiso, G., Garattini, S., and Caccia, S. (1988). Disposition of (−)fenfluramine and its active metabolite, (−)-norfenfluramine in rat: A single dose-proportionality study. *Xenobiotica* **18**, 573–584.

Steele, T. D., Brewster, W. K., Johnson, M. P., Nichols, D. E., and Yim, G. K. (1991). Assessment of the role of α-methylepinine in the neurotoxicity of MDMA. *Pharmacol. Biochem. Behav.* **38**, 345–351.

Strobel, H. W., Cattaneo, E., Adesnik, M., and Maggi, A. (1989). Brain cytochrome P-450 are responsive to phenobarbital and tricyclic amines. *Pharmacol. Res.* **21**, 169–175.

Sum, C. Y., and Cho, A. K. (1977). The N-hydroxylation of phentermine by rat liver microsomes. *Drug Metab. Disp.* **5**, 464–468.

Thoenen, H., Hurlimann, A., Gey, K. F., and Haefely, W. (1966). Liberation of p-hydroxynorephedrine from cat spleen by sympathetic nerve stimulationn after pretreatment with amphetamine. *Life Sci.* **5**, 1715–1722.

Tyndale, R. F., Sunahara, R., Inaba, T., Kalow, W., Gonzalez, F. J., and Niznik, H. B. (1991). Neuronal cytochrome P450IID1 (debrisoquine/sparteine-type): Potent inhibition of activity by (−)-cocaine and nucleotide sequence identity to human hepatic P450 gene CYP2D6. *Mol. Pharmacol.* **40**, 63–68.

Tynes, R. E., and Hodgson, E. (1983). Oxidation of thiobenzamide by the FAD-containing and cytochrome P-450-dependent monooxygenases of liver and lung microsomes. *Biochem. Pharmacol.* **32**, 3419–3428.

Van Lier, J. E., and Rousseu, J. (1976). Mechanism of cholesterol side-chain cleavage: Enzymatic rearrangement of 20-beta-hydroperoxy-20-isocholesterol to 20-beta,21-dihydroxy-20-isocholesterol. *FEBS Lett.* **70**, 23–27.

Vree, T. B. (1973). "Pharmacokinetics and Metabolism of Amphetamines." Ph.D. Thesis. Catholic University of Nijmegen, The Netherlands.

Walsh, C. (1979). "Enzymatic Reaction Mechanisms." Freeman, San Francisco.

Walther, B., Ghersi-Egea, J. F., Minn, A., and Siest, G. (1986). Subcellular distribution of cytochrome P450 in the brain. *Brain Res.* **375**, 338–344.

Warner, M., Kohler, C., Hansson, T., and Gustafsson, J. (1988). Regional distribution of cytochrome P450 in the rat brain: Spectral quantitation and contribution of P450 b,e and P450c,d. *J. Neurochem.* **50**, 1057–1065.

Weli, A. M., Hook, B. B., and Lindeke, B. (1985). N-Dealkylation and N-oxidation of two α-methyl-substituted pargyline analogues in rat liver microsomes. *Acta Pharm. Suec.* **22**, 1–16.

Wilkinson, G. R., and Beckett, A. H. (1968). Absorption, metabolism and excretion of the ephedrine in man. II. Pharmacokinetics. *J. Pharm. Sci.* **57**, 1933–1938.

Wright, J., Cho, A. K., and Gal, J. (1977a). The metabolism of amphetamine in vitro by rabbit liver preparations: A comparison of $R(-)$ and $S(+)$ enantiomers. *Xenobiotica* **7**, 257–266.

Wright, J., Cho, A. K., and Gal, J. (1977b). The role of N-hydroxyamphetamine in the metabolic deamination of amphetamine. *Life Sci.* **20**, 467–474.

Yamada, H., Baba, T., Hirata, Y., Oguri, K., and Yoshimura, H. (1984). Studies on N-demethylation of methamphetamine by liver microsomes of guinea-pigs and rats; The role of flavin-containing mono-oxygenase and cytochrome P450 systems. *Xenobiotica* **14**, 861–866.

Yamada, H., Oguri, K., and Yoshimura, H. (1986). Effects of several factors on urinary excretion of methamphetamine and its metabolites in rats. *Xenobiotica* **16**, 137–141.

Yamada, H., Honda, S., Oguri, K., and Yoshimura, H. (1989). A rabbit liver constitutive form of cytochrome P450 responsible for amphetamine deamination. *Arch. Biochem. Biophys.* **273**, 26–33.

Yamada, H., Ohkuma, T., Urata, K., Watabe, E., Oguri, K., and Yoshimura, H. (1992). On the formation of benzoic acid from amphetamine—A possible mechanism of this metabolism including C-C bond cleavage. *In* "Abstracts of the 23rd Symposium on Drug Metabolism and Action," Octoboer 1992. Kyoto, Japan.

Yamamoto, T., Takano, R., Egashira, T., and Yamanaka, Y. (1984). Metabolism of methamphetamine, amphetamine and *p*-hydroxymethamphetamine by rat-liver microsomal preparations *in vitro*. *Xenobiotica* **14,** 867–875.

Yoshida, T., Yamada, Y., Yamamoto, T., and Kuroiwa, Y. (1986). Metabolism of deprenyl, a selective monoamine oxidase (MAO) B inhibitor in rat: Relationship of metabolism to MAO-B inhibitory potency. *Xenobiotica* **16,** 129–136.

Yoshida, T., Oguro, T., and Kuroiwa, Y. (1987). Hepatic and extrahepatic metabolism of deprenyl, a selective monoamine oxidase (MAO) B inhibitor, of amphetamines in rats: sex and strain differences. *Xenobiotica* **17,** 957–963.

Zhiyang, Z., Castagnoli, N., Jr., Ricaurte, G. A., Steele, T., and Martello, M. (1992). Synthesis and neurotoxicological evaluation of putative metabolites of the serotonergic neurotoxin 2-(methylamino)-1-[3,4-(methylenedioxy)phenyl]propane [(methylenedioxy)methamphetamine]. *Chem. Res. Toxicol.* **5,** 89–94.

Ziegler, D. M. (1991). Mechanism, multiple forms and substrate specificities of flavin-containing mono-oxygenases. *In* "N-Oxidation of Drugs" (P. Hlavica and L. A. Damani, eds.), pp. 59–68. Chapman and Hall, London.

II
NEUROPSYCHOPHARMACOLOGY

Ronald Kuczenski
David S. Segal

3
Neurochemistry of Amphetamine

I. INTRODUCTION

Amphetamine (AMPH) produces a broad spectrum of central and peripheral effects. However, compelling evidence derived from a variety of different sources suggests that central nervous system (CNS) catecholamine (CA) neurons represent crucial elements through which AMPH and AMPH-like stimulants exert their behavioral activating effects. Enhanced neurotransmitter activity, particularly at dopamine (DA) receptors has been implicated in the locomotor and stereotypy responses (Creese and Iversen, 1974; Kelly, 1977; Cole, 1978; Sessions et al., 1980), as well as in the rewarding properties of these drugs (see Chapter 8). Although a preeminent role for various DA systems in these behaviors has been well established, several issues remain to be elucidated fully, including (1) the mechanisms by which AMPH facilitates neurotransmission and (2) the relative contribution of the broad spectrum of neurochemical actions of AMPH to the various components of the response profile of the stimulant.

This chapter focuses on the mechanisms underlying the acute neurotransmitter response to AMPH. An understanding of these actions is a necessary prerequisite to elucidating the consequences of chronic AMPH administration, which may be of more direct relevance to issues of stimulant abuse. In Chapter 4, the AMPH response with respect to possible neurochemical–behavioral relationships will be considered, with particular emphasis on changes that follow chronic AMPH administration.

II. MECHANISMS OF INTERACTION OF AMPHETAMINE WITH CATECHOLAMINE NEURONS

Early studies using *in vitro* methodologies revealed three sites on the catecholaminergic nerve terminal at which AMPH could interact: the uptake carrier, the storage vesicle, and the mitochondrial monoamine oxidase (Fig. 1). In each case, the anticipated consequence of the interaction would be enhancement of the intraterminal availability and/or intrasynaptic concentration of the neurotransmitter, which could augment CA transmission.

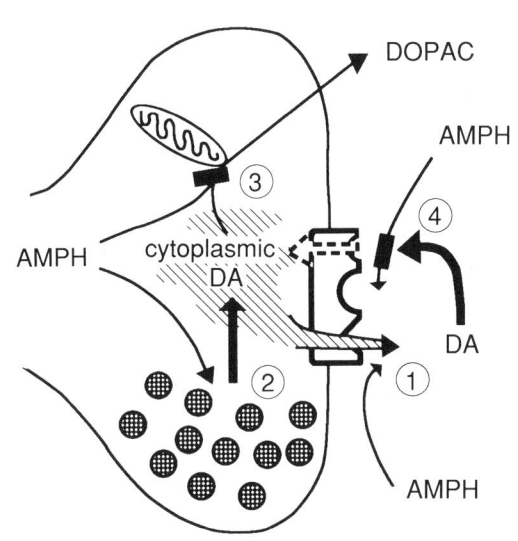

FIGURE 1 Sites of action of amphetamine (AMPH) at the dopaminergic nerve terminal. (1) AMPH interacts with the dopamine (DA) transport carrier to facilitate DA release from the cytoplasm through an exchange diffusion mechanism (see text and Fig. 2). At higher intracellular concentrations, AMPH also can (2) disrupt vesicular storage of DA and (3) inhibit monoamine oxidase. Both these actions will augment cytoplasmic DA concentrations. (4) AMPH also inhibits DA uptake by virtue of its binding to and transport by the DA carrier. DOPAC, Dihydroxyphenylacetic acid.

A. Uptake Carrier

Since the early 1960s (Ross and Renyi, 1964; Glowinski and Axelrod, 1965) AMPH has been well documented to interfere with the neuronal retention of [^{3}H]CAs. Initially, this effect was attributed primarily to AMPH inhibition of CA uptake across the neuronal membrane, but subsequently several investigators provided evidence that, through its interaction with the uptake carrier, AMPH also participates in a nonexocytotic transmitter release process. Heikkila *et al.* (1975) and Raiteri and co-workers (1975) first suggested that AMPH-induced release, rather than uptake blockade, played the more important role in enhancing extrasynaptosomal DA. Additional studies revealed that AMPH-induced DA release, in contrast to exocytotic DA release, could be prevented by DA uptake blockers (Raiteri *et al.*, 1979), exhibited little or no calcium requirement (Meyerhoff and Kant, 1978; Kamal *et al.*, 1981), and was insensitive to DA autoreceptor activity (Kamal *et al.*, 1981). These observations suggested that AMPH-induced DA release proceeded through a mechanism alternative to exocytosis, and led to the hypothesis that DA was released by means of an exchange diffusion mechanism through the DA uptake carrier (Fischer and Cho, 1979; Raiteri *et al.*, 1979). This process (Fig. 2), which involves a reversal of the normal operation of the neuronal DA uptake process, requires the DA uptake carrier to bind AMPH and transport it down the concentration gradient into the nerve terminal. Subsequently, when AMPH dissociates from the carrier, the carrier becomes available to bind DA, which then is transported down the AMPH concentration gradient into the neuron. That AMPH-induced DA release could be prevented by uptake blockers suggested that the uptake carrier played a key role in this process (Raiteri *et al.*, 1979). Fischer and Cho (1979) presented convincing indirect evidence that this role included the transport of AMPH into the neuron. Although the lipophilic AMPH molecule can enter the neuron readily and achieve equilibrium concentrations by diffusion (Obianwu *et al.*, 1968; Wong *et al.*, 1972), Fischer and Cho (1979) argued that the intraneuronal accumulation of AMPH is not essential to the release process. Rather, DA release is enhanced because the transport of AMPH into the neuron increases the availability of the transport carrier inside the neuron to bind and transport DA to the extraneuronal space. Although distinguishing carrier-mediated intraneuronal accumulation of AMPH from its passive diffusion has been difficult, Zaczek *et al.* (1991) have characterized the active transport of [^{3}H]AMPH into striatal synaptosomes with characteristics consistent with a DA carrier-mediated process and a K_m near 100 nM.

Other data from *in vitro* studies support the view that carrier-mediated outward movement of DA represents a viable mechanism of DA release, and provide additional insight into the dynamics of the carrier-mediated release process. For example, a similar carrier-dependent mechanism is apparently

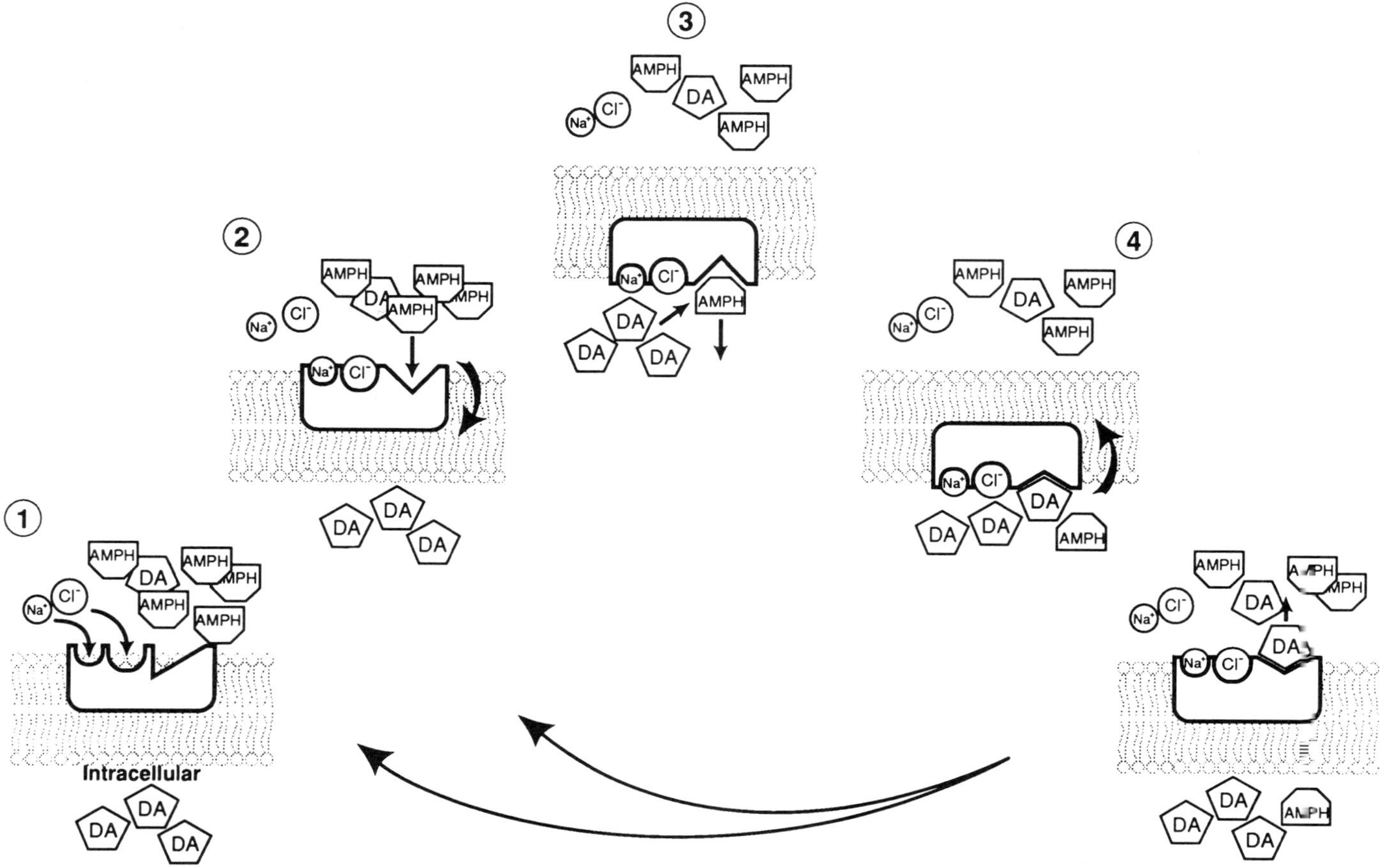

1
2
3
4
Intracellular

responsible for the release of synaptosomal DA observed in Na^+-free medium and after inhibition of Na^+,K^+-ATPase by ouabain (Raiteri et al., 1979). Kinetic studies of uptake carriers operating in the normal direction have led to a model in which the carrier binds Na^+, Cl^-, and the protonated substrate to move transmitter down the Na^+ and electrical gradients into the neuron (Holz and Coyle, 1974; Blaustein and King, 1976; Sanchez-Armass and Orrego, 1977, 1978; Sammet and Graefe, 1979; Wheeler, 1980; Krueger, 1990; Trendelenburg, 1991). In the absence of bound ions or substrate, the carrier is free to move within the membrane. The initial binding of extracellular Na^+ immobilizes the carrier on the outside of the neuronal membrane to permit Cl^- and substrate binding; in addition, immobilization facilitates substrate binding. Once substrate is bound, the carrier is again free to move. Therefore, the model would predict that removal of extracellular Na^+ would permit continued carrier movement, thereby enhancing the probability of the appearance of the carrier at the membrane internal surface to bind intracellular DA. In the second situation, ouabain-induced inhibition of Na^+,K^+-ATPase, by allowing intracellular Na^+ to increase, not only would augment the frequency of immobilization of the carrier at intracellular sites but also would facilitate binding of intracellular DA. In both cases, the loss of the normal inward-directed Na^+ gradient that drives DA uptake would enhance the net outward movement of DA, down its concentration gradients. Within the context of such a model, a concentration of ouabain that, by itself, is ineffective in releasing DA should potentiate AMPH-induced DA release (Rutledge, 1978; Connor and Kuczenski, 1986), supporting the view that carrier-mediated DA release is a function of the intraneuronal concentration of both Na^+ and carrier binding sites.

During their initial studies of AMPH-induced CA release from brain tissue preparations, neither Heikkila et al. (1975) nor Raiteri and co-workers (1975) observed a substantial releasing effect of AMPH on norepinephrine (NE) under conditions under which DA release was evident. In contrast, AMPH-induced release of NE from peripheral sympathetic NE nerve terminals has been well documented (for reviews, see Bonisch and Trendelenburg, 1988; Trendelenburg, 1991). Further, ouabain increases the release of NE through a desipramine (DMI)-sensitive mechanism (Sweadner,

---

FIGURE 2 Schematic of the exchange–diffusion process. (1) Sodium and chloride bind to the dopamine (DA) transport carrier to immobilize it at the extracellular surface of the neuronal membrane and to alter the conformation of the DA binding site to facilitate substrate binding. (2) Amphetamine (AMPH), in competition with extracellular DA, binds to the carrier. Substrate binding allows movement of the carrier to the intracellular surface of the neuronal membrane, driven by the favorable sodium and AMPH concentration gradients. (3) AMPH dissociates from the carrier, making the binding site available for cytoplasmic DA. (4) DA binding to the carrier enables movement of the carrier to the extracellular surface of the neuronal membrane, driven by the favorable DA concentration gradient. (5) DA dissociates from the carrier, making the carrier available for another cycle of transport.

1985); DMI also blocks the ability of a Na$^+$-free medium to release NE from heart slices (Paton, 1973) and hypothalamic synaptosomes (Raiteri et al., 1977). Thus, carrier-mediated NE release apparently can occur. In addition, some evidence suggests that AMPH can be transported into NE synaptosomes through a mechanism sensitive to uptake blockers (Azzaro et al., 1974; Rutledge and Vollmer, 1979). Bonisch (1984) showed that the uptake$_1$ transporter can actively transport AMPH into PC12 cells through a Na$^+$-, Cl$^-$-, and cocaine-sensitive process. Since carrier-mediated release requires a cytoplasmic pool of transmitter, researchers have argued (Kuczenski, 1983; Trendelenburg, 1991) that the failure to detect substantial AMPH-induced NE release in brain preparations might reflect the absence of a significant cytoplasmic pool of NE because of the vesicular location of dopamine β-hydroxylase (Lundberg et al., 1977), the final enzyme in the NE biosynthetic pathway. In fact, some authors (Bonisch and Trendelenburg, 1988; Knepper et al., 1988; Trendelenburg, 1991) have argued that a robust NE-releasing effect by AMPH may depend on a concomitant mobilization of vesicular NE into the cytoplasm.

B. Vesicle

A dynamic equilibrium is likely to exist between cytoplasmic and vesicular CA, and a decrease in cytoplasmic concentration might promote leakage of the vesicular transmitter. In addition, AMPH can affect vesicular processes directly. First, AMPH is an effective competitive inhibitor for the uptake of both DA and NE into dopaminergic and noradrenergic vesicles, with a K_i near 1–2 μM (Schumann and Philippu, 1962; Philippu and Beyer, 1973; Ferris and Tang, 1979; Knepper et al., 1988). Thus, concentrations of AMPH typically used in *in vitro* CA release studies can be expected to interfere with the vesicular accumulation of the neurotransmitter. Second, high concentrations of AMPH may promote the efflux of transmitter from the vesicle by abolishing the pH gradient across the vesicular membrane (Johnson et al., 1982; Phillips, 1982; Sulzer and Rayport, 1990). First, vesicular uptake of CAs is driven by the proton gradient (Phillips and Allison, 1978; Toll and Howard, 1978; Johnson et al., 1979; Erickson et al., 1990). Therefore, increasing the intravesicular pH will inhibit CA vesicular uptake. In addition, a sufficient increase in intravesicular pH will facilitate leakage of the vesicular transmitter by reducing its protonation. Thus, actions of AMPH on the vesicle may supplement the cytoplasmic CA pool to contribute to the carrier-mediated release process. In this regard, some evidence suggests that reserpine alters AMPH release of DA and NE *in vitro*. Fischer and Cho (1979) noted a biphasic effect of AMPH on the release of DA from synaptosomes. The second component, which occurred at high (>100 μM) AMPH concentrations, was prevented by reserpine pretreatment and could reflect AMPH-mediated alterations in vesicular pH. Likewise, the release of NE from rat vas deferens, particularly at higher AMPH concentrations, was diminished by reserpine pretreatment (Bonisch and Trendelenburg, 1988). Some authors (Parker and Cubeddu, 1986a,b) reported significantly less DA release from striatal slices after reserpine pretreatment at all concentrations of AMPH, whereas others (Niddam et al., 1985) failed to observe a reserpine effect. Thus, the currently available data from *in vitro* studies are not entirely consistent regarding a contribution from the vesicular pool to AMPH-induced CA release.

C. Monoamine Oxidase

Monoamine oxidase (MAO) represents a third potential site of interaction of AMPH with DA and NE dynamics. Amphetamine inhibits MAO in broken cell preparations (Mann and Quastel, 1940); inhibition of this pathway of CA degradation in intact tissue preparations could enhance the AMPH-releasable CA pool. In this regard, the release of DA and NE by AMPH is accompanied by a dramatic reduction in their respective metabolites, an effect that could be attributed to inhibition of MAO. However, carrier-mediated release of CAs after other *in vitro* manipulations, such as addition of ouabain (Stute and Trendelenburg, 1984; Schomig and Trendelenburg, 1987), also decreases CA metabolites. Researchers have argued (Stute and Trendelenburg, 1984; Zetterstrom et al., 1986; Kuczenski and Segal, 1989; see subsequent discussion) that this effect primarily reflects a decrease in the availability of the transmitter for metabolism by MAO. More direct assessment of the ability of AMPH to inhibit MAO *in vivo* suggests that such an effect occurs only at relatively high AMPH concentrations (Green and El Hait, 1980; Miller et al., 1980). As a consequence, AMPH inhibition of MAO generally is not considered to be pharmacologically important. Nevertheless, definitive evidence ruling out a significant action of AMPH at this site is not yet available.

Based on *in vitro* studies, then, AMPH-induced CA release appears to proceed through a series of steps that begins with the binding of AMPH to the uptake carrier at the external surface of the neuronal membrane. Na^+ facilitates this binding, both by immobilizing the carrier at the external surface and by increasing its affinity for AMPH. The Na^+–AMPH–carrier complex moves down the Na^+ and AMPH gradients to the inside of the nerve terminal, where cytoplasmic CA can replace AMPH on the carrier to be transported down the CA gradient to the synaptic space. Such a mechanism clearly depends on the availability of a substantial cytoplasmic CA pool to maintain a favorable outwardly directed CA gradient. DA neurons appear to maintain such a pool, whereas NE neurons may not contain sufficient cytoplasmic transmitter to sustain a robust NE release. However, AMPH accumulation within the nerve terminal, by its active uptake and through its passive diffusion across the cellular membrane, may achieve sufficiently high concentrations to facilitate movement of CA from vesicles into the cytoplasm and to inhibit MAO so nonvesicular CA can accumulate for transport from the nerve terminal.

III. POSTMORTEM STUDIES OF SYNTHESIS AND METABOLISM

Although AMPH *in vitro* interacts with both DA and NE dynamics, over the past two decades most systematic studies of *in vivo* neurotransmitter synthesis and metabolism in response to AMPH have focused on DA, particularly caudate DA. These studies showed that AMPH promotes a complex time- and dose-dependent pattern of effects on DA dynamics which, in retrospect, is consistent with its proposed mechanism of interaction with the DA nerve terminal. These effects have been reviewed (Kuczenski, 1983) and will be considered only briefly here.

A. Dopamine Synthesis and Turnover

Low doses of AMPH (0.25–1.5 mg/kg), which increase locomotor activity but do not promote a focused stereotypy phase, increase caudate DA synthesis when the incorporation of [^{3}H]-labeled tyrosine into DA is used as the index of synthesis (Costa *et al.*, 1972; Carenzi *et al.*, 1975; Kuczenski, 1977, 1979). The magnitude and duration of the increase are proportional to the AMPH dose. Likewise, the accumulation of dihydroxyphenylalanine (DOPA) after aromatic amino acid decarboxylase (AADC) inhibition also is increased over this dose range (Kehr *et al.*, 1977; Pearl and Seiden, 1979). Increased synthesis likely reflects the alleviation of end-product inhibition of tyrosine hydroxylase (TH) as the regulatory cytoplasmic pool of the transmitter is released from the nerve terminal into the synaptic space. Because synthesis is increased despite substantially elevated synaptic DA concentrations (see subsequent discussion), synthesis-modulating autoreceptors on the DA nerve terminal apparently are not capable of overcoming the effect of the loss of the cytoplasmic pool of DA. This suggestion is consistent with the mechanism of action of synthesis-modulating DA autoreceptors, which appear to prevent or reverse phosphorylation-dependent activation of TH rather than inhibit TH activity (Strait and Kuczenski, 1986).

As the dose of AMPH is increased further, the incorporation of [^{3}H]-labeled tyrosine into DA begins to decline and falls below baseline at higher doses (5–10 mg/kg) (Kuczenski, 1977). The decline in synthesis may reflect increases in cytoplasmic DA as AMPH concentrations become high enough to inhibit MAO and to interfere with vesicular storage of the transmitter. DOPA accumulation after decarboxylase inhibition also falls from peak levels at higher AMPH doses (Pearl and Seiden, 1979), although this decline after decarboxylase inhibition occurs at higher AMPH doses than the decline in [^{3}H]DA formation. Inhibition of DA synthesis at the decarboxylase step may deprive the cytoplasm of newly synthesized transmitter and contribute to a rightward shift in AMPH-induced inhibition of synthesis. In this regard, except after decarboxylase inhibition (Pearl and Seiden, 1979), AMPH increases tissue levels of DA (Kuczenski, 1977, 1980; Hartman and Halaris, 1980). Thus, synthesis must continue to contribute significantly to the overall tissue DA level. Although the nature of this increase is somewhat obscure, it simply may reflect the summation of both intra- and extracellular transmitter or may include a component consequent to MAO inhibition.

B. Dopamine Metabolites

Amphetamine promotes a rapid and substantial dose-dependent decrease in the DA metabolite dihydroxyphenyl acetic acid (DOPAC) (Bunney *et al.*, 1973; Roth *et al.*, 1976; Westerink and Korf, 1976) to about 40% of predrug levels (for reviews, see Kuczenski, 1983; Westerink, 1985). Levels are reduced maximally with AMPH doses near 1–2 mg/kg, and remain depressed as the AMPH dose is increased further (Kuczenski, 1980). This effect is unlikely to reflect inhibition of MAO since AMPH, particularly at lower doses, appears to be a relatively weak MAO inhibitor. Rather, since DOPAC derives from the actions of MAO on cytoplasmic DA, AMPH-induced release of DA from this pool should deprive MAO of its substrate and lead to a decrease in the formation of DOPAC. A similar explanation has been proffered for the decline in dihydroxyphenylethylene glycol (DOPEG) levels accompanying the AMPH-induced increase in NE release from sympathetically innervated tissues (Stute and Trendelenburg, 1984). To the extent that DOPAC reflects the availability of cytoplasmic DA, the failure of higher doses of AMPH to decrease DOPAC further while continuing to increase DA release (see subsequent discussion) would suggest, as discussed earlier, that the cytoplasm is supplemented continually with newly synthesized and/or vesicular DA that can sustain AMPH-induced release.

Concomitant with the decline in DOPAC, homovanillic acid (HVA) levels also initially decline. However, in contrast to DOPAC, HVA eventually increases as a function of increasing dose of AMPH (Braestrup, 1977; Kuczenski, 1980). The initial parallel decline in DOPAC and HVA reflects the sequential relationship of these metabolites, that is, most brain HVA is formed from the metabolism of DOPAC by extraneuronal catechol O-methyl transferase (COMT) (Westerink, 1979b). The divergence of DOPAC and HVA at higher doses of AMPH reflects the shift in DA metabolism from intra- to extraneuronal as cytoplasmic DA is depleted, as the nerve terminal uptake process is inhibited more completely and as extracellular DA accumulates. Under these conditions, more DA first is O-methylated to 3-methoxytyramine (3-MT) and then is metabolized to HVA, thus contributing more substantially to the overall HVA content. Tissue levels of 3-MT do rise, especially after higher doses of AMPH (Westerink, 1979a; Kehr, 1981; Ponzio *et al.*, 1981; Waldmeier *et al.*, 1981), but for methodological reasons a more exact relationship between dose of AMPH and level of 3-MT formation has been difficult to determine using postmortem techniques. More recent studies using microdialysis procedures in awake animals (see subsequent discussion) have confirmed that 3-MT formation after AMPH administration rises dramatically.

Generally researchers have assumed that AMPH promotes a progressive dose-dependent enhancement of dopaminergic transmission. However, none of the postmortem measures of DA dynamics just described, with the possible exception of tissue levels of 3-MT, can be interpreted readily to reflect a progressive increase in synaptic DA. In the absence of drug administration, changes in DA synthesis and metabolism often appear to parallel increases and decreases in dopaminergic neuronal activity and, by inference, synaptic DA concentrations. However, the unique manner in which AMPH releases DA into the synapse dissociates synaptic DA content from the regulatory events that normally link DA synthesis and metabolism to levels of DA transmission. Further, AMPH can alter DA synthesis and metabolism directly through a variety of mechanisms, thus further confounding interpretation in terms of DA function. In addition to the indirect nature of these biochemical data, two other shortcomings of postmortem techniques are particularly conspicuous with respect to establishing neurochemistry–behavior relationships. First, the methods used tend to obscure individual variations in responsivity to drugs. In addition, whereas drug-induced changes are dynamic and can be multiphasic, the postmortem strategy provides a static perspective on neurotransmitter processes. As a consequence, although converging evidence has supported the general hypothesis that changes in DA function contribute significantly to the behavioral effects of AMPH, the variety of attempts to relate the AMPH-induced changes in DA synthesis and metabolism to synaptic DA concentrations and to specific components of the drug-induced behavioral profile have been relatively unsuccessful.

IV. EXTRACELLULAR FLUID SAMPLING IN AWAKE ANIMALS

The intracerebral microdialysis techniques (see Benveniste and Hüttemeier, 1990, for a review) samples the extracellular space for neurotransmitter substances that have escaped the synaptic cleft (Fig. 3). Converging evidence suggests that changes in extracellular transmitter concentrations parallel changes in synaptic transmitter concentrations (Westerink *et al.*, 1987, 1988; Kuczenski *et al.*, 1991; Manley *et al.*, 1992; see Di Chiara, 1991, for a review). Therefore, this approach appears to provide a more direct measure of DA dynamics with which to test hypotheses regarding the mechanisms of AMPH-induced DA release. In addition, this procedure allows for a

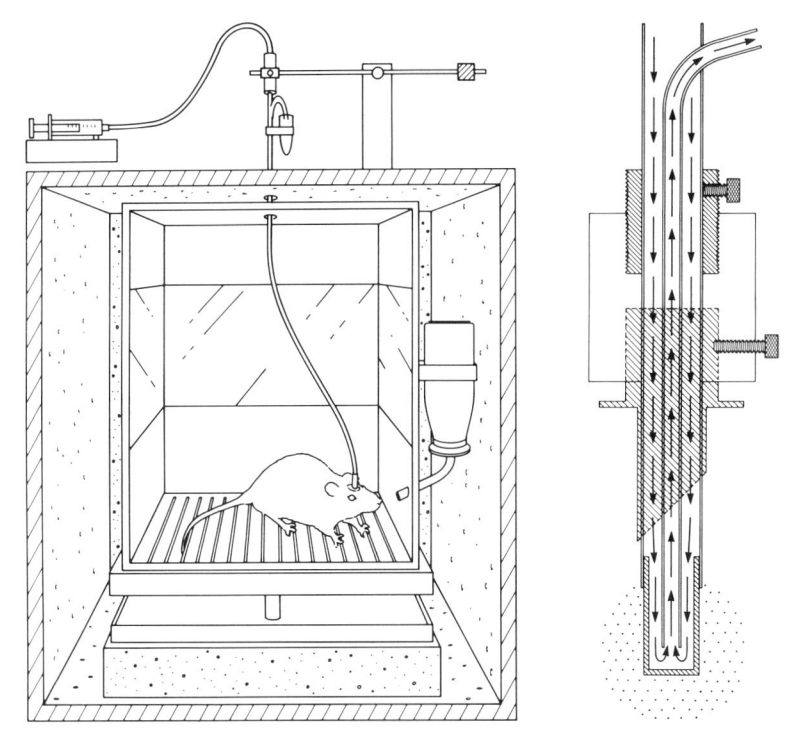

FIGURE 3 (*Left*) Schematic of a freely moving animal undergoing microdialysis in a soundproofed behavioral activity chamber. Dialysis samples are collected outside the chamber to avoid disrupting ongoing behaviors. (*Right*) Microdialysis probe with a 250-μm diameter tip is attached to the awake animal via a permanently implanted stainless steel guide cannula. Reprinted by permission of the *Journal of Neuroscience* from Kuczenski and Segal, 1989.

continuous measure of DA function in freely moving animals, facilitating the identification of behavior–biochemistry relationships.

A. Extracellular Dopamine

Zetterstrom *et al.* (1983) first described the effects of AMPH on extracellular DA and its metabolites in the caudate of the awake rat. Subsequently, various investigators utilized this methodology to characterize the mechanisms of interaction of AMPH with the dynamics of the DA nerve terminal and to assess quantitative relationships between the AMPH-induced alterations in DA function and the various features of the stimulant behavioral response profile.

These studies have shown that AMPH produces a rapid and profound dose-dependent increase in caudate extracellular DA. For example, in response to 2 mg/kg AMPH, DA concentrations rise 15- to 20-fold within the

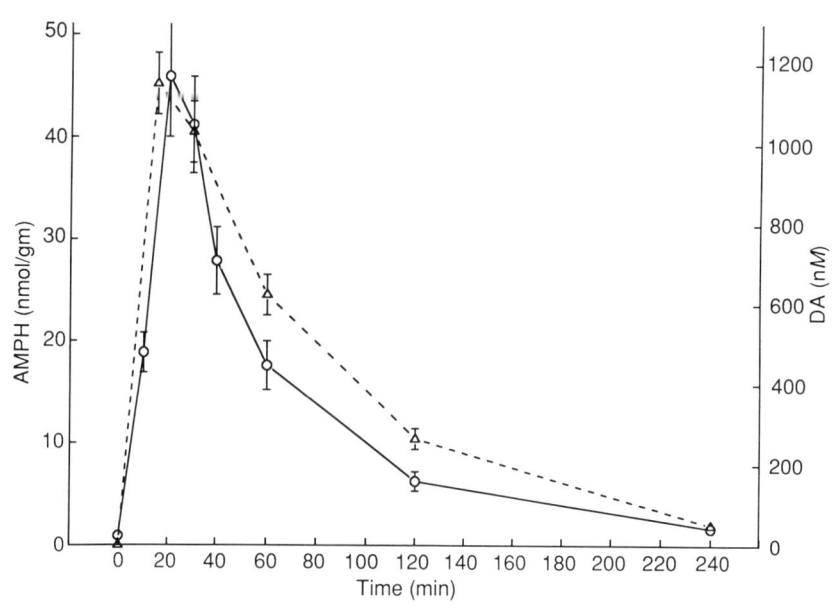

FIGURE 4 Comparison of the caudate extracellular dopamine (DA, ○) concentrations and tissue levels of amphetamine (AMPH, △) as a function of time after the subcutaneous administration of 2.5 mg/kg AMPH. AMPH concentrations were determined using gas chromatography–mass spectrometry techniques (D. S. Segal and A. Cho, unpublished results).

first 20–40 min of drug administration, from basal levels typically near 25 nM. DA concentrations then gradually return to baseline over the next 2–3 hr (Zetterstrom et al., 1983; Sharp et al., 1987; Kuczenski and Segal, 1989). Doses of AMPH as low as 0.1–0.25 mg/kg significantly increase extracellular DA (Zetterstrom et al., 1986; Carboni et al., 1989; Segal et al., 1992); extracellular DA concentrations as high as 1500 nM, or >30-fold above baseline, have been observed at high AMPH doses (5 mg/kg) (Sharp et al., 1987; Kuczenski and Segal, 1989, 1992a). In general, the temporal pattern of the caudate DA response to AMPH appears to parallel the pharmacokinetics of AMPH in the caudate (Fig. 4). These data suggest that the quantitative features of the presynaptic DA response are linked to the presence of AMPH, and are unlikely to reflect a significant residual component that is initiated by AMPH but does not require the continued presence of the drug to be maintained. Importantly, as discussed in Chapter 4, the intensity of the behavioral response as a function of time, particularly at higher doses of AMPH, does not parallel extracellular DA or the pharmacokinetics of the drug.

B. Extracellular 3-Methoxytyramine

Amphetamine also dose-dependently increases caudate extracellular 3-MT concentrations proportional to the increase in DA (Wood and Altar, 1988; Brown et al., 1991; Kuczenski and Segal, 1992a). However, although some disagreement exists (Brown et al., 1991), the rise in 3-MT appears to be delayed temporally relative to the increase in DA (Wood and Altar, 1988; Kuczenski and Segal, 1992a) and may reflect a more rapid diffusion of DA from the synaptic cleft relative to the several steps leading to 3-MT accumulation in the extracellular fluid, that is, DA uptake into the nondopaminergic sites containing COMT, its subsequent conversion to 3-MT, and 3-MT release and diffusion into the extracellular fluid (Kaplan et al., 1979; Rivett et al., 1983; Kaakkola et al., 1987).

C. Extracellular Dopamine Acid Metabolites

In contrast to the AMPH-induced increase in extracellular DA and 3-MT, and consistent with an exchange diffusion mechanism of AMPH-induced release of cytoplasmic DA, the acid metabolites DOPAC and HVA decline (Zetterstrom et al., 1983, 1986; Imperato and Di Chiara, 1984; Kuczenski and Segal, 1989, 1992a; Kuczenski et al., 1991). Although qualitatively similar metabolite profiles are observed in postmortem studies (see preceding discussion), the microdialysis technique enables quantitative comparisons of the dose and temporal features of the response profiles for DA and its metabolites, and can provide insight into their possible interrelationships. For example, the maximum decline in extracellular DOPAC is observed at AMPH doses of 0.5–1 mg/kg (Zetterstrom et al., 1986; Kuczenski and Segal, 1989, 1992a), whereas DA concentrations continue to increase as the dose of the drug is increased. In addition, the temporal pattern of DOPAC decline does not parallel the DA rise (Zetterstrom et al., 1986; Kuczenski and Segal, 1989, 1992a; Kuczenski et al., 1991), that is, levels of DOPAC achieve their minimum after DA peaks and remain depressed longer than DA increases. Thus, changes in AMPH-induced DA release apparently can occur without parallel changes in the amount of cytoplasmic DA, as reflected by DOPAC concentrations (Kuczenski and Segal, 1989; Kuczenski et al., 1990). Such changes could occur because the dynamics of cytoplasmic DA are altered through enhanced DA synthesis and a shift of DA from vesicular storage, to sustain not only additional AMPH-induced release but also the continuing production of DOPAC via MAO (Kuczenski and Segal, 1989; Kuczenski et al., 1990). However, that the recovery of DOPAC to predrug levels is retarded relative to DA is difficult to reconcile with a model that attributes the AMPH-induced decline in DOPAC solely to a decrease in cytoplasmic DA content. One could argue that, during recovery of the DA nerve terminal from the releasing effect of AMPH, the transmitter biosynthetic capacity is not sufficient to overcome the depletion of the cytoplasmic amine immediately, resulting in a delayed restoration of intraneuronal DA and a retarded recovery of DOPAC. However, such a depletion of cytoplasmic and/or vesicular DA is not consistent with the elevated tissue levels of DA observed in postmortem studies after AMPH administration. Further, the DA uptake blocker amperozide (Eriksson and Christensson, 1990) has been shown to effect a dissociation of AMPH-induced DA release and the decline in DOPAC (Ichikawa and Meltzer, 1992), that is, a dose of amperozide that, by itself, increased extracellular DOPAC and DA prevented the AMPH-induced increase in extracellular DA by 95% without blocking the AMPH-induced decline in DOPAC. However, a contrasting AMPH–amperozide profile did not include a dissociation of the DA and DOPAC responses has been reported (Kimura et al., 1993). In addition, the uptake blocker nomifensine produces a parallel inhibition of the AMPH-induced changes in DA and DOPAC (Butcher et al., 1988). Although other factors also may contribute significantly to the AMPH-induced decline in DOPAC concentrations, the release of cytoplasmic DA is likely to predominate in this effect (Zetterstrom et al., 1986).

Caudate extracellular HVA concentrations respond to AMPH in a pattern qualitatively similar to postmortem tissue levels. At low doses, extracellular HVA declines, paralleling the DOPAC decline. As discussed earlier in the context of postmortem data, the initial parallel decline in DOPAC and HVA concentrations presumably reflects the sequential relationship between DOPAC and HVA; the decline in HVA is a direct consequence of AMPH-induced disruption of DOPAC formation. Consistent with this idea, significant positive correlations are obtained between the AMPH-induced decreases in these two DA metabolites with doses of AMPH up to 2 mg/kg (Kuczenski and Segal, 1989). At higher AMPH doses, the decline in dialysate HVA is attenuated or reversed, whereas DOPAC concentrations remain maximally suppressed (Kuczenski and Segal, 1989, 1992a). In the absence of additional changes in DOPAC formation, the increase in HVA observed at higher AMPH doses probably arises through 3-MT.

D. Exchange–Diffusion Model

The dialysate DA metabolite response profile to AMPH doses, like its postmortem counterpart, can be interpreted readily in the context of an exchange diffusion mechanism of AMPH-induced DA release. The factors regulating the AMPH-induced increase in extracellular DA differ from those that regulate depolarization-induced DA release and are more consistent with the exchange diffusion mechanism. Thus, the removal of Ca^{2+} from the dialysis perfusion medium and the addition of Ca^{2+} antagonists decrease basal, presumably impulse-dependent, excocytotic DA release but have little effect on the AMPH-induced increase in extracellular DA (Westerink et al., 1988, 1989b; Carboni et al., 1989; Pani et al., 1990). Likewise, the infusion of micromolar quantities of tetrodotoxin (TTX), which blocks neuronal impulse conductance (Narahashi, 1974), eliminates basal DA (Westerink et al., 1987; Osborne et al., 1991) but has little effect on the AMPH-induced DA response (Westerink et al., 1987). Similarly, pretreatment with γ-butyrolactone (Carboni et al., 1989) or apomorphine (APO) (Kuczenski et al., 1990), both of which inhibit neuronal firing activity in nigrostriatal DA neurons, has no effect on AMPH-induced DA release. These observations suggest that the AMPH-induced inhibition of neuronal activity in dopaminergic neurons (Bunney et al., 1973) has no direct consequences for the quantitative features of the nerve terminal DA response profile. Further, that APO pretreatment fails to affect the AMPH-induced DA response (Kuczenski et al., 1990) also supports the suggestion, derived from *in vitro* studies, that AMPH-induced DA release *in vivo* is not influenced by autoreceptor activation (Shenoy and Ziance, 1979; Kamal et al., 1981).

Various pharmacological manipulations of intraneuronal DA pools affect the AMPH-induced DA response in a manner predictable from an exchange diffusion mechanism. First, 24 hr after pretreatment with 2.5 mg/kg reserpine, basal dialysate DA derived from exocytotic release was decreased by 80%, but neither the quantitative nor the temporal features of the caudate DA response to low and intermediate doses of AMPH (up to 2.5 mg/kg) was altered (Callaway et al., 1989). Thus, the DA response to AMPH over this dose range does not require intact functional vesicles and can be sustained, presumably by ongoing DA synthesis, from the cytoplasm in the absence of a substantial contribution from vesicles. However, in the nonreserpinized animal, redistribution of vesicular DA through equilibration with ongoing synthesis may maintain cytoplasmic DA. Further, whether mobilization of vesicular DA contributes to the response to doses of AMPH that are sufficiently high to disrupt vesicular function remains to be determined.

In contrast to the absence of an effect of reserpine, manipulations that more directly affect cytoplasmic DA do alter the DA response to AMPH. Pretreatment with the MAO-A inhibitor clorgyline, which should enhance the availability of cytoplasmic DA by preventing its degradation, more than doubled the increase in extracellular DA in response to 0.25 mg/kg AMPH (Segal et al., 1992). Conversely, although the effects of α-methyltyrosine (α-MT) on AMPH-induced increases in extracellular DA have not been assessed in awake animals, pretreatment of anesthetized rats with this TH inhibitor attenuated the caudate DA response to 4 mg/kg AMPH by almost 90% (Butcher et al., 1988). Importantly, even when α-MT was administered simultaneously with AMPH to preclude substantial depletion of vesicular DA, the DA response to AMPH still was reduced by about 60%. These data support the contention that AMPH-induced increases in extracellular DA depend on cytoplasmic transmitter, sustained by ongoing synthesis, and are consistent with the hypothesized exchange diffusion mechanism of AMPH-induced DA release.

That uptake blockers inhibit the AMPH-induced increase in extracellular DA (Butcher *et al.*, 1988; Hurd and Ungerstedt, 1989a) is consistent with the putative role of the carrier in exchange diffusion. However, interpretation of this observation in the context of an exchange diffusion mechanism is confounded. *In vivo,* uptake blockers and releasing agents both increase extracellular DA; thus, uptake blockers should inhibit the DA response not only of releasing agents but also of other uptake blockers. For example, the high affinity DA uptake inhibitor GBR12909 antagonizes the ability of cocaine to increase extracellular levels of DA (Rothman *et al.*, 1991). In contrast, *in vitro,* uptake blockers alone do not increase DA in superfusates; the superfusion technique readily distinguishes these drugs from releasing agents such as AMPH. Nevertheless, note that *in vivo,* Ca^{2+} removal and TTX infusion, in contrast to their inability to affect the AMPH-induced DA response, prevent the rise in extracellular DA associated with uptake blockers such as cocaine (Westerink *et al.*, 1989b; Nomikos *et al.*, 1990; Pani *et al.*, 1990). Further, the increase in extracellular DA associated with AMPH is 3- to 5-fold greater than that observed with behaviorally equivalent doses of uptake blockers (Kuczenski *et al.*, 1991; see Chapter 4 for further discussion). Thus, a reasonable conclusion from these microdialysis studies is that the role of the DA carrier in the AMPH-induced increase in extracellular DA is different than its role in the action of uptake blockers.

In contrast to these supporting data, other agents that have been used successfully *in vitro* to document carrier-mediated DA release, such as Na^+-free medium or ouabain (see previous discussion), have led to results that are not entirely consistent with carrier-mediated release when applied *in vivo* in microdialysis studies. For example, replacement of normal (155 mM) Na^+ in the dialysis medium with 50 mM Na^+ (choline added to maintain tonicity) resulted in an 80-fold increase in extracellular DA (Hurd and Ungerstedt, 1989b). The increase could be blocked by coadministration of either of the DA uptake blockers nomifensine or LU-19005. These observations are consistent with *in vitro* studies, and implicate a carrier-mediated DA release mechanism in the low Na^+ effect. However, whereas the model for carrier-mediated release of cytoplasmic DA predicts a concomitant decrease in DOPAC, the low Na^+-induced increase in DA was accompanied by an increase in DOPAC. Likewise, ouabain (1–100 μM) added to the dialysis medium increased not only extracellular DA as much as 50-fold but also DOPAC as much as 2-fold (Westerink *et al.*, 1989a). Note, however, that the ouabain effects were blocked completely by TTX and were partially Ca^{2+} sensitive. Thus, if the ouabain-induced DA release *in vivo* occurs through an impulse-dependent exocytotic mechanism rather than a carrier-mediated process, a decrease in DOPAC would not be expected. However, although data regarding the TTX and Ca^{2+} sensitivity of the low Na^+-induced increase in DA were not presented (Hurd and Ungerstedt, 1989b), that the release of DA under these conditions was sensitive to uptake blockers implicates the DA carrier and not an exocytotic release process. Nevertheless, the unexpected increases in DOPAC, the magnitude of the low Na^+- and ouabain-induced increases in DA, and the sensitivity of the ouabain effects to TTX and Ca^{2+} suggest that, *in vivo*, these manipulations introduce other effects that add to and obscure the consequences of exchange diffusion release of DA.

In summary, the AMPH-induced increase in extracellular DA is independent of impulse-mediated nerve terminal depolarization and normal exocytotic release since it is insensitive to TTX and Ca^{2+}, in contrast to the effects of uptake blockers such as cocaine. In addition, AMPH-induced release is insensitive to reserpine, γ-butyrolactone, and APO. The increase can be inhibited by uptake blockers and appears to depend on cytoplasmic DA, since it is blocked by α-MT and enhanced by MAO inhibition. Most evidence generated from *in vivo* studies is reconciled readily with a carrier-mediated DA release mechanism: *in vitro* data support the potential for such release and the characteristics of the AMPH-induced increase in extracellular DA *in vivo* are most consistent with at least some variant of such a mechanism.

V. REGIONAL CHARACTERISTICS OF THE EXTRACELLULAR DOPAMINE RESPONSE

As discussed earlier, the administration of increasing doses of AMPH [0.1–5.0 mg/kg, subcutaneously (sc)] promotes a monotonic increase in caudate extracellular DA, proportional in both magnitude and duration to the dose of the drug. Maximum DA concentrations are achieved 20–40 min after drug administration and then rapidly decline, falling to 50% of the maximum within 50–90 min of drug administration. Similar effects are observed in the nucleus accumbens, the only other DA-innervated brain region that has been examined systematically (Sharp *et al.*, 1987; Carboni *et al.*, 1989; Robinson and Camp, 1990; Kuczenski *et al.*, 1991; Kuczenski and Segal, 1992b). Maximum AMPH-induced extracellular DA concentrations observed in the accumbens are only about two-thirds of caudate values, presumably reflecting the less dense DA innervation in this brain region (Doucet *et al.*, 1986). However, in studies in which the regional DA responses to AMPH were assessed simultaneously (Robinson and Camp, 1990; Kuczenski *et al.*, 1991; Kuczenski and Segal, 1992a; Segal and Kuczenski, 1992a), the temporal and dose-related features of the caudate and accumbens DA responses were otherwise comparable. Further, the DA metabolites DOPAC and HVA exhibit at least qualitatively similar response profiles to AMPH in the two brain regions (Robinson and Camp, 1990; Kuczenski et al., 1991, Kuczenski and Segal, 1992a).

Carboni et al. (1989) reported that the mesolimbic DA system exhibited a more robust response than the mesostriatal system to psychomotor stimulants including AMPH, when the maximal DA response was expressed as a percentage change from baseline to normalize for differences in the density of DA innervation. For example, in response to 0.25 mg/kg AMPH, these investigators observed a greater than 5-fold increase in accumbens DA but only a 2.5-fold increase in caudate DA. Likewise, 5 mg/kg cocaine increased accumbens DA 3-fold over predrug baseline values but caudate DA only 2-fold. Although a greater accumbens than caudate DA response to the uptake blockers cocaine and nomifensine has been confirmed (Kuczenski and Segal, 1992a; Segal and Kuczenski, 1992b), most investigators (Hernandez et al., 1987; Sharp et al., 1987; Maisonneuve et al., 1990, 1992; Pehek et al., 1990; Robinson and Camp, 1990; Kuczenski et al., 1991; Nomikos et al., 1991; Kuczenski and Segal, 1992a) have not observed a preferential accumbens response to a wide range of AMPH doses. In addition, the ratio of caudate to accumbens DA appears to remain constant throughout the peak DA response to AMPH, and to be equivalent to the predrug ratio (Segal and Kuczenski, 1992a). Thus, the weight of evidence suggests that these two brain regions respond to AMPH in a quantitatively similar manner.

AMPH also increases extracellular DA in frontal cortex, substantia nigra, and ventral tegmentum, although to date most of these studies have been performed in anesthetized rats at relatively short intervals after dialysis probe implantation. Several investigators (Maisonneuve et al., 1990; Moghaddam et al., 1990; During et al., 1992) have reported that moderate doses of AMPH (1–2 mg/kg) increased frontal cortex DA 3- to 8-fold above basal concentrations (typically near 2 nM; about 10% of caudate levels), a somewhat smaller change than typically is observed in caudate and accumbens. Moghaddam and co-workers (1990) simultaneously examined the accumbens and frontal cortex DA responses to 1 mg/kg AMPH [intravenously (iv)] and observed similar regional changes in DA in terms of both magnitude and duration. However, the magnitude of DA increase in the accumbens (300% above basal concentrations) was substantially less than typically is observed in awake animals. Kalivas et al. (1989) simultaneously assessed the DA response in both ventral tegmentum and accumbens to a range of AMPH doses (0.67–6.7 mg/kg). Their data suggested a DA response in ventral tegmentum that was substantially less in magnitude than in the accumbens, even when the data were expressed relative to basal values (Kalivas et al., 1989). In contrast, Robertson et al. (1991) examined the caudate and substantia nigra DA response to 2 mg/kg AMPH in awake animals. These investigators reported no significant differences in the magnitude or duration of the response when the data were expressed as percentage of predrug baseline DA levels, although the absolute basal and AMPH-induced DA concentrations in the substantia nigra were only 10% of the caudate values. Thus, based on these comparisons, the temporal pattern of the DA response appears not to exhibit substantial regional variations and, therefore, probably parallels AMPH concentrations in the brain. However, a variety of mechanisms, in addition to pharmacokinetic factors, could contribute to regional differences in the magnitude of the percentage change in DA concentration in response to AMPH. For example, the relative number of transport carriers at the site of release and the amount and dynamics of the available cytoplasmic DA play critical roles in the rate and amount of DA released. Additionally, microheterogeneity of DA transporters, possibly reflecting differences in glycosylation (Lew *et al.*, 1992) or sialic acid content (Zaleska and Erecinska, 1987), have been noted that could influence the ability of the carrier system to transport both AMPH and DA. Other evidence suggests regional differences in the relative degree of metabolism of released DA to 3-MT, which also could contribute to a differential response to stimulants (Kuczenski and Segal, 1992a). Since the quantitative features of the DA response to a given dose of AMPH are determined by these and other factors, additional data are required for meaningful predictions and conclusions to be made regarding regional variations in the DA response.

VI. EFFECTS OF AMPHETAMINE ON EXTRACELLULAR CONCENTRATIONS OF OTHER NEUROTRANSMITTERS

The mesolimbic and mesostriatal DA pathways have been implicated in many of the AMPH-induced behavioral responses. Efforts to define the role of DA further using microdialysis have focused on DA projections to the nucleus accumbens and caudate. The results of these studies, discussed in detail in Chapter 4, have revealed important dissociations of the quantitative aspects of the AMPH-induced DA and behavioral response profiles.

1. Individual animals within a dose and across doses exhibit DA responses that are not consistent with the specific elicited behaviors.
2. The temporal patterns of the DA and behavioral responses are not parallel.
3. Pharmacological manipulations differentially alter the DA and behavioral response profiles.

On the basis of these observations, other effects of AMPH are likely also to contribute significantly to the behavioral response. In this regard, AMPH has been reported to alter the dynamics of a variety of substances in the extracellular space—for example, γ-aminobutyric acid (Bourdelais and Kalivas, 1990) and neurotensin (During *et al.*, 1992)—but these effects have not yet been well characterized. The effects of AMPH on several additional substances, including NE, serotonin, acetylcholine, and ascorbic acid, are being studied more extensively.

A. Norepinephrine

Evidence indicates that AMPH increases extracellular levels of NE in both anesthetized (L'Heureux *et al.*, 1986) and awake (Kuczenski and Segal, 1992b) rats. Although this result should not be surprising, since AMPH is a potent blocker of NE uptake *in vitro*, earlier postmortem studies had failed to detect any substantial AMPH-induced changes in NE biochemistry. Indeed, the lack of effect of the drug on NE turnover at low locomotor-stimulating doses that increased DA turnover (Costa *et al.*, 1972) was instrumental in focusing attention on DA as a potentially more significant transmitter with respect to stimulant-induced behaviors. However, with the advent of more direct indices of *in vivo* CNS NE function in awake animals, AMPH has been shown to promote a dose-dependent increase in both hippocampus and frontal cortex extracellular NE (Kuczenski and Segal, 1992b). The magnitude of the increase (10- to 30-fold above baseline at 0.5–2.5 mg/kg AMPH) is comparable to the caudate DA response. Note, however, that in terms of their temporal patterns, the NE response is prolonged relative to the DA response (Kuczenski and Segal, 1992b), perhaps reflecting differences in neurotransmitter dynamics within the respective neurons.

Although early efforts to detect AMPH-induced NE release from brain slices and synaptosomes were unsuccessful, the profound increase in brain extracellular NE could reasonably result from release of the transmitter rather than simply through uptake blockade. On one hand, as discussed earlier, the NE nerve terminal uptake carrier in sympathetically innervated tissues and in *in vitro* brain preparations exhibits many characteristics that suggest that it can participate in an exchange diffusion mechanism. Further, whereas the increase in extracellular transmitter through uptake blockade requires the continuous impulse-dependent release of NE, AMPH profoundly inhibits the activity of NE neurons (Graham and Aghajanian, 1971; Engborg and Svensson, 1979; Ramirez and Wang, 1986; Pitts and Marwah, 1987) even more effectively than it inhibits DA neurons. On the other hand, although AMPH promotes a greater NE response than does a behaviorally similar dose of the uptake blocker cocaine (S. Florin, R. Kuczenski, and D. S. Segal, unpublished observations), the differential observed for NE is much smaller than that obtained for DA. For example, the increase in caudate DA following 2.5 mg/kg AMPH is 4- to 5-fold greater than the increase obtained with 30–40 mg/kg cocaine (Kuczenski *et al.*, 1991; Kuczenski and Segal, 1992a). In contrast, the AMPH-induced increase in NE in hippocampus and frontal cortex is only 1.5- to 2-fold greater. Although, if AMPH is releasing both DA and NE, these contrasting patterns are likely to reflect differences in neurotransmitter dynamics within DA and NE neurons, further studies of the interaction of AMPH with brain NE are required to specify the mechanism involved more conclusively.

B. Serotonin

Systemic AMPH also increases caudate extracellular serotonin (5-HT) in awake animals, although this effect is relatively transient and appears to be restricted to higher doses of the drug (Kuczenski and Segal, 1989). In early *in vitro* studies, AMPH had been shown to release [^{3}H]5-HT from synaptosomes into superfusates (Raiteri *et al.*, 1975). More recently, a variety of substituted amphetamines, including *p*-chloroamphetamine, methamphetamine, 3,4-methylenedioxymethamphetamine (MDMA), and fenfluramine, was shown to release 5-HT from brain tissue preparations. This release was sensitive to the 5-HT uptake blockers cocaine and fluoxetine (Berger *et al.*, 1992; Rudnick and Wall, 1992). In addition, *in vivo* fluoxetine attenuated the fenfluramine-induced increase in 5-HT in hippocampal dialysates (Sabol *et al.*, 1992). Thus, although the mechanism underlying the AMPH-induced increase in dialysate 5-HT has not been examined systematically, 5-HT also appears to be released via an exchange diffusion mechanism as do DA and NE. This AMPH effect may occur directly through an interaction with the 5-HT transport carrier.

C. Acetylcholine

The direct infusion of AMPH into the caudate via a microdialysis probe rapidly decreases the dialysate levels of acetylcholine (ACh), whereas the DA receptor antagonists haloperidol and sulpiride increase ACh levels (De Boer *et al.*, 1990,1992). These observations are consistent with a variety of converging evidence (McGeer *et al.*, 1974; Sethy and van Woert, 1974; Guyenet *et al.*, 1975; DeBelleroche *et al.*, 1982; Scatton, 1982; Kubota *et al.*, 1987; Chang, 1988) that suggests an inhibitory interaction between DA and caudate cholinergic neurons mediated by D_2 DA receptors. This model predicts that systemic AMPH, by increasing synaptic DA, should decrease caudate ACh through inhibitory D_2 receptors. However, although systemic administration of relatively specific D_2 agonists and antagonists produces effects on caudate ACh that are predictable from such a model (Bertorelli and Consolo, 1990; Damsma *et al.*, 1990a; Marien and Richard, 1990; Florin *et al.*, 1992), the systemic administration of AMPH increases rather than decreases caudate extracellular ACh (Damsma *et al.*, 1991; Consolo *et al.*, 1992; Florin *et al.*, 1992; Guix *et al.*, 1992). Since systemic D_1 agonists increase (Damsma *et al.*, 1990b, 1991) and D_1 antagonists decrease (Consolo *et al.*, 1987; Bertorelli and Consolo, 1990; Damsma *et al.*, 1991) caudate extracellular ACh, researchers have suggested that D_1 DA receptor activation may predominate in the AMPH effect (Damsma *et al.*, 1991; Consolo *et al.*, 1992; Florin *et al.*, 1992; but see De Boer *et al.*, 1992). In fact, both systemic and intrastriatal administration of the D_1 antagonist SCH23390 have been shown to prevent the AMPH-induced increase in caudate ACh (Damsma *et al.*, 1991; Consolo *et al.*, 1992).

The dose–response characteristics of the AMPH-induced increase in caudate extracellular ACh also are not entirely clear. Florin *et al.* (1992) reported a significant increase (maximum 125% above baseline) in caudate ACh after administration of 5.0 mg/kg AMPH, but failed to observe any effect at lower doses. However, other investigators observed significant (40–60%) increases in caudate ACh after administration of lower AMPH doses (1.5–2.0 mg/kg) (Damsma *et al.*, 1991; Consolo *et al.*, 1992; Guix *et al.*, 1992). Additional studies are required to define the dose–response characteristics of AMPH effects on extracellular ACh further before potential relationships between the AMPH-induced increase in caudate ACh and specific components of the behavioral response profile can be established.

D. Ascorbic Acid

Although ascorbic acid (AA) is present in high concentrations (near 200 μM) in the extracellular fluid of mammalian brain (Basse-Tomusk and Rebec, 1990), its functional relationship to neurotransmission has not been elucidated fully. However several investigators, using *in vivo* voltammetry in awake animals, have shown that AMPH increases caudate extracellular AA as much as 100% over basal levels in a dose- and time-dependent manner, generally paralleling the DA response (Oh *et al.*, 1989; Mueller, 1990; Mueller and Kunko, 1990; Pierce and Rebec, 1990,1992; Yount *et al.*, 1991). Whether similar effects occur in other brain regions is controversial (Mueller, 1990; Mueller and Kunko, 1990; Yount *et al.*, 1991). In caudate, the AA response can be blocked by the combined administration of D_1 and D_2 DA receptor antagonists (Oh *et al.*, 1989) and can be mimicked by combined D_1 and D_2 agonists (Pierce and Rebec, 1990; Zetterström *et al.*, 1991). Thus, the AMPH-induced increase in AA appears to be indirect and dependent on DA receptor activation. However, intrastriatal AMPH does not increase extracellular AA (Wilson *et al.*, 1986); investigators have suggested that the changes in caudate AA following systemic AMPH administration are mediated by DA receptors outside the caudate (Desole *et al.*, 1992; Koob, 1992; Pierce and Rebec, 1992; Woolverton and Johnson, 1992). Some evidence suggests that extracellular AA in caudate may be linked to the neuronal release of glutamate (O'Neill *et al.*, 1983,1984; Grunewald and Fillenz, 1984; Fillenz *et al.*, 1986; Basse-Tomusk and Rebec, 1990). Thus, the AA response to AMPH could reflect a DA receptor-dependent activation of corticostriate glutamatergic neurons. Although possible effects of AMPH on extracellular glutamate have not been documented, high repeated doses of methamphetamine have been reported to increase the caudate extracellular concentration of this amino acid (Nash and Yamamoto, 1992).

VII. CONCLUSIONS AND SPECULATIONS

The preponderance of evidence from *in vitro* and *in vivo* studies supports the hypothesis that AMPH increases dopaminergic transmission in the CNS through an exchange diffusion process mediated by the DA transport carrier. Data also have begun to accumulate that document the likelihood that AMPH interacts with analogous NE and 5-HT transport sites in a similar manner, although additional studies comparable to those characterizing AMPH-DA interactions are necessary to confirm this suggestion. However, although delineation of the exchange diffusion mechanism has provided important insight into the interactions of AMPH with biogenic amine neurons, the physiological relevance of this action is not resolved entirely. As noted earlier, some pronounced inconsistencies exist between the quantitative features of the DA and behavioral responses to AMPH. The most striking inconsistency is the observation that, at behaviorally similar doses, AMPH promotes 3- to 5-fold higher extracellular DA concentrations than the psychomotor stimulants that exert their effects through the blockade of DA uptake (Kuczenski *et al.*, 1991; Kuczenski and Segal, 1992a). Thus, to the extent that synaptic DA concentrations are related critically to the behavioral actions of AMPH-like drugs, researchers must account for these large differences in extracellular DA levels.

Several explanations for this apparent dissociation are possible. First, the extracellular DA that is subject to microdialysis is not identical to synaptic DA and, therefore, may not provide an accurate indication of synaptic events. However, although caution must be exercised in interpreting dialysis data, most evidence supports the assumption that the extracellular pool provides a reasonable estimate of synaptic transmitter dynamics.

However, the differences between AMPH and the uptake blockers are more likely to derive from the unique manner in which AMPH releases DA. Unlike AMPH, uptake blocking agents increase synaptic activity at DA sites by interfering with the inactivation of transmitter that has entered the synapse through ongoing impulse-dependent exocytotic release. Thus, the effects of uptake blockers are linked directly to those processes that typically modulate dopamine release, including heteroreceptor- and autoreceptor-mediated changes in impulse traffic and vesicular transmitter release, whereas the effects of AMPH are not influenced by these regulatory mechanisms. As a consequence, whereas uptake blockers will increase DA in proportion to the level of impulse flow in active dopaminergic neurons, AMPH should increase transmitter at all dopaminergic synapses. Therefore, a significant portion of the AMPH-induced elevation of DA may occur at synapses that do not contribute substantially to the motor-activating effects of these drugs, but contribute to those aspects of the behavioral profiles that differ between AMPH and the uptake blockers instead (for example, see Segal and Kuczenski, 1987).

However, only some portion of the AMPH-induced increases in extracellular DA may represent elevated synaptic DA accurately. For example, AMPH may release DA through transport carriers located at nonsynaptic sites that normally do not participate in the impulse-dependent vesicular release process. Alternatively, the subsynaptic ultrastructure into which AMPH releases DA may be different from that in impulse-dependent release, and may place unique directional restraints on transmitter diffusion. Such constraints may limit the progression of DA toward postsynaptic sites and perhaps facilitate its movement out of the synapse. One observation in support of the possibility that the increases in extracellular DA in response to AMPH and uptake blockers may be dissimilar is that DA released by AMPH appears to be converted to 3-MT less efficiently than DA that accumulates in the synapse through uptake blockade (Kuczenski and Segal, 1992a).

On the other hand, if synaptic levels of DA in response to AMPH and uptake blockers are as different as demonstrated by the extracellular measures, then those differences must be nullified largely by other effects of these drugs to produce behaviorally similar profiles. One possibility is that other DA effects of AMPH may attenuate the response to the exaggerated DA release. For example, because of its structural similarity to DA, AMPH may interfere with DA–receptor interactions. Alternatively, differential effects of AMPH and uptake blockers on other neurotransmitter systems may effectively equalize the consequences of differential DA receptor activation.

In summary, research over the past two decades has yielded significant progress in defining the mechanisms by which AMPH interacts with several neurotransmitter systems in the brain, particularly the DA system. However, a more complete characterization of the CNS effects of this drug is required before all its behaviorally relevant mechanisms are elucidated fully.

ACKNOWLEDGMENTS

Preparation of this manuscript was supported in part by United States Public Health Service Grants DA-04157 and DA-01568 and by a Research Scientist Award MH-70183 to D. S. Segal.

REFERENCES

Azzaro, A. J., Ziance, R. J., and Rutledge, C. O. (1974). The importance of neuronal uptake of amines for amphetamine-induced release of ^{3}H-norepinephrine from isolated brain tissue. *J. Pharmacol. Exp. Ther.* **189**, 110–118.

Basse-Tomusk, A., and Rebec, G. V. (1990). Corticostriatal and thalamic regualtion of amphetamine-induced ascorbate release in the neostriatum. *Pharmacol. Biochem. Behav.* **35**, 55–60.

Benveniste, H., and Hïtemeier, P. C. (1990). Microdialysis—Theory and application. *Progr. Neurobiol.* **35**, 195–215.

Berger, U. V., Gu, X. F., and Azmitia, E. C. (1992). The substituted amphetamines 3,4-methylenedioxymethamphetamine, methamphetamine, p-chloroamphetamine and fenfluramine induce 5-hydroxytryptamine release via a common mechanism blocked by fluoxetine and cocaine. *Eur. J. Pharmacol.* **215**, 153–160.

Bertorelli, R., and Consolo, S. (1990). D_1 and D_2 dopaminergic regulation of acetylcholine release from striata of freely moving rats. *J. Neurochem.* **54**, 2145–2148.

Blaustein, M. P., and King, A. C. (1976). Influence of membrane potential on the sodium-dependent uptake of gamma-aminobutyric acid by presynaptic nerve terminals: Experimental observations and theoretical considerations. *J. Membr. Biol.* **30**, 153–173.

Bonisch, H. (1984). The transport of (+)-amphetamine by the neuronal noradrenaline carrier. *Naunyn-Schmiedeberg's Arch. Pharmacol.* **327**, 267–272.

Bonisch, H., and Trendelenburg, U. (1988). The mechanism of action of indirectly acting sympathomimetic amines. *In* "Handbook of Experimental Pharmacology" (U. Trendelenburg and N. Weiner, eds.), pp. 247–277. Springer-Verlag, Berlin.

Bourdelais, A., and Kalivas, P. W. (1990). Amphetamine lowers extracellular GABA concentration in the ventral pallidum. *Brain Res.* **516**, 132–136.

Braestrup, C. (1977). Biochemical differentiation of amphetamine vs. methylphenidate and nomifensine in rats. *J. Pharm. Pharmacol.* **29**, 463–472.

Brown, E. E., Damsma, G., Cumming, P., and Fibiger, H. C. (1991). Interstitial 3-methoxytyramine reflects striatal dopamine release: An in vivo microdialysis study. *J. Neurochem.* **57**, 701–707.

Bunney, B. S., Walters, J. R., Roth, R. H., and Aghajanian, G. K. (1973). Dopaminergic neurons: Effect of antipsychotic drugs and amphetamine on single cell activity. *J. Pharmacol. Exp. Ther.* **208**, 560–571.

Butcher, S. P., Fairbrother, I. S., Kelly, J. S., and Arbuthnott, G. W. (1988). Amphetamine-induced dopamine release in the rat striatum: An in vivo microdialysis study. *J. Neurochem.* **50**, 346–355.

Callaway, C. W., Kuczenski, R., and Segal, D. S. (1989). Reserpine enhances amphetamine stereotypies without increasing amphetamine-induced changes in striatal dialysate dopamine. *Brain Res.* **505**, 83–90.

Carboni, E., Imperato, A., Perezzani, L., and Di Chiara, G. (1989). Amphetamine, cocaine, phencyclidine and nomifensine increase extracellular dopamine concentrations preferentially in the nucleus accumbens of freely moving rats. *Neuroscience* **28**, 653–661.

Carenzi, A., Guidotti, A., Revuelta, A., and Costa, E. (1975). Molecular mechanisms in the action of morphine and viminol (R2) on rat striatum. *J. Pharmacol. Exp. Ther.* **194**, 311–318.

Chang, H. T. (1988). Dopamine-acetylcholine interaction in the rat striatum: A dual-labeling immunocytochemical study. *Brain Res. Bull.* **21**, 295–304.

Cole, S. O. (1978). Brain mechanisms of amphetamine-induced anorexia, locomotion, and stereotypy: A review. *Neurosci. Biobehav. Rev.* **2**, 89–100.

Connor, C. E., and Kuczenski, R. (1986). Evidence that amphetamine and Na^+ gradient reversal increase striatal synaptosomal dopamine synthesis through carrier-mediated efflux of dopamine. *Biochem. Pharmacol.* **35**, 3123–3130.

Consolo, S., Wu, F. C., and Fusi, R. (1987). D-1 receptor-linked mechanism modulates cholinergic neurotransmission in rat striatum. *J. Pharmacol. Exp. Ther.* **242**, 300–305.

Consolo, S., Girotti, P., Russi, G., and Di Chiara, G. (1992). Endogenous dopamine facilitates striatal in vivo acetylcholine release by acting on D_1 receptors localized in the striatum. *J. Neurochem.* **59**, 1555–1557.

Costa, E., Groppetti, A., and Naimzada, M. K. (1972). Effects of amphetamine on the turnover rate of brain catecholamines and motor activity. *Br. J. Pharmacol.* **44,** 742–751.

Creese, I., and Iversen, S. D. (1974). The role of forebrain dopamine systems in amphetamine induced stereotypy in the adult rat following neonatal treatment with 6-hydroxydopamine. *Psychopharmacology* **39,** 345–357.

Damsma, G., De Boer, P., Westerink, B. H. C., Fibiger, H. C. (1990a). Dopaminergic regulation of striatal cholinergic interneurons: An in vivo microdialysis study. *Naunyn-Schmiedeberg's Arch. Pharmacol.* **342,** 523–527.

Damsma, G., Tham, C.-S., Robertson, G. S., and Fibiger, H. C. (1990b). Dopamine D_1 receptor stimulation increases striatal acetylcholine release in the rat. *Eur. J. Pharmacol.* **186,** 335–338.

Damsma, G., Robertson, G. S., Tham, C.-S., and Fibiger, H. C. (1991). Dopaminergic regulation of striatal acetylcholine release: Importance of D_1 and NMDA receptors. *J. Pharmacol. Exp. Ther.* **259,** 1064–1072.

DeBelleroche, J., Coutinho-Netto, J., and Bradford, H. F. (1982). Dopamine inhibition of the release of endogenous acetylcholine from corpus striatum and cerebral cortex in tissue slices and synaptosomes: A presynaptic response? *J. Neurochem.* **39,** 217–222.

De Boer, P., Damsma, G., Fibiger, H. C., Timmerman, W., de Vries, J. B., and Westerink, B. H. C. (1990). Dopaminergic-cholinergic interactions in the striatum: The critical significance of calcium concentrations in brain microdialysis. *Naunyn-Schmiedeberg's Arch. Pharmacol.* **342,** 528–534.

De Boer, P., Damsma, G., Schram, Q., Stoof, J. C., Zaagsma, J., and Westerink, B. H. C. (1992). The effect of intrastriatal application of directly and indirectly acting dopamine agonists and antagonists on the in vivo release of acetylcholine measured by brain microdialysis. The importance of the post-surgery interval. *Naunyn-Schmiedeberg's Arch. Pharmacol.* **345,** 144–152.

Desole, M. S., Miele, M., Enrico, P., Fresu, L., Esposito, G., De Natale, G., and Miele, E. (1992). The effects of cortical ablation on *d*-amphetamine-induced changes in striatal dopamine turnover and ascorbic acid catabolism in the rat. *Neurosci. Lett.* **139,** 29–33.

Di Chiara, G. (1991). Brain dialysis of monoamines. *In* "Microdialysis in the Neurosciences" (T. E. Robinson and J. B. Justice, Jr., eds.), pp. 175–187. Elsevier, New York.

Doucet, G., Descarries, L., and Garcia, S. (1986). Quantification of the dopamine innervation in adult rat neostriatum. *Neuroscience* **19,** 427–445.

During, M. J., Bean, A. J., and Roth, R. H. (1992). Effects of CNS stimulants on the in vivo release of the colocalized transmitters, dopamine and neurotensin, from rat prefrontal cortex. *Neurosci. Lett.* **140,** 129–133.

Engborg, G., and Svensson, T. H. (1979). Amphetamine-induced inhibition of central noradrenergic neurons: A pharmacological analysis. *Life Sci.* **24,** 2245–2254.

Erickson, J. D., Masserano, J. M., Barnes, E. M., Ruth, J. A., and Weiner, N. (1990). Chloride ion increases [^{3}H]dopamine accumulation by synaptic vesicles purified from rat striatum: Inhibition by thiocyanate ion. *Brain Res.* **516,** 155–160.

Eriksson, E., and Christensson, E. (1990). The effect of amperozide on uptake and release of [^{3}H]-dopamine in vitro from perfused rat striatal and limbic brain areas. *Pharmacol. Toxicol. (Suppl.)* **66,** 45–48.

Ferris, R. M., and Tang, L. M. (1979). Comparison of the effects of the isomers of amphetamine, methylphenidate and deoxipradrol on the uptake of 1-[^{3}H]norepinephrine and [^{3}H]dopamine by synaptic vesicles from rat whole brain, striatum and hypothalamus. *J. Pharmacol. Exp. Ther.* **210,** 422–428.

Fillenz, M., O'Neill, R. D., and Grunewald, R. A. (1986). Changes in extraceullular brain ascorbate concentrations as an index of excitatory amino acid release. *In* "Monitoring Neurotransmitter Release during Behavior" (M. H. Joseph, M. Fillenz, I. A. Macdonald, and C. A. Marsden, eds.), pp. 144–163. Ellis Horwood, Chichester.

Fischer, J. F., and Cho, A. K. (1979). Chemical release of dopamine from striatal homogenates: Evidence for an exchange diffusion model. *J. Pharmacol. Exp. Ther.* **192**, 642–653.

Florin, S. M., Kuczenski, R., and Segal, D. S. (1992). Amphetamine-induced changes in behavior and caudate extracellular acetylcholine. *Brain Res.* **581**, 53–58.

Glowinski, J., and Axelrod, J. (1965). Effects of drugs on the uptake release and metabolism of ^{3}H-norepinephrine in the rat brain. *J. Pharmacol. Exp. Ther.* **149**, 43–49.

Graham, A., and Aghajanian, G. K. (1971). Effects of amphetamine on single unit activity in a catecholamine nucleus, the locus coeruleus. *Nature (London)* **234**, 100–102.

Green, A. L., and El Hait, A. S. (1980). A new approach to the assessment of the potency of reversible monoamine oxidase inhibitors in vivo, and its application to (+)-amphetamine, *p*-methoxyamphetamine and harmaline. *Biochem. Pharmacol.* **29**, 2781–2789.

Grunewald, R. A., and Fillenz, M. (1984). Release of ascorbate form a synaptosomal fraction from rat brain. *Neurochem. Int.* **6**, 491–500.

Guix, T., Hurd, Y. L., and Ungerstedt, U. (1992). Amphetamine enhances extracellular concentrations of dopamine and acetylcholine in dorsolateral striatum and nucleus accumbens of freely moving rats. *Neurosci. Lett.* **138**, 137–140.

Guyenet, P. G., Agid, Y., Javoy, F., Beaujouan, J. C., Rossier, J., and Glowinski, J. (1975). Effects of dopaminergic receptor agonists and antagonists on the activity of the neo-striatal cholinergic system. *Brain Res.* **84**, 227–244.

Hartman, J., and Halaris, A. E. (1980). Compartmentation of catecholamines in rat brain: Effects of agonists and antagonists. *Brain Res.* **200**, 421–436.

Heikkila, R. E., Orlansky, H., Mytilineou, C., and Cohen, G. (1975). Amphetamine: Evaluation of *d*- and *l*-isomers as releasing agents and uptake inhibitors for ^{3}H-dopamine and ^{3}H-norepinephrine in slices of rat neostriatum and cerebral cortex. *J. Pharmacol. Exp. Ther.* **193**, 47–56.

Hernandez, L., Lee, F., and Hoebel, B. G. (1987). Simultaneous microdialysis and amphetamine infusion in the nucleus accumbens and striatum of freely moving rats: Increase in extracellular dopamine and serotonin. *Brain Res. Bull.* **19**, 623–628.

Holz, R. W., and Coyle, J. (1974). The effects of various salts, temperature, and the alkaloids veratridine and tetrodotoxin on the uptake of ^{3}H-dopamine into synaptosomes from rat striatum. *Mol. Pharmacol.* **10**, 746–758.

Hurd, Y. L., and Ungerstedt, U. (1989a). Ca^{2+} dependence of the amphetamine, nomifensine, and Lu 19-005 effect on in vivo dopamine transmission. *Eur. J. Pharmacol.* **166**, 261–269.

Hurd, Y. L., and Ungerstedt, U. (1989b). Influence of a carrier transport process on *in vivo* release and metabolism of dopamine: Dependence on extracellular Na^+. *Life Sci.* **45**, 283–293.

Ichikawa, J., and Meltzer, H. Y. (1992). Amperozide, a novel antipsychotic drug, inhibits the ability of *d*-amphetamine to increase dopamine release in vivo in rat striatum and nucleus accumbens. *J. Neurochem.* **58**, 2285–2291.

Imperato, A., and Di Chiara, G. (1984). Trans-striatal dialysis coupled to reverse phase high performance liquid chromatography with electrochemical detection: A new method for the study of the in vivo release of endogenous dopamine and metabolites. *J. Neurosci.* **4**(4), 966–977.

Johnson, R. G., Pfister, D., Carty, S. E., and Scarpa, A. (1979). Biological amine transport in chromaffin ghosts: Coupling to the transmembrane proton and potential gradients. *J. Biol. Chem.* **254**, 10963–10972.

Johnson, R. G., Carty, S. E., Hayflick, S., and Scarpa, A. (1982). Mechanism of accumulation of tyramine, metaraminol, and isoproterenol in isolated chromaffin granules. *Biochem. Pharmacol.* **31**, 815–823.

Kaakkola, S., Mannisto, P. T., and Nissinen, E. (1987). Striatal membrane-bound and soluble catechol-O-methyl-transferase after selective neuronal lesions in the rat. *J. Neural Transm.* **69**, 221–228.

Kalivas, P. W., Bourdelais, A., Abhold, R., and Abbott, L. (1989). Somatodendritic release of endogenous dopamine: In vivo dialysis in the A10 dopamine region. *Neurosci. Lett.* **100**, 215–220.

Kamal, L., Arbilla, S., and Langer, S. (1981). Presynaptic modulation of the release of dopamine from the rabbit caudate nucleus: Differences between electrical stimulation, amphetamine and tyramine. *J. Pharmacol. Exp. Ther.* **216**, 592–598.

Kaplan, G. P., Hartman, B. K., and Creveling, C. R. (1979). Immunohistochemical demonstration of catechol-O-methyltransferase in mammalian brain. *Brain Res.* **167**, 241–250.

Kehr, W. (1981). 3-Methoxytyramine and normetanephrine as indicators of dopamine and norepinephrine release in mouse brain in vivo. *J. Neural Transm.* **50**, 165–178.

Kehr, W., Speckenbach, W., and Zimmerman, R. (1977). Interaction of haloperidol and gamma-butyrolactone with (+)-amphetamine-induced changes in monoamine synthesis and metabolism in rat brain. *J. Neural Transm.* **40**, 129–147.

Kelly, P. H. (1977). Drug-induced motor behavior. In "Handbook of Psychopharmacology" (L. Iversen, S. Iversen, and S. H. Snyder, eds.), Vol. 8, pp. 295–331. Plenum Press, New York.

Kimura, K., Nomikos, G. G., and Svensson, T. H. (1993). Effects of amperozide on psychostimulant-induced hyperlocomotion and dopamine release in the nucleus accumbens. *Pharmacol. Biochem. Behav.* **44**, 27–36.

Knepper, S. M., Grunewald, G. L., and Rutledge, C. O. (1988). Inhibition of norepinephrine transport into synaptic vesicles by amphetamine analogs. *J. Pharmacol. Exp. Ther.* **247**, 487–494.

Koob, G. F. (1992). Drugs of abuse: Anatomy, pharmacology and function of reward pathways. *Trends Pharmacol. Sci.* **13**, 177–184.

Krueger, B. K. (1990). Kinetics and block of dopamine uptake in synaptosomes from rat caudate nucleus. *J. Neurochem.* **55**, 260–267.

Kubota, Y., Inagaki, S., Shimada, S., Kito, S., Eckenstein, F., and Tohyama, M. (1987). Neostriatal cholinergic neurons recieve direct synaptic inputs from dopaminergic axons. *Brain Res.* **413**, 179–184.

Kuczenski, R. (1977). Biphasic effects of amphetamine on striatal dopamine dynamics. *Eur. J. Pharmacol.* **46**, 249–257.

Kuczenski, R. (1979). Effects of para-chlorophenylalanine on amphetamine and haloperidol-induced changes in striatal dopamine turnover. *Brain Res.* **164**, 217–225.

Kuczenski, R. (1980). Amphetamine–haloperidol interactions on striatal and mesolimbic tyrosine hydroxylase activity and dopamine metabolism. *J. Pharmacol. Exp. Ther.* **215**, 135–142.

Kuczenski, R. (1983). Biochemical actions of amphetamine and other stimulants. In "Stimulants: Neurochemical, Behavioral, and Clinical Perspectives" (I. Creese, ed.), pp. 31–61. Raven Press, New York.

Kuczenski, R., and Segal, D. S. (1989). Concomitant characterization of behavioral and striatal neurotransmitter response to amphetamine using in vivo microdialysis. *J. Neurosci.* **9**, 2051–2065.

Kuczenski, R., and Segal, D. S. (1992a). Differential effects of amphetamine and dopamine uptake blockers (cocaine, nomifensine) on caudate and accumbens dialysate dopamine and 3-methoxytyramine. *J. Pharmacol. Exp. Ther.* **262**, 1085–1094.

Kuczenski, R., and Segal, D. S. (1992b). Regional norepinephrine response to amphetamine using dialysis: Comparison to caudate dopamine. *Synapse* **11**, 164–169.

Kuczenski, R., Segal, D. S., and Manley, L. D. (1990). Apomorphine does not alter amphetamine-induced dopamine release measured in striatal dialysates. *J. Neurochem.* **54**, 1492–1499.

Kuczenski, R., Segal, D. S., and Aizenstein, M. L. (1991). Amphetamine, fencamfamine, and cocaine: Relationships between locomotor and stereotypy response profiles and caudate and accumbens dopamine dynamics. *J. Neurosci.* **11**, 2703–2712.

Lew, R., Patel, A., Vaughan, R. A., Wilson, A., and Kuhar, M. J. (1992). Microheterogeneity of dopamine transporters in rat striatum and nucleus accumbens. *Brain Res.* **584,** 266–271.

L'Heureux, R., Dennis, T., Curet, O., and Scatton, B. (1986). Measurement of endogenous noradrenaline release in the rat cerebral cortex in vivo by transcortical dialysis: Effects of drugs affecting noradrenergic transmission. *J. Neurochem.* **46,** 1794–1801.

Lundberg, J., Bylock, A., Goldstein, M., Hansson, H., and Dahlstrom, A. (1977). Ultrastructural localization of dopamine-β-hydroxylase in nerve terminals of the rat brain. *Brain Res.* **120,** 549–552.

Maisonneuve, I. M., Keller, R. W., and Glick, S. D. (1990). Similar effects of D-amphetamine and cocaine on extracellular dopamine levels in medial prefrontal cortex of rats. *Brain Res.* **535,** 221–226.

Maisonneuve, I. M., Keller, R. W., Jr., and Glick, S. D. (1992). Interactions of ibogaine and D-amphetamine: In vivo microdialysis and motor behavior in rats. *Brain Res.* **579,** 87–92.

Manley, L. D., Kuczenski, R., Segal, D. S., Young, S. J., and Groves, P. M. (1992). Effects of frequency and pattern of medial forebrain bundle stimulation on caudate dialysate dopamine and serotonin. *J. Neurochem.* **58,** 1491–1498.

Mann, P., and Quastel, J. (1940). Benzedrine (β-phenylisopropylamine) and brain metabolism. *Biochem. J.* **34,** 414–431.

Marien, M. R., and Richard, J. W. (1990). Drug effects on the release of endogenous acetylcholine in vivo: Measurement by intracerebral dialysis and gas chromatography–mass spectrometry. *J. Neurochem.* **54,** 2016–2023.

McGeer, P. L., Grewaal, D. S., and McGeer, E. G. (1974). Influence of non-cholinergic drugs on rat striatal acetylcholine levels. *Brain Res.* **80,** 211–217.

Meyerhoff, J. L., and Kant, G. J. (1978). Release of endogenous dopamine from corpus striatum. *Life Sci.* **23,** 1481–1486.

Miller, H. H., Shore, P. A., and Clarke, D. E. (1980). In vivo monoamine oxidase inhibition by d-amphetamine. *Biochem. Pharmacol.* **29,** 1347–1354.

Moghaddam, R., Roth, R. H., and Bunney, B. S. (1990). Characterization of dopamine release in the rat medial prefrontal cortex as assessed by *in vivo* microdialysis: Comparison to the striatum. *Neurosci.* **36,** 669–676.

Mueller, K. (1990). The effects of haloperidol and amphetamine on ascorbic acid and uric acid in caudate and nucleus accumbens of rats as measured by voltammetry in vivo. *Life Sci.* **47,** 735–742.

Mueller, K., and Kunko, P. M. (1990). The effects of amphetamine and pilocarpine on the release of ascorbic and uric acid in several rat brain areas. *Pharmacol. Biochem. Behav.* **35,** 871–876.

Narahashi, T. (1974). Chemical tools in the study of excitable membranes. *Physiol. Rev.* **54,** 813–889.

Nash, J. F., and Yamamoto, B. K. (1992). Methamphetamine neurotoxicity and striatal glutamate release: Comparison to 3,4-methylenedioxymethamphetamine. *Brain Res.* **581,** 237–243.

Niddam, R., Abrila, S., Scatton, B., Dennis, T., and Langer, S. Z. (1985). Amphetamine-induced release of endogenous dopamine in vitro is not reduced following pretreatment with reserpine. *Naunyn-Schmiedeberg's Arch. Pharmacol.* **329,** 123–127.

Nomikos, G. G., Damsma, G., Wenkstern, D., and Fibiger, H. C. (1990). In vivo characterization of locally applied dopamine uptake inhibitors by striatal microdialysis. *Synapse* **6,** 106–112.

Nomikos, G. G., Damsma, G., Wenkstern, D., and Fibiger, H. C. (1991). Chronic desipramine enhances amphetamine-induced increases in interstitial concentrations of dopamine in the nucleus accumbens. *Eur. J. Pharmacol.* **195,** 63–73.

Obianwu, H. O., Stitzel, R., and Lundborg, P. (1968). Subcellular distribution of ³H-amphetamine and ³H-guanethidine and their interactions with adrenergic neurons. *J. Pharm. Pharmacol.* **20**, 585–594.

Oh, C., Gardiner, T. W., and Rebec, G. V. (1989). Blockade of both D_1- and D_2-dopamine receptors inhibits amphetamine-induced ascorbate release in the neostriatum. *Brain Res.* **480**, 184–189.

O'Neill, R. D., Grunewald, R. A., Fillenz, M., and Albery, W. J. (1983). The effect of unilateral cortical lesions on the circadian changes in rat striatal ascorbate and homovanillic acid levels measured in vivo using voltammetry. *Neurosci. Lett.* **42**, 105–110.

O'Neill, R. D., Fillenz, M., Sundstrom, L., and Rawlins, J. N. P. (1984). Voltammetrically monitored brain ascorbate as an index of excitatory amino acid release in the unrestrained rat. *Neurosci. Lett.* **52**, 227–233.

Osborne, P. G., O'Connor, W. T., Kehr, J., and Ungerstedt, U. (1991). In vivo characterization of extracellular dopamine, GABA and acetylcholine from the dorsolateral striatum of awake freely moving rats by chronic microdialysis. *J. Neurosci. Meth.* **37**, 93–102.

Pani, L., Kuzmin, A., Diana, M., De Montis, G., Gessa, G. L., and Rossetti, Z. L. (1990). Calcium receptor antagonists modify cocaine effects in the central nervous system differently. *Eur. J. Pharmacol.* **190**, 217–221.

Parker, E. M., and Cubeddu, L. X. (1986a). Effects of d-amphetamine and dopamine synthesis inhibitors on dopamine and acetylcholine neurotransmission in the striatum. I. Release in the absence of vesicular transmitter stores. *J. Pharmacol. Exp. Ther.* **237**, 179–192.

Parker, E. M., and Cubeddu, L. X. (1986b). Effects of d-amphetamine and dopamine synthesis inhibitors on dopamine and acetylcholine neurotransmission in the striatum. II. Release in the presence of vesicular transmitter stores. *J. Pharmacol. Exp. Ther.* **237**, 193–203.

Paton, D. M. (1973). Mechanism of efflux of noradrenaline from adrenergic nerves in rabbit atria. *Br. J. Pharmacol.* **49**, 614–627.

Pearl, R. G., and Seiden, L. S. (1979). D-Amphetamine-induced increase in catecholamine synthesis in the corpus striatum of the rat: Persistence of the effect after tolerance. *J. Neural Transm.* **44**, 21–38.

Pehek, E. A., Schechter, M. D., and Yamamoto, B. K. (1990). Effects of cathinone and amphetamine on the neurochemistry of dopamine *in vivo*. *Neuropharmacology* **29**, 1171–1176.

Philippu, A., and Beyer, J. (1973). Dopamine and noradrenaline transport in subcellular vesicles of the striatum. *Naunyn-Schmiedeberg's Arch. Pharmacol.* **278**, 387–402.

Phillips, J. H. (1982). Dynamic aspects of chromaffin granule structure. *Neuroscience* **7**, 1595–1609.

Phillips, J. H., and Allison, Y. P. (1978). Proton translocation by the bovine chromaffin granule membrane. *Biochem. J.* **170**, 661–672.

Pierce, R. C., and Rebec, G. V. (1990). Stimulation of both D_1 and D_2 dopamine receptors increases behavioral activation and ascorbate release in the neostriatum of freely moving rats. *Eur. J. Pharmacol.* **191**, 295–302.

Pierce, R. C., and Rebec, G. V. (1992). Dopamine-, NMDA- and sigma-receptor antagonists exert differential effects on basal and amphetamine-induced changes in neostriatal ascorbate and DOPAC in awake, behaving rats. *Brain Res.* **579**, 59–66.

Pitts, D. K., and Marwah, J. (1987). Electrophysiological actions of cocaine on noradrenergic neurons in rat locus ceruleus. *J. Pharmacol. Exp. Ther.* **240**, 345–351.

Ponzio, F., Achilli, G., and Algeri, S. (1981). A rapid and simple method for the determination of picogram levels of 3-methoxytyramine in brain tissue using liquid chromatography with electrochemical detection. *J. Neurochem.* **36**, 1361–1367.

Raiteri, M., Bertollini, A., Angelini, F., and Levi, G. (1975). d-Amphetamine as a releaser or reuptake inhibitor of biogenic amines in synaptosomes. *Eur. J. Pharmacol.* **34**, 189–195.

Raiteri, M., Del Carmine, R., Bertollini, A., and Levi, G. (1977). Effect of desmethylimipramine on the release of ³H-norepinephrine induced by various agents in hypothalamic synaptosomes. *Mol. Pharmacol.* **13**, 746–758.

Raiteri, M., Cerrito, F., Cervoni, A., and Levi, G. (1979). Dopamine can be released by two mechanisms differentially affected by the dopamine transport inhibitor nomifensine. *J. Pharmacol. Exp. Ther.* **208**, 195–202.

Ramirez, O. A., and Wang, R. Y. (1986). Locus coeruleus norepinephrine-containing neurons: Effects produced by acute and subchronic treatment with antipsychotic drugs and amphetamine. *Brain Res.* **362**, 165–170.

Rivett, A. J., Francis, A., and Roth, J. A. (1983). Distinct cellular localization of membrane-bound and soluble forms of catechol-O-methyltransferase in brain. *J. Neurochem.* **40**, 215–219.

Robertson, G. S., Damsma, G., and Fibiger, H. C. (1991). Characterization of dopamine release in the substantia nigra by *in vivo* microdialysis in freely moving rats. *J. Neurosci.* **11**, 2209–2216.

Robinson, T. E., and Camp, D. M. (1990). Does amphetamine preferentially increase the extracellular concentration of dopamine in the mesolimbic system of freely moving rats? *Neuropsychopharmacology* **3**, 163–173.

Ross, S. B., and Renyi, A. L. (1964). Blocking action of sympathomimetic amines on the uptake of tritiated noradrenaline by mouse cerebral cortex tissues in vitro. *Acta Pharmacol. Toxicol.* **21**, 226–239.

Roth, R. H., Murrin, T. C., and Walters, J. R. (1976). Central dopaminergic neurons: Effects of alterations in impulse flow on the accumulation of dihydroxyphenylacetic acid. *Eur. J. Pharmacol.* **36**, 163–171.

Rothman, R. B., Mele, A., Reid, A. A., Akunne, H. C., Greig, N., Thurkauf, A., De Costa, B. R., Rice, K. C., and Pert, A. (1991). GBR12909 antagonizes the ability of cocaine to elevate extracellular levels of dopamine. *Pharmacol. Biochem. Behav.* **40**, 387–397.

Rudnick, G., and Wall, S. C. (1992). p-Chloroamphetamine induces serotonin release through serotonin transporters. *Biochemistry* **31**, 6710–6718.

Rutledge, C. O. (1978). Effect of metabolic inhibitors and ouabain on amphetamine- and potassium-induced release of biogenic amines from isolated brain tissue. *Biochem. Pharmacol.* **27**, 511–516.

Rutledge, C. O., and Vollmer, S. (1979). Evidence for carrier-mediated efflux of norepinephrine displayed by amphetamine. *In* "Catecholamines: Basic and Clinical Frontiers" (E. Usdin, I. J. Kopin, and J. Barchas, eds.), pp. 304–306. Pergamon Press, New York.

Sabol, K. E., Richards, J. B., and Seiden, L. S. (1992). Fluoxetine attenuates the D,L-fenfluramine-induced increase in extracellular serotonin as measured by in vivo dialysis. *Brain Res.* **585**, 421–424.

Sammet, S., and Graefe, K.-H. (1979). Kinetic analysis of the interaction between noradrenaline and Na^+ in neuronal uptake: Kinetic evidence for co-transport. *Naunyn-Schmiedeberg's Arch. Pharmacol.* **309**, 99–107.

Sanchez-Armass, S., and Orrego, F. (1977). A major role for chloride in (^{3}H)-noradrenaline transport by rat heart adrenergic nerves. *Life Sci.* **20**, 1829–1838.

Sanchez-Armass, S., and Orrego, F (1978). Noradrenaline transport by rat heart sympathetic nerves: A re-examination of the role of sodium ions. *Naunyn-Schmiedeberg's Arch. Pharmacol.* **302**, 255–261.

Scatton, B. (1982). Further evidence for the involvement of D_2, but not D_1 dopamine receptors in dopaminergic control of striatal cholinergic transmission. *Life Sci.* **31**, 2883–2890.

Schomig, E., and Trendelenburg, U. (1987). Simulation of outward transport of neuronal ^{3}H-noradrenaline with the help of a two-compartment model. *Naunyn-Schmiedeberg's Arch. Pharmacol.* **336**, 631–640.

Schumann, H. J., and Philippu, A. (1962). Release of catecholamines from isolated medullary granules by sympathetic amines. *Nature (London)* **193**, 890–891.

Segal, D. S., and Kuczenski, R. (1987). Behavioral and neurochemical characteristics of stimulant-induced augmentation. *Psychopharmacol. Bull.* **23**, 417–424.

Segal, D. S., Kuczenski, R. (1992a). In vivo microdialysis reveals a diminished amphetamine-induced dopamine response corresponding to behavioral sensitization produced by repeated amphetamine pretreatment. *Brain Res.* **571**, 330–337.

Segal, D. S., and Kuczenski, R. (1992b). Repeated cocaine administration induces behavioral sensitization and corresponding decreased extracellular dopamine responses in caudate and accumbens. *Brain Res.* **577**, 351–355.

Segal, D. S., Kuczenski, R., and Okuda, C. (1992). Clorgyline-induced increases in presynaptic DA: Changes in the behavioral and neurochemical effects of amphetamine using in vivo microdialysis. *Pharmacol. Biochem. Behav.* **42**, 421–429.

Sessions, G., Meyerhoff, J., Kant, G. J., and Koob, G. F. (1980). Effects of lesions of the ventral medial tegmentum on locomotor activity, biogenic amines and response to amphetamine in rats. *Pharmacol. Biochem. Behav.* **12**, 603–608.

Sethy, V. H., and van Woert, M. H. (1974). Modification of striatal acetylcholine concentration by dopamine receptor agonists and antagonists. *Res. Commun. Chem. Pathol. Pharmacol.* **8**, 13–28.

Sharp, T., Zetterstrom, T., Ljungberg, T., and Ungerstedt, U. (1987a). A direct comparison of amphetamine-induced behaviours and regional brain dopamine release in the rat using intracerebral dialysis. *Brain Res.* **401**, 322–330.

Shenoy, A., and Ziance, R. (1979). Comparative regulation of potassium and amphetamine induced release of ^{3}H-norepinephrine from rat brain via presynaptic mechanisms. *Life Sci.* **24**, 255–264.

Strait, K. A., and Kuczenski, R. (1986). Dopamine autoreceptor regulation of the kinetic state of striatal tyrosine hydroxylase. *Mol. Pharmacol.* **29**, 561–569.

Stute, N., and Trendelenburg, U. (1984). The outward transport of axoplasmic noradrenaline induced by a rise of the sodium concentration in the adrenergic nerve endings of the rat vas deferens. *Naunyn-Schmiedeberg's Arch. Pharmacol.* **327**, 124–132.

Sulzer, D., and Rayport, S. (1990). Amphetamine and other psychostimulants reduce pH gradients in midbrain dopaminergic neurons and chromaffin granules: A mechanism of action. *Neuron* **5**, 797–808.

Sweadner, K. J. (1985). Ouabain-evoked noradrenaline release from intact rat sympathetic neurons: Evidence for carrier-mediated release. *J. Neurosci.* **5**, 2397–2406.

Toll, L., and Howard, B. D. (1978). Role of Mg^{2+}-ATPase and pH gradient in the storage of catecholamines in synaptic vesicles. *Biochem.* **17**, 2517–2523.

Trendelenburg, U. (1991). Functional aspects of the neuronal uptake of noradrenaline. *Trends Pharmacol. Sci.* **12**, 334–337.

Waldmeier, P. C., Lauber, J., Blum, W., and Richter, W. (1981). 3-Methoxytyramine: Its suitability as an indicator of synaptic dopamine release. *Naunyn-Schmiedeberg's Arch. Pharmacol.* **315**, 219–225.

Westerink, B. H. C. (1979a). Effects of drugs on the formation of 3-methoxy-tyramine, a dopamine metabolite, in the substantia nigra, striatum, nucleus accumbens and tuberculum olfactorium of the rat. *J. Pharm. Pharmacol.* **31**, 94–99.

Westerink, B. H. C. (1979b). Further studies on the sequence of dopamine metabolism in the rat brain. *Eur. J. Pharmacol.* **56**, 313–322.

Westerink, B. H. C. (1985). Sequence and significance of dopamine metabolism in the rat brain. *Neurochem. Int.* **7**, 221–227.

Westerink, B. H. C., and Korf, J. (1976). Turnover of acid dopamine metabolites in striatal and mesolimbic tissue of the rat brain. *Eur. J. Pharmacol.* **37**, 249–255.

Westerink, B. H. C., Tuntler, J., Damsma, G., Rollema, H., and de Vries, J. B. (1987). The use of tetrodotoxin for the characterization of drug-enhanced dopamine release in conscious rats studied by brain dialysis. *Arch. Pharmacol.* **336**, 502–507.

Westerink, B. H., Hofsteede, H. M., Damsma, G., and de Vries, J. B. (1988). The significance of extracellular calcium for the release of dopamine, acetylcholine and amino acids in conscious rats, evaluated by brain microdialysis. *Naunyn-Schmiedeberg's Arch. Pharmacol.* **337**, 373–378.

Westerink, B. H. C., Damsma, G., and de Vries, J. B. (1989a). Effect of ouabain applied by intrastriatal microdialysis on the in vivo release of dopamine, acetylcholine, and amino acids in the brain of conscious rats. *J. Neurochem.* **52**, 705–712.

Westerink, B. H. C., Hofsteede, R. M., Tuntler, J., and de Vries, J. B. (1989b). Use of calcium antagonism for the characterization of drug-evoked dopamine release from the brain of conscious rats determined by microdialysis. *J. Neurochem.* **52**, 722–729.

Wheeler, D. D. (1980). A model for GABA and glutamic acid transport by cortical synaptosomes. *Pharmacology* **21**, 141–152.

Wilson, R. L., Kamata, K., Wightman, R. M., and Rebec, G. V. (1986). Unilateral infusions of amphetamine produce differential, bilateral changes in unit activity and extracellular levels of ascorbate in the neostriatum of the rat. *Brain Res.* **384**, 342–347.

Wong, T. W., Van Frank, R. M., and Horng, J. (1972). Accumulation of amphetamine and p-chloroamphetamine into synaptosomes of rat brain. *J. Pharm. Pharmacol.* **24**, 171–173.

Wood, P. L., and Altar, C. A. (1988). Dopamine release in vivo from nigrostriatal, mesolimbic, and mesocortical neurons: Utility of 3-methoxytyramine measurements. *Pharmacol. Rev.* **40**, 163–188.

Woolverton, W. L., and Johnson, K. M. (1992). Neurobiology of cocaine abuse. *Trends Pharmacol. Sci.* **13**, 193–200.

Yount, S. E., Kraft, M. E., Pierce, R. C., Langley, P. E., and Rebec, G. V. (1991). Acute and long-term amphetamine treatments alter extracellular ascorbate in neostriatum but not nucleus accumbens of freely moving rats. *Life Sci.* **49**, 1237–1244.

Zaczek, R., Culp, S., and De Souza, E. B. (1991). Interactions of [^{3}H]amphetamine with rat brain synaptosomes. II. Active transport. *J. Pharmacol. Exp. Ther.* **257**, 830–835.

Zaleska, M. M., and Erecinska, M. (1987). Involvement of sialic acid in high-affinity uptake of dopamine by synaptosomes from rat brain. *Neurosci. Lett.* **82**, 107–112.

Zetterstrom, T., Sharp, T., Marsden, C. A., and Ungerstedt, U. (1983). In vivo measurement of dopamine and its metabolites by intracerebral dialysis: Changes after d-amphetamine. *J. Neurochem.* **41**, 1769–1773.

Zetterstrom, T., Sharp, T., and Ungerstedt, U. (1986). Further evaluation of the mechanism by which amphetamine reduces striatal dopamine metabolism: A brain dialysis study. *Eur. J. Pharmacol.* **132**, 1–9.

Zetterström, T., Wheeler, D. B., Boutelle, M. G., and Fillenz, M. (1991). Striatal ascorbate and its relationship to dopamine receptor stimulation and motor activity. *Eur. J. Neurosci.* **3**, 940–946.

David S. Segal
Ronald Kuczenski

4
Behavioral Pharmacology of Amphetamine

I. INTRODUCTION

Amphetamine (AMPH) and related psychomotor stimulants produce a broad spectrum of behavioral effects. Although the specific response profile can be influenced profoundly by constitutional, experiential, and situational factors, the psychomotor activating and rewarding/euphorigenic effects of these agents have been identified in all mammals tested, despite obvious species-specific differences. A vast array of studies has been done, utilizing different behavioral paradigms designed to assess the various aspects of the stimulant response. Considering this extremely large and diverse literature is, of course, beyond the scope of this chapter. Instead, the primary emphasis is on the changes in the motor activating effects associated with various patterns of repeated AMPH administration in rodents. In addition to the reinforcing effects of these drugs (discussed in Chapter 8 of this volume), various features of the behavioral activation have been implicated in the changes produced by the abuse of stimulants (Segal and Mandell, 1974; Segal and Schuckit, 1983; Robinson, 1991b). Although similar response patterns have been observed in a variety of species (for reviews, see Segal and Janowsky, 1978; Segal and Schuckit, 1983), rodents have been examined most thoroughly and systematically with respect to these effects and their potential underlying mechanisms. Researchers have argued that evidence of behavioral–neurochemical relationships in this species is relevant to the effects in humans (Segal and Janowsky, 1978, Segal and Schuckit, 1983; Segal and Geyer, 1985; Robinson and Becker, 1986; Robinson, 1991a). Finally, chronic AMPH exposure of moderate to high doses appears to be related most closely to the patterns of abuse typically associated with addiction and relapse (Segal *et al.*, 1981; Robinson and Becker, 1986; Robinson, 1991b) as well as with the induction of various forms of psychopathology although, with respect to the induction of AMPH psychosis, the relative importance of dose and chronicity are difficult to assess (Segal and Schuckit, 1983; see also Chapter 13).

Repeated administration of AMPH and related stimulants results in a complex multiphasic spectrum of behavioral changes including (1) a progressive behavioral augmentation or sensitization to drug challenge that persists after prolonged periods of abstinence (Segal and Mandell, 1974; Segal, 1975a) and (2) a poststimulant withdrawal syndrome (PSWS) that also exhibits progressive and dynamic changes as a function of dose and withdrawal interval (Segal, 1975a; Gawin and Kleber, 1986; Gawin and Ellinwood, 1988; Teicher *et al.*, 1989; Paulson *et al.*, 1991). Variations in the characteristics of the altered responsiveness are associated with different dosage regimens and may be relevant to different patterns of AMPH abuse. In addition to possible implications for stimulant abuse, sensitization in particular may reflect an important form of neuronal plasticity in the central nervous system (CNS).

Several aspects of the altered responsiveness have become increasingly apparent as the spectrum of changes involved has been characterized more thoroughly. Perhaps the most important observation, particularly with respect to identifying underlying mechanisms, is that the behavioral changes can vary substantially, depending on several different factors. For these and other reasons (discussed subsequently), the alterations that result with repeated AMPH administration are likely not to represent a unitary process but to reflect multiple behavioral and neurotransmitter mechanisms.

II. FACTORS THAT INFLUENCE THE EMERGENT BEHAVIORAL PROFILES

Accumulating evidence now indicates that a variety of interrelated factors influences the behavioral response profiles that emerge with repeated AMPH administration. An understanding of the role of such factors is necessary to facilitate the identification of the psychological processes and the neurotransmitter systems and mechanisms responsible for these behavioral changes. Among the most important factors responsible for the expression of the chronic response are (1) dose and time parameters, (2) situational and experiential variables, and (3) individual differences.

A. Dosage Parameters

1. Acute Response

Accurate conclusions regarding the behavioral changes that result with repeated AMPH administration require a thorough understanding of the acute dose–response characteristics (Lat, 1965; Segal, 1975a; Schiorring, 1979; Segal et al., 1981; Segal and Schuckit, 1983). In general, low doses in rats produce locomotor activation including horizontal and vertical movements (Fig. 1). As the dose is increased, the locomotion becomes less varied and more perseverative. This behavior, especially during its early phase, occasionally is interrupted by episodes of stereotyped sniffing and/or repetitive head and limb movements. A distinct stereotypy phase emerges with higher doses. During this phase, which is preceded and followed by periods of locomotion, repetitive movements and/or intense oral stereotypies may be displayed for prolonged periods of time, depending on dose. Other temporal features of this multiphasic response also change as a function of increasing dose; the latency to onset of the stereotypy is decreased and the duration of the stereotypy phase and the subsequent period of locomotor activation is prolonged. A similar pattern of effects has been observed in most mammals tested; in fact, comparable behavioral characteristics can be identified in primates (Ellinwood et al., 1973; Garver et al., 1975; Utena et al., 1975; Post et al., 1992; for humans, see Chapters 13 and 14). As illustrated in subsequent sections, qualitative as well as quantitative features of the dose–response profiles should be considered to characterize chronic AMPH-induced changes accurately, especially with respect to assessing possible underlying mechanisms.

2. Chronic Response

Dose profoundly influences the behavioral characteristics that result from repeated intermittent AMPH administration, determining not only the specific response characteristics but also the rate and, in fact, the direction of the behavioral change (Segal and Mandell, 1974; Segal et al., 1981; Segal and Schuckit, 1983; Robinson and Becker, 1986, Mittleman et al., 1991). In general, the locomotor response to low doses is enhanced progressively (Fig. 2.). With increasing dose, a more rapid onset and intensification of the stereotypy phase and an increase in the subsequent period of locomotor activation are seen. Importantly, the duration of the stereotypy phase is not

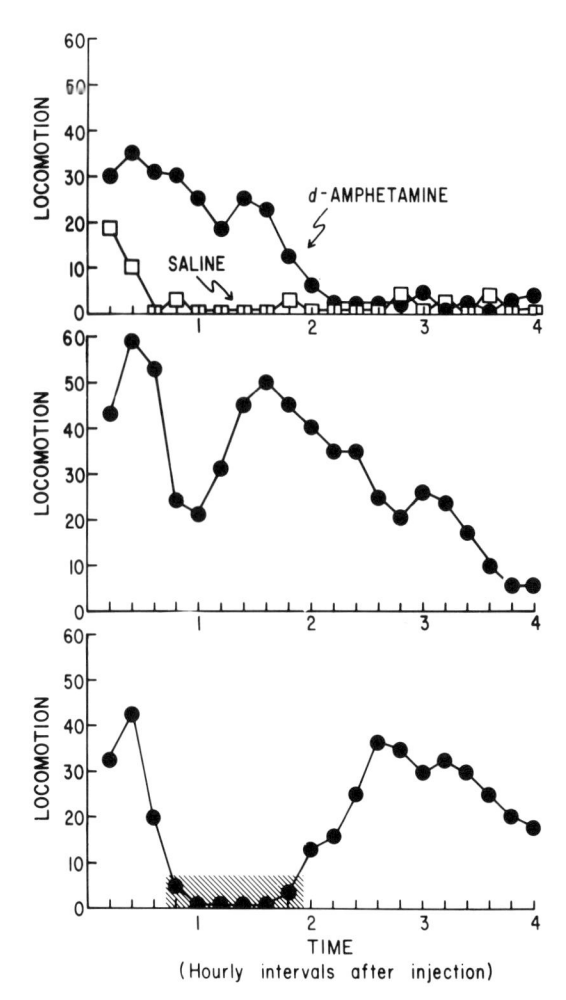

FIGURE 1 Temporal pattern of amphetamine (AMPH)-induced behavioral effects. Response to single injections with various doses of d-AMPH as represented by the patterns of locomotor activity over time. (*Top*) For low doses (0.5–1.5 mg/kg), locomotion is the predominant response. (*Middle*) for intermediate close range (1.5–2.5 mg/kg), locomotion is interrupted by brief episodes of focused stereotyped behaviors (e.g., sniffing, repetitive head and limb movements). (*Bottom*) With high doses (2.5–7.5 mg/kg), multiphasic response pattern emerges consisting of early and late periods of locomotion and an intermediate phase characterized by continuous stereotypy. Reprinted by permission of Raven Press from Segal and Schuckit (1983).

prolonged correspondingly, as would occur with increasing dose (Browne and Segal, 1977; Segal *et al.*, 1980). Administration of relatively high doses may result in a combination of augmentation and tolerance to different features of the response (see subsequent discussion). Note that both qualitative and quantitative changes are associated with the emergent response

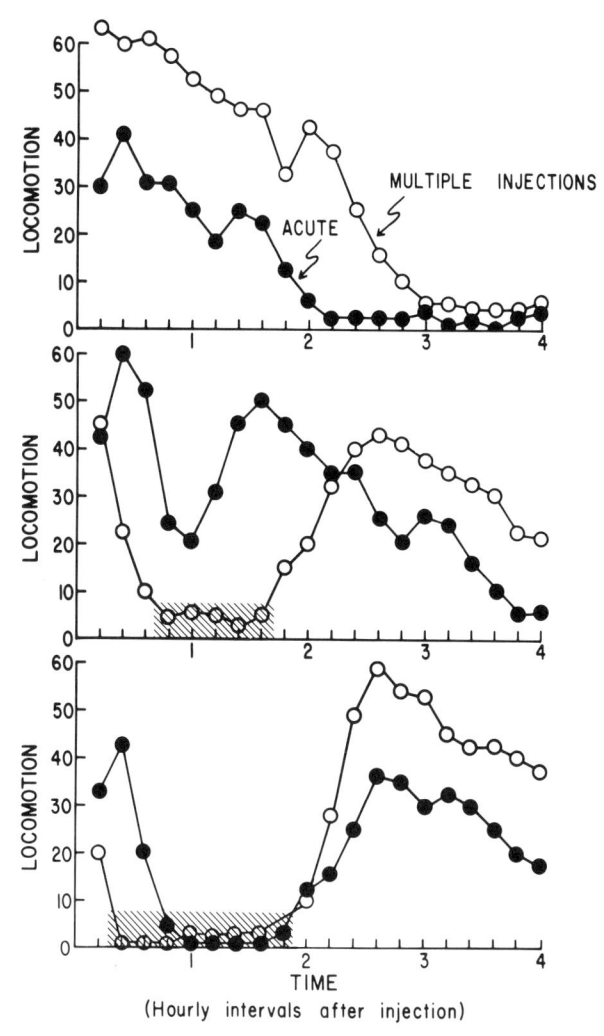

FIGURE 2 *d*-Amphetamine (AMPH) causes alterations in response pattern with repeated daily injections of constant doses. (*Top*) For low doses (0.5–1.5 mg/kg), 5–10 days of administration results in a progressive increase in locomotion, as reflected in the magnitude and duration of the response. (*Middle*) For intermediate doses (1.5–2.5 mg/kg), the duration of stereotypy episodes are increased and, by 3–5 days, a continuous stereotypy phase (shading) emerges. Post-stereotypy hyperactivity is enhanced also. (*Bottom*) For the high-dose range (2.5–7.5 mg/kg), stereotypy (shading) appears more rapidly and is intensified by the second injection, although the duration of the stereotypy phase is not prolonged correspondingly. Locomotion after the stereotypy phase is increased progressively. Reprinted by permission of Raven Press from Segal and Schuckit (1983).

profiles. For example, with lower doses, locomotion may be altered by the intrusion of other behaviors including episodes of stereotypy. In addition, horizontal and vertical activity may change differentially (Mazurski and Beninger, 1987; Segal and Kuczenski, 1987b; Kuczenski and Segal, 1988; Stewart and Vezina, 1991). Thus, assessment of total activity alone may be misleading. Further, a more detailed analysis of the spatial pattern of locomotion is required to distinguish between possible alterations in the relative degree of varied (exploratory-like) and perseverative patterns of ambulation (Lat, 1965; Schiorring, 1979; Geyer et al., 1987; Eilam et al., 1989; Eilam and Golani, 1990; Nicholls et al., 1992). Likewise, with higher doses, qualitative changes may be reflected in the type of stereotypy and in the composition of behaviors during the poststereotypy phase of the response. Importantly, these various behavioral alterations may occur at different rates (Leith and Kuczenski, 1982; Segal and Kuczenski, 1987b; Kuczenski and Segal, 1988).

Temporal parameters also can influence the behavioral characteristics of the chronic response profile significantly. With respect to interinjection interval, evidence has accumulated indicating that the general pattern of sensitization is not qualitatively different over a wide range of repeated intermittent dosage regimens (Segal et al., 1980; Segal and Geyer, 1985; Paulson et al., 1991), that is, regimens of either constant or escalating doses, spaced at either long (days) or short (multiple daily injections) intervals, produce relatively similar patterns of behavioral sensitization. Evidence does suggest, however, that the rate of development of at least some aspects of behavioral change are facilitated with longer interinjection intervals (Post, 1980; Robinson and Becker, 1986; Segal and Kuczenski, 1987b; Chaudhry et al., 1988; Kuczenski and Segal, 1988; Paulson et al., 1991). In contrast, with continuous AMPH exposure (using implanted pellets or minipumps), a different ("novel") behavioral profile emerges that includes behaviors not typically observed with acute doses (Ellison et al., 1978a; Lyon and Nielsen, 1979; Nielsen et al., 1980). This response may result, at least in part, as a consequence of the neurotoxicity associated with continuous exposure to AMPH (Segal et al., 1981; Segal and Geyer, 1985; Robinson and Becker, 1986). Investigators have suggested that continuous exposure may simulate more closely at least some patterns of self-administration in stimulant abusers (i.e., "binge" or "speed run") (Ellison et al., 1978b; Neilsen et al., 1980; King et al., 1992). Therefore, further examination of the mechanisms underlying the behavioral response to this treatment might provide a more complete understanding of the whole range of stimulant abuse.

Another important temporal factor is the time between discontinuation of repeated treatment and subsequent challenge with AMPH. Most observations indicate that altered responsiveness persists for prolonged periods of time and, in fact, can be demonstrated for at least several weeks, even after a single injection (Browne and Segal, 1977; Leith and Kuczenski, 1982; Segal and Schuckit, 1983; Robinson and Becker, 1986). Other evidence has revealed a relatively dynamic process during the withdrawal period, especially with respect to underlying mechanisms (see subsequent discussion), suggesting the need for a more detailed behavioral analysis of the response to AMPH at various times after withdrawal.

In this regard, the poststimulant depression (PSD) of motor activity is most pronounced early after withdrawal from repeated high-dose pretreatment (Segal, 1975a; Paulson *et al.*, 1991), but is also apparent even after the initial exposure to relatively moderate doses (Segal, 1975a). The characteristics of the PSWS must be considered for a number of reasons. First, the presence of various degrees of behavioral depression may influence the expression and/or development of sensitization differentially (Paulson *et al.*, 1991). Such interactions could contribute to the differences in the behavioral and neurochemical responses that have been reported to occur as a function of time between successive stimulant administrations and after withdrawal (see subsequent discussion). Further, PSD also exhibits sensitization-like changes (Segal, 1975a), that is, PSD may not be apparent after an acute AMPH administration but may emerge gradually with repeated drug injections. In addition, as for the sensitization effect, several days after discontinuation of AMPH treatment a single challenge results in a significant PSD. Therefore, future studies should characterize both phenomena concurrently to assess their possible interrelationships, particularly with respect to determining how the emergence of depression might affect the development and/or expression of sensitization. These considerations further emphasize the importance of obtaining a complete characterization of the behavioral response alterations associated with repeated stimulant administration as a prerequisite to mechanistic studies.

B. Situational and Experiential Factors

Responsiveness to AMPH appears to be influenced significantly by the state of stress present at the time of drug administration (Antelman *et al.*, 1980). Further, repeated exposures to a stressor results in a progressively enhanced stress reaction and in an augmented responsiveness to AMPH (Herman *et al.*, 1984; Robinson *et al.*, 1985; Robinson and Becker, 1986). Neurochemical data also support an important interrelationship between AMPH and stressors (for review, see Kalivas and Stewart, 1991). In fact, researchers have suggested that the sensitization associated with repeated AMPH administration may be the result of AMPH-induced stress (Antelman and Chiodo, 1983). Since the degree of stress reactivity to a specific stimulus depends on a variety of experiential and situational factors, stress-related variations in experimental design may have pronounced effects on the AMPH response (see also Section II,C).

Another potentially important experiential factor in the development of

AMPH sensitization is context-dependent conditioning (for example, see Stewart, 1991,1992; Nader et al., 1992; Post et al., 1992; Stewart and Eikelboom, 1987). The degree to which this process influences the response to repeated administration depends most critically on the experimental paradigm employed, especially the relationship of the stimulus environment during pretreatment and during testing. Drug dose, as well as other variables, also appears to have a significant impact on the relative role of conditioning in the development of sensitization (Post et al., 1992). However, despite the importance of learning factors, accumulating evidence indicates that sensitization can develop in the absence of environment-specific conditioning (Segal, 1975a; Browne and Segal, 1977; Leith and Kuczenski, 1982; Robinson and Becker, 1986; Vezina and Stewart, 1990; Stewart and Vezina, 1991; Damianopoulos and Carey, 1992). The role of state-dependent conditioning has not been as well defined (Browne and Segal, 1977; Kuczenski et al., 1982; Drew and Glick, 1988; Levy et al., 1988; Post et al., 1992).

C. Individual Differences

The administration of AMPH to rats results in a relatively wide range of individual differences in responsiveness that is also apparent in the clinical literature (see Chapter 13). Such differences in the initial behavioral response profile can be related to the pattern of sensitization that emerges with repeated administration (Segal and Kuczenski, 1987a; Kuribara and Tadokoro, 1989). For example, animals administered a dose of AMPH that is transitional relative to the induction of locomotion and continuous stereotypy could be separated into two subgroups, displaying either continuous locomotion (S1) or a multiphasic pattern including a prolonged focused stereotypy phase (S2). With repeated administration, two prominent features of the augmentation, the emergence of stereotypy and the enhancement of locomotion during the last half of the response, were dissociable in the subgroups (Figs. 3, 4). Progressive changes in the stereotypy phase were confined primarily to S1 animals, whereas only the S2 subgroup displayed the augmentation of locomotion. This dissociation of stereotypy from poststereotypy locomotion changes also suggests that these two features of the

---

FIGURE 3 (*Top*) Temporal pattern of crossover response for S1 (*left*) and S2 (*right*) subgroups to daily injections of 1.75 mg/kg amphetamine (AMPH). Rats were administered AMPH sc daily for 7 days. Data represent response patterns on days 1, 2, 4, and 7. (*Bottom*) Comparison of crossovers cumulated over the specified intervals for subgroups S1 ($n = 16$) and S2 ($n = 12$) with the combined subgroups ($n = 28$) on the days indicated. $P < .01$ (**) and $P < .001$ (***) on paired t-test with Bonferroni adjustment after significant 1-day analysis of variance with repeated measures [S1, 10–60 min: $F(3,42) = 26.54$, $P < .001$; S1 + S2, 10–60 min: $F(3,75) = 9.08$, $P < .001$; S2, 60–150 min: $F(3,30) = 22.17$, $P < .001$; S1 + S2, 60–150 min: $F(3,75) = 14.05$, $P < .001$]. Significant difference between S1 and S2 subgroups of $P < .001$ in an unpaired t-test (+++). Reprinted with permission from Segal and Kuczenski (1987a).

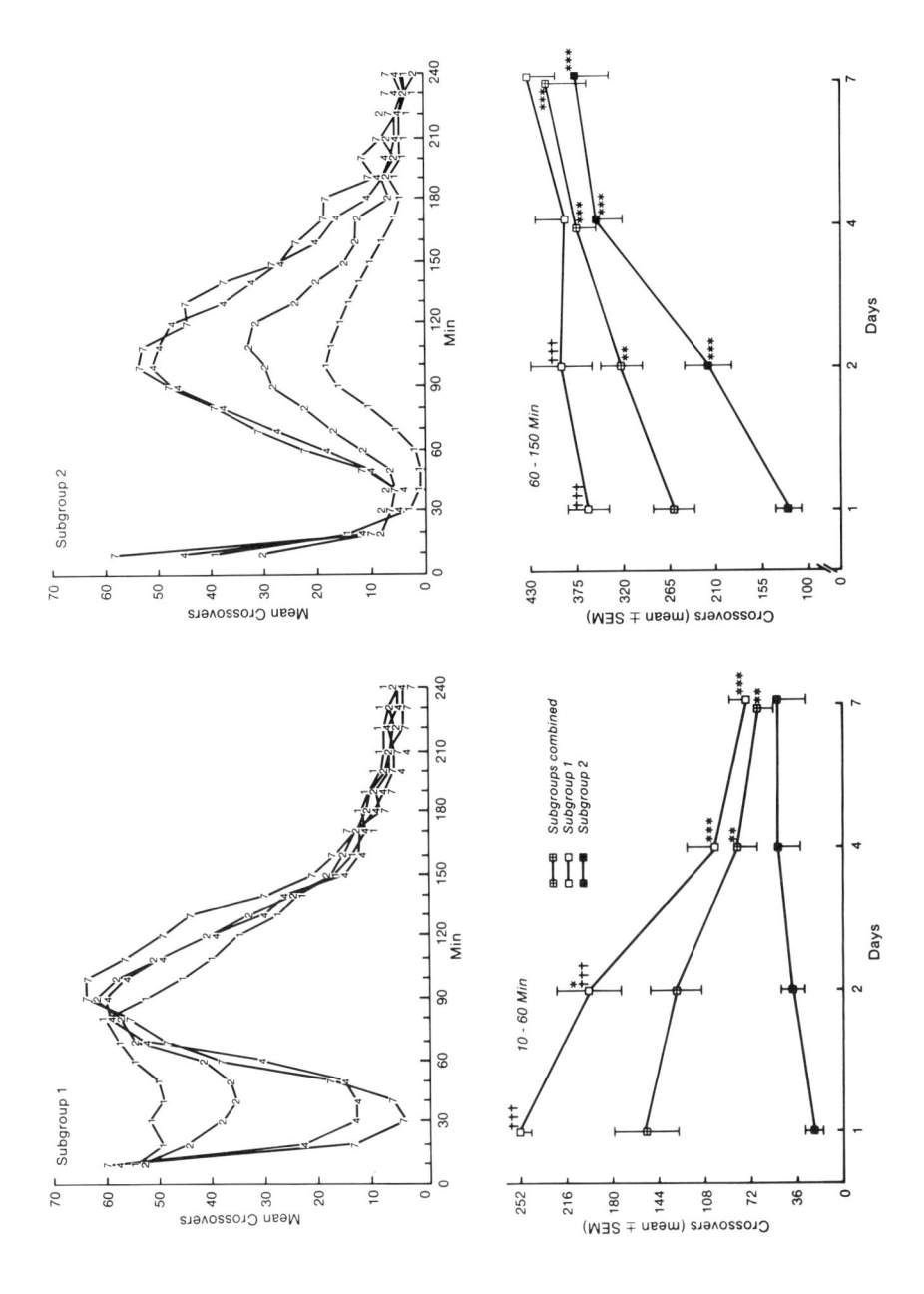

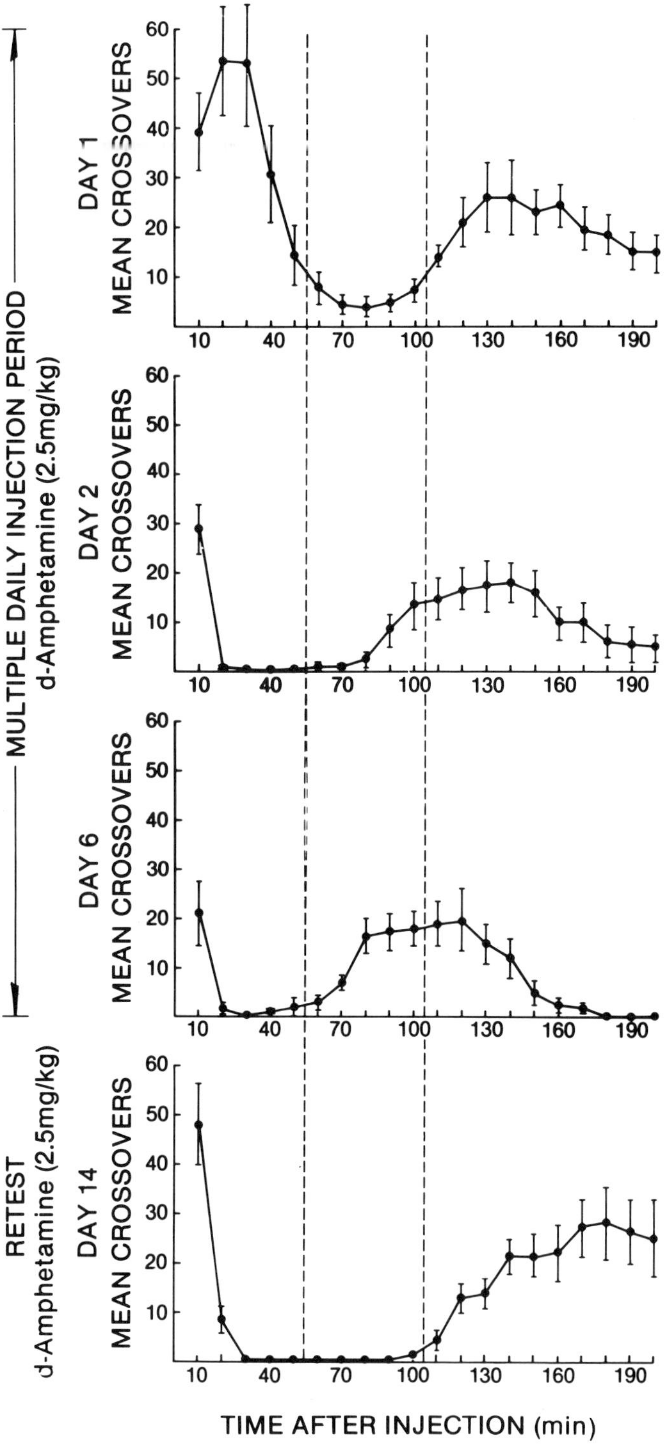
MULTIPLE DAILY INJECTION PERIOD
d-Amphetamine (2.5mg/kg)
RETEST
d-Amphetamine (2.5mg/kg)
DAY 1
DAY 2
DAY 6
DAY 14
MEAN CROSSOVERS
60
50
40
30
20
10
10
40
70
100
130
160
190
TIME AFTER INJECTION (min)

response are mediated by different underlying mechanisms. These results indicate that characterization of individual responses may be required to determine accurately the neurochemical mechanisms involved in the behavioral effects of stimulants. These observations also underscore the potential value of *in vivo* procedures in which individual variations in behavior and neurochemistry can be assessed using each animal as its own control. (See Chapter 3 for a more complete description of *in vivo* procedures.)

Evidence indicates the existence of a relationship between individual variation in response to a novel environment, the behavioral and nucleus accumbens dopamine (DA) response to AMPH or cocaine, and the rate and degree of sensitization with repeated drug administration (Bradberry *et al.*, 1991a,b; Hooks *et al.*, 1991a,b,c,1992). However, the predictive value of the various measures of spontaneous activity used appears to be limited to the locomotor activating effects of relatively low doses of stimulants, that is, correlations are not apparent thus far for doses that produce stereotypy (Hooks *et al.*, 1991c). A more detailed characterization of spontaneous activity, with particular emphasis on indices of perseverative tendency, may reveal relationships to stimulant-induced stereotypy that are not evident with measures of total locomotion or exploration. The available evidence does, however, indicate that individual differences in predrug responsiveness must be characterized to facilitate identification of mechanisms underlying the development of at least low-dose sensitization. Also in this regard, some evidence now suggests that stress reactivity may be a critical factor linking individual differences in acute and chronic stimulant-response profiles to a corresponding vulnerability to developing stimulant self-administration (Deminiere *et al.*, 1989; Piazza *et al.*, 1989,1990; Deroche *et al.*, 1992). This proposed relationship is consistent with the large body of converging evidence (discussed earlier) that suggests a significant interrelationship between the effects of stressors and stimulants. Thus, failure to consider individual differences may obscure important behavioral and neurochemical features associated with both acute and chronic drug administration.

In summary, consideration of these various issues emphasizes the obvious complexity of the so-called sensitization response. In particular, investigators must recognize that repeated AMPH administration does not result in a unitary response modification (Segal and Kuczenski, 1987b; Kuczenski and Segal, 1988). This interpretation is consistent with a number of observations, some of which have been discussed. First, various features and ele-

---

FIGURE 4 Mean crossovers (± standard errors) during successive 10-min intervals following the 10 A.M. *d*-amphetamine (AMPH) injection on days 1, 2, and 6, and again 8 days after the discontinuation of long-term treatment (retest day). Dashed lines indicate the period of focused stereotypy produced by the first AMPH injection. $N \geq 11$ in each group. Reprinted with permission from Segal *et al.* (1980). Copyright © 1980 by the AAAS.

ments or components of the AMPH behavioral response change at different rates, for example, the more rapid onset of the stereotypy phase appears sooner and persists longer than does the increase in poststereotypy hyperactivity (Leith and Kuczenski, 1982). In addition, these two features of the augmentation are differentially sensitive to the interval between successive injections. Extending the interval from 24 to 72 hr or 14 days increased the rate at which focused stereotypy emerged but not the rate of locomotor activation during the poststereotypy phase (Segal and Kuczenski, 1987b). The duration of the stereotypy phase also can vary as a function of withdrawal interval; in fact, the duration actually may be reduced in response to AMPH challenge soon after chronic pretreatment with moderate to high doses (Fig. 5), although the more rapid stereotypy onset remains relatively constant (Segal *et al.*, 1980; Segal and Kuczenski, 1987b; Kuczenski and Segal, 1988). In this regard note that, even at relatively long withdrawal intervals, the duration of neither the stereotypy phase nor the total response is prolonged markedly. Thus, the temporal pattern of the response that emerges with repeated administration cannot be reproduced fully by any acute dose of the drug. Therefore, the behavioral alterations associated with repeated AMPH administration do not reflect simply a "shift to the left" of the dose–response relationship. In contrast, an apparent shift to the left of the AMPH dose–response curve does occur, for example, following the DA receptor supersensitivity alleged to result from repeated haloperidol administration (Rebec *et al.*, 1982) or after administration of drugs that inhibit AMPH metabolism (D. S. Segal and R. Kuczenski, unpublished observation; see also Chapter 2).

Also consistent with a multimodal model of sensitization is the observation that, depending on dosage parameters, some behaviors exhibit tolerance with repeated stimulant administration (Eichler *et al.*, 1980; Rebec and Segal, 1980). For instance, the oral stereotypy elicited by acute high doses is reduced with repeated administration; however, this shift to a less intense form of stereotypy coexists with a more rapid onset of stereotypy. These observations indicate the potential for response competition to underlie at least some of the changes with repeated AMPH administration (Segal and Mandell, 1974; Segal, 1975a; Segal and Schuckit, 1983; Whishaw *et al.*, 1992). Thus, for instance, an apparent increase in some behaviors (inferred to be sensitization) could result from decrements in competing response tendencies (tolerance). These considerations suggest that terms such as sensitization or augmentation may be somewhat misleading in guiding the search for underlying mechanisms. Instead, the behavioral changes that result with repeated AMPH administration apparently do not represent a unitary phenomenon, but reflect the dynamic interaction of a number of different stimulant effects leading to dose- and time-related multiphasic sequences of behavioral alterations.

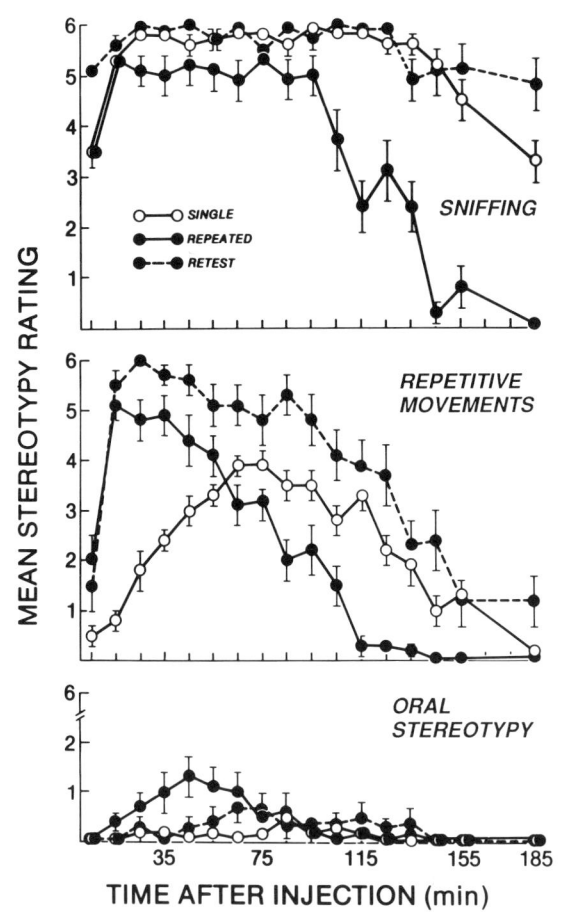

FIGURE 5 Mean stereotypy scores (11) during successive 10-min intervals after the injection of *d*-amphetamine (AMPH) (2.5 mg/kg) 4 hr after long-term preliminary treatment with saline (single AMPH group, ○) or *d*-AMPH (repeated AMPH group, ●), or 8 days after the long-term preliminary treatment with AMPH was discontinued (retest group, ●--●). $N \geq 11$ in each group. Reprinted with permission from Segal *et al.* (1980). Copyright © 1980 by the AAAS.

III. AMPHETAMINE-INDUCED BEHAVIOR–NEUROTRANSMITTER RELATIONSHIPS

A. Dopamine

1. Acute Response

Most research aimed at elucidating neurobiological mechanisms underlying the altered responsiveness to AMPH and related stimulants has focused on DA dynamics, since converging evidence suggests that DA systems play a critical role in acute and chronic stimulant responses (Segal and Schuckit, 1983; Robinson and Becker, 1986; Kuczenski and Segal, 1988; Kalivas and Stewart, 1991; Rebec, 1991). The development of *in vivo* neurochemical methodologies such as microdialysis has enabled a more direct evaluation of stimulant-induced changes in DA function in freely moving animals (for a more detailed discussion, see Chapter 3). Because the mesolimbic and mesostriatal DA pathways have been implicated in these behavioral responses, efforts to establish quantitative relationships between AMPH-induced changes in behavior and DA function have focused on DA projections to the nucleus accumbens and caudate. Using these procedures, researchers have confirmed that AMPH-induced changes in behavior and the DA response generally are related (Sharp *et al.*, 1987; Kuczenski and Segal, 1989). However, demonstrating a quantitative relationship between the DA response and any specific behavior within a given dose of the drug has been more difficult (Fig. 6). In fact, across doses, individual animals exhibiting similar DA responses could express qualitatively different behaviors (Kuczenski and Segal, 1989,1990). Efforts to correlate behavior with DA responses to other AMPH-like stimulants such as cocaine also have led to inconsistent results; some data suggest significant correlations between cocaine-induced locomotor activation and nucleus accumbens DA levels whereas others fail to demonstrate such a relationship (Kalivas and Duffy, 1990; Bradberry *et al.*, 1991a; Hooks *et al.*, 1991c,1992).

The absence or presence of correlations of individual variations in specific behaviors with the magnitude of the regional DA response critically depends not only on the degree to which a specific brain region contributes to the behavior, but also on the manner in which both the behavior and DA response are sampled. Currently, the extent to which these variables contribute to the lack of consistent results is not clear. However, other characteristics of the AMPH-induced changes in behavior and extracellular DA argue against a simple relationship (Kuczenski and Segal, 1990). First, a pronounced temporal dissociation of the pattern of the DA response from the expression of specific stimulant-induced behaviors is observed (Sharp *et al.*, 1987; Kuczenski and Segal, 1989; Kuczenski *et al.*, 1991). For example, after administration of 5 mg/kg AMPH, maximal caudate DA concentrations and the onset of oral stereotypies occurred at the 20- to 40-min interval after drug administration (Fig. 7). However, whereas the magnitude of the stereotypy response was maintained during the subsequent 2 hr, DA concentrations declined 7-fold during this same time period. In fact, oral stereotypies continued during these later intervals when the extracellular DA concentrations were less than peak levels typically obtained with lower doses of the drug that promote no focused stereotypies (Kuczenski and Segal, 1989). A similar temporal dissociation of other stereotyped behaviors and caudate

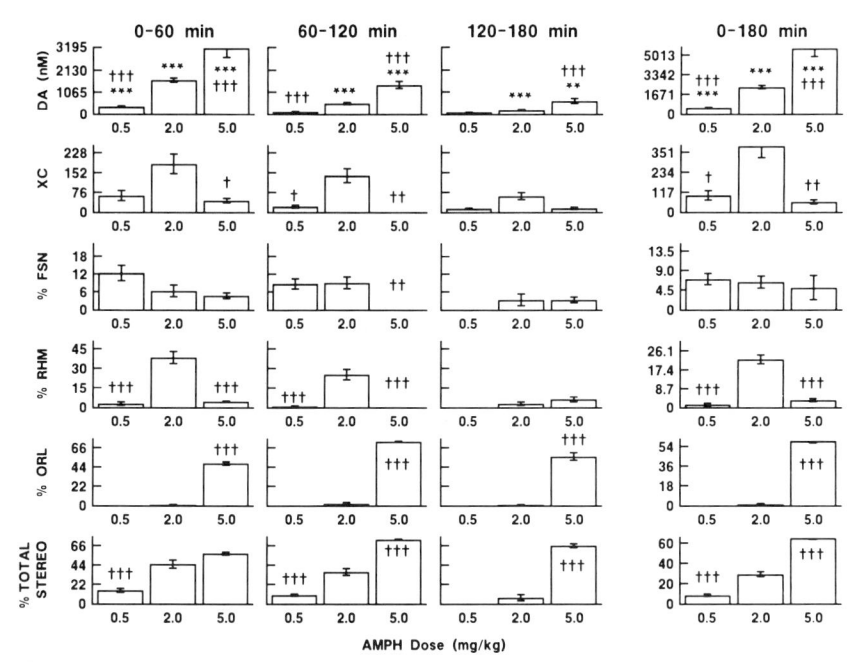

FIGURE 6 Dose–response relationships of amphetamine (AMPH)-induced changes in striatal dopamine (DA) compared with the presence of specific drug-induced behaviors over the specified time interval. *, $P < 0.05$; **, $P < 0.01$; ***, $P < 0.001$ compared with within-group predrug baseline concentrations. †, $P < 0.05$; ††, $P < 0.01$; †††, $P, < 0.001$ compared with the 2.0 mg/kg AMPH response. Reprinted by permission of the *Journal of Neuroscience* from Kuczenski and Segal (1989).

DA levels was evident at lower doses of AMPH as well (Sharp *et al.*, 1987; Kuczenski and Segal, 1989; Kuczenski *et al.*, 1991). Apparently, whereas the magnitude of the extracellular DA response parallels AMPH pharmacokinetics, the magnitude of the stereotypy response does not (see Chapter 3). Further, whereas some investigators have noted parallel locomotor and DA temporal profiles (Sharp *et al.*, 1987), others have not (Kuczenski *et al.*, 1990,1991; Nomikos *et al.*, 1991). That a temporal dissociation of behavior and extracellular DA derives from inappropriate or inconsistent sampling of brain regions is unlikely, since the temporal pattern of the DA response appears to be similar in all brain regions tested.

Additional data support the contention that the expression of specific stimulant-induced behaviors is not simply a reflection of the magnitude of the DA response. First, behaviorally equivalent doses of AMPH-like stimulants with different molecular mechanisms of action exhibit quantitatively

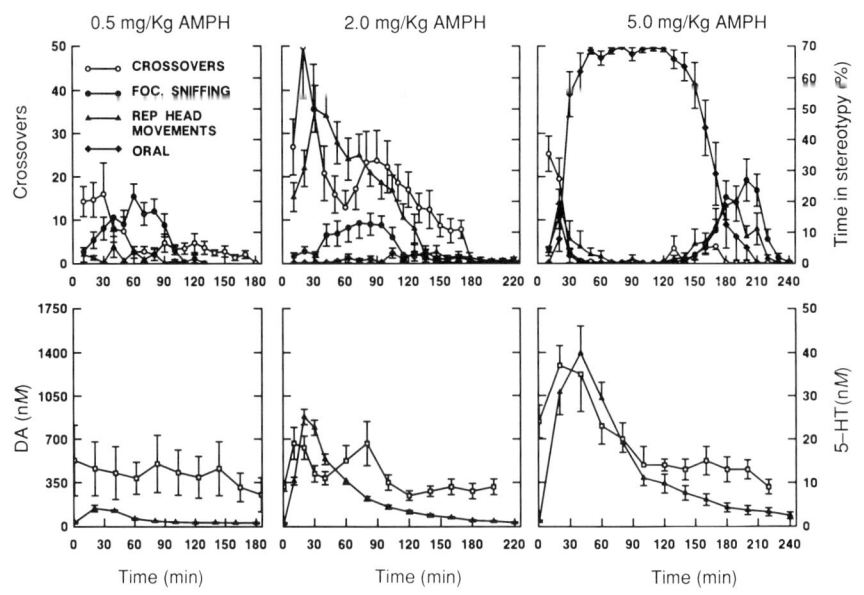

FIGURE 7 Temporal pattern of behavioral and concomitant neurotransmitter response to single injection of the indicated dose of amphetamine (AMPH). Each value represents the mean ± SEM. Crossovers (○), focused sniffing (●), repetitive head movements (▲), oral stereotypy (♦). DA, dopamine (▲), 5-HT, 5-hydroxytryptamine (□). Reprinted by permission of the *Journal of Neuroscience* from Kuczenski and Segal (1989).

different DA response profiles (Kuczenski *et al.*, 1991). For example, when direct comparisons have been made, drugs that act primarily by blocking the uptake of DA promote a much smaller increase in extracellular DA concentrations than behaviorally equivalent doses of AMPH. Thus, 2.5 mg/kg AMPH and 6 mg/kg fencamfamine (FCF), a DA uptake blocker (Seyfried, 1983; De Lucia *et al.*, 1984), produced qualitatively and quantitatively similar locomotor activation and focused stereotypy, but the DA response to FCF in both caudate and nucleus accumbens was only about one-fifth the DA response to AMPH. Likewise, the regional DA response to 40 mg/kg cocaine was markedly less than the response to AMPH and more similar to the response to FCF. Contrasting patterns of behavior and DA responses among the three drugs were also evident following doses that increased locomotor activity in the absence of focused stereotypies. A low dose of AMPH (0.5 mg/kg) produced significantly lower levels of locomotor activation than did 10 mg/kg cocaine or 1.7 mg/kg FCF, but nearly 3- to 4-fold more extracellular DA in both caudate and nucleus accumbens than did the two uptake blockers. Further, as for AMPH, the temporal patterns of the caudate and nucleus accumbens DA responses to the uptake blockers were dissociable from their behavioral response patterns.

Additional support for this dissociation stems from studies of pharmacological manipulations that produce differential effects on the AMPH-induced DA and behavioral responses. For example, doses of apomorphine (APO) that are alleged to activate DA autoreceptors preferentially significantly inhibit the locomotor stimulant effects of AMPH (Creese et al., 1982; Strombom and Liedman, 1982; Kuczenski et al., 1990). However, although APO decreases basal DA release in both caudate and nucleus accumbens, this compound has no effect on AMPH-induced DA release (Zetterstrom and Ungerstedt, 1984; Imperato et al., 1988; Kuczenski et al., 1990). Thus, the behavioral effects of AMPH are inhibited in the absence of a decreased DA response. Conversely, the behavioral response to AMPH can be enhanced without altering the DA response. For example, pretreatment of animals with 2.5 mg/kg reserpine decreases tissue DA levels by about 90% and decreases basal caudate extracellular DA by about 85%, but has no effect on either the magnitude or the duration of the caudate DA response to 1.25 or 2.5 mg/kg AMPH (Callaway et al., 1989). However, the behavioral effects of AMPH are intensified markedly by reserpine pretreatment. A dissociation of the effects of amperozide on stimulant-induced locomotion and on nucleus accumbens DA also has been reported (Kimura et al., 1993). Thus, although DA clearly plays a critical role in AMPH-induced behaviors, no simple relationship appears to exist between the quantitative features of the DA response to AMPH and the appearance of specific drug-induced behaviors. On the basis of these observations, other effects of AMPH are likely to contribute significantly to the behavioral response.

2. Repeated Response

a. Augmented DA Response Despite the apparent dissociation of the acute DA response and the behavioral response profiles, activation of DA systems appears to be required for sensitization to occur (Kuczenski and Leith, 1981; Karler et al., 1990,1991; Post et al., 1992; Ujike et al., 1992; however, see Mittleman et al., 1991). In fact, in agreement with some *in vitro* data, several investigators have reported significant increases in the caudate or nucleus accumbens dialysate DA response after repeated AMPH or cocaine administration (Robinson et al., 1988; Akimoto et al., 1989,1990; Kazahaya et al., 1989; Kalivas and Duffy, 1990; Pettit et al., 1990; Kalivas and Stewart, 1991; Patrick et al., 1991; Robinson, 1991b). These increases appear to occur in the absence of a chronic AMPH-induced elevation of basal levels. Researchers have concluded, therefore, that an augmented DA release might be responsible for behavioral sensitization. Various presynaptic mechanisms have been proposed to mediate the enhanced DA response, including alterations in (1) the terminal and somatodendritic autoreceptors

(Kalivas and Stewart, 1991), (2) the distribution of intracellular DA (Robinson and Becker, 1986; Robinson, 1991a), and (3) the DA uptake carrier (Kazhaya et al., 1989; Alzenstein et al., 1990).

i. Autoreceptors Some attention has been directed toward the possible role of altered autoreceptor sensitivity and/or reduced somatodendritic release of DA in the enhanced nerve terminal DA response after chronic AMPH pretreatment (for review, see Kalivas and Stewart, 1991). Note, however, that AMPH-induced DA release appears to be independent of autoreceptor function as well as of other mechanisms implicated in the regulation of normal impulse-dependent DA release (see Chapter 3). Thus, changes in these functions, at least over relatively short intervals (see subsequent discussion), might not be expected to influence the DA response to the drug directly. In fact, studies have shown that the DA response to AMPH is not changed when neuronal activity is inhibited experimentally by APO (Kuczenski et al., 1990) or γ-butyrolactone (Carboni et al., 1989). These observations support the idea that the acute AMPH response is not dependent on factors that influence neuronal activity. However, both nerve terminal synthesis-modulating autoreceptors and cell body impulse-modulating autoreceptors could contribute to the level of DA synthesis at least indirectly through their influence on tyrosine hydroxylase (TH) (Mestikawy et al., 1986; Strait and Kuczenski, 1986; Thompson et al., 1990; see Zigmond et al., 1989, for review). In this regard, alterations in ongoing DA synthesis rather than in vesicular function are likely to underlie changes in the size of the AMPH-releasable pool, since DA synthesis inhibition with α-methyltyrosine alters the DA response to AMPH (Butcher et al., 1988) whereas reserpine pretreatment does not (Callaway et al., 1989). Thus, relatively persistent increases in impulse traffic consequent to a decrease in the sensitivity of cell body autoreceptors might be expected to activate TH. Likewise, a decrease in the sensitivity of nerve terminal autoreceptors would alleviate their inhibitory influence on TH. As a consequence, relatively prolonged desensitization of autoreceptor function could lead to an enhanced AMPH-releasable DA pool and corresponding changes in the behavioral response.

However, several observations are not readily compatible with such a model. First, DA cell body autoreceptors have been reported to undergo marked time-dependent changes during withdrawal from chronic stimulants (White and Wang, 1984; Henry et al., 1989; Jeziorski and White, 1989; Ackerman and White, 1990; Henry and White, 1992), whereas the enhanced behavioral response typically persists throughout this period. However, although somatodendritic effects may not be involved in the expression of the chronic effect, some evidence implicates somatodendritic regions, especially the ventral tegmentum, in the induction of sensitization (Stewart and Vezina, 1989; Vezina and Stewart, 1990; Kalivas and Stewart, 1991). Thus, for example, activation of somatodendritic D_1 receptors may initiate various changes in gene expression at the cell body level that are expressed at

DA terminals at different times after the start of chronic stimulant treatment (Robinson, 1991a).

ii. Intracellular DA Pools The possibility that intraneuronal changes in compartmentalization of DA might alter the DA and behavioral responses to AMPH has been examined by pretreating animals with various drugs that are alleged to change intracellular DA distribution. In this context, converging evidence indicates that the AMPH response is influenced most critically by the dynamics of the cytoplasmic DA pool (reviewed in Chapter 3). Therefore, in one study, microdialysis procedures were used to determine how monamine oxidase (MAO) inhibition, which should increase cytoplasmic DA, affects AMPH-induced changes in behavioral and extracellular DA dynamics (Segal *et al.*, 1992). Consistent with its effects as an irreversible MAO inhibitor, clorgyline (4.0 mg/kg), which produced prolonged increases in caudate and nucleus accumbens extracellular DA and 3-methoxytyramine (3-MT) and corresponding decreases in homovanillic acid (HVA) and dihydroxyphenylacetic acid (DOPAC), enhanced the behavioral and DA responses to AMPH. However, a potentially important difference between the effects of clorgyline and those of repeated AMPH administration is that only the MAO inhibitor enhances basal DA levels. Basal DA, which presumably reflects impulse-dependent vesicular DA release, is increased persistently after clorgyline pretreatment (Segal *et al.*, 1992), whereas repeated AMPH administration does not appear to alter baseline DA levels significantly (Robinson *et al.*, 1988; Akimoto *et al.*, 1990; Patrick *et al.*, 1991; Segal and Kuczenski, 1992b; however, see Weiss *et al.*, 1992). These results indicate that the MAO inhibitor-induced increase in cytoplasmic DA results in an elevation of vesicular DA through dynamic re-equilibration between these pools. Thus, these data are difficult to reconcile with a mechanism for an enhanced DA response to repeated AMPH administration that is mediated by changes in cytoplasmic DA.

The effects of reserpine on the AMPH-induced behavioral and DA responses are also relevant in this regard. The results of one such study (described earlier) showed that, whereas reserpine pretreatment enhanced certain features of the behavioral effects of AMPH, the amount and duration of AMPH-induced DA release were not altered. These data suggest that mechanisms other than enhanced DA release can underlie sensitization-like changes in the AMPH response. In conjunction with the MAO inhibition results, these data question the concept that a shift in either pool mediates the enhanced DA response to AMPH challenge following repeated AMPH pretreatment.

iii. Uptake Carrier Studies aimed at determining the generality of the chronic AMPH effect have led to the hypothesis that increased affinity of the DA uptake carrier may account for the enhanced DA response. For example, cross-sensitization occurs between AMPH and various other stimulants including cocaine, FCF, and methylphenidate but not caffeine (Segal and

Kuczenski, 1987b; Kuczenski and Segal, 1988; Kazahaya *et al.,* 1989; Aizenstein *et al.,* 1990). Because the transporter represents a common site of interaction with the DA nerve terminal for all these drugs, chronic AMPH-induced changes in its properties could be reflected in an altered response to both releasing agents and uptake blockers. For example, an increased affinity of the transporter for AMPH would facilitate the DA-releasing potential of this drug, and also could be translated into increased affinity for uptake blockers. However, efforts to establish direct evidence for chronic stimulant-induced changes in the DA transporter have yielded inconsistent results (for example, Missale *et al.,* 1985; Izenwasser and Cox, 1990; Kula and Baldessarini, 1991; Sharpe *et al.,* 1991). On the other hand, investigators have reported that chronic methamphetamine treatment alters the responsivity of caudate DA nerve terminals to the DA-releasing actions of ouabain (Kanzaki *et al.,* 1992). Thus, factors such as the regulation of transmembrane Na^+ gradients, which indirectly affect the consequences of AMPH or cocaine interaction with the DA transporter, may be modified after chronic stimulant administration to affect the DA response profile to subsequent stimulant administration.

b. Reduced DA Response Additional research obviously is required to determine the mechanisms underlying the chronic stimulant-induced increase in DA release. However, some observations question the essential role of an enhanced DA response in the sensitization effect since, particularly during short withdrawal intervals, an increased DA response has not always been found (Peris *et al.,* 1990; Kalivas and Stewart, 1991). In fact, investigators have shown (Figs. 8, 9) that behavioral sensitization produced by AMPH or cocaine can occur in the presence of decreased caudate and nucleus accumbens DA responses (Guix *et al.,* 1992; Segal and Kuczenski, 1992a,b). Thus, the relationship between the altered behavioral and DA responses appears to be relatively complex; the degree or direction of association depends on factors such as pretreatment dosage parameters and withdrawal time. These results suggest that an enhanced DA response may not be required for the expression of behavioral sensitization and, therefore, that other DA mechanisms and/or other neurotransmitter systems significantly contribute to the AMPH response profile.

The decreased extracellular DA response to stimulant challenge during short withdrawal intervals may reflect compensatory adaptational processes that are triggered by prolonged AMPH-induced hyperdopaminergic activity. For example, researchers have suggested that chronic stimulants may promote a compensatory increase in the DA uptake carrier that could lead to a decreased extracellular DA response to stimulant challenge (Hurd *et al.,* 1990; Ng *et al.,* 1991; however, see Sharpe *et al.,* 1991). However, since behavioral sensitization in the form of a more rapid onset and intensification

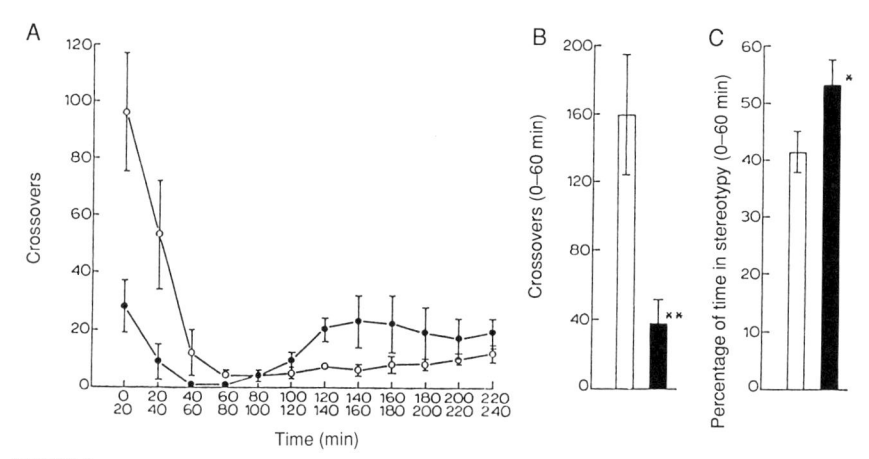

FIGURE 8 (A) Locomotor response (crossovers) to amphetamine (AMPH, 2.5 mg/kg) 48 hr to 6 days (data combined; see text) after 4 daily injections of either saline (○, N = 13) or AMPH (2.5 mg/kg, ●, N = 16). The pattern of augmentation includes the more rapid onset of stereotypy reflected by a decrease in locomotion during the 0- to 60-min interval (B), and a corresponding increase in the time spent engaged in repetitive head and limb movements (C). $P < 0.05$ (*) and $P < 0.01$ (**) relative to saline pretreated controls. Reprinted with permission from Segal and Kuczenski (1992b).

of stereotypy is apparent despite a diminished extracellular DA response (Segal and Kuczenski, 1992a,b), this decrease in extracellular DA may be overcome by other effects of repeated AMPH treatment that are more directly responsible for the enhanced behavioral responsiveness. Several possible alternative DA mechanisms have been considered, including (1) alterations in postsynaptic DA receptors and (2) changes in the ratio of mesolimbic–mesostriatal activity.

i. Postsynaptic DA Receptors Although the role of DA receptor changes in stimulant-induced sensitization remains somewhat controversial (Robinson and Becker, 1986; Robinson, 1991a), some evidence supports an up-regulation of postsynaptic receptors in the expression of sensitization, especially during the early withdrawal phase. For example, Peris *et al.* (1990) reported that an initial increase in DA D_2 receptor sensitivity preceded an elevated nucleus accumbens extracellular DA response after repeated AMPH administration (see also Hurd *et al.*, 1992). Changes in D_1 postsynaptic receptor sensitivity also may contribute to the early behavioral response pattern. Accumulating evidence indicates that the expression of specific stimulant-induced behaviors may be influenced by an interaction of D_1 and D_2 receptor activation (Arnt *et al.*, 1987; Clark and White, 1987; Waddington and O'Boyle, 1987,1989; Eilam *et al.*, 1991; Hamamura *et al.*,

FIGURE 9 (A) Caudate and (B) nucleus accumbens dopamine (DA) response patterns for baseline (BL) and for 4 hr after injection of amphetamine (AMPH, 2.5 mg/kg). (C) Ratio of caudate to accumbens DA is presented for baseline (BL) and for the first 100 min after AMPH injection. Data are presented in 20-min intervals as either absolute values (*top*) or percentage of baseline (*bottom*). Histograms (0–60 min response) illustrate the DA response during the period of most profound behavioral augmentation. $P < 0.05$ (*) and $P < 0.01$ (**) relative to saline pretreated controls. Reprinted with permission from Segal and Kuczenski (1992b).

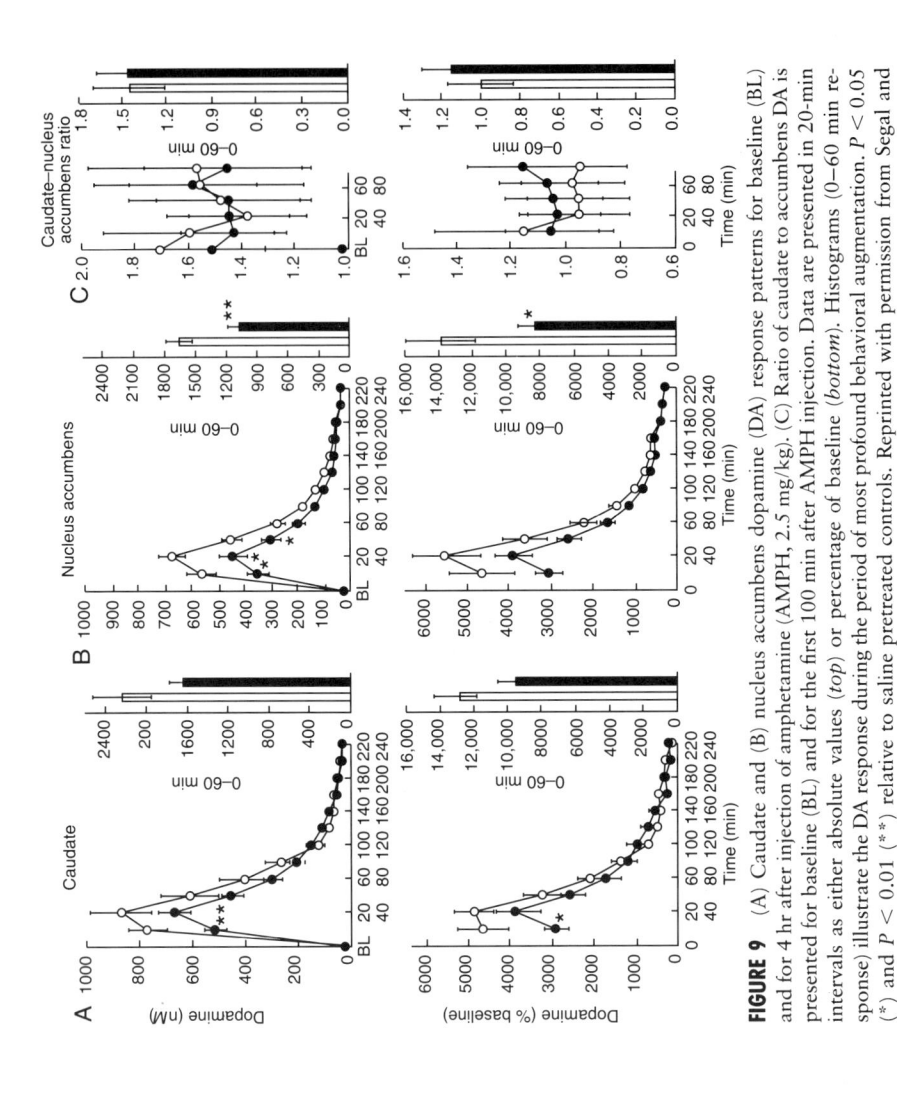

1991). Although the exact nature of this relationship remains the subject of some controversy, alterations in behavior that are associated with repeated AMPH treatment may reflect a change in the balance of D_2 and D_1 activation, perhaps because the two receptor subtypes are differentially responsive to compensatory adjustments. Thus, for example, D_1 desensitization may lead to a predominance of D_2 activation (Barnett and Kuczenski, 1986); this shift in balance could mediate the more rapid appearance and intensification of stereotypy (Roberts-Lewis et al., 1986; Barnett et al., 1987; Roseboom et al., 1990).

ii. Mesolimbic and Mesostriatal Ratio Rather than a simple increase or decrease of extracellular DA, a change in the relative activation of the mesolimbic and mesostriatal DA systems may determine the expression of the stimulant response. Such a switching mechanism may be particularly important to the transition from locomotion to stereotypy and to the more rapid onset of the stereotypy phase that occurs as a prominent feature of sensitization at moderate to high AMPH doses (Segal and Kuczenski, 1987a). In fact, some observations are consistent with an important role for a mesostriatal–mesolimbic balance (Gately et al., 1987; Segal and Kuczenski, 1987a). However, no differences were found in the ratio of caudate and nucleus accumbens DA responses as a function of AMPH dose (Kuczenski and Segal, 1992), nor did repeated AMPH alter the relative striatal–accumbens extracellular DA levels (Segal and Kuczenski, 1992a). Thus, the balance between the regional DA responses does not appear to regulate the expression of the behavior after either acute or chronic AMPH treatment.

B. Other Neurotransmitter Systems

That neurotransmitter systems other than DA significantly contribute to the acute and the chronic effects is indicated by converging evidence from a variety of sources.

1. *In vivo* microdialysis studies have revealed dissociations in the DA–behavioral response relationship.
2. Increases *and* decreases in regional extracellular DA are associated with AMPH sensitization as a function of dose and withdrawal time.
3. *In vivo* microdialysis data show that AMPH produces pronounced dose- and time-related alterations in regional norepinephrine (NE), serotonin (5-HT), and acetylcholine (ACh).

The effects of AMPH on non-DA systems may interact with the DA response through one or more different mechanisms, ranging from modulatory influences on DA transmission to independent actions that compete with AMPH-induced DA effects for expression.

1. Norepinephrine

The interest in the role of NE pathways in the acute and chronic effects of stimulants has been long standing (Segal et al., 1974; Geyer et al., 1986; Kokkinidis, 1988; Kuczenski and Segal, 1988; Harris and Williams, 1992). The dose–response effects of AMPH were examined using *in vivo* microdialysis procedures to monitor extracellular NE levels in prefrontal cortex and hippocampus (Kuczenski and Segal, 1992). NE exhibited a pronounced and rapid dose- and time-dependent increase to AMPH, which corresponded closely to the time course of the behavioral response. Further, to the extent that stress is implicated significantly in the stimulant response (see previous discussion), the observation that stressful stimulation increases regional extracellular NE levels (Abercrombie et al., 1988; Rossetti et al., 1990; Nisenbaum et al., 1991; Britton et al., 1992) additionally supports a role for NE systems in the effects of AMPH on behavior. Comparison of the NE and DA responses to AMPH reveals that the temporal pattern of NE response significantly differs from that of DA. One difference, with potentially important behavioral implications, is that the decline of extracellular NE to half-maximal concentrations is substantially longer than for DA, that is, whereas the NE response continues to be maintained during intervals of persisting behavioral activation, the temporal pattern for DA more closely parallels the pharmacokinetics of AMPH (see Chapter 3).

2. Serotonin

5-HT systems also appear to play a role in the acute and chronic effects of AMPH (Segal, 1976,1977; Kuczenski et al., 1987; Kokkinidis, 1988; Kuczenski and Segal, 1988,1989; Cunningham et al., 1992). Relatively high doses of AMPH (> 2.5 mg/kg) were found to be required to elevate 5-HT concentration in caudate dialysate significantly. Further, the magnitude and persistence of this effect was less than for the catecholamines (Kuczenski and Segal, 1989). In addition, in contrast to AMPH, one study that examined the effects of cocaine using *in vivo* voltammetry reported a decrease in nucleus accumbens 5-HT (Broderick, 1991). Although further assessment of the effects of stimulants on 5-HT dynamics is necessary, the dialysis results suggest that, since the 5-HT response corresponds to the induction of focused stereotypies, 5-HT systems may contribute to the transition between locomotion and stereotypy that occurs as a function of dose and chronicity. Future studies should compare 5-HT and DA to test the possibility that a DA–5-HT interaction is involved in the behavioral transition.

3. Acetylcholine

Changes in ACh may be of particular importance in revealing the integrated effects of AMPH actions on DA and other neurotransmitter systems that converge on cholinergic neurons (Guyenet et al., 1975; Damsma et al., 1990; Drukarch et al., 1990; Williams and Millar, 1990; Wickens et al.,

1991). Therefore, using *in vivo* microdialysis, the effects of AMPH on caudate extracellular ACh were compared with the drug-induced behavioral response profile (Florin *et al.*, 1992). Whereas an intermediate dose of AMPH (1.75 mg/kg) did not alter ACh levels significantly, a higher dose of AMPH (5.0 mg/kg) promoted a 2-fold increase in ACh. This increase paralleled the appearance of oral stereotypies in these animals. These results, in conjunction with other findings (Damsma *et al.*, 1991; Consolo *et al.*, 1992; Guix *et al.*, 1992), suggest that the response of caudate cholinergic interneurons to AMPH may be modulated by various DA receptors inside and outside the caudate, as well as by other neurotransmitter systems. Further, these results suggest that ACh release might contribute to the emergence of intense stereotypies (particularly oral behaviors) and, therefore, that changes in ACh might be associated with specific behavioral features of sensitization or tolerance that results from repeated administration of moderate or high doses of AMPH.

In contrast to the effects of AMPH on 5-HT and ACh, the DA and NE responses exhibit broad dose- and time-dependent characteristics that more generally parallel the level of stimulant-induced behavioral activation, including both locomotion and stereotypy. Therefore, whereas 5-HT and ACh as well as other neurotransmitter systems may play a role in modulating specific components of the stimulant-induced behavioral response (e.g., focused stereotypy), DA and NE may be involved more generally in the overall behavioral activation associated with psychomotor stimulants. Future studies should assess more systematically the effect of repeated AMPH treatment on the various neurotransmitters to test the possibility that behavioral expression may be regulated by *relative* alterations in these systems. In this regard, the chronic stimulant-induced decrease in the DA response may be related functionally to the behavioral sensitization that appears during the early withdrawal phase. We have argued previously that selective tolerance and corresponding changes in various competing response tendencies may contribute significantly to the progressive pattern of behavioral alterations that results with repeated stimulant administration (Segal, 1975a; Kuczenski *et al.*, 1982; Segal and Schuckit, 1983; Segal and Geyer, 1985; see also Mittleman *et al.*, 1991). Therefore, if non-DA mechanisms play an important role in the stimulant behavioral profile, a relatively selective reduction in the DA component of the response may disinhibit or unmask behavioral features of the sensitized response that are mediated by other neurotransmitter systems.

4. Glutamate

Neurotransmitter systems that modulate the DA response at the somatodendritic or terminal regions may play a role in the initiation or the expression of sensitization (for review, see Kalivas and Stewart, 1991). In this regard, evidence implicates glutamatergic mechanisms in the induction of the sensitized response (Karler *et al.*, 1989,1990,1991; Wolf and Khansa, 1991). Of particular interest is the finding that MK-801 (a noncompetitive N-methyl-D-asparate receptor antagonist) appears to prevent the induction of chronic stimulant-induced sensitization selectively without altering the expression of either the acute or the augmented response (however, see Ujike *et al.*, 1992). Although some evidence suggests a glutamate–DA interaction (see, for example, Carter *et al.*, 1988; Arias-Montano *et al.*, 1992; Keefe *et al.*, 1992; Westerink *et al.*, 1992), studies of the effects of MK-801 on dialysate DA have yielded inconsistent results (Imperato *et al.*, 1990; Kashihara *et al.*, 1990; Weihmuller *et al.*, 1991; Whitton *et al.*, 1992). However, one report states that MK-801 attenuated the caudate DA response to methamphetamine (Weihmuller *et al.*, 1991). Current evidence is not adequate to define a role for glutamate in the actions of AMPH, nor a locus for the putative effects of MK-801 on the development of sensitization. However, excitatory amino acids as well as other neurotransmitter systems are likely to be involved significantly in the behavioral changes associated with repeated AMPH administration.

IV. CONCLUSIONS AND SPECULATIONS

That multiple mechanisms and processes and not simply a shift in the DA response are likely to be involved in the alterations that result from repeated administration of AMPH and related stimulants may be inferred from the following behavioral observations.

1. The temporal pattern of the response that emerges cannot be reproduced fully by an acute dose of the stimulant.
2. The various response components can be dissociated—changes occur at different rates.
3. Some characteristics of the response increase while others decrease, thus raising the possibility that an apparent increase in some behaviors may result from decreases in competing response tendencies.
4. Evaluation of patterns of responsiveness to repeated AMPH suggests broad variations in the behavioral profiles of individual animals.

The chronic response to AMPH is complicated further by the presence of a withdrawal syndrome, the dose- and time-related characteristics of which remain to be determined. However, the magnitude and duration of PSD are likely to influence the behavioral and neurochemical responses to AMPH significantly. Therefore, sensitization and PSD likely reflect an evolving multiphasic sequence of neurochemical and behavioral alterations. DA effects appear to represent only one, albeit critical, aspect of this complex spectrum of changes. The DA neurochemical adaptations that do occur may involve pre- *and* postsynaptic mechanisms that are expressed or predominate at different times after withdrawal. In addition, on the basis of *in vivo* dialysis results, the effects of AMPH on other neurotransmitter systems apparently either modulate dopaminergic transmission or independently influence the behavioral response profile that emerges with repeated AMPH administration.

The importance of recognizing the complexity of the altered behavioral responsiveness to repeated AMPH should not be minimized, since elucidation of the mechanisms underlying the various changes that can occur may be a prerequisite to fully understanding stimulant addiction and relapse, as well as the apparent persistent hypersensitivity to the psychotoxic effects of these drugs. Additional characterization of the behavioral–neurochemical relationships associated with chronic AMPH exposure in animals also may contribute to the development of novel therapies for the treatment of stimulant abuse.

ACKNOWLEDGMENTS

Preparation of this manuscript was supported in part by U.S. Public Health Service Grants DA-01568 and DA-04157, and by Research Scientist Award MH-70183 to D. S. Segal.

REFERENCES

Abercrombie, E. D., Keller, R. W., Jr., and Zigmond, M. J. (1988). Characterization of hippocampal norepinephrine release as measured by microdialysis perfusion: Pharmacological and behavioral studies. *Neuroscience* 27, 897–904.

Ackerman, J. M., and White, F. J. (1990). A10 somatodendritic dopamine autoreceptor sensitivity following withdrawal from repeated cocaine treatment. *Neurosci. Lett.* 117, 181–187.

Aizenstein, M. L., Segal, D. S., and Kuczenski, R. (1990). Repeated amphetamine and fencamfamine: Sensitization and reciprocal cross-sensitization. *Neuropsychopharmacology* 3, 283–290.

Akimoto, K., Hamamura, T., and Otsuki, S. (1989). Subchronic cocaine treatment enhances cocaine-induced dopamine efflux, studied by in vivo intracerebral dialysis. *Brain Res.* 490, 339–344.

Akimoto, K., Hamamura, T., Kazahaya, Y., Akiyama, K., and Otsuki, S. (1990). Enhanced extracellular dopamine level may be the fundamental neuropharmacological basis of cross-behavioral sensitization between methamphetamine and cocaine—An in vivo dialysis study in freely moving rats. *Brain Res.* 507, 344–346.

Antelman, S. M., and Chiodo, L. A. (1983). Amphetamine as a stressor. *In* "Stimulants: Neurochemical, Behavioral and Clinical Perspectives" (I. Creese, ed.), pp. 269–299. Raven Press, New York.

Antelman, S. M., Eichler, A. J., Black, C. A., and Kocan, D. (1980). Interchangeability of stress and amphetamine in sensitization. *Science* 207, 329–331.

Arias-Montaño, J. A., Martínez-Fong, D., and Aceves, J. (1992). Glutamate stimulation of tyrosine hydroxylase is mediated by NMDA receptors in the rat striatum. *Brain Res.* 569, 317–322.

Arnt, J., Hyttel, J., and Perregaard, J. (1987). Dopamine D-1 receptor agonists combined with the selective D-2 agonist quinpirole facilitate the expression of oral stereotyped behaviour in rats. *Eur. J. Pharmacol.* **133**, 137–145.

Barnett, J. V., and Kuczenski, R. (1986). Desensitization of rat striatal dopamine-stimulated adenylate cyclase after acute amphetamine administration. *J. Pharmacol Exp. Therap.* **237**, 820–825.

Barnett, J. V., Segal, D. S., and Kuczenski, R. (1987). Repeated amphetamine pretreatment alters the responsiveness of striatal dopamine-stimulated adenylate cyclase to amphetamine-induced desensitization. *J. Pharmacol. Exp. Ther.* **242**, 40–47.

Bradberry, C. W., Gruen, R. J., Berridge, C. W., and Roth, R. H. (1991a). Individual differences in behavioral measures: Correlations with nucleus accumbens dopamine measured by microdialysis. *Pharmacol. Biochem. Behav.* **39**, 877–882.

Bradberry, C. W., Lory, J. D., and Roth, R. H. (1991b). The anxiogenic β-carboline FG 7142 selectively increases dopamine release in rat prefrontal cortex as measured by microdialysis. *J. Neurochem.* **56**, 748–752.

Britton, K. T., Segal, D. S., Kuczenski, R., and Hauger, R. (1992). Dissociation between in vivo hippocampal norepinephrine response and behavioral/neuroendocrine responses to noise stress in rats. *Brain Res.* **574**, 125–130.

Broderick, P. A. (1991). Cocaine: On-line analysis of an accumbens amine neural basis for psychomotor behavior. *Pharmacol. Biochem. Behav.* **40**, 959–968.

Browne, R. G., and Segal, D. S. (1977). Metabolic and experiential factors in the behavioral response to repeated amphetamine. *Pharmacol. Biochem. Behav.* **6**, 545–552.

Butcher, S. P., Fairbrother, I. S., Kelly, J. S., and Arbuthnott, G. W. (1988). Amphetamine-induced dopamine release in the rat striatum: An in vivo microdialysis study. *J. Neurochem.* **50**, 346–355.

Callaway, C. W., Kuczenski, R., and Segal, D. S. (1989). Reserpine enhances amphetamine stereotypies without increasing amphetamine-induced changes in striatal dialysate dopamine. *Brain Res.* **505**, 83–90.

Carboni, E., Acquas, E., Frau, R., and Di Chiara, G. (1989). Differential inhibitory effects of a 5-HT_3 antagonist on drug-induced stimulation of dopamine release. *Eur. J. Pharmacol.* **164**, 515–519.

Carter, C. J., L'Heureux, R., and Scatton, B. (1988). Differential control by N-methyl-D-aspartate and kainate of striatal dopamine release in vivo: A trans-striatal dialysis study. *J. Neurochem.* **51**, 462–468.

Chaudhry, I. A., Turkanis, S. A., and Karler, R. (1988). Characteristics of "reverse" tolerance to amphetamine-induced locomotor stimulation in mice. *Neuropharmacology* **27**, 777–781.

Clark, D. and White, F. J. (1987). D_1 Dopamine receptor—The search for a function: A critical evaluation of the D_1/D_2 dopamine receptor classification and its functional implications. *Synapse* **1**, 347–388.

Consolo, S., Girotti, P., Russi, G., and Di Chiara, G. (1992). Endogenous dopamine facilitates striatal in vivo acetylcholine release by acting on D_1 receptors localized in the striatum. *J. Neurochem.* **59**, 1555–1557.

Creese, I., Kuczenski, R., and Segal, D. S. (1982). Lack of behavioral evidence for dopamine autoreceptor subsensitivity after acute electroconvulsive shock. *Pharmacol. Biochem. Behav.* **17**, 375–376.

Cunningham, K. A., Paris, J. M., and Goeders, N. E. (1992). Serotonin neurotransmission in cocaine sensitization. *Ann. N.Y. Acad. Sci.* **654**, 117–127.

Damianopoulos, E. N., and Carey, R. J. (1992). Conditioning, habituation and behavioral reorganization factors in chronic cocaine effects. *Behav. Brain Res.* **49**, 149–157.

Damsma, G., De Boer, P., Westerink, B. H. C., and Fibiger, H. C. (1990). Dopaminergic regulation of striatal cholinergic interneurons: An in vivo microdialysis study. *Naunyn-Schmiedeberg's Arch. Pharmacol.* **342**, 523–527.

Damsma, G., Robertson, G. S., Tham, C. -S., and Fibiger, H. C. (1991). Dopaminergic regulation of striatal acetylcholine release: Importance of D_1 and NMDA receptors. *J. Pharmacol. Exp. Ther.* **259**, 1064–1072.
De Lucia, R., Bernardi, M. M., Scavone, C., and Aizenstein, M. L. (1984). On the mechanism of central stimulation action of fencamfamine. *Gen. Pharmacol.* **15**, 407–410.
Deminière, J. -M., Piazza, P. V., Le Moal, M., and Simon, H. (1989). Experimental approach to individual vulnerability to psychostimulant addiction. *Neurosci. Biobehav. Rev.* **13**, 141–147.
Deroche, V., Piazza, P. V., Maccari, S., Le Moal, M., and Simon, H. (1992). Repeated corticosterone administration sensitizes the locomotor response to amphetamine. *Brain Res.* **584**, 309–313.
Drew, K. L., and Glick, S. D. (1988). Characterization of the associate nature of sensitization to amphetamine-induced circling behavior and of the environment dependent placebo-like response. *Psychopharmacology* **95**, 482–487.
Drukarch, B., Schepens, E., and Stoof, J. C. (1990). Muscarinic receptor activation attenuates D_2 dopamine receptor mediated inhibition of acetylcholine release in rat striatum: Indications for a common signal transduction pathway. *Neuroscience* **37**, 1–9.
Eichler, A. J., Antelman, S. M., and Black, C. A. (1980). Amphetamine stereotypy is not a homogenous phenomenon: Sniffing and licking show distinct profiles of sensitization and tolerance. *Psychopharmacology* **68**, 287–290.
Eilam, D., and Golani, I. (1990). Home base behavior in amphetamine-treated tame wild rats (*Rattus norvegicus*). *Behav. Brain Res.* **36**, 161–170.
Eilam, D., Golani, I., and Szechtman, H. (1989). D_2-Agonist quinprole induces perseveration of routes and hyperactivity but no perseveration of movements. *Brain Res.* **490**, 255–267.
Eilam, D., Clements, K. V. A., and Szechtman, H. (1991). Differential effects of D_1 and D_2 dopamine agonists on stereotyped locomotion in rats. *Behav. Brain Res.* **45**, 117–124.
Ellinwood, E. H., Sudilovsky, A., and Nelson, L. (1973). Evolving behavior in the clinical and experimental amphetamine (model) psychoses. *Am. J. Psychiatry* **130**, 1088–1093.
Ellison, G. D., Eison, M., Huberman, H., and Daniel, F. (1978a). Long-term changes in dopaminergic innervation of caudate nucleus after continuous amphetamine administration. *Science* **201**, 276–278.
Ellison, G. D., Eison, M., and Huberman, H. (1978b). Stages of constant amphetamine intoxication: Delayed appearance of abnormal social behaviors in rat colonies. *Psychopharmacology* **56**, 293–299.
Florin, S. M., Kuczenski, R., and Segal, D. S. (1992). Amphetamine-induced changes in behavior and caudate extracellular acetylcholine. *Brain Res.* **581**, 53–58.
Garver, D. S., Schlemmer, R. F., Maas, J. W., and Davis, J. M. (1975). A schizophreniform behavioral psychosis mediated by dopamine. *Am. J. Psychiatry* **133**, 33–38.
Gately, P. F., Segal, D. S., and Geyer, M. A. (1987). Sequential changes in behavior induced by continuous infusions of amphetamine in rats. *Psychopharmacology* **91**, 217–220.
Gawin, F. H., and Ellinwood, E. J. (1988). Cocaine and other stimulants. Actions, abuse, and treatment. *N. Engl. J. Med.* **318**, 1173–1182.
Gawin, F. H., and Kleber, H. D. (1986). Abstinence symptomatology and psychiatric diagnosis in cocaine abusers. *Arch. Gen. Psychiatry* **43**, 107–113.
Geyer, M. A., Masten, G., and Segal, D. S. (1986). Behavioral effects of xylamine-induced depletions of brain norepinephrine: Interactions with amphetamine. *Behav. Brain Res.* **21**, 55–64.
Geyer, M. A., Russo, P. V., Segal, D. S., and Kuczenski, R. (1987). Effects of apomorphine and amphetamine on patterns of locomotor and investigatory behavior in rats. *Pharmacol. Biochem. Behav.* **28**, 393–399.
Guix, T., Hurd, Y. L., and Ungerstedt, U. (1992). Amphetamine enhances extracellular concen-

trations of dopamine and acetylcholine in dorsolateral striatum and nucleus accumbens of freely moving rats. *Neurosci. Lett.* **138**, 137–140.

Guyenet, P. G., Agid, Y., Javoy, F., Beaujouan, J. C., Rossier, J., and Glowinski, J. (1975) Effects of dopaminergic receptor agonists and antagonists on the activity of the neo-striatal cholinergic system. *Brain Res.* **84**, 227–244.

Hamamura, T., Akiyama, K., Akimoto, K., Kashihara, K., Okumura, K., Ujike, H., and Otsuki, S. (1991). Co-administration of either a selective D_1 or D_2 dopamine antagonist with methamphetamine prevents methamphetamine-induced behavioral sensitization and neurochemical change, studied by in vivo intracerebral dialysis. *Brain Res.* **546**, 40–46.

Harris, G. C., and Williams, J. T. (1992). Sensitization of locus coeruleus neurons during withdrawal from chronic stimulants and antidepressants. *J. Pharmacol. Exp. Therap.* **261**, 476–483.

Henry, D. J., and White, F. J. (1992). Electrophysiological correlates of psychomotor stimulant-induced sensitization. *Ann. N.Y. Acad. Sci.* **654**, 88–100.

Henry, D. J., Greene, M. A., and White, F. J. (1989). Electrophysiological effects of cocaine in the mesoaccumbens dopamine system: Repeated administration. *J. Pharmacol. Exp. Ther.* **251**, 833–839.

Herman, J. P., Stinus, L., and LeMoal, M. (1984). Repeated stress increases locomotor response to amphetamine. *Psychopharmacology* **84**, 431–435.

Hooks, M. S., Jones, G. H., Neill, D. B., and Justice, J. B., Jr. (1991a). Individual differences in amphetamine sensitization: Dose-dependent effects. *Pharmacol. Biochem. Behav.* **41**, 203–210.

Hooks, M. S., Jones, G. H., Smith, A. D., Neill, D. B., and Justice, J. B., Jr. (1991b). Individual differences in locomotor activity and sensitization. *Pharmacol. Biochem. Behav.* **38**, 467–470.

Hooks, M. S., Jones, G. H., Smith, A. D., Neill, D. B., and Justice, J. B., Jr. (1991c). Response to novelty predicts the locomotor and nucleus accumbens dopamine response to cocaine. *Synapse* **9**, 121–128.

Hooks, M. S., Colvin, A. C., Juncos, J. L., and Justice, J. B., Jr. (1992). Individual differences in basal and cocaine-stimulated extracellular dopamine in the nucleus accumbens using quantitative microdialysis. *Brain Res.* **587**, 306–312.

Hurd, Y. L., Weiss, F., Koob, G., and Ungerstedt, U. (1990). The influence of cocaine self-administration on in vivo dopamine and acetylcholine neurotransmission in rat caudate-putamen. *Neurosci. Lett.* **109**, 227–233.

Imperato, A., Tanda, G., Frau, R., and Di Chiara, G. (1988). Pharmacological profile of dopamine receptor agonists as studied by brain dialysis in behaving rats. *J. Pharmacol. Exp. Ther.* **245**, 257–264.

Imperato, A., Honoré, T., and Jensen, L. H. (1990). Dopamine release in the nucleus caudatus and in the nucleus accumbens is under glutamatergic control through non-NMDA receptors: A study in freely-moving rats. *Brain Res.* **530**, 223–228.

Izenwasser, S., and Cox, B. M. (1990). Daily cocaine treatment produces a persistent reduction of [^{3}H]dopamine uptake in vitro in rat nucleus accumbens but not in striatum. *Brain Res.* **531**, 338–341.

Jeziorski, M., and White, F. J. (1989). Dopamine agonists at repeated "autoreceptor-selective" doses: Effects upon the sensitivity of A10 dopamine autoreceptors. *Synapse* **4**, 267–280.

Kalivas, P. W., and Duffy, P. (1990). Effect of acute and daily cocaine treatment on extracellular dopamine in the nucleus accumbens. *Synapse* **5**, 48–58.

Kalivas, P. W., and Stewart, J. (1991). Dopaminergic transmission in the initiation and expression of drug- and stress-induced sensitization of motor activity. *Brain Res. Rev.* **16**, 223–244.

Kanzaki, A., Akiyama, K., and Otsuki, S. (1992). Subchronic methamphetamine treatment enhances ouabain-induced striatal dopamine efflux in vivo. *Brain Res.* **569**, 181–188.

Karler, R., Calder, L. D., Chaudhry, I. A., and Turkanis, S. A. (1989). Blockade of "reverse tolerance" to cocaine and amphetamine by MK-801. *Life Sci.* **45**, 599–606.

Karler, R., Chaudhry, I. A., Calder, L. D., and Turkanis, S. A. (1990). Amphetamine behavioral sensitization and the excitatory amino acids. *Brain Res.* **537**, 76–82.

Karler, R., Calder, L. D., and Turkanis, S. A. (1991). DNQX blockade of amphetamine behavioral sensitization. *Brain Res.* **552**, 295–300.

Kashihara, K., Hamamura, T., Okumura, K., and Otsuki, S. (1990). Effect of MK-801 on endogenous dopamine release in vivo. *Brain Res.* **528**, 80–82.

Kazahaya, Y., Kiyoshi, A., and Otsuki, S. (1989). Subchronic methamphetamine treatment enhances methamphetamine- or cocaine-induced dopamine efflux in vivo. *Biol. Psychiatry* **25**, 903–912.

Keefe, K. A., Zigmond, M. J., and Abercrombie, E. D. (1992). Extracellular dopamine in striatum: Influence of nerve impulse activity in medial forebrain bundle and local glutamatergic input. *Neuroscience* **47**, 325–332.

Kimura, K., Nomikos, G. G., and Svensson, T. H. (1993). Effects of amperozide on psychostimulant-induced hyperlocomotion and dopamine release in the nucleus accumbens. *Pharmacol. Biochem. Behav.* **44**, 27–36.

King, G., Joyner, C., Lee, C., Kuhn, C., and Ellinwood, Jr., E. H. (1992). Intermittent and continuous cocaine administration: Residual behavioral states during withdrawal. *Pharmacol. Biochem. Behav.* **43**, 243–248.

Kokkinidis, L. (1988). Animal models of psychiatric disorders. *In* "An Inquiry into Schizophrenia and Depression" (P. Simon, P. Soubrie, and D. Widlocher, eds.), pp. 148–173. Karger, Basel.

Kuczenski, R., and Leith, N. J. (1981). Chronic amphetamine: Is dopamine a link in or a mediator of the development of tolerance and reverse tolerance. *Pharmacol. Biochem. Behav.* **15**, 405–413.

Kuczenski, R., and Segal, D. S. (1988) Psychomotor stimulant-induced sensitization: Behavioral and neurochemical correlates. *In* "Sensitization in the Nervous System" (P. Kalivas and T. Barnes, eds.), pp. 175–205. Telford Press, Caldwell, New Jersey.

Kuczenski, R., and Segal, D. S. (1989). Concomitant characterization of behavioral and striatal neurotransmitter response to amphetamine using in vivo microdialysis. *J. Neurosci.* **9**, 2051–2065.

Kuczenski, R., and Segal, D. S. (1990). In vivo measures of monoamines during amphetamine-induced behaviors in rats. *Prog. Neuro-psychopharmacol. Biol. Psychiat.* **14** (suppl.), S37–S50.

Kuczenski, R., and Segal, D. S. (1992). Regional norepinephrine response to amphetamine using dialysis: Comparison to caudate dopamine. *Synapse* **11**, 164–169.

Kuczenski, R., Segal, D. S., Weinberger, S. B., and Browne, R. G. (1982). Evidence that a behavioral augmentation following repeated amphetamine administration does not involve peripheral mechanisms. *Pharmacol. Biochem. Behav.* **17**, 547–553.

Kuczenski, R., Segal, D. S., Leith, N. J., and Applegate, C. D. (1987). Effects of amphetamine, methylphenidate, and apomorphine on regional brain serotonin and 5-hydroxyindole acetic acid. *Psychopharmacology (Berlin)* **93**, 329–335.

Kuczenski, R., Segal, D. S., and Manley, L. D. (1990). Apomorphine does not alter amphetamine-induced dopamine release measured in striatal dialysates. *J. Neurochem.* **54**, 1492–1499.

Kuczenski, R., Segal, D. S., and Aizenstein, M. L. (1991). Amphetamine, fencamfamine, and cocaine: Relationships between locomotor and stereotypy response profiles and caudate and accumbens dopamine dynamics. *J. Neurosci.* **11**, 2703–2712.

Kula, N. S., and Baldessarini, R. J. (1991). Lack of increase in dopamine transporter binding or function in rat brain tissue after treatment with blockers of neuronal uptake of dopamine. *Neuropharmacology* **30**, 89–92.

Kuribara, H., and Tadokoro, S. (1989). Reverse tolerance to ambulation-increasing effects of methamphetamine and morphine in 6 mouse strains. *Jpn. J. Pharmacol.* **49**, 197–203.

Lat, J. (1965). The spontaneous exploratory reactions as a tool for psychopharmacological studies. In "Proceedings of the Second International Pharmacological Meeting" (M. Mikhelson and V. Longo, eds.), pp. 47–66. Pergamon Press, London.

Leith, N. J., and Kuczenski, R. (1982). Two dissociable components of behavioral sensitization following repeated amphetamine administration. *Psychopharmacology (Berlin)* **76**, 310–315.

Levy, A. D., Kim, J. J., and Ellison, G. D. (1988). Chronic amphetamine alters D-2 but not D-1 agonist-induced behavioral response in rats. *Life Sci.* **43**, 1207–1213.

Lyon, M., and Nielsen, E. (1979). Drug-induced stereotypies and psychosis. In "Psychopathology in Animals" (J. Keehn, ed.), pp. 103–142. Academic Press, New York.

Mazurski, E. J., and Beninger, R. J. (1987). Environment-specific conditioning and sensitization with (+)-amphetamine. *Pharmacol. Biochem. Behav.* **27**, 61–65.

Mestikawy, S. E., Glowinski, J., and Hamon, M. (1986). Presynaptic dopamine autoreceptors control tyrosine hydroxylase activation in depolarized striatal dopaminergic terminals. *J. Neurochem.* **46**, 12–22.

Missale, C., Castelletti, L., Govoni, S., Spano, P. F., Trabucchi, M., and Hanbauer, I. (1985). Dopamine uptake is differentially regulated in rat striatum and nucleus accumbens. *J. Neurochem.* **45**, 51–56.

Mittleman, G., Jones, G. H., and Robbins, T. W. (1991). Sensitization of amphetamine-stereotypy reduces plasma corticosterone: Implications for stereotypy as a coping response. *Behav. Neural Biol.* **56**, 170–182.

Nader, M. A., Tatham, T. A., and Barrett, J. E. (1992). Behavioral and pharmacological determinants of drug abuse. *Ann. N.Y. Acad. Sci.* **654**, 368–385.

Ng, J. P., Hubert, G. W., and Justice, J. B., Jr. (1991). Increased stimulated release and uptake of dopamine in nucleus accumbens after repeated cocaine administration as measured by in vivo voltammetry. *J. Neurochem.* **56**, 1485–1492.

Nicholls, B., Springham, A., and Mellanby, J. (1992). The playground maze: A new method for measuring directed exploration in the rat. *J. Neurosci. Methods* **43**, 171–180.

Nielsen, E. B., Lee, T. H., and Ellison, G. (1980). Following several days of continuous administration amphetamine acquires hallucinogen-like properties. *Psychopharmacology* **4**, 17–20.

Nisenbaum, L. K., Zigmond, M. J., Sved, A. F., and Abercrombie, E. D. (1991). Prior exposure to chronic stress results in enhanced synthesis and release of hippocampal norepinephrine in response to a novel stressor. *J. Neurosci.* **11**, 1478–1484.

Nomikos, G. G., Damsma, G., Wenkstern, D., and Fibiger, H. C. (1991). Chronic desipramine enhances amphetamine-induced increases in interstitial concentrations of dopamine in the nucleus accumbens. *Eur. J. Pharmacol.* **195**, 63–73.

Patrick, S. L., Thompson, T. L., Walker, J. M., and Patrick, R. L. (1991). Concomitant sensitization of amphetamine-induced behavioral stimulation and in vivo dopamine release from rat caudate nucleus. *Brain Res.* **538**, 343–346.

Paulson, P. E., Camp, D. M., and Robinson, T. E. (1991). Time course of transient behavioral depression and persistent behavioral sensitization in relation to regional brain monoamine concentrations during amphetamine withdrawal in rats. *Psychopharmacology* **103**, 480–492.

Peris, J., Boyson, S. J., Cass, W. A., Curella, P., Dwoskin, L. P., Larson, G., Lin, L.-H., Yasuda, R. P., and Zahniser, N. R. (1990). Persistence of neurochemical changes in dopamine systems after repeated cocaine administration. *J. Pharmacol. Exp. Ther.* **253**, 38–44.

Pettit, H. O., Pan, H.-T., Parsons, L. H., and Justice, J. B., Jr. (1990). Extracellular concentrations of cocaine and dopamine are enhanced during chronic cocaine administration. *J. Neurochem.* **55**, 798–804.

Piazza, P. V., Deminière, J.-M., Le Moal, M., and Simon, H. (1989). Factors that predict individual vulnerability to amphetamine self-administration. *Science* **245**, 1511–1513.

Piazza, P. V., Deminière, J.-M., Le Moal, M., and Simon, H. (1990). Stress- and

pharmacologically-induced behavioral sensitization increases vulnerability to acquisition of amphetamine self-administration. *Brain Res.* **514**, 22–26.
Post, R. (1980). Intermittent versus continuous stimulation: Effect of time interval on the development of sensitization or tolerance. *Life Sci.* **26**, 1275–1282.
Post, R. M., Weiss, S. R. B., and Pert, A. (1992). Sensitization and kindling effects of chronic cocaine administration. *In* "Cocaine: Pharmacology, Physiology, and Clinical Strategies" (J. M. Lakoski, M. P. Galloway, and F. J. White, eds.), pp. 115–161. CRC Press, Boca Raton, Florida.
Rebec, G. V. (1991). Changes in brain and behavior produced by amphetamine: A perspective based on microdialysis, voltammetry, and single-unit electrophysiology in freely moving animals. *In* "Biochemistry and Physiology of Substance Abuse" (R. R. Watson, ed.), pp. 93–115. CRC Press, Boston.
Rebec, G. V., and Segal, D. S. (1980). Apparent tolerance to some aspects of amphetamine stereotypy with long-term treatment. *Pharmacol. Biochem. Behav.* **13**, 793–979.
Rebec, G. V., Peirson, E. E., McPherson, F. A., and Brugge, K. (1982). Differential sensitivity to amphetamine following long-term treatment with clozapine or haloperidol. *Psychopharmacology* **77**, 360–366.
Roberts-Lewis, J. M., Roseboom, P. H., Iwaniec, L. M., and Gnegy, M. E. (1986). Differential down-regulation of D_1-stimulated adenylate cyclase activity in rat forebrain after in vivo amphetamine treatments. *J. Neurosci.* **6**, 2245–2251.
Robinson, T. E. (1991a). Persistent sensitizing effects of drugs on brain dopamine systems and behavior: Implications for addiction and relapse. *In* "The Biological Basis of Substance Abuse" (J. Barchas and S. Korenman, eds.) pp. 373–402. Oxford University Press, Oxford.
Robinson, T. E. (1991b). The neurobiology of amphetamine psychosis: Evidence from studies with an animal model. *Taniguchi Symp. Brain Sci.* **14**, 185–201.
Robinson, T. E., and Becker, J. B. (1986). Enduring changes in brain and behavior produced by chronic amphetamine administration: A review and evaluation of animal models of amphetamine psychosis. *Brain Res. Rev.* **11**, 157–198.
Robinson, T. E., Angus, A. L., and Becker, J. B. (1985). Sensitization to stress: The enduring effects of prior stress on amphetamine-induced rotational behavior. *Life Sci.* **37**, 1039–1042.
Robinson, T. E., Jurson, P. A., Bennett, J. A., and Bentgen, K. M. (1988). Persistent sensitization of dopamine neurotransmission in ventral striatum (nucleus accumbens) produced by prior experience with (+)-amphetamine: A microdialysis study in freely moving rats. *Brain Res.* **462**, 211–222.
Roseboom, P. H., Hewlett, G. H. K., and Gnegy, M. E. (1990). Repeated amphetamine administration alters the interaction between D_1-stimulated adenylyl cyclase activity and calmodulin in rat striatum. *J. Pharmacol. Exp. Ther.* **255**, 197–203.
Rossetti, Z. L., Portas, C., Pani, L., Carboni, S., and Gessa, G. L. (1990). Stress increases noradrenaline release in the rat frontal cortex: Prevention by diazepam. *Eur. J. Pharmacol.* **176**, 229–231.
Schiorring, E. (1979). An open field study of stereotyped locomotor activity in amphetamine treated rats. *Psychopharmacology* **66**, 281–287.
Segal, D. S. (1975a). Behavioral and neurochemical correlates of repeated *d*-amphetamine administration. *In* "Advances in Biochemical Psychopharmacology" (A. J. Mandell, ed.), pp. 247–266. Raven Press, New York.
Segal, D. S. (1975b). Behavioral characterization of *d*- and *l*-amphetamine: Neurochemical implications. *Science* **190**, 475–477.
Segal, D. S. (1976). Differential effects of *para*-chlorophenylalanine on amphetamine-induced locomotion and stereotypy. *Brain Res.* **116**, 267–276.
Segal, D. S. (1977). Differential effects of serotonin depletion on amphetamine-induced locomotion and stereotypy. *In* "Cocaine and Other Stimulants" (E. H. Ellinwood and M. J. Kilbey, eds.), pp. 431–443. Raven Press, New York.

Segal, D. S., and Geyer, M. A. (1985) Animal models of psychopathology. *In* "Psychiatry" (J. O. Cavenar, Jr., ed.), pp. 1–18. Lippincott, Philadelphia.

Segal, D. S., and Janowsky, D. S. (1978) Psychostimulant-induced behavioral effects: Possible models of schizophrenia. *In* "Psychopharmacology: A Generation of Progress" (M. A. Lipton, A. MiMascio, and K. F. Killman, eds.), pp. 1113–1124. Raven Press, New York.

Segal, D. S., and Kuczenski, R. (1987a). Individual differences in responsiveness to single and repeated amphetamine administration: Behavioral characteristics and neurochemical correlates. *J. Pharmacol. Exp. Ther.* **242**, 917–926.

Segal, D. S., and Kuczenski, R. (1987b). Behavioral and neurochemical characteristics of stimulant-induced augmentation. *Psychopharmacol. Bull.* **23**, 417–424.

Segal, D. S., and Kuczenski, R. (1992a). Repeated cocaine administration induces behavioral sensitization and corresponding decreased extracellular dopamine responses in caudate and accumbens. *Brain Res.* **577**, 351–355.

Segal, D. S., and Kuczenski, R. (1992b). In vivo microdialysis reveals a diminished amphetamine-induced dopamine response corresponding to behavioral sensitization produced by repeated amphetamine pretreatment. *Brain Res.* **571**, 330–337.

Segal, D. S., and Mandell, A. J. (1974). Long-term administration of *d*-amphetamine: Progressive augmentation of motor activity and stereotypy. *Pharmacol. Biochem. Behav.* **2**, 249–255.

Segal, D. S., and Schuckit, M. A. (1983) Animal models of stimulant-induced psychosis. *In* "Stimulants: Neurochemical, Behavioral and Clinical Perspectives" (I. Creese, ed.), pp. 131–167. Raven Press, New York.

Segal, D. S., McAllister, C., and Geyer, M. A. (1974). Ventricular infusion of norepinephrine and amphetamine: Direct versus indirect action. *Pharmacol. Biochem. Behav.* **2**, 79–86.

Segal, D. S., Weinberger, S., Cahill, J., and McCunney, S. (1980). Multiple daily amphetamine administration: Behavioral and neurochemical alterations. *Science* **207**, 904–907.

Segal, D. S., Geyer, M. A., and Schuckit, M. A. (1981). Stimulant-induced psychosis: An evaluation of animals models. *In* "Essays in Neurochemistry and Neuropharmacology" (M. B. H. Youdim, W. Lovenberg, D. F. Sharman, and J. R. Lagnado, eds.), pp. 95–129. John Wiley & Sons, Sussex, England.

Segal, D. S., Kuczenski, R., and Okuda, C. (1992). Clorgyline-induced increases in presynaptic DA: Changes in the behavioral and neurochemical effects of amphetamine using in vivo microdialysis. *Pharmacol. Biochem. Behav.* **42**, 421–429.

Seyfried, C. A. (1983). Dopamine uptake inhibiting versus dopamine releasing properties of fencamfamine: An in vitro study. *Biochem. Pharmacol.* **32**, 2329–2331.

Sharp, T., Zetterstrom, T., Ljungberg, T., and Ungerstedt, U. (1987). A direct comparison of amphetamine-induced behaviours and regional brain dopamine release in the rat using intracerebral dialysis. *Brain Res.* **401**, 322–330.

Sharpe, L. G., Pilotte, N. S., Mitchell, W. M., and De Souza, E. B. (1991). Withdrawal of repeated cocaine decreases autoradiographic [^{3}H]mazindol-labelling of dopamine transporter in rat nucleus accumbens. *Eur. J. Pharmacol.* **203**, 141–144.

Stewart, J. (1991). Conditioned stimulus control of the expression of sensitization of the behavioral activating effects of opiate and stimulant drugs. *In* "Learning and Memory: The Behavioral and Biological Substrates" (I. Gormezano and E. A. Wasserman, eds.), 410–437. Erlbaum, Hillsdale, New Jersey.

Stewart, J. (1992). Neurobiology of conditioning to drugs of abuse. *Ann. N.Y. Acad. Sci.* **654**, 335–346.

Stewart, J., and Eikelboom, R. (1987). Conditioned drug effects. In "Handbook of Psychopharmacology" (L. L., Iversen, S. D. Iversen, and S. H. Snyder, eds.), Vol. 19, pp. 1–57. Plenum Press, New York.

Stewart, J., and Vezina, P. (1989). Microinjections of Sch-23390 into the ventral tegmental area and substantia nigra pars reticulata attenuate the development of sensitization to the locomotor activating effects of systemic amphetamine. *Brain Res.* **495**, 401–406.

Stewart, J., and Vezina, P. (1991). Extinction procedurs abolish conditioned stimulus control but spare sensitized responding to amphetamine. *Behav. Pharmacol.* **2**, 65–71.

Strait, K. A., and Kuczenski, R. (1986). Dopamine autoreceptor regulation of the kinetic state of striatal tyrosine hydroxylase. *Mol. Pharmacol.* **29**, 561–569.

Strombom, U. H., and Liedman, B. (1982). Role of dopaminergic neurotransmission in locomotor stimulation by dexamphetamine and ethanol. *Psychopharmacology* **78**, 271–276.

Teicher, M. H., Barber, N. I., Lawrence, J. M., and Baldessarini, R. J. (1989). Motor activity and antidepressant drugs: A proposed approach to categorizing depression syndromes and their animal models. *In* "Animal Models of Depression" (G. F. Koob, C. Ehlers, and D. J. Kupfer, eds.), pp. 135–157. Birkhauser, Boston.

Thompson, T. L., Colby, K. A., and Patrick, R. L. (1990). Activation of striatal tyrosine hydroxylase by in vivo electrical stimulation: Comparison with cyclic AMP mediated activation. *Neurochem. Res.* **15**, 1159–1166.

Ujike, H., Tsuchida, H., Kanzaki, A., Akiyama, K., and Otsuki, S. (1992). Competitive and non-competitive N-methyl-D-aspartate antagonists fail to prevent the induction of methamphetamine-induced sensitization. *Life Sci.* **50**, 1673–1681.

Utena, H., Machiyama, Y., Hsu, S. C., Katagiri, M., and Hirata, A. (1975). A monkey model for schizophrenia produced by methamphetamine. *In* "Contemporary Primatology" (S. Kondo, M. Kawai, and A. Ehara, eds.), pp. 502–507. Karger, Bassel.

Vezina, P., and Stewart, J. (1990). Amphetamine administered to the ventral tegmental area but not to the nucleus accumbens sensitizes rats to systemic morphine: Lack of conditioned effects. *Brain Res.* **516**, 99–106.

Waddington, J. L., and O'Boyle, K. M. (1987). The D-1 dopamine receptor and the search for its functional role: From neurochemistry to behaviour. *Rev. Neurosci.* **1**, 157–184.

Waddington, J. L., and O'Boyle, K. M. (1989). Drugs acting on brain dopamine receptors: A conceptual re-evaluation five years after the first selective D-1 antagonist. *Pharmacol. Ther.* **43**, 1–52.

Weihmuller, F. B., O'Dell, S. J., Cole, B. N., and Marshall, J. F. (1991). MK-801 attenuates the dopamine-releasing but not the behavioral effects of methamphetamine: An in vivo microdialysis study. *Brain Res.* **549**, 230–235.

Weiss, F., Paulus, M. P., Lorang, M. T., and Koob, G. F. (1992). Increases in extracellular dopamine in the nucleus accumbens by cocaine are inversely related to basal levels: Effects of acute and repeated administration. *J. Neurosci.* **12**, 4372–4380.

Westerink, B. H. C., Santiago, M., and de Vries, J. B. (1992). The release of dopamine from nerve terminals and dendrites of nigrostriatal neurons induced by excitatory amino acids in the conscious rat. *Naunyn-Schmiedeberg's Arch. Pharmacol.* **345**, 523–529.

Whishaw, I. Q., Fiorino, D., Mittleman, G., and Castañeda, E. (1992). Do forebrain structures compete for behavioral expression? Evidence from amphetamine-induced behavior, microdialysis, and caudate-accumbens lesions in medial frontal cortex damaged rats. *Brain Res.* **576**, 1–11.

White, F. J., and Wang, R. Y. (1984). Electrophysiological evidence for A10 dopamine autoreceptor subsensitivity following chronic d-amphetamine treatment. *Brain Res.* **309**, 283–292.

Whitton, P. S., Biggs, C. S., Pearce, B. R., and Fowler, L. J. (1992). Regional effects of MK-801 on dopamine and its metabolites studied by in vivo microdialysis. *Neurosci. Lett.* **142**, 5–8.

Wickens, J. R., Alexander, M. E., and Miller, R. (1991). Two dynamic modes of striatal function under dopaminergic–cholinergic control: Simulation and analysis of a model. *Synapse* **8**, 1–12.

Williams, G. V., and Millar, J. (1990). Concentration-dependent actions of stimulated dopamine release on neuronal activity in rat striatum. *Neuroscience* **39**, 1–16.

Wolf, M. E., and Khansa, M. R. (1991). Repeated administration of MK-801 produces sensitization to its own locomotor stimulant effects but blocks sensitization to amphetamine. *Brain Res.* **562,** 164–168.

Zetterstrom, T., and Ungerstedt, U. (1984). Effects of apomorphine on the in vivo release of dopamine and its metabolites, studied by brain dialysis. *Eur. J. Pharmacol.* **97,** 29–36.

Zigmond, R. E., Schwarzschild, M. A., and Rittenhouse, A. R. (1989). Acute regulation of tyrosine hydroxylase by nerve activity and by neurotransmitters via phosphorylation. *Ann. Rev. Neurosci.* **12,** 415–461.

Christopher J. Schmidt

5
Neurochemistry of Ring-Substituted Amphetamine Analogs

I. INTRODUCTION

The ring-substituted amphetamines belong to a broad class of chemical compounds more appropriately referred to as the phenylisopropylamines. Despite the overall structural similarity of these compounds, simple additions to the aromatic ring can change the pharmacology of this group of drugs from that of central stimulants such as *d*-amphetamine to that of potent hallucinogens such as 2,5-dimethoxy-4-bromophenylisopropylamine (DOB) and 2,5-dimethoxy-4-methylphenylisopropylamine (DOM) (Glennon, 1989). In parallel with this shift in behavioral effects is a change in the neurochemical activities of the agents from that of indirect agonists acting primarily via the release of endogenous neurotransmitters to that of direct agonists acting at specific monoamine receptors. Between these two extremes exist compounds with varying degrees of each type of activity. This chapter primarily focuses on the indirect agonists since the amphetamines generally are regarded as the classical examples of such agents. However, because of the continuum between direct and indirect agonists, and for the sake of completeness, we begin with a very brief discussion of the neurochemistry of the hallucinogenic phenylisopropylamines.

Substitutions on the phenyl ring of amphetamine are generally of four types: halogens or methoxy, methylenedioxy, and alkyl groups. The number and the positioning of these ring substituents produce the wide range of neurochemical effects of these agents. Of these various substitutions, halogenated amphetamines are considered in Chapter 7. Simple ring-alkylated amphetamines have received little attention, particularly in terms of their central nervous system (CNS) activity, and will not be discussed here, with one exception. The acute neurochemistry of fenfluramine, a trifluoromethyl ring-substituted amphetamine, will be described as a member of this group, notwithstanding the fact that this compound is also a halogenated amphetamine. Although substitution at other positions in the phenylisopropylamine structure is possible, only a few N-alkyl analogs are considered here since only these agents still retain the amphetamine backbone. For a more comprehensive review of the structure–activity relationships of the phenylisopropylamines, see Shulgin (1978).

II. METHOXYLATED AMPHETAMINES (HALLUCINOGENS)

The sequential addition of methoxy groups to the phenyl ring of amphetamine generates hallucinogenic phenylisopropylamines and culminates in agents such as DOB and DOM (although these compounds also possess a halogen and an alkyl substituent, respectively). In general, the serial addition of ring substituents to amphetamine tends to reduce the "dopaminergic" actions of the drug while enhancing its direct or indirect "serotonergic" effects. Although this statement appears to be true for several different substituents (e.g., Cl, methylenedioxy, trifluoromethyl), this behavior is demonstrated most clearly by the addition of a single *para*-methoxy group to amphetamine to yield *p*-methoxyamphetamine (PMA). Behaviorally, PMA possesses some properties of both the parent molecule amphetamine and the more typical hallucinogens. In humans, the hallucinogenic potency of PMA is reported to be 5 times that of mescaline (Shulgin *et al.*, 1969), yet the drug is also a potent central stimulant. Behaviorally, this action can be demonstrated using rats trained to discriminate amphetamine from saline (Glennon *et al.*, 1986; Young and Glennon, 1986). In such animals, PMA generalizes to amphetamine, suggesting a similarity in the discriminative cue. The addition of a second methoxy group to the aromatic ring of amphetamine further reduces its central stimulant activity, as shown by the limited generalization observed for any of the six positional isomers of dimethoxyamphetamine (DMA). Predictably, trimethoxyamphetamines show even less generalization to amphetamine whereas compounds such as DOM and

DOB are potent 5-HT$_2$ receptor agonists with virtually no dopaminergic character. Thus, as the number of methoxy groups increases, the behavioral actions of the drug switch from amphetamine-like to hallucinogen-like (Young and Glennon, 1986; Glennon, 1989).

These changes in the behavioral effects of the amphetamine analogs with the addition of methoxy groups is a reflection of changes in the neurochemical activity of the drugs. Whereas methoxylation of amphetamine to PMA has little effect on the ability to release [^{3}H]norepinephrine from cortical slices *in vitro*, a reduction of approximately two orders of magnitude occurs in its potency for the release of [^{3}H]dopamine from striatal slices (Tseng *et al.*, 1976). With this reduction in potency for dopamine release comes an increase in the potency of the compound for inducing [^{3}H]5-HT (serotonin) release *in vitro* (Tseng *et al.*, 1976) and *in vivo* (Loh and Tseng, 1979). Despite this change in selectivity, the mechanism of neurotransmitter release for PMA is the same carrier-dependent process described for amphetamine (Fischer and Cho, 1979). This mechanism is demonstrated by the ability of the 5-HT uptake inhibitor chlorimipramine to reduce PMA-induced [^{3}H]5-HT release both *in vitro* and *in vivo* (Loh and Tseng, 1979).

Although the behavioral effects of PMA may be accounted for by its actions as an indirect agonist, the mechanism of action of agents such as 2,5-DMA is clearly different. PMA releases transmitter from the brains of rats loaded intracerebroventricularly (icv) with [^{3}H]5-HT in a fluoxetine-sensitive manner (Tseng, 1979), consistent with a carrier-dependent mechanism of release. In the same system however, 2,5-DMA has been shown to reduce release (Tseng, 1978), suggesting a direct receptor effect. This effect can be demonstrated in ligand binding studies in which the affinity of PMA (K_i) for the 5-HT$_2$ binding site in rat cortex is approximately 33,600 nM (versus [^{3}H]ketanserin) whereas the same value for 2,5-DMA is approximately 5200 nM (Titeler *et al.*, 1990). Although these values appear to be too high to have any physiological relevance, studies in rats using the 5-HT$_2$-specific ligand [^{3}H]DOB to label the high affinity state of the 5-HT$_2$ receptor show that agonists have an approximately 5-fold greater potency for displacement of agonist than antagonists (Titeler *et al.*, 1990). This result may indicate that the affinity of agents such as 2,5-DMA for the 5-HT$_2$ receptor is greater than was predicted based on antagonist displacement, and may be of potential physiological relevance. A report using the novel ligand [^{77}Br]DOB suggests that the difference between displacement of agonist and antagonist binding may be larger than 5-fold in human tissue (Pierce and Peroutka, 1988).

III. METHYLENEDIOXYAMPHETAMINES

The methylenedioxy derivatives of amphetamine have received a great deal of attention because of the popularity of 3,4-methylenedioxymethamphetamine (MDMA) as an illicit recreational agent. These agents also possess a number of unique behavioral and neurochemical effects that have made them popular research tools. With its N-desmethyl (MDA) and N-ethyl (MDE) analogs, MDMA provides a distinctive example of the intermediate state between central stimulants and hallucinogens. Among the many interesting features of these compounds is the fact that the methylenedioxy-substituted amphetamine agents represent an exception to the general rule that the R-(−) enantiomer of phenethylamine and phenylisopropylamine psychotomimetics is the active conformation of the drug (Anderson et al., 1978; Marquardt et al., 1978; Nichols et al., 1982; Glennon, 1984; Shannon et al., 1984). Although this rule seems to hold for MDA, N methylation to MDMA results in a change of some stereochemical requirement for activity so S-(+)-MDMA is the more active isomer in humans (Anderson et al., 1978; Braun et al., 1980; Nichols et al., 1982; Nichols and Glennon, 1984). This explanation is probably an oversimplification since researchers have by no means determined that MDA and MDMA produce their behavioral effects by a common mechanism (see Nichols and Oberlender, 1989). However, this unusual feature of the structure–activity relationship for these compounds has prompted a great deal of neurochemical and behavioral work on the stereoisomers of MDA and MDMA. A second feature of the methylenedioxyamphetamines that differentiates them from other phenylisopropylamine psychotomimetics is that N alkylation of MDA does not decrease its potency as generally is observed for other amphetamines (Anderson et al., 1978; Braun et al., 1980; Nichols et al., 1982; Nichols and Glennon, 1984). This statement is also an oversimplification. Although MDA and MDMA have similar potencies for producing behavioral effects, a qualitative change from hallucinogen-like to central stimulant-like activity does occur in the conversion of MDA to MDMA (see Glennon, 1989).

A. *In Vitro* Neurochemistry

1. Ligand Binding Studies

Initial efforts at defining any direct receptor affinity of the methylenedioxyamphetamines examined the interaction of the stereoisomers of MDA and MDMA with the 5-HT$_2$ receptor as defined by the binding of [^{3}H]ketanserin. All four agents displaced ketanserin with micromolar potency; the R-(−) enantiomers of both MDA and MDMA were 4–5 times more potent than the S-(+) stereoisomers (Lyon et al., 1986). This result reflects the proper stereoselectivity for binding to the 5-HT$_2$ receptor and is consistent with the observation that the R-(−) stereoisomer of MDA is the more potent hallucinogen in humans (Shulgin, 1978). However, as already discussed, this result is inconsistent with the greater activity of the S-(+) stereoisomer of MDMA in humans (Shulgin, 1978). These stereochemical considerations call into question the importance of direct receptor activity in the behavioral effects of MDMA. In another study already referred to, the stereoisomers of MDA and MDMA were compared for their ability to displace the binding of the selective 5-HT$_2$ agonist [^{3}H]DOB (Titeler et al., 1990). DOB binds to the high-affinity, agonist-preferring state of the 5-HT$_2$ receptor and, hence, is displaced more potently by agonists than by antagonists. As in the studies with ketanserin, the R-(−) isomers of both MDA and MDMA most potently displaced DOB. From these data, one could conclude that MDA may produce its behavioral effects by a direct action at the 5-HT$_2$ receptor. Indeed, both racemic and R-(−)-MDA generalize to DOM in drug discrimination studies, whereas neither racemic MDMA nor either of its optical isomers produces such generalization (Glennon, 1989). Further, whereas R-(−)-MDA shows generalization to LSD, neither R-(−)- nor S-(+)-MDMA does (Nichols and Oberlender, 1989). These results suggest that the mechanism of action of MDMA may have some unique features that differentiate it from its desmethyl precursor.

Although concerns about the stereochemical requirements for MDMA binding to the 5-HT$_2$ receptor may call into question the role of any direct receptor effects in its mechanism of action, the results of the displacement studies with [^{3}H]DOB indicate that the drug may have agonist activity. Racemic MDMA displaced [^{3}H]ketanserin with a K_i of 8300 nM, compared with a K_i of 214 nM for the displacement of [^{3}H]DOB. The higher affinity for the DOB site is consistent with agonist activity of the drug at the 5-HT$_2$ receptor. Studies conducted in human cortical membranes using [^{77}Br]DOB also indicate that MDA, MDMA, and MDE bind to 5-HT$_2$ receptors with IC$_{50}$ values of approximately 1 μM. In agreement with the results observed using [^{3}H]ketanserin in rat membranes, all three agents were far less potent at displacing [^{3}H]spiperone indicating that the agents bound primarily to the high-affinity, agonist-preferring conformation of the 5-HT$_2$ receptor. The extent to which this direct agonist activity of the agents contributes to their neurochemical and behavioral effects remains to be determined, although Zaczek et al. (1989) demonstrated brain concentrations of MDMA in the rat as high as 165 μM after a dose of 20 mg/kg. If these concentrations are present at the receptor site, the contribution of direct agonist activity may have to be considered in the neurochemical and behavioral effects of the methylenedioxyamphetamines.

Battaglia et al. (1988) determined the affinity of MDMA for a number of other brain neurotransmitter recognition sites in displacement studies. The affinity of MDMA for the monoaminergic uptake sites was 0.6 μM for 5-HT sites, 15.8 μM for norepinephrine sites, and 24.4 μM for dopamine sites. The general selectivity of the ring-substituted amphetamines for the serotonergic system is illustrated again by the approximately 40-fold higher affinity of MDMA for the 5-HT uptake site relative to the dopamine uptake site. Note, however, that activity at all these sites is possible based on the central MDMA concentrations measured by Zaczek et al. (1989). The affinity of MDMA for a number of neurotransmitter receptors also falls below or within the 100-μM range. The rank order of this affinity is $\alpha_2 > 5\text{-}HT_2 > M_1 > H_1 > M_2 > \alpha_1 > \beta > 5\text{-}HT_1 > D_2 > D_1$.

2. Uptake and Release Studies

The modest affinity of the methylenedioxyamphetamines for the $5\text{-}HT_2$ receptor and their central stimulant activity point to the importance of indirect agonism in their behavioral effects. This aspect of the pharmacology of MDA first was examined by Marquardt (1978) in studies of norepinephrine uptake by hypothalamic synaptosomes. Both $S\text{-}(+)$-amphetamine and $S\text{-}(+)$-MDA exhibited submicromolar IC_{50}s for the inhibition of [^{3}H]norepinephrine uptake whereas $R\text{-}(-)$-MDA was slightly less potent. Steele et al. (1987) found that only the $S\text{-}(+)$ stereoisomers of MDA and MDMA inhibited the uptake of [^{3}H]dopamine into rat striatal synaptosomes and both were less potent than d-amphetamine. In contrast, although equipotent, the stereoisomers of MDA and MDMA all were more potent than d-amphetamine in inhibiting the uptake of [^{3}H]5-HT by hippocampal synaptosomes. Although MDMA and its analogs have reasonable affinity for the various monoamine uptake carriers (Battaglia et al., 1988), two studies have been unable to find evidence of any uptake of [^{3}H]MDMA into rat brain synaptosomes (Schmidt et al., 1987; Wang et al., 1987), perhaps because of the lipophilicity of the drug and its rapid diffusion across synaptosomal membranes.

In studies on 5-HT transport by synaptic vesicles, MDMA has been shown both to block accumulation of 5-HT and to increase its efflux. These phenomena occur by a direct effect at the ATP-dependent transporter and by a disruption of the pH gradient responsible for concentrating 5-HT (Rudnick and Wall, 1992). Such a displacement of intraneuronal stores may provide the favorable concentration gradient that allows MDMA and related drugs to cause transmitter efflux from the nerve terminal by way of the plasma membrane carrier.

Most of the behavioral and neurochemical effects of the methylenedioxyamphetamines are probably attributable to their effects on neurotransmitter release, particularly of monoamines. All three methylenedioxyamphetamines are similarly potent at releasing [^{3}H]5-HT from striatal slices *in vitro* (Schmidt, 1987a) and are more potent than either amphetamine or methamphetamine (Schmidt et al., 1987,1991a). In keeping with the effect of ring substitutions on the dopaminergic activity, MDMA is much less potent than either methamphetamine or amphetamine at releasing [^{3}H]dopamine from striatal slices (Schmidt et al., 1987,1991). Figure 1 shows that, although increasing the size of the N-alkyl substituent has little effect on 5-HT release, the progression from MDA to MDE reduces the potency of the compound for the release of [^{3}H]dopamine *in vitro* (Schmidt, 1987). Similar results have been demonstrated using *in vivo* microdialysis (Nash and Nichols, 1991).

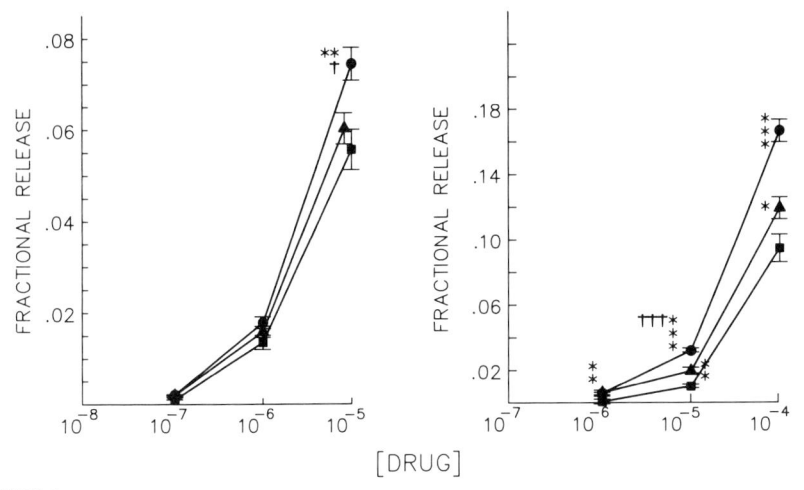

FIGURE 1 Comparison of [^{3}H]serotonin (*left*) and [^{3}H]dopamine (*right*) release by methylenedioxymethamphetamine (MDMA, ▲) and its *N*-desmethyl (MDA, ●) and *N*-ethyl (MDE, ■) analogs from superfused striatal slices of the rat. Release was expressed as a fraction of the radioactivity in the slices during a 5-min pulse of the drug. Results are means ± SEM. Reprinted with permission from Schmidt (1987a).

Release of 5-HT and dopamine are Ca^{2+}-independent carrier-mediated processes consistent with their amphetamine structure (Schmidt *et al.*, 1987). As predicted by the exchange diffusion model of Fisher and Cho (1979), MDMA-induced release of 5-HT from striatal slices was blocked by the selective 5-HT uptake inhibitor citalopram, whereas DA release was blocked by amfonelic acid (Schmidt *et al.*, 1987). Similar results have been achieved in synaptosomes using fluoxetine, cocaine, and desmethylimipramine to block the MDMA-induced release of 5-HT, dopamine, and norepinephrine, respectively (Fitzgerald and Reid, 1990). In a more elaborate study, the affinity of several uptake inhibitors for the 5-HT uptake site as defined by the displacement of [^{3}H]paroxetine binding was shown to correlate with the IC_{50} of these compounds for the inhibition of MDMA-induced 5-HT release from synaptosomes (Hekmatpanah and Peroutka, 1990). Nash and Brodkin (1991) showed that peripherally administered uptake blockers such as mazindol and GBR12909 can antagonize the dopamine release produced by intrastriatal infusions of MDMA.

The stereoselectivity of transmitter release also has been examined. Nichols *et al.* (1982) studied the effect of both stereoisomers of MDA and MDMA on the release of [^{3}H]5-HT from whole brain synaptosomes in the rat. Their results showed that these agents were potent releasers of 5-HT, with a tendency for the *S*-(+) isomer to be more potent. Johnson *et al.* (1986) and Schmidt *et al.* (1987) compared the stereoisomers of MDMA for their effects on the release of 5-HT from hippocampal and striatal slices, respectively, and found no differences in potency. In contrast to the results for 5-HT, the S-(+) stereoisomers of both drugs were significantly more potent than the R (−)isomer at producing dopamine release. The greater potency of S-(+)- over R-(−)-MDMA in increasing dopamine release has been confirmed in microdialysis studies *in vivo* (Hiramatsu and Cho, 1990).

In one study that combined electrophysiological activity and 5-HT release as end points, Sprouse *et al.* (1989) demonstrated that the S-(+) isomer of MDMA was 2- to 3-fold more potent than the R-(−) stereoisomer in inhibiting serotonergic cell firing in a dorsal raphe slice preparation. Collection of 5-HT directly from the surface of the slice using a modified microdialysis probe showed that the inhibition of cell firing correlated with the MDMA-enhanced efflux of the transmitter. The addition of the 5-HT uptake inhibitor fluoxetine to the medium bathing the slice almost completely prevented the effect of MDMA on cell firing, whereas the norepinephrine uptake inhibitor desmethylimipramine was without effect.

B. *In Vivo* Neurochemistry

All three methylenedioxy analogs of amphetamine produce dramatic neurochemical alterations in laboratory animals *in vivo*. The acute (and many of the long-term) neurochemical effects of MDMA and its analogs resemble those described for the well-known serotonergic neurotoxin *p*-chloroamphetamine (PCA; see Chapter 7) and the anorectic agent fenfluramine (see subsequent discussion). The persistent or neurotoxic effects of these agents, particularly MDMA, have received the greatest attention. These effects will be addressed elsewhere in this book. We focus on the acute neurochemical effects of this group of drugs with an emphasis on MDMA. As might be expected from the *in vitro* activities of MDMA already discussed, the primary effects of MDMA *in vivo* occur in the serotonergic nervous system.

1. Acute Changes in Tissue Concentrations of 5-HT and Dopamine

Figure 2 shows results from one of the initial experiments examining postmortem neurochemistry in the rat brain following acute MDMA administration. A single high dose of MDMA can produce dramatic and dose-dependent changes in striatal 5-HT concentrations within hours of drug administration (Schmidt *et al.*, 1986). Examination of the time course of this phenomenon indicated that transmitter concentrations reached a nadir 3–6 hr after drug administration, with a variable recovery evident by 24 hr. This recovery was virtually complete in the cortex at doses of 10 mg/kg MDMA or less, suggesting that these acute changes were temporary (Schmidt, 1987). In contrast to these changes in the serotonergic system, only slight alterations were observed in neurochemical markers of dopaminergic function following MDMA administration. The concentrations of dopamine and ho-

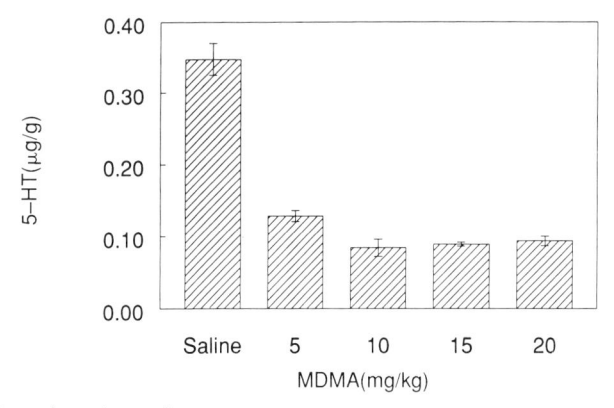

FIGURE 2 Dose-dependent reduction in rat neostriatal 5-hydroxytryptamine (5-HT) concentrations 3 hr after the administration of methylenedioxymethamphetamine (MDMA) (sc). Results are expressed as the mean ± SEM. All four doses produced a significant depletion in the tissue content of transmitter. Reprinted with permission from Schmidt et al. (1986).

movanillic acid (HVA) showed small but significant elevations, although dihydroxyphenylacetic acid (DOPAC) levels were reduced markedly as is observed commonly with most amphetamines. All these changes observed in this study were transient and disappeared by 12 hr. These changes, particularly the elevation of dopamine levels, are consistent with an MDMA-induced increase in transmitter efflux. Such an acute rise in dopamine concentrations has been reported with other releasing agents including amphetamine and PCA (Schmidt et al., 1991a). Investigators generally agree that the release of newly synthesized transmitter by such agents relieves feedback inhibition of tyrosine hydroxylase (TH) and results in a transient increase in synthesis. In agreement with this explanation, blocking MDMA-induced dopamine release with the dopamine uptake inhibitor nomifensine also was shown to block the increase in tissue concentrations of striatal dopamine produced by MDMA administration (Schmidt et al., 1991b). Nash et al. (1990) used the accumulation of DOPA following decarboxylase inhibition to demonstrate directly that MDMA increases dopamine synthesis *in vivo* in the rat striatum and nucleus accumbens.

2. Tryptophan Hydroxylase

Stone et al. (1986) showed that, in addition to the widespread changes in 5-HT concentrations, both MDMA and MDA produced an acute decrease in the activity of the rate limiting enzyme for 5-HT synthesis—tryptophan hydroxylase (TPH). In contrast, neither drug had any effect on TH activity. As shown in Fig. 3, this loss of TPH activity actually preceded the decline in 5-HT concentrations and could be observed within 15 min of drug administration (Stone et al., 1987; Schmidt and Taylor, 1988). Note

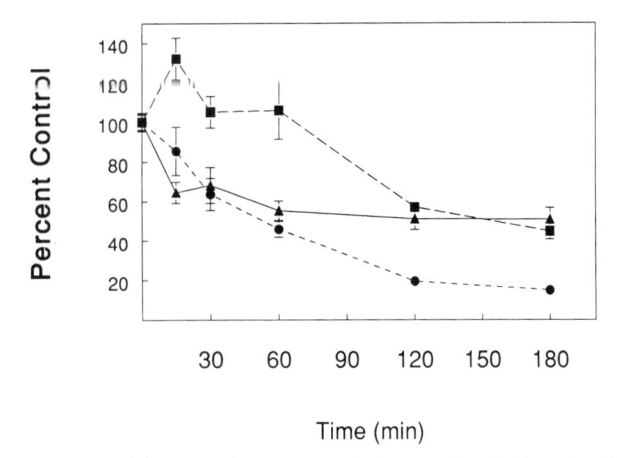

FIGURE 3 Time course of the acute changes in cortical tryptophan hydroxylase (TPH) activity (▲) and the concentrations of 5-hydroxytryptamine (5-HT, ●) and 5-hydroxyindoleacetic acid (5-HIAA, ■) after the administration of a single 10 mg/kg dose of methylenedioxymethamphetamine (MDMA) in the rat. Results are expressed as the mean ± SEM. Reprinted with permission from Schmidt and Taylor (1988).

that 5-hydroxyindoleacetic acid (5-HIAA) concentrations rose temporarily shortly after MDMA administration, suggesting that an increase in 5-HT release may precede all the changes that follow. This conclusion also follows from the observation that inhibitors of the 5-HT uptake carrier could block MDMA-induced release, as well as the reduction in 5-HT concentrations and the acute loss of TPH activity produced by the administration of MDMA (Schmidt and Taylor, 1987; Schmidt et al., 1987).

As observed for the release of 5-HT *in vitro*, these acute neurochemical effects of MDMA did not show any stereoselectivity. Both stereoisomers produced similar depletions of 5-HT (Schmidt, 1987b; Schmidt et al., 1987) and decreases in TPH activity (Schmidt and Taylor, 1988). This characteristic distinguishes the acute effects of MDMA from the long-term deficits in 5-HT parameters since the latter have been shown to be produced primarily by the S-(+) stereoisomer. The same stereoselectivity was demonstrated later for MDA (Schmidt, 1987a). Interestingly and again consistent with both *in vivo* and *in vitro* release data, the acute elevation of dopamine observed with MDMA is also primarily due to the S(+) isomer (Schmidt et al., 1987; Schmidt et al., 1991b).

The precise mechanism responsible for the acute effect of MDMA and similar agents on TPH activity remains unknown. Studies showing that the acute loss of TPH activity occurred independently of the more slowly developing neurotoxic effect of MDMA differentiated these early neurochemical effects from the later changes. Further, investigators demonstrated that the serotonergic system could be spared the neurotoxic effects of MDMA even after the acute loss of TPH activity by the delayed administration of the 5-HT uptake inhibitor fluoxetine (Schmidt and Taylor, 1987). Despite the rapid loss of TPH activity following MDMA administration, neurotoxicity appears to have some requirement for prolonged exposure to the drug, as can be inferred from a number of observations. First, acute administration of a single dose of the parent drug in this series, amphetamine, does not reduce TPH activity in the rat significantly. However, multiple injections of amphetamine closely spaced together will reduce TPH activity (C. J. Schmidt and V. L. Taylor, unpublished observation). A similar effect can be achieved by the addition of a *para*-chloro group to amphetamine. Substitution at this position prolongs the half-life of amphetamine by interfering with its primary route of metabolism in rats (*p*-hydroxylation). PCA produces significant reductions in both 5-HT concentrations and TPH activity on acute administration (see Fig. 6). The methylenedioxy group of MDMA probably functions in a similar manner to slow metabolism of the drug. These studies are consistent with results from experiments described subsequently in which bolus injections of MDMA directly into the rat brain did not affect TPH whereas slow infusions of comparable doses did lower enzyme activity.

Studies on the effect of MDMA on TPH activity *ex vivo* have demonstrated that the loss of enzyme activity is the result of a decrease in the V_{max} of the enzyme and suggested a loss of the functional protein (Schmidt and Taylor, 1987). The effect of MDMA on enzyme activity is not caused by a direct interaction of the drug with the enzyme. TPH activity was not affected by the direct addition of MDMA to brain homogenates (Schmidt and Taylor, 1987) nor was the enzyme activity of the mouse mast cell line P-815 altered by exposure to high concentration of the drug (Schmidt and Taylor, 1988). Direct icv injections of MDMA in awake, freely moving rats had no effect on hippocampal or cortical enzyme activity acutely. Interestingly, when these injections were made into the substantia nigra, striatal concentrations of dopamine, DOPAC, and HVA increased dramatically on the ipsilateral side without any changes in serotonergic parameters being observed. Only slow icv infusions of MDMA resulted in significant reductions in cortical and hippocampal TPH activity (Schmidt and Taylor, 1988), whereas the same dose administered peripherally was without effect. Consistent with the first effect of MDMA occurring on TPH activity, a 1 mg/kg infusion of MDMA for 1 hr was sufficient to reduce TPH activity without altering regional 5-HT concentrations. A higher infusion dose of 2 mg/kg did reduce both enzyme activity and transmitter levels. Thus, although the acute effect of MDMA is not caused by a direct interaction with TPH, data suggest that the effect does appear to be centrally mediated and does not require any peripheral metabolism of the drug.

The three methylenedioxyamphetamine derivatives appear to be similar in terms of their ability to reduce regional 5-HT concentrations and TPH activity acutely. This result again mirrors their similar effects on 5-HT release *in vitro*. This comparison also highlights the distinction between the acute and long-term neurochemical effects of these agents, since marked differences appear in their potency for producing the long-term deficits (Johnson *et al.*, 1987; Schmidt, 1987a). Indeed, the doses of MDE that must be given to produce evidence of neurotoxicity (50–100 mg/kg) are one order of magnitude greater than those needed to produce the acute effects on transmitter levels and TPH activity (Ricaurte *et al.*, 1987).

The observation by Stone *et al.* (1989) that TPH is inactivated oxidatively after MDMA administration is one of the most revealing findings made in this field in recent years. These authors showed that TPH activity in cortical homogenates from rats treated acutely with MDMA could be recovered completely by incubation of the homogenates in an anaerobic atmosphere for 20–24 hr in the presence of dithiothreitol (DTT) and Fe^{2+}. This regeneration of enzyme activity *in vitro* is shown in Table I for cortical homogenates from rats treated with saline or MDMA [10 mg/kg, subcutaneously (sc)] 3 hr prior to sacrifice (C. J. Schmidt and V. L. Taylor, unpublished observations). Incubation of control homogenates with DTT alone resulted in the loss of control enzyme activity and additional loss of activity in the MDMA homogenates. However, the addition of Fe^{2+} plus DTT to the homogenates maintained control enzyme activity and regenerated activity from the MDMA-treated rats. These results clearly show that the enzyme protein remains within the serotonergic terminal for some time after its inactivation. Stone *et al.* (1989) showed that active TPH could not be regenerated 3 days after MDMA administration, suggesting that the inactive enzyme is degraded eventually. These results give some insight into the very early neurochemical events that occur in the serotonergic neuron on exposure to MDMA *in vivo*. Clearly TPH is inactivated oxidatively very rapidly after

TABLE I *In Vitro* Regeneration of Tryptophan Hydroxylase Activity in Cortical Homogenates of Methylenedioxymethamphetamine-Treated Rats Using Dithiothreitol and FE^{2+} [a]

Treatment	TPH activity	
	Before incubation	After incubation
Saline	33.1 ± 1.3	28.9 ± 1.4
MDMA	11.1 ± 1.2	26.1 ± 1.2

[a] Rats were treated with 10 mg/kg MDMA and sacrificed at 2 hr. Cortical homogenates were incubated with 6 mM DTT and 50 μM FE^{2+} for 24 hr under nitrogen using the conditions described by Stone *et al.* (1989a). Enzyme activity was assayed according to the $^{14}CO_2$ trapping method described by Schmidt and Taylor (1988) and is presented as nmol tryptophan oxidized/g tissue/hr.

MDMA administration. At lower doses of MDMA, or if a 5-HT uptake inhibitor is added shortly after MDMA administration, endogenous reducing systems apparently can reactivate the enzyme *in vivo*. This hypothesis is based on the rapid recovery of TPH activity under both conditions (Stone *et al.*, 1989a; Schmidt and Taylor, 1990). This recovery is not the result of the synthesis of new enzyme, since the partial recovery of enzyme activity observed 24 hr after MDMA administration was not altered by the administration of cycloheximide (a protein synthesis inhibitor). Under the high doses of MDMA that typically lead to the long-term serotonergic deficits, however, these endogenous reducing systems may be compromised or severely depleted themselves. A complete loss of endogenous antioxidants may be the prerequisite for the development of MDMA-induced neurotoxicity. The lack of effect of MDMA on TPH activity in the raphe region may relate to the presence of greater stores of antioxidants in the cell bodies relative to the terminals (Schmidt and Taylor, 1988). Note that the relatively nontoxic analog MDE produces similar acute effects including the oxidative inactivation of TPH *in vivo*. This result provides additional evidence that the early biochemical changes in the serotonergic neuron, although possibly indicative of oxidative stress, are not irreversible.

3. Dopamine System *In Vivo*

Although the information presented to this point would suggest that the effects of MDMA and its analogs on the dopamine system are inconsequential, this is, in fact, not the case. Although only small postmortem changes in dopaminergic parameters are observed after MDMA administration, considerable data now support the hypothesis that the acute interactions of MDMA with the dopaminergic system are somehow responsible or required for the long-term effects of the drug on the serotonergic system. Although these data are discussed elsewhere in this book, we address here the acute responses of the dopaminergic system to MDMA.

The first demonstration of MDMA-induced dopamine release *in vivo* was by Yamamoto and Spanos (1988) using voltammetry in anesthetized rats. Nash (1990) subsequently demonstrated that MDMA produced a dose-dependent increase in extracellular concentrations of striatal dopamine in the awake, freely moving rat. Dialysis studies have shown (+)-MDMA to be a more potent releaser of DA than the (−) isomer (Hiramatsu and Cho, 1990) and have demonstrated that such release is sensitive to dopamine uptake blockers such as mazindol and nomifensine (Nash and Brodkin, 1991).

One interesting characteristic of MDMA-induced dopamine release *in vivo* is its sensitivity to 5-HT_2 receptor antagonists (Nash, 1990; Schmidt *et al.*, 1992b). Figure 4 provides data showing the elevation of extracellular dopamine concentrations in the striatum of the freely moving rat after MDMA administration. Also shown is the attenuation of MDMA-induced

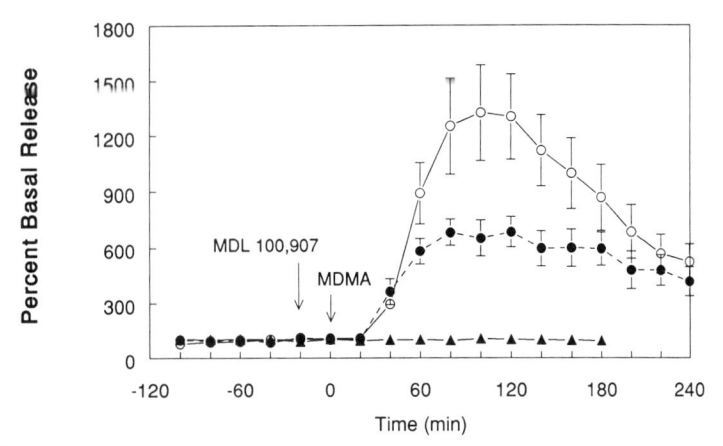

FIGURE 4 Increase in extracellular concentrations of striatal dopamine after methylenedioxymethamphetamine (MDMA) administration (20 mg/kg, ○) as measured by *in vivo* microdialysis in the awake, freely moving rat. Also shown is the attenuation of MDMA-induced release after pretreatment with the selective 5-HT$_2$ antagonist MDL 100,907 (MDMA + MDL 100,907, ●). Note that, although the antagonist had no effect on basal dopamine efflux (1 mg/kg, ▲), it reduced release caused by MDMA by approximately 40%. Reprinted with permission from Schmidt *et al.* (1992b).

release following pretreatment with the selective 5-HT$_2$ receptor antagonist MDL 100,907. This effect on MDMA-induced dopamine release results from the ability of such antagonists to interfere with MDMA-stimulated dopamine synthesis (Nash *et al.*, 1990; Schmidt *et al.*, 1992a,b) and probably explains why agents such as ketanserin and MDL 100,907 prevent the long-term or neurotoxic effects of MDMA (Nash *et al.*, 1990; Schmidt and Kehne, 1990; Schmidt *et al.*, 1990a,1990b).

C. Electrophysiology

Of the three methylenedioxyamphetamines discussed, only MDMA has been characterized in terms of its electrophysiological effects. As would be expected for an agent that increases extracellular concentrations of dopamine, MDMA administration reduces the rate of dopaminergic cell firing (Kelland *et al.*, 1989; Matthews *et al.*, 1989; Schmidt *et al.*, 1992a) although it is much less potent than amphetamine in this regard (Piercey *et al.*, 1990). Like that of amphetamine, the MDMA-induced reduction in dopamine cell firing rates is sensitive to the inhibition of dopamine synthesis with α-methyl-*p*-tyrosine or 5-HT synthesis with PCA (Kelland *et al.*, 1989; Sorenson *et al.*, 1992). As might be anticipated from the preceding discussion, the effects of MDMA on dopaminergic cell firing also can be blocked

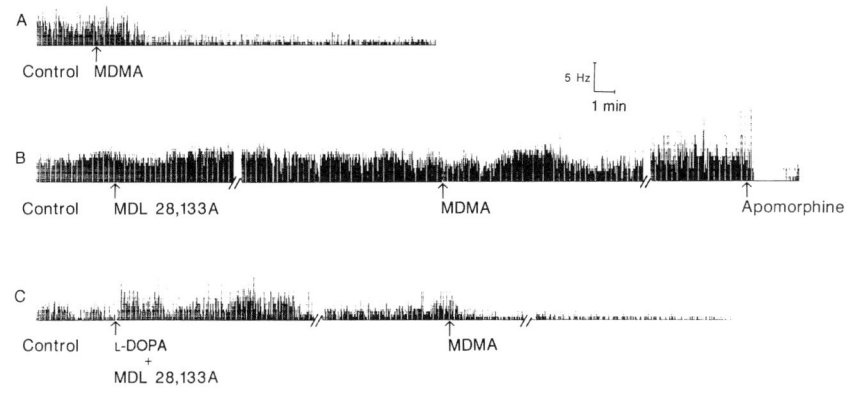

FIGURE 5 (A) Effect of methylenedioxymethamphetamine (MDMA, 15 mg/kg, iv) on the firing rate of a single dopaminergic neuron in the substantia nigra compacta. (B) Administration of the 5-HT$_2$ antagonist MDL 28,133A (1 mg/kg) 20 min prior to MDMA administration completely prevents the response to a subsequent injection of MDMA. (C) Coadministration of the dopamine precursor L-DOPA with the antagonist preserves the inhibitory effect of MDMA on cell firing. Reprinted with permission from Schmidt et al. (1992a).

by pretreatment with 5-HT$_2$ receptor antagonists (Schmidt et al., 1990a). Each of these observations is consistent with the known dependency of carrier-mediated release on the newly synthesized pool of transmitter and by the requirement for 5-HT$_2$ receptor activation in MDMA-induced dopamine synthesis and release. Figure 5 provides the results from an experiment examining the effects of the 5-HT$_2$ receptor antagonist MDL 28,133 on the MDMA-induced slowing of A9 dopamine neurons. The effect of the receptor antagonist was negated when dopamine synthesis was maintained by administration of the dopamine precursor L-DOPA (Schmidt et al., 1992a). Similar results were observed using amphetamine to slow the firing rate of A10 dopaminergic neurons (Sorensen et al., 1992), suggesting in both cases that the site of action of the 5-HT$_2$ receptor antagonists was at the level of dopamine synthesis.

Only one study to date has examined the effects of MDMA on the activity of the serotonergic cell groups in the raphe nuclei or of the norepinephrine-containing cells of the locus coeruleus (Piercey et al., 1990). Both classes of neurons were inhibited potently by peripheral administration of MDMA. This effect was observed at doses lower than those required to affect dopaminergic activity, suggesting a greater sensitivity of these neurons to the transmitter released by MDMA. Interestingly, the dorsal raphe nucleus contained two groups of cells with a 14-fold difference in sensitivity to MDMA. In contrast, the median raphe nucleus contained only the high-sensitivity cell type.

IV. RING-ALKYLATED AMPHETAMINES (FENFLURAMINE)

For the sake of completeness, the anorectic fenfluramine (trifluoromethyl-N-ethylamphetamine) will be discussed as a ring-alkylated amphetamine, although this compound is grouped most often with the halogenated amphetamines. Not surprisingly, many of the acute neurochemical effects of fenfluramine are similar to those of the halogenated and methylenedioxy-substituted amphetamines. However, fenfluramine is a unique drug if only because, of all the indirect agonists discussed, only fenfluramine does not possess the stimulant properties of the parent molecule amphetamine.

The acute neurochemical effects of fenfluramine have been well studied in an attempt to understand its anorectic activity. Although this activity is associated primarily with the d stereoisomer, the drug is used therapeutically as a racemate and both stereoisomers have received extensive study. This fact is particularly important in the case of fenfluramine because the two stereoisomers of the drug appear to have different pharmacological properties. The metabolism of fenfluramine is also an important consideration. The half-life of fenfluramine in the rat is approximately 2 hr, compared with 12 hr for its primary metabolite norfenfluramine, which is also active (Rowland and Carlton, 1986). Both fenfluramine and norfenfluramine are believed to exert their anorectic activity by increasing central serotonergic activity; this interaction has been the focus of most studies. For a comprehensive review of the neurochemical and behavioral effects of fenfluramine, see Rowland and Carlton (1986).

A. *In Vitro* Neurochemistry

Fenfluramine is a potent 5-HT-releasing agent *in vitro* (Maura *et al.*, 1982; Gobbi *et al.*, 1992). This compound is extremely selective in this regard, particularly at the low brain concentrations of 1–10 μM that are believed to be achieved at anorectic doses in rats (Rowland and Carlton, 1986). Concentrations greater than 10 μM have been reported to produce dopamine release *in vitro* (Duhault *et al.*, 1980; Liang and Rutledge, 1982), although fenfluramine is by far the most selective of all the releasing agents discussed (Schmidt *et al.*, 1991a). This lack of effect on dopamine release is probably responsible for the absence of stimulant activity noted for fenfluramine. Typical of carrier-mediated release *in vitro*, fenfluramine-induced release is Ca^{2+} independent (Langer and Moret, 1982) and sensitive to 5-HT uptake inhibitors such as chlorimipramine (Maura *et al.*, 1982). d-Fenfluramine is reported to be more potent than the l isomer at releasing 5-HT from synaptosomes (Garattini *et al.*, 1979).

Although both fenfluramine and norfenfluramine produce their behavioral effects by releasing 5-HT, data suggest that the two drugs may accomplish this end by different mechanisms. Borroni *et al.* (1983) used synaptosomal systems to provide evidence that norfenfluramine acts primarily by releasing 5-HT whereas fenfluramine increases synaptic transmitter concentrations by inhibiting the 5-HT uptake carrier and through displacement of transmitter from a reserpine-sensitive pool. Fuller *et al.* (1988) concluded that fenfluramine acts primarily as a direct releasing agent *in vivo* rather than an uptake inhibitor, based on its inability to block PCA-induced depletions of 5-HT. However, the relatively short half-life of fenfluramine in comparison to that of norfenfluramine should be taken into account when interpreting these studies.

B. *In Vivo* Neurochemistry

Fenfluramine-induced 5-HT release first was monitored *in vivo* by examining the elevation of the 5-HIAA:5-HT ratio as a function of time after drug administration. Changes in this ratio indicate that transmitter efflux is enhanced for approximately 2 hr following drug administration (Rowland and Carlton, 1986). This result correlates extremely well with results from microdialysis studies (Laferrere and Wurtman, 1989; Sabol *et al.*, 1992), although such a short duration of action is curious given the long half-life of norfenfluramine in the rat brain. However, fenfluramine-induced 5-HT release may be a self-limiting process because of the rapid decline in brain 5-HT concentrations that occurs following administration of the drug (Fig. 6). The ability of fenfluramine to reduce brain concentrations of 5-HT acutely in laboratory animals was reported first almost three decades ago (see review by Rowland and Carlton, 1986). Although the metabolism of fenfluramine to norfenfluramine is a confounding feature of such experiments, the acute decrease in 5-HT concentrations is observed after the administration of either agent. Fuller *et al.* (1978) reported that norfenfluramine produced a more rapid depletion of 5-HT than did the parent drug. Invernizzi *et al.* (1991) showed that direct icv infusion of either *d*-fenfluramine or *d*-norfenfluramine could produce acute depletion in brain 5-HT concentrations by 4 hr. However, either treatment yielded similar brain concentrations of *d*-norfenfluramine, indicating that peripheral metabolism of fenfluramine was occurring despite the route of administration.

As already noted, many of the neurochemical characteristics of acute fenfluramine administration are identical to those already described for MDMA. *d*- and *l*-Fenfluramine were reported to be equipotent in producing 5-HT depletions (Duhault *et al.*, 1979), although Invernizzi *et al.* (1986) noted that the *d* isomer may be more potent in reducing 5-HT concentrations and in inhibiting the accumulation of 5-hydroxytryptophan (5-HTP) after decarboxylase inhibition. Predictably, the effect of fenfluramine on 5-HT concentrations is blocked by inhibitors of the 5-HT uptake carrier (Steranka and Sanders-Bush, 1979). Delayed administration of the uptake inhibitor also reverses fenfluramine-induced 5-HT depletion (Clineschmidt

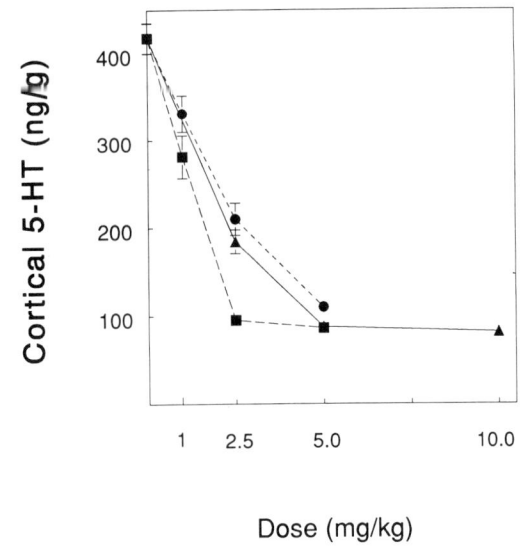

FIGURE 6 Comparison of the dose-dependent depletion of cortical 5-hydroxytryptamine (5-HT) produced by methylenedioxymethamphetamine (MDMA, ▲), fenfluramine (●), p-chloroamphetamine (PCA, ■) 3 hr after drug administration. Data are expressed as the mean ± SEM (C. J. Schmidt and V. L. Taylor, unreported observation).

et al., 1978), as demonstrated for the deficits produced by either MDMA (Schmidt and Taylor, 1990) or PCA (Fuller *et al.*, 1975). This result suggests that the acute effect of all three agents depends on the continuous release of transmitter or uptake of drug. The decrease in transmitter levels after fenfluramine administration results not only from the increase in 5-HT efflux but from a simultaneous decrease in synthesis as well. The latter has been demonstrated as a fenfluramine-induced reduction in the accumulation of 5-HTP in NSD 1015-treated rats (Fuller and Perry, 1983). This acute loss of TPH activity can be measured *ex vivo* as well, and appears to be caused by a reversible oxidation of the enzyme similar to that demonstrated after PCA, MDMA, or methamphetamine administration (Stone *et al.*, 1989a,1989b). Although at low doses (< 15 mg/kg) this effect appears to be reversible (Duhault *et al.*, 1975), higher doses of fenfluramine produce the same long-term neurochemical changes in the serotonergic system observed with high doses of PCA and MDMA. These long-term effects will not be discussed here, but have been reviewed elsewhere (Johnson and Nichols, 1990).

Despite the specificity of fenfluramine for the serotonergic system *in vitro*, considerable biochemical data now indicate that high doses of fenfluramine do have effects on the dopaminergic system *in vivo*. The first such effect to be observed was an acute increase in the concentration of the dopamine metabolite HVA after the administration of fenfluramine (DOPAC levels increased as well). In contrast to the biochemical effects of fenfluramine on the serotonergic system and the behavioral effects such as anorexia, the *l* isomer of fenfluramine is primarily responsible for the effects of the drug on dopamine metabolism. The available data suggest that the *d* and *l* isomers of fenfluramine may affect the dopaminergic system by different mechanisms. The modest effects of *d*-fenfluramine on striatal HVA concentrations are prevented by the 5-HT uptake inhibitor LM 5008 or by prior depletion of 5-HT with PCA. In contrast, the effect of *l*-fenfluramine on striatal dopamine metabolism is not blocked by the 5-HT uptake inhibitor or by 5-HT depletion but is antagonized by the dopamine agonist piribedil (Crunelli *et al.*, 1980). Ligand binding studies indicate that fenfluramine has virtually no affinity for the D_2 receptor (Burt *et al.*, 1976; Invernizzi *et al.*, 1989). These observations have led to the suggestion that the effect of *l*-fenfluramine on dopamine metabolism is not due to 5-HT release but to a "functional" blockade of dopamine receptors. *In vivo* dialysis studies confirm that *l*-fenfluramine (10 mg/kg) increases striatal dopamine efflux by a firing rate-dependent mechanism similar to that of haloperidol (Bettini *et al.*, 1987). Further, the effect of *l*-fenfluramine on striatal HVA concentrations (Invernizzi *et al.*, 1989) and on dopamine release *in vivo* (Bettini *et al.*, 1987) shows cross-tolerance with haloperidol following chronic treatment with the D_2 antagonist. Finally, antagonism of some (but not all) aspects of the stereotyped behavior produced by amphetamine or apomorphine also has been observed with fenfluramine. The *l* isomer of fenfluramine is more potent in this effect, as might be expected. Note, however, that although haloperidol was equipotent at blocking the behavior induced by either apomorphine or amphetamine, fenfluramine was more potent against behaviors induced by amphetamine, a result that may indicate a presynaptic effect of fenfluramine on the dopaminergic neuron (Bendotti *et al.*, 1980).

Rowland and Carlton (1986) reported that fenfluramine produces a small decrease in telencephalic dopamine synthesis, as measured by DOPA accumulation, whereas the remainder of the brain shows a slight increase in synthesis. In our own studies of fenfluramine, we observed an increase in dopamine synthesis in both the striatum and the nucleus accumbens that appears to be caused only by the *l* isomer (C. J. Schmidt and V. L. Taylor, unpublished observations). Although these results may explain the increase in DOPAC and HVA concentrations also observed with fenfluramine, additional studies are necessary to confirm this relationship.

Very little work has been done on the electrophysiological effects of fenfluramine. Scuvee-Moreau and Dreese (1990) compared the effects of *d*- and *l*-fenfluramine on the firing rate of neurons in the dorsal raphe nucleus, the locus coeruleus, and the A10 region of the midbrain. Although neither stereoisomer affected the firing rate of dopaminergic neurons in the A10 region at cumulative doses up to 5 mg/kg, serotonergic firing rates were inhibited by both isomers; *d*-fenfluramine was more potent. Only the *d* isomer of fenfluramine reduced cell firing in the noradrenergic neurons of the locus coeruleus.

V. FUTURE DIRECTIONS

The indirect agonist properties of amphetamine and its ring-substituted analogs provide a means of assessing the neurochemical and behavioral effects of increased transmitter release from one or more of the monoaminergic systems. The availability of agents with differing degrees of specificity for these neurotransmitters allows some degree of selection for the transmitter system that is affected. For example, although amphetamine certainly releases 5-HT, most of its effects are attributable to its release of dopamine. To a similar extent, the effects of increased serotonin release can be examined by administering d-fenfluramine or low doses of MDE. Agents such as MDMA facilitate assessing the effects of increases in the release of both 5-HT and dopamine. This range of responses makes the amphetamines invaluable tools for studying the biochemistry, physiology, and behavioral pharmacology of the monoaminergic systems and, to some extent, the transmitter systems on which they act.

One of the most intriguing questions regarding the pharmacology of the ring-substituted amphetamines is what role their acute effects play in the development of the long-term neurochemical deficits they produce in the dopaminergic and serotonergic systems. Clearly some relationship must exist based on the observation that any manipulation that interferes with the releasing action of the amphetamines also prevents or attenuates their long-term effects. The selectivity of the drug for neurotransmitter release appears to correlate with the specificity of its long-term effects. For example, the long-term neurochemical deficits produced by amphetamine are restricted largely to the dopaminergic system. Methamphetamine, which appears to produce somewhat more serotonin release than amphetamine, produces long-term deficits in both the dopaminergic and the serotonergic systems of the rat. For the methylenedioxyamphetamines, a clear bias exists toward the serotonergic system, in terms of both transmitter release and long-term effects. Although the basis of this relationship is not currently clear, it certainly will be critical to understanding the mechanism of the long-term effects as well as the selective vulnerability of these transmitter systems to neurotoxic insult.

Finally, the acute effect of agents such as MDMA, PCA, and fenfluramine on TPH activity is an area requiring further study. The selective oxidative inactivation of TPH by these agents is a unique phenomenon. That this response is reversible both *in vitro* and *in vivo*, as well as the rate of TPH inactivation following drug administration, suggests that the indirect agonists may be interfering with an endogenous process. The redox cycling of enzymes between active and inactive states has been proposed as a regulatory mechanism in some biochemical systems (Holmgren, 1989). Perhaps such a mechanism operates to regulate the activity of TPH. The sensitivity of TPH to oxidative inactivation and reactivation certainly suggests that the enzyme is susceptible to regulation by such a mechanism. Determining which intracellular reducing systems could serve to maintain TPH in its active state would be of interest. Glutathione represents the major intracellular source of reducing equivalents whereas the thioredoxin system has been shown to be very effective at reducing protein disulfides (Holmgren, 1989). Even if such a system can be identified or implicated, the primary question remains—What causes these systems to be depleted or inoperative in rats treated with ring-substituted amphetamine analogs?

REFERENCES

Anderson, G. M., Braun, G., Braun, U., Nichols, D. E., and Shulgin, A. T. (1978). Absolute configuration and psychotomimetic activity. *In* "Quasar Research Monograph 22" (G. Barnett, M. Trsic, and R. Willette, eds.), pp. 8–15. National Institute on Drug Abuse, Washington, D.C.

Battaglia, G., Brooks, B. P., Kulsakdinun, C., and De Souza, E. B. (1988). Pharmacologic profile of MDMA (3,4-methylenedioxymethamphetamine) at various brain recognition sites. *Eur. J. Pharmacol.* **149**, 159–163.

Bendotti, C., Borsini, F., Zanini, M. G., Samanin, R., and Garattini, S. (1980). Effect of fenfluramine and norfenfluramine stereoisomers on stimulant effects of *d*-amphetamine and apomorphine in the rat. *Pharmacol. Res. Commun.* **12**, 567–575.

Bettini, E., Ceci, A., Spinelli, R., and Samanin, R. (1987). Neuroleptic-like effects of the *l*-isomer of fenfluramine on striatal dopamine release in freely moving rats. *Biochem. Pharmacol.* **14**, 2387–2391.

Borroni, E., Ceci, A., Garattini, S., and Mennini, T. (1983). Differences between *d*-fenfluramine and *d*-norfenfluramine in serotonin presynaptic mechanisms. *J. Neurochem.* **40**, 891–893.

Braun, U., Shulgin, A. T., and Braun, G. (1980). Centrally active N-substituted analogs of 3,4-methylenedioxyphenylisopropylamine (3,4 methylenedioxyamphetamine). *J. Pharm. Sci.* **69**, 192–195.

Burt, D. R., Creese, I., and Snyder, S. H. (1976). Properties of [^{3}H]haloperidol and [^{3}H]dopamine binding associated with dopamine receptors in calf brain membranes. *Mol. Pharmacol.* **12**, 800–812.

Clineschmidt, B. V., Zacchei, A. G., Totaro, J. A., Pflueger, A. B., McGuffin, J. C., and Wishousky, T. I. (1978). Fenfluramine and brain serotonin. *Ann. N.Y. Acad. Sci.* **308**, 222–241.

Crunelli, V., Bernasconi, S., and Samanin, R. (1980). Effects of *d*- and *l*-fenfluramine on striatal homovanillic acid concentrations in rats after pharmacological manipulation of brain serotonin. *Pharmacol. Res. Commun.* **12**, 215–223.

Duhault, J., Malen, C., Boulanger, M., Voisin, C., Beregi, V., and Schmidt, H. (1975). Fenfluramine and 5-hydroxytryptamine. Part 1. *Arzneim. Forsch.* **25**, 1755–1762.

Duhault, J., Beregi, V., and Du Boistesselin, R. (1979). General and comparative pharmacology of fenfluramine. *Curr. Med. Res. Opin. (Suppl.)* **6**, 3–14.

Duhault, J. Beregi, V., and Roman, R. (1980). Substituted phenylethylamines and anorexia. *Prog. Neuropsychopharmacol.* **4**, 341–349.

Fischer, J. F., and Cho, A. K. (1979). Chemical release of dopamine from striatal homogenates: Evidence for an exchange-diffusion model. *J. Pharmacol. Exp. Ther.* **208**, 203–209.

Fitzgerald, J. L., and Reid, J. J. (1990). Effects of methylenedioxymethamphetamine on the release of monoamines from rat brain slices. *Eur. J. Pharmacol.* **191**, 217–220.

Fuller, R. W., and Perry, K. W. (1983). Decreased accumulation of brain 5-hydroxytryptophan after decarboxylase inhibition in rats treated with fenfluramine, norfenfluramine or *p*-chloroamphetamine. *J. Pharm. Pharmacol.* **35**, 597–598.

Fuller, R. W., Perry, K. W., and Molloy, B. B. (1975). Reversible and irreversible phases of serotonin depletion by 4-chloroamphetamine. *Eur. J. Pharmacol.* **33**, 119–124.

Fuller, R. W., Snoddy, H. D., and Hemrick, S. K. (1978). Effects of fenfluramine and norfenfluramine on brain serotonin metabolism in rats. *Proc. Soc. Exp. Biol. Med.* **157**, 202–205.

Fuller, R. W., Snoddy, H. D., and Robertson, D. W. (1988). Mechanisms of effects of *d*-fenfluramine on brain serotonin metabolism in rats: Uptake inhibition versus release. *Pharmacol. Biochem. Behav.* **30**, 715–721.

Garattini, S., Caccia, S., Mennini, T., Samanin, R., Consolo, S., and Ladinsky, H. (1979). Biochemical pharmacology of the anorexic drug fenfluramine: A review. *Curr. Med. Res. Opin. (Suppl.)* **6**, 15–27.

Glennon, R. A. (1984). Hallucinogen phenylisopropylamines: Stereochemical aspects. In "CRC Handbook of Stereoisomers: Drugs in Psychopharmacology" (D. F. Smith, ed.), pp. 327–368. CRC Press, Boca Raton, Florida.

Glennon, R. A. (1989). Stimulus properties of hallucinogenic phenylalkylamines and related designer drugs: Formulation of structure activity relationships. *NIDA Res. Monogr.* **94**, 43–67.

Glennon, R. A., Titeler, M., and Young, R. (1986). Structure activity relationships and mechanisms of actions of hallucinogenic agents based on drug discrimination and radioligand binding studies. *Psychopharmacol. Bull.* **22**, 953–958.

Gobbi, M., Frittoli, E., Mennini, T., and Garattini, S. (1992). Releasing activities of d-fenfluramine and fluoxetine on rat hippocampal synaptosomes preloaded with [^{3}H]serotonin. *Naunyn-Schmiedeberg's Arch. Pharmacol.* **345**, 1–6.

Hekmatpanah, C. R., and Peroutka, S. J. (1990). 5-Hydroxytryptamine uptake blockers attenuate the 5-hydroxytryptamine-releasing effect of 3,4-methylenedioxymethamphetamine and related agents. *Eur. J. Pharmacol.* **177**, 95–98.

Hiramatsu, M., and Cho, A. K. (1990). Enantiomeric differences in the effects of 3,4-methylenedioxymethamphetamine on extracellular monoamines and metabolites in the striatum of freely-moving rats: An *in vivo* microdialysis study. *Neuropharmacology* **29**, 269–275.

Holmgren, A. (1989). Thioredoxin and glutaredoxin systems. *J. Biol. Chem.* **264**, 13963–13966.

Invernizzi, R., Berettera, C., Garattini, S., and Samanin, R. (1986). D- and L-isomers of fenfluramine differ markedly in their interaction with brain serotonin and catecholamines in the rat. *Eur. J. Pharmacol.* **120**, 9–15.

Invernizzi, R., Bertorelli, R., Consolo, S., Garattini, S., and Samanin, R. (1989). Effects of the *l*-isomer of fenfluramine on dopamine mechanisms in rat brain: Further studies. *Eur. J. Pharmacol.* **164**, 241–248.

Invernizzi, R., Fracasso, C., Caccia, S., Garattini, S., and Samanin, R. (1991). Effects of intracerebroventricular administration of *d*- fenfluramine and *d*-norfenfluramine, as a single injection or 2-HR infusion, on serotonin in brain: Relationship to concentrations of drugs in brain. *Neuropharmacology* **30**, 119–123.

Johnson, M. P., and Nichols, D. E. (1990). Comparative serotonin neurotoxicity of the stereoisomers of fenfluramine and norfenfluramine. *Pharmacol. Biochem. Behav.* **36**, 105–109.

Johnson, M. P., Hoffman, A. J., and Nichols, D. E. (1986). Effects of the enantiomers of MDA,

MDMA and related analogues on [³H]serotonin and [³H]dopamine release from superfused rat brain slices. *Eur. J. Pharmacol.* **132**, 269–276.

Johnson, M., Hanson, G. R., and Gibb, J. W. (1987). Effects of N-ethyl-3,4-methylenedioxyamphetamine (MDE) on central serotonergic and dopaminergic systems of the rat. *Biochem. Pharmacol.* **36**, 4085–4093.

Kelland, M. D., Freeman, A. S., and Chiodo, L. A. (1989). (±)3,4-Methylenedioxymethamphetamine-induced changes in the basal activity and pharmacological responsiveness of nigrostriatal dopamine neurons. *Eur. J. Pharmacol.* **169**, 11–21.

Laferrere, B., and Wurtman, R. J. (1989). Effect of D-fenfluramine on serotonin release in brains of anaesthetized rats. *Brain Res.* **504**, 258–263.

Langer, S. Z., and Moret C. (1982). Citalopram antagonizes the stimulation by lysergic acid diethylamide of presynaptic inhibitory serotonin autoreceptors in the rat hypothalamus. *J. Pharmacol. Exp. Ther.* **222**, 220–226.

Liang, N. Y., and Rutledge, C. O. (1982). Comparison of the release of [³H]dopamine from isolated corpus striatum by amphetamine, fenfluramine and unlabelled dopamine. *Biochem. Pharmacol.* **31**, 983–992.

Loh, H. H., and Tseng, L. (1979). Role of biogenic amines in the actions of monomethoxyamphetamine. In "Psychopharmacology of Hallucinogens" (R. C. Stillman and R. E. Willete, eds.), pp. 13–22. Pergamon Press, New York.

Lyon, R. A., Glennon, R. A., and Titeler, M. (1986). 3,4-Methylenedioxymethamphetamine (MDMA): Stereo-selective interactions at brain 5-HT$_1$ and 5-HT$_2$ receptors. *Psychopharmacology* **88**, 525–526.

Marquardt, G. M., DiStefano, U., and Ling, L. L. (1978). Pharmacological effects of (±), (S)- and (R)-MDA. In "The Psychopharmacology of Hallucinogens" (R. C. Stillman and R. E. Willette, eds.), pp. 84–103. Pergamon Press, New York.

Matthews, R. T., Champney, T. H., and Frye, G. D. (1989). Effects of (±)3,4-methylenedioxymethamphetamine (MDMA) on brain dopaminergic activity in rats. *Pharmacol. Biochem. Behav.* **33**, 741–747.

Maura, G., Gemignani, A., Versace, P., Martire, M., and Raiteri, M. (1982). Carrier-mediated and carrier-independent release of serotonin from isolated central nerve endings. *Neurochem. Int.* **4**, 219–224.

Nash, J. F. (1990). Ketanserin pretreatment attenuates MDMA-induced dopamine release in the striatum as measured by *in vivo* microdialysis. *Life Sci.* **47**, 2401–2408.

Nash, J. F., and Brodkin, J. (1991). Microdialysis studies on 3,4-methylenedioxymethamphetamine-induced dopamine release: Effect of dopamine uptake inhibitors. *J. Pharmacol. Exp. Ther.* **259**, 820–825.

Nash, J. F., and Nichols, D. E. (1991). Microdialysis studies on 3,4-methylenedioxyamphetamine and structurally related analogues. *Eur. J. Pharmacol.* **200**, 53–58.

Nash, J. F., Meltzer, H. Y., and Gudelsky, G. A. (1990). Effect of 3,4-methylenedioxymethamphetamine on 3,4-dihydroxyphenylalanine accumulation in the striatum and nucleus accumbens. *J. Neurochem.* **54**, 1062–1067.

Nichols, D. E., and Glennon, R. A. (1984). Medicinal chemistry and structure activity relationships of hallucinogens. In "Hallucinogens: Neurochemical, Behavioral and Clinical Perspectives" (B. L. Jacobs, ed.), pp. 95–142. Raven Press, New York.

Nichols, D. E., and Oberlender, R. (1989). Structure–activity relationships of MDMA-like substances. *NIDA Res. Monogr.* **94**, 1–29.

Nichols, D. E., Lloyd, D. H., Hoffman, A. J., Nichols, M. B., and Yim, G. K. W. (1982). Effect of certain hallucinogenic amphetamine analogues on the release of [³H]serotonin from rat brain synaptosomes. *J. Med. Chem.* **25**, 530–535.

Pierce, P. A., and Peroutka, S. J. (1988). Ring-substituted amphetamine interactions with neurotransmitter receptor binding sites in human cortex. *Neurosci. Lett.* **95**, 208–212.

Piercey, M. F., Lum, J. T., and Palmer, J. R. (1990). Effects of MDMA ('ecstasy') on firing rates

of serotonergic, dopaminergic, and noradrenergic neurons in the rat. *Brain Res.* **526,** 203–206.
Ricaurte, G. A., Finnegan, K. F., Nichols, D. F., Delanney, L. E., Irwin, I., and Langston, J. W. (1987). 3,4-Methylenedioxyethylamphetamine (MDE), a novel analogue of MDMA, produces long-lasting depletions of serotonin in the rat brain. *Eur. J. Pharmacol.* **137,** 265–268.
Rowland, N. E., and Carlton, J. (1986). Neurobiology of an anorectic drug: Fenfluramine. *Progr. Neurobiol.* **27,** 13–62.
Rudnick, G., and Wall, S. C. (1992). The molecular mechanism of "ecstasy" [3,4-methylenedioxymethamphetamine (MDMA)]: Serotonin transporters are targets for MDMA-induced serotonin release. *Proc. Natl. Acad. Sci. U.S.A.* **89,** 1817–1821.
Sabol, K. E., Richards, J. B., and Seiden, L. S. (1992). Fenfluramine-induced increases in extracellular hippocampal serotonin are progressively attenuated *in vivo* during a four-day fenfluramine regimen in rats. *Brain Res.* **571,** 64–72.
Schmidt, C. J. (1987a). Acute administration of methylenedioxymethamphetamine: Comparison with the neurochemical effects of its N-desmethyl and N-ethyl analogs. *Eur. J. Pharmacol.* **136,** 81–88.
Schmidt, C. J. (1987b). Neurotoxicity of the psychedelic amphetamine, methylenedioxymethamphetamine. *J. Pharmacol. Exp. Ther.* **240,** 1–7.
Schmidt, C. J., and Kehne, J. H. (1990). Neurotoxicity of MDMA: Neurochemical effects. *Ann. N.Y. Acad. Sci.* **600,** 665–681.
Schmidt, C. J., and Taylor, V. L. (1987). Depression of rat brain tryptophan hydroxylase activity following the acute administration of methylenedioxymethamphetamine. *Biochem. Pharmacol.* **36,** 4095–4102.
Schmidt, C. J., and Taylor, V. L. (1988). Direct central effects of acute methylenedioxymethamphetamine on serotonergic neurons. *Eur. J. Pharmacol.* **156,** 121–131.
Schmidt, C. J., and Taylor, V. L. (1990). Reversal of the acute effects of 3,4-methylenedioxymethamphetamine by 5-HT uptake inhibitors. *Eur. J. Pharmacol.* **181,** 133–136.
Schmidt, C. J., Wu, L., and Lovenberg W. (1986). Methylenedioxymethamphetamine: A potentially neurotoxic amphetamine analogue. *Eur. J. Pharmacol.* **124,** 175–178.
Schmidt, C. J., Levin, J. A., and Lovenberg, W. (1987). *In vitro* and *in vivo* neurochemical effects of methylenedioxymethamphetamine on striatal monoaminergic systems in the rat brain. *Biochem. Pharmacol.* **36,** 747–755.
Schmidt, C. J., Black, C. K., and Taylor, V. L. (1990a). Antagonism of the neurotoxicity due to a single administration of methylenedioxymethamphetamine. *Eur. J. Pharmacol.* **181,** 59–70.
Schmidt, C. J., Abbate, G. M., Black, C. K., and Taylor, V. L. (1990b). Selective 5-hydroxytryptamine$_2$ receptor antagonists protect against the neurotoxicity of methylenedioxymethamphetamine in rats. *J. Pharmacol. Exp. Ther.* **255,** 478–483.
Schmidt, C. J., Black, C. K., and Taylor, V. L. (1991a). L-DOPA potentiation of the serotonergic deficits due to a single administration of 3,4-methylenedioxymethamphetamine, p-chloroamphetamine or methamphetamine to rats. *Eur. J. Pharmacol.* **203,** 41–49.
Schmidt, C. J., Taylor, V. L., Abbate, G. M., and Niedzak, T. R. (1991b). 5-HT$_2$ Antagonists stereoselectively prevent the neurotoxicity of 3,4-methylenedioxymethamphetamine by blocking the acute stimulation of dopamine synthesis: Reversal by L-DOPA. *J. Pharmacol. Exp. Ther.* **256,** 230–235.
Schmidt, C. J., Black, C. K., Taylor, V. L., Fadayel, G. M., Humphreys, T. M., Niedzak, T. R., and Sorensen, S. M. (1992a). THe 5-HT$_2$ receptor antagonist, MDL 28,133A, disrupts the serotonergic-dopaminergic interaction mediating the neurochemical effects OF MDMA. *Eur. J. Pharmacol.* **220,** 151–159.
Schmidt, C. J., Fadayel, G. M., Sullivan, C. K., and Taylor, V. L. (1992b). 5-HT$_2$ Receptors exert a state dependent regulation of dopaminergic function: Studies with MDL 100,907 and the amphetamine analogue, 3,4-methylenedioxymethamphetamine. *Eur. J. Pharmacol.* **223,** 65–74.

Scuvee-Moreau, J., and Dreese, A. (1990). Influence of fenfluramine and norfenfluramine stereoisomer on the firing rate of central monoaminergic neurons in the rat. *Eur. J. Pharmacol.* **179**, 211–215.

Shannon, M., Battaglia, G., Glennon, R. A., and Titeler, M. (1984). 5-HT$_1$ and 5-HT$_2$ Binding properties of derivatives of the hallucinogen 1-(2,5-dimethoxyphenyl)-2-aminopropane (2,5-DMA). *Eur. J. Pharmacol.* **102**, 23–29.

Shulgin, A. T. (1978). Psychotomimetic drugs: Structure–activity relationships. In "Handbook of Psychopharmacology VII" (L. L. Iversen, S. D. Iversen, and S. H. Snyder, eds.), pp. 243–333. Plenum Press, New York.

Shulgin, A. T., Sargent, T., and Narango, C. (1969). Structure activity relationships of one-ring psychotomimetics. *Nature (London)* **216**, 537–541.

Sorensen, S. M., Humphreys, T. M., Taylor, V. L., and Schmidt, C. J. (1992). 5-HT$_2$ receptor antagonists reverse amphetamine-induced slowing of dopaminergic neurons by interfering with stimulated dopamine synthesis. *J. Pharmacol. Exp. Ther.* **260**, 872–878.

Sprouse, J. S., Bradberry, C. W., Roth, R. H., and Aghajanian, G. K. (1989). MDMA (3,4-methylenedioxymethamphetamine) inhibits the firing of dorsal raphe neurons in brain slices via release of serotonin. *Eur. J. Pharmacol.* **167**, 375–383.

Steele, T. D., Nichols, D. E., and Yim, G. K. W. (1987). Stereochemical effects of 3,4-methylenedioxymethamphetamine (MDMA) and related amphetamine derivatives on inhibition of uptake of [^{3}H]monoamines into synaptosomes from different regions of rat brain. *Biochem. Pharmacol.* **36**, 2297–2303.

Steranka, L. R., and Sanders-Bush, E. (1979). Long-term effects of fenfluramine on central serotonergic mechanisms. *Psychopharmacology* **18**, 895–903.

Stone, D. M., Stahl, D. C., Hanson, G. R., and Gibb, J. W. (1986). The effects of 3,4-methylenedioxymethamphetamine (MDMA) and 3,4-methylenedioxyamphetamine (MDA) on monoaminergic systems in the rat brain. *Eur. J. Pharmacol.* **128**, 41–48.

Stone, D. M., Merchant, K. M., Hanson, G. R., and Gibb, J. W. (1987). Immediate and long-term effects of 3,4-methylenedioxymethamphetamine on serotonin pathways in brain of rat. *Neuropharmacology* **26**, 1677–1683.

Stone, D. M., Hanson, G. R., and Gibb, J. W. (1989a). *In vitro* reactivation of rat cortical tryptophan hydroxylase following *in vivo* inactivation by methylenedioxymethamphetamine. *J. Neurochem.* **53**, 572–581.

Stone, D. M., Johnson, M., Hanson, G. R., and Gibb, J. W. (1989b). Acute inactivation of tryptophan hydroxylase by amphetamine analogs involves the oxidation of sulfhydryl sites. *Eur. J. Pharmacol.* **172**, 93–97.

Titeler, M., Leonhardt, S., Appel, N. M., DeSouza, E. B., and Glennon, R. A. (1990). Receptor pharmacology of MDMA and related hallucinogens. *Ann. N.Y. Acad. Sci.* **600**, 626–639.

Tseng, L. (1978). Effects of para-methoxyamphetamine and 2,5-dimethoxyamphetamine on serotonergic mechanisms. *Naunyn-Schmiedeberg's Arch. Pharmacol.* **304**, 101–105.

Tseng, L. (1979). 5-Hydroxytryptamine uptake inhibitors block *para*-methoxyamphetamine-induced 5-HT release. *Br. J. Pharmacol.* **66**, 185–190.

Tseng, L., Menon, M. K., and Loh, H. H. (1976). Comparative actions of monomethoxy-amphetamines on the release and uptake of biogenic amines in brain tissue. *J. Pharmacol. Exp. Ther.* **197**, 263–271.

Wang, S. S., Ricaurte, G. A., and Peroutka, S. J. (1987). [^{3}H]3,4-Methylenedioxymethamphetamine (MDMA) interactions with brain membranes and glass fiber filter paper. *Eur. J. Pharmacol.* **138**, 439–443.

Yamamoto, B. K., and Spanos, L. J. (1988). The acute effects of methylenedioxymethamphetamine on dopamine release in the awake-behaving rat. *Eur. J. Pharmacol.* **148**, 195–203.

Young, R., and Glennon, R. A. (1986). Discriminative stimulus properties of amphetamine and structurally related phenylalkylamines. *Med. Res. Rev.* **6**, 99–130.

Zaczek, R., Hart, S., Culp, S., and DeSouza, E. B. (1989). Characterization of brain interactions with methylenedioxyamphetamine and methylenedioxymethamphetamine. *NIDA Res. Monogr.* **94**, 223–239.

Mark A. Geyer
Clifton W. Callaway

6

Behavioral Pharmacology of Ring-Substituted Amphetamine Analogs

I. INTRODUCTION

Acute administrations of phenalkylamine drugs produce psychopharmacological effects that can be divided into at least three categories. First, the primary behavioral effects of the psychostimulant drugs exemplified by amphetamine consist of increased arousal and motor activation. Many studies suggest that these effects involve the drug-induced release of catecholamines (Creese and Iversen, 1975; Kelly et al., 1975). Second, ring-substituted derivatives of amphetamine, such as mescaline or 2,5-dimethoxy-4-methylphenylisopropylamine (DOM), produce hallucinogen-like perceptual distortions without significant motor activation (Shulgin, 1978). Hallucinogenic phenalkylamines are believed to produce their characteristic behavioral changes by direct interactions at serotonin (5-HT) receptors, particularly receptors of the 5-HT$_2$ subtype (i.e., the class that now includes receptors called 5-HT$_{2A}$ and 5-HT$_{2C}$, which previously were called 5-HT$_2$ and 5-HT$_{1C}$, respectively) (Glennon, 1985; Wing et al., 1990). Third, phenalkylamines resembling 3,4-methylenedioxymethamphetamine (MDMA) produce alterations in affect and in the perception of emotions that are unlike the behavioral effects of either amphetamine or DOM (Anderson et al., 1978; Shulgin, 1978; Nichols et al., 1982). Several findings indicate that the primary behavioral effects of MDMA-like drugs are attributable to their prominent serotonin-releasing properties. This chapter focuses on the behavioral alterations produced by the pharmacological actions of methylenedioxy-substituted amphetamines and examines the relationship between the unique behavioral effects of these drugs and release of presynaptic serotonin.

In part because of the phenalkylamine structure common to amphetamine-like psychostimulants and mescaline-like hallucinogens, the pharmacological actions of the MDMA-like drugs were thought to be similar to both the typical amphetamines and the traditional hallucinogens (Nozaki et al., 1977; Dimpfel et al., 1989). Clearly, however, variations in the location and identity of substituent groups profoundly alter the effects of these drugs (Glennon, 1989). The unusual behavioral properties of 3,4-methylenedioxyamphetamine (MDA), a popular drug of abuse for about 20 years (Nichols et al., 1989), and of its more recently abused congeners MDMA ("Ecstasy") and MDEA (3,4-methylenedioxy-N-ethylamphetamine; MDE or "Eve") do not involve the profound sensory disruptions characteristic of the hallucinogens (Naranjo et al., 1980), but include powerful alterations in emotions, empathy, and affiliative bonds with other persons instead. These effects are accompanied by side effects such as tachycardia, dry mouth, bruxism, and trismus. In humans, the α-ethyl homolog of MDMA, N-methyl-1-(1,3-benzodioxol-5-yl)-2-butanamine (MBDB), has effects that are qualitatively similar to, although less potent than, those of MDMA (Nichols, 1986). The behavioral effects of these drugs are distinctly different from the effects of amphetamine, as supported by the reports of experienced humans (Anderson et al., 1978; Greer and Strassman, 1985; Shulgin, 1986) and by the different discriminative stimulus properties of these drugs in drug discrimination paradigms (Oberlender and Nichols, 1988; Glennon, 1989; Nichols et al., 1989). As reviewed in this chapter, systematic behavioral studies in animals have confirmed the differences reported in humans by demonstrating that the methylenedioxy-substituted phenalkylamines cannot be characterized simply as hallucinogens or stimulants.

The structural manipulations based on the phenalkylamine skeleton also confer differential neurochemical actions on these compounds (Gehlert et al., 1985; Battaglia et al., 1988). Neurochemically, serotonin release from whole brain rat synaptosomes is induced by MDA, MDMA, and MDE; in all cases, the S-(+) isomers generally are more potent (Nichols et al., 1982; Johnson et al., 1986; Nichols, 1986). MDMA and related drugs, including p-chloroamphetamine (PCA), MDE, MBDB, and MDA, are now well-established to be potent releasers of presynaptic serotonin and to have varying potencies as dopamine releasers (Johnson et al., 1986; Hekmatpanah and Peroutka, 1990; Berger et al., 1992). In addition, these drugs have varying potencies as monoamine uptake blockers (Nichols, 1986; Johnson et al., 1987). In rats, MDMA reduces tissue levels of serotonin, its metabolite, and its synthetic enzyme in brain (Stone et al., 1986; McCann and Ricaurte, 1991). In vitro, MDMA potently releases serotonin from rat striatal slices but is less effective at increasing dopamine release (Schmidt, 1987). Whereas R-(−)-MDMA possesses greater serotonin receptor binding affinity than the S-(+) enantiomer (Lyon et al., 1986; Teitler et al., 1990), the S-(+) enantiomer is a more potent psychoactive compound in humans (Adler, 1985). Hence, the psychoactive effects of MDMA appear to be correlated positively with the serotonin-releasing properties of the S-(+) enantiomer rather than with the receptor binding properties of the R-(−) enantiomer.

II. BEHAVIORAL EFFECTS AFTER ACUTE ADMINISTRATION

A. Effects in Unconditioned Behavior Paradigms

1. "Serotonin Syndrome"

Drugs resembling MDMA produce dramatic and unusual effects on unconditioned behavior of rodents. At lower doses, the predominant behavioral effect of MDMA and related drugs is increasing locomotor activity and concomitantly decreasing investigatory activity (Gold et al., 1988). At higher doses (> 5.0 mg/kg), MDMA elicits some behaviors that are included in the so-called "serotonin syndrome" (Hiramatsu et al., 1989; Slikker et al., 1989; Spanos and Yamamoto, 1989; Millan and Colpaert, 1991). The elements of this syndrome that are increased most dramatically by MDMA are low body posture, forepaw treading, and tail flicks. In addition, some head-weaving is observed. Both increases and decreases have been observed in the effectiveness of MDMA in producing the syndrome after repeated daily injections (Slikker et al., 1989; Spanos and Yamamoto, 1989). Although the elicitation of this behavioral profile is consistent with an indirect activation of 5-HT$_{1A}$ receptors, the nonspecificity of this syndrome precludes any definitive conclusions about the monoaminergic receptors involved. The hyperactivity, decreased investigatory responding, and increased syndrome-like behavior all are produced most potently by the S-(+) isomer of MDMA (Hiramatsu et al., 1989; Callaway et al., 1990), suggesting that these behaviors are attributable to the serotonin-releasing properties of the drug rather than to direct serotonin receptor activation. The only serotonin syndrome component that has been studied systematically with respect to its neuropharmacological substrates is the increases in tail flicks (Millan and Colpaert,

1991). The dependence of this effect on the release of serotonin but not catecholamines has been established by demonstrating that selective serotonin uptake blockers, paroxetine and citalopram, block the MDMA-induced increases in tail flicks. In contrast, neither the dopamine uptake blocker bupropion nor the norepinephrine uptake blocker maprotiline had any influence on the effect of MDMA. Further, the apparent importance of 5-HT_{1A} receptors in the indirect mediation of this effect of MDMA was confirmed by demonstrating that appropriate antagonists, including methiothepin, $(-)$-alprenolol, spiperone, BMY 7378, and NAN-190, blocked this effect of MDMA. Antagonists at 5-HT_2, 5-HT_3, D_1, D_2, α_1, α_2, β_1, or β_2 receptors had little effect on the tail-flick response elicited by MDMA (Millan and Colpaert, 1991). Thus, the increases in tail flicks observed after acute administrations of MDMA appear to be attributable to the release of presynaptic serotonin acting on 5-HT_{1A} receptors.

2. Stereotyped Behavior

Although the increases in locomotor activity produced by MDMA in rats (Gold et al., 1988) have been suggested to be caused by an amphetamine-like activation of catecholaminergic systems (Spanos and Yamamoto, 1989), several observations have distinguished clearly the locomotor activation produced by MDMA from that produced by amphetamine. For example, increasing the dose of MDMA or its close congeners produces graded increases in locomotor hyperactivity (Gold et al., 1988; Callaway et al., 1990,1991a). In contrast, increasing the dose of amphetamine, cocaine, and other psychostimulants that are potent dopamine-releasing agents results in the emergence of focused stereotypies to the exclusion of locomotor activity (Lyon and Robbins, 1975; Segal, 1975). Systematic assessments have confirmed that even high doses of MDMA fail to produce an amphetamine-like pattern of stereotyped behavior (Gold et al., 1988). Indeed, throughout the MDMA dose range, direct observation confirms that the focused sniffing, rapid head movements, and frequent changes in direction of movement that accompany amphetamine-induced hyperactivity are absent in MDMA-treated animals. Instead, MDMA-treated animals exhibit a low body posture and plodding straightforward locomotion that is interrupted only when the animal encounters an obstruction.

3. Locomotor Activating Effects of MDMA and Congeners

In rats, MDMA produces dramatic increases in locomotor activity (Gold et al., 1988). Extensive studies now have been conducted using a Behavioral Pattern Monitor (BPM) to explore the characteristics of and the mechanisms contributing to the locomotor hyperactivity induced by MDMA. The BPM combines the features of activity and holeboard chambers and measures individual response frequencies and durations. The BPM chamber is a 30×60 cm box that is criss-crossed with infrared beams and contains 3 holes in the floor and 7 in the walls, each equipped with an infrared beam. A microcomputer records and stores the successive holepokes, rearings, and positions of the animal in an x–y coordinate system with a resolution of 55 msec in time and 3.8 cm in space. The sequential pattern of movements, holepokes, and rearings can be displayed at variable rates on a video terminal or can be plotted on paper. Statistical assessments of these spatiotemporal sequences have proven very useful in discriminating drug effects (Geyer *et al.*, 1986,1987; Gold *et al.*, 1988; Geyer and Paulus, 1992). In addition to monitoring both locomotor and investigatory responses, the BPM permits assessment of changes in both the geometrical and the dynamical structures of motor behavior. Using this approach, the effects of various stimulants can be differentiated statistically at doses that produce comparable increases in the amount of locomotion but marked differences in qualitative aspects of behavior involving spatiotemporal patterns of locomotion and investigatory responses directed at specific environmental stimuli (Gately *et al.*, 1985; Geyer *et al.*, 1986,1987; Geyer and Paulus, 1992).

4. MDMA-Induced Changes in Behavioral Profiles

The analysis of the behavioral profiles of MDMA and MDE provided one of the first clues that the behavioral effects of these drugs were not simply the result of their weak ability to induce an amphetamine-like release of dopamine. The hyperactivity produced by MDMA and its congeners is accompanied by a profound reduction of investigatory responses. Both rearings and exploratory holepokes are suppressed almost completely by MDMA and MDE (Gold *et al.*, 1988). In contrast, amphetamine and other dopamine-releasing agents produce concomitant increases in locomotor movements, rearings, and holepokes (Geyer *et al.*, 1986,1987). Similarly, anticholinergic drugs such as scopolamine, which produce comparable increases in locomotor activity, also increase the absolute numbers of investigatory responses (Geyer *et al.*, 1986). Thus, the multiresponse profile characteristic of MDMA is most similar to that of the direct dopamine agonist apomorphine because both suppress holepokes and rearings while increasing locomotion. However, apomorphine produces relatively weak increases in locomotion and dramatic increases in sniffing behavior, whereas MDMA-induced hyperactivity is accompanied by reductions in sniffing.

5. MDMA-Induced Changes in Patterns of Locomotion

The dose-dependent and stereoselective increases in locomotor activity produced by MDMA or MDE are accompanied by marked alterations in the spatial patterning of locomotion (Gold *et al.*, 1988). As illustrated in Fig. 1, rats treated with MDMA consistently ambulated around the perimeter of the BPM chamber in unusually straight paths. Similar patterns are observed after administration of a variety of other serotonin-releasing drugs such as MDE, MBDB, MDA, and PCA (Callaway *et al.*, 1991a; Paulus and Geyer,

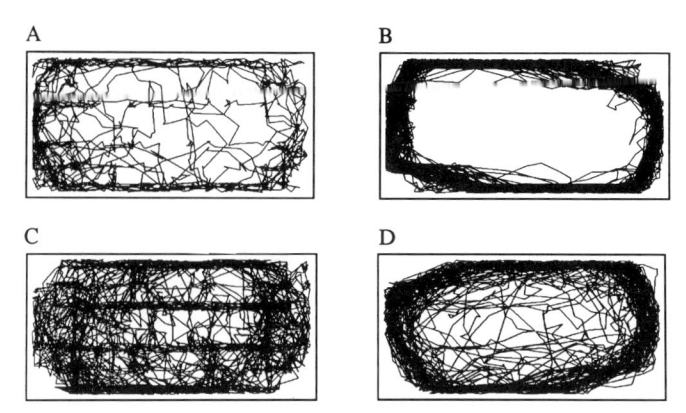

FIGURE 1 The spatial patterns of activity exhibited by individual rats in the 30 × 60 cm Behavioral Pattern Monitor (BPM) chambers are shown for 60-min sessions. Representative animals are shown following treatment with (A) saline, (B) 2.5 mg/kg RU24969, (C) 1.0 mg/kg d-amphetamine, or (D) 10.0 mg/kg R,S-methylenedioxymethamphetamine (MDMA). **Significantly different from saline controls.

1992). This behavior differed from the rotational behavior associated with drugs such as apomorphine (Geyer *et al.*, 1987). First, MDMA- or MDE-treated rats frequently reversed directions. Second, these animals moved in straight paths even when they were not walking close to a wall. These changes in patterns of locomotor activity have been quantitated using the BPM (Gold *et al.*, 1988). Specifically, the coefficient of variation (CV) assesses the predictability of spatial patterns of locomotion by measuring the consistency with which the animal repeats movements from one part of the chamber to another (Geyer, 1982,1990). The increases in straight and predictable path patterns produced by MDMA and related drugs are reflected in increases in the CV statistic (Gold *et al.*, 1988). In contrast, across a wide range of doses, amphetamine produces highly varied patterns of activity that either decrease or fail to alter the CV (Geyer *et al.*, 1986; Gold *et al.*, 1989).

Similarly, more definitive and powerful measures of the differences between MDMA-like drugs and the traditional psychostimulants have come from the use of scaling and complexity measures (Paulus and Geyer, 1991). The spatial scaling exponent d is based conceptually on fractal geometry and assesses the roughness or smoothness of the spatial path traversed by the animal (Paulus and Geyer, 1991). Briefly, the length of the path taken by a rat is calculated using several different spatial resolutions. Using scaling arguments, the rate at which the observed path length decreases as a function of decreasing spatial resolution is fitted to an exponential curve. The fitted coefficient for the exponent of this function is the variable d. The d measure increases when the locomotor path is rougher and includes many changes of direction. As did the CV analysis of the predictability of the spatial paths taken by the rats, the spatial scaling exponent d revealed that MDMA and its congeners increase straight movements, as reflected in decreases in d. In contrast, amphetamine has little influence on d. These findings are summarized in Fig. 2. Other independent analyses using newly developed measures of complexity such as the entropy statistic confirmed these differences between the MDMA family of drugs and the amphetamine-like stimulants (Paulus et al., 1990). Over dose ranges at which amphetamine and MDMA produce comparable increases in locomotor activity, amphetamine increased the number of probable subsets of behavioral sequences and the total number of different sequences, dose dependently, whereas low doses of MDMA increased and high doses of MDMA decreased the probable subsets of behavioral sequences (Paulus et al., 1990). After 5 mg/kg MDMA, some animals exhibited unpredictable sequences of behavior whereas others showed more predictable patterns. Thus, high doses of MDMA produce very constrained sequences of behavior that are readily distinguishable from the behaviors associated with classical stimulants such as amphetamine or cocaine (Paulus et al., 1990,1991).

Further, the characteristic patterns of locomotion produced by MDMA-like drugs can be distinguished from the effects of hallucinogens, despite the fact that they produce a similar tendency for rat locomotor paths to be constrained to the periphery of the BPM (Adams and Geyer, 1985). First, the hallucinogenic serotonin agonists decrease both locomotor and exploratory behavior when animals are tested in a novel environment (Adams and Geyer, 1985; Wing et al., 1990). In the case of hallucinogens, suppressions of locomotor activity in general and of entries into the center of the BPM in particular are hypothesized to reflect a potentiation of the normal avoidance of novel open areas by rats, as a result of the well-documented anxiogenic effects of these drugs. However, the suppression of locomotor activity and investigatory responses by hallucinogens is attenuated by familiarity with the testing environment (Adams and Geyer, 1985; Wing et al., 1990). In contrast, MDMA-induced increases in locomotor hyperactivity are not diminished when animals are tested in a familiar environment, an observation that further distinguishes these drugs from the traditional hallucinogens (Callaway et al., 1991b). In addition, the tendency for locomotion induced by MDMA-like drugs to consist of straight paths is a much more robust effect than the tendency to avoid entries into the center of the BPM (Callaway et al., 1991a,b). Perhaps avoidance of the center region is an epiphenomenon of the straight paths followed by MDMA-treated animals rather than a specific avoidance of a novel or open area, because the periphery of the BPM represents the longest path in which to engage in forward locomotion without encountering any obstructions. Quantitative measures of the effects of these drugs on the spatial structure of the locomotion confirm these observations. Specifically, the decreases in locomotor activity produced by hallucinogens such as LSD and 2,5-dimethoxy-4-iodophenylisopropylamine

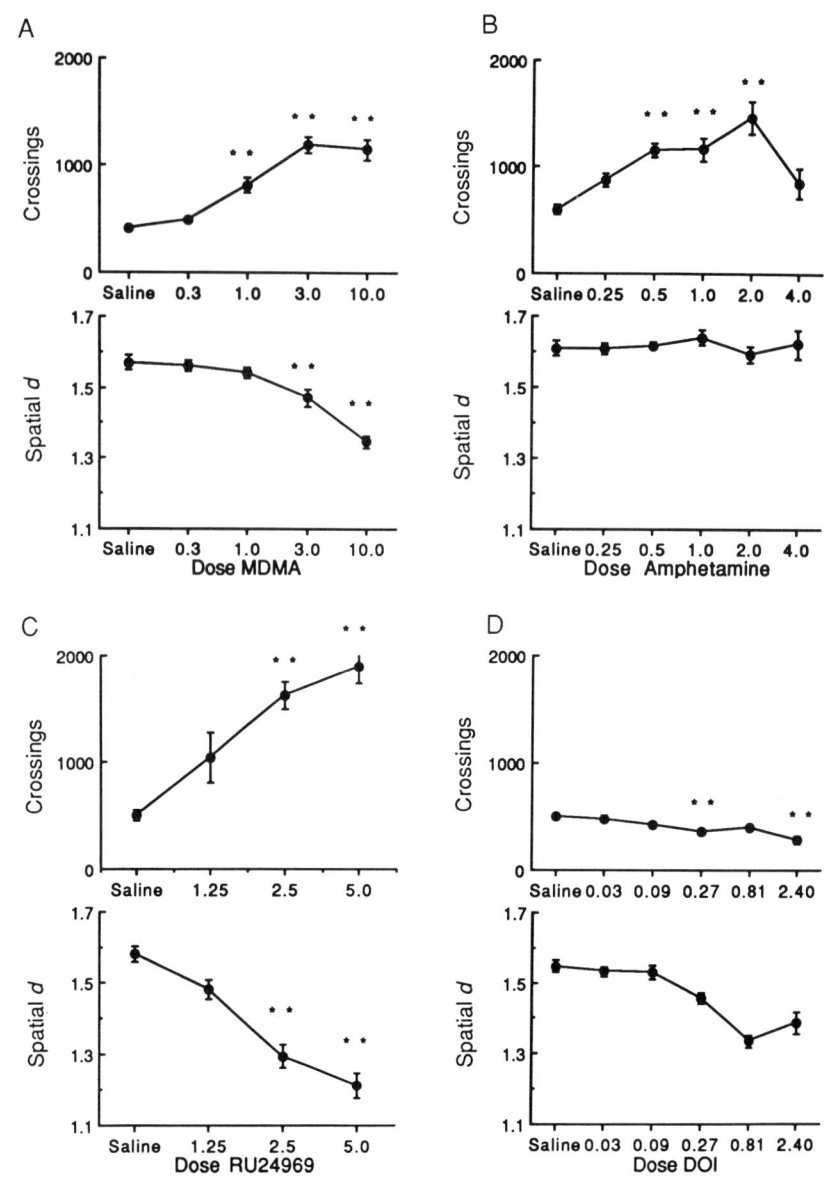

FIGURE 2 Primary effects of four phenalkylamines in the Behavioral Pattern Monitor (BPM) paradigm. Locomotor activity is quantified by the number of crossings between 15-cm square sectors per 60 min. The geometrical structure of the locomotion is quantified using the spatial scaling exponent d, calculated for the same interval. S-Amphetamine (B) increases locomotor activity without affecting the geometrical structure of that activity. In contrast, S-methylenedioxymethamphetamine (MDMA) (A) and RU24969 (C) increase locomotor activity and increase the contribution of straight distance-covering movements, as reflected by decreases in d. DOI (D) significantly decreases locomotor activity while also decreasing d.

(DOI) are associated with decreases or no change in the CV measure, because the movements of the animals are somewhat less predictable (Adams and Geyer, 1982), and with no significant change in the d statistic, as shown in Fig. 2. Collectively, these data suggest that the locomotor activating effects of MDMA differ from the effects of both classical psychostimulants and hallucinogens.

6. Contributions of Serotonin Release

Because the most prominent neurochemical effect of MDMA is the ability to increase serotonin release, several studies have evaluated the contribution of serotonin release to the behavioral profile of MDMA. By a hypothesized competition for binding to neuronal membrane carrier proteins, selective serotonin uptake inhibitors can decrease MDMA-induced serotonin release (Schmidt et al., 1987; Hekmatpanah and Peroutka, 1990; McKenna et al., 1991). Capitalizing on the fact that these uptake inhibitors produce few behavioral changes in the BPM paradigm, investigators have employed pretreatment with serotonin uptake inhibitors to evaluate the contribution of serotonin release to the behavioral effects of MDMA. Inhibitors of serotonin uptake such as fluoxetine potently antagonized MDMA-induced locomotor hyperactivity (Callaway et al., 1990). Further, the spatial scaling exponent revealed that pretreatment with fluoxetine also antagonized MDMA-induced changes in the spatial patterning of exploration (Fig. 3). Because of its selectivity for serotonergic systems, fluoxetine was believed to antagonize MDMA-induced behavior by preventing drug-induced serotonin release. Indeed, studies confirmed that fluoxetine pretreatment antagonized a decrease in tissue serotonin levels that was thought to reflect increased serotonin release (Callaway et al., 1991a). Similar antagonism of MDMA-induced hyperlocomotion was observed after a p-chlorophenylalanine (PCPA) pretreatment that produced a selective depletion of brain serotonin. These data thus indicated that serotonin release was necessary for the particular pattern of hyperactivity induced by MDMA in this paradigm.

The behavioral effects of other drugs that increase serotonin release provided converging evidence that serotonergic mechanisms mediate the locomotor activating effects of MDMA. For example, systemic administration of MBDB produced a fluoxetine-sensitive locomotor hyperactivity with spatial and temporal organization identical to that produced by MDMA (Callaway et al., 1991a). This observation confirmed the results of drug-discrimination studies in which MBDB was found to substitute completely for MDMA (Oberlender and Nichols, 1988) and of human psychopharmacological studies in which the primary subjective effects of MBDB were reported to be the same as those of MDMA (Nichols et al., 1986). The efficacy of MBDB is particularly important in ruling out the contribution of direct dopamine release to the primary behavioral effects of MDMA, because MBDB is virtually inactive as a dopamine-releasing agent *in vitro*

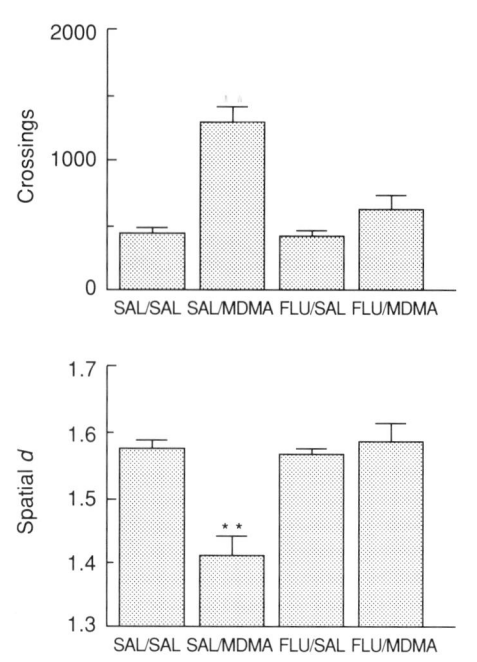

FIGURE 3 Antagonism of the response to S-methylenedioxymethamphetamine (MDMA) by fluoxetine (FLU). The quantity and spatial structure of locomotor activity are expressed by the number of crossings per 60 min and spatial d, respectively. Animals were pretreated with saline (SAL) or 10.0 mg/kg fluoxetine 50 min prior to receiving saline or 3.0 mg/kg S-MDMA. Data from Callaway et al. (1990).

(Johnson et al., 1986). In addition to that produced by MBDB, locomotor hyperactivity that was qualitatively similar to MDMA-induced locomotor hyperactivity was observed after the administration of other serotonin-releasing agents including PCA, MDA (Callaway et al., 1991a), and 5,6-(methylenedioxy)-2-aminoindan (MDAI) (Callaway et al., 1991b). These studies thus supported the hypothesis that the unique behavioral effects of MDMA are mediated via serotonin release.

7. Serotonin Receptor Involvement in Activating Effects of Serotonin Releasers

Several studies have examined the hypothesis that specific serotonin receptor subtypes contribute to the serotonergic effects of MDMA on unconditioned motor behavior. First, the behavioral effects of indirect serotonin agonists were compared with the effects of direct serotonin agonists that have some subtype specificity. The results of several dose–response studies indicate that neither 5-HT_{1A} nor 5-HT_2 agonists produce a comparable behavioral profile in the same paradigm (Mittman and Geyer, 1989;

Wing et al., 1990). However, previous studies indicate that the $5\text{-HT}_{1A/1B}$ agonist RU24969 increases locomotor activity in rats (Oberlander et al., 1986,1987). In the BPM, the behavioral effects of RU24969 resembled those of MDMA or MBDB, dose-dependently increasing locomotor activity and decreasing investigatory rearings and holepokes. In particular, the hyperactivity was characterized by repetitive spatial patterns of locomotion. The activating effects of RU24969 are attributable to postsynaptic serotonin receptors, since these are potentiated by neurotoxin lesions of serotonin terminals (Nisbet and Marsden, 1984; Oberlander et al., 1986,1987; Tricklebank et al., 1986). Hence, the $5\text{-HT}_{1A/1B}$ agonist RU24969 produces a behavioral activation that is similar to that of MDMA and dissimilar from the effects of 8-hydroxy-2(di-n-propylamino) tetralin (8-OH-DPAT), suggesting that 5-HT_{1B} receptors may be critical mediators of the activating effects of released serotonin (Rempel et al., 1993). These results suggest that MDMA and related drugs act via indirect agonist actions at 5-HT_{1B} receptors to produce behavioral activation similar to that produced by the direct agonist RU24969.

The contributions of various serotonin receptor subtypes to the effects of MDMA on rat exploration in the BPM also have been assessed by pretreating rats with selective and nonselective serotonin receptor antagonists (Callaway et al., 1992). Both propranolol and pindolol, β-noradrenergic antagonists with affinity for 5-HT_1 receptors, antagonized S-MDMA-induced locomotor hyperactivity. Further, pretreatment with (−)-propranolol but not (+)-propranolol antagonized the hyperactivity induced by S-MDMA (Rempel et al., 1993). Among nonselective serotonin antagonists, methiothepin was effective whereas methysergide and cyproheptadine were ineffective in antagonizing S-MDMA-induced hyperactivity. In other experimental paradigms, methiothepin has been found to be a good antagonist at 5-HT_{1B} receptors whereas methysergide, cyproheptadine, and ritanserin were ineffective. These studies support the hypothesis that locomotor hyperactivity produced by S-MDMA in rats depends on activation of 5-HT_1-like receptors, possibly of the 5-HT_{1B} subtype.

8. Contribution of Different Serotonin Terminal Regions

Confirming reports that many effects observed after systemic administration of psychostimulants can be reproduced by direct administration of the same drugs into the ventral striatum (Pijnenburg et al., 1976; Delfs et al., 1990; Swerdlow et al., 1990), injections of MDMA into the nucleus accumbens produce locomotor hyperactivity (Callaway and Geyer, 1992a) (see Fig. 4). Injections of MDMA into the dorsolateral caudate did not produce any hyperactivity, confirming the anatomical specificity of this effect. However, the behavioral stimulation observed after intra-accumbens MDMA administration differs from the effects of systemically administered MDMA. First, MBDB did not share the ability of MDMA to induce hyperactivity when

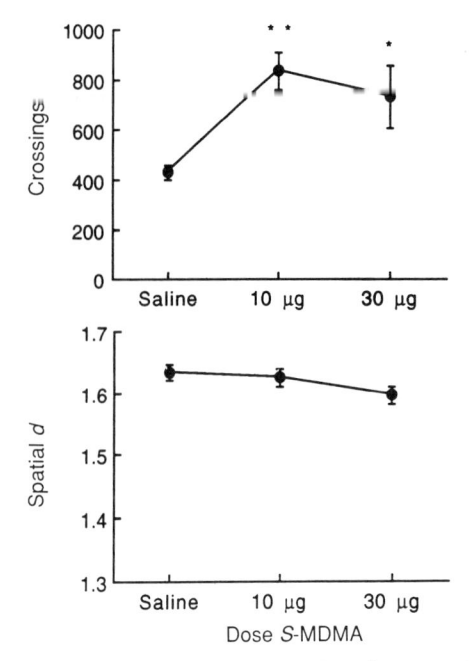

FIGURE 4 Behavioral response to S-methylenedioxymethamphetamine (MDMA) administration into the nucleus accumbens. In contrast to systemic administration, injection of S-MDMA into the nucleus accumbens produces an amphetamine-like increase in locomotor activity (crossings) without altering the spatial structure of activity (spatial d). Data from Callaway and Geyer (1992a).

injected into the nucleus accumbens. Second, the hyperactivity induced by injection of MDMA into the nucleus accumbens was not antagonized by fluoxetine pretreatment. Further, the pattern of activity induced by MDMA administered into the nucleus accumbens resembled the effects of amphetamine rather than the effects of MDMA (Fig. 5). These data indicate that the amphetamine-like dopamine-releasing properties of MDMA account for the activating effects of this drug when injected into the nucleus accumbens. This amphetamine-like action of MDMA may explain why drug discrimination studies find that amphetamine sometimes substitutes for MDMA (Glennon and Young, 1984) but not for MBDB (Nichols and Oberlender, 1990). Unfortunately, MDMA actions in the nucleus accumbens do not explain the unique behavioral effects of MDMA and MBDB that are believed to depend on serotonin release.

9. Fenfluramine—The Red Herring of Serotonin Releasers?

The activating effects of indirect serotonin agonists stand in apparent contradiction to some historical notions that central serotonin has primarily inhibitory effects on behavior (Gerson and Baldessarini, 1980; Soubrie,

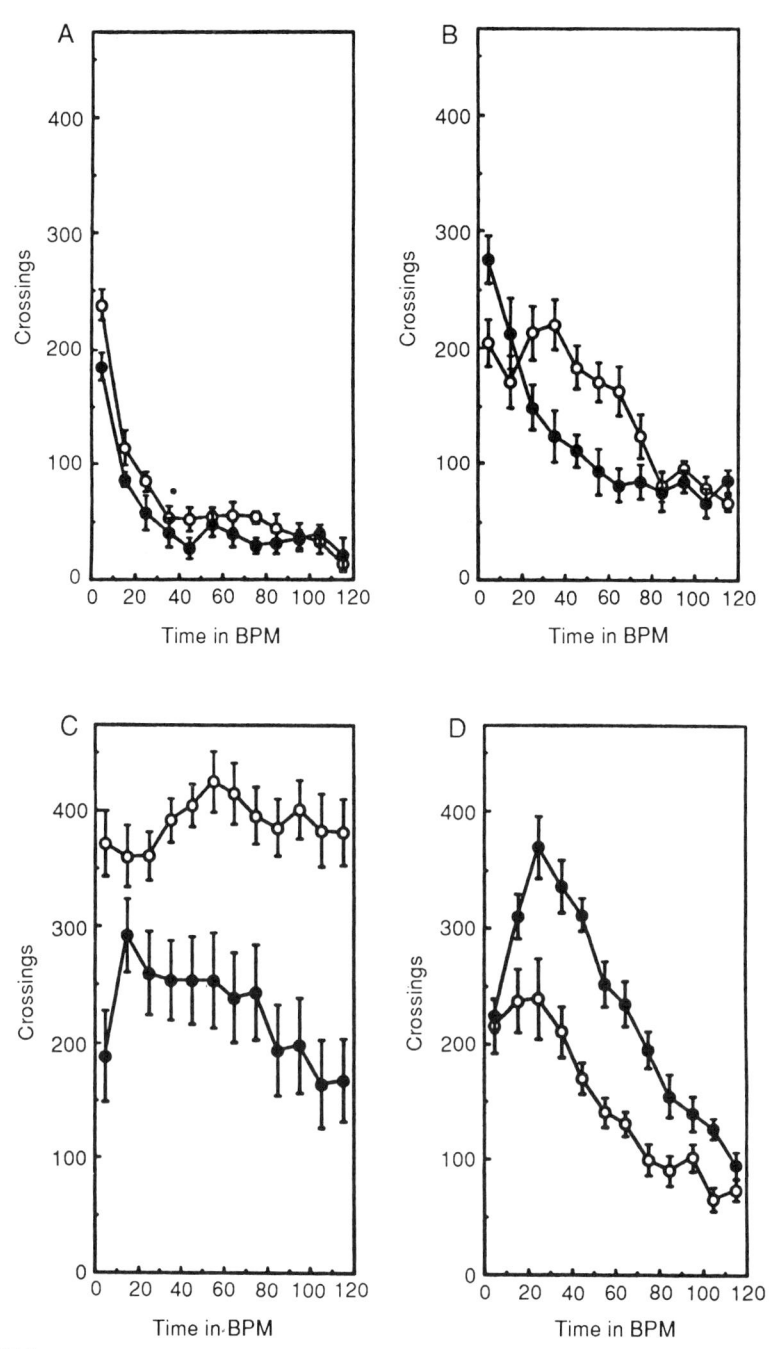

FIGURE 5 Effects of repeated R,S-methylenedioxymethamphetamine (MDMA) pretreatment on the activating effects of 3.0 mg/kg S-MDMA (B), 2.5 mg/kg RU24969 (C), and 1.0 mg/kg S-amphetamine (D) are shown as group means of crossings for twelve 10-min intervals. Saline (A) was used as a control. Saline (○) or R,S-MDMA (●) was administered (sc) every 12 hr for 4 days. Animals were tested 36 hr after the last pretreatment and 10 min after the challenge treatment. BPM, Behavioral Pattern Monitor.

1986). Such notions have persisted despite the accumulation of considerable evidence to the contrary. First, overwhelming evidence suggests that the multiple serotonergic pathways in the brain subserve different functional influences (Jacobs et al., 1974; Lorens and Guldberg, 1974; Geyer, 1978). Second, although the idea that serotonin inhibits motor activity originated with the observation that animals were hyperactive after lesions of the median raphe nucleus (Kostowski et al., 1968; Jacobs et al., 1974; Lorens and Guldberg, 1974; Geyer et al., 1976), selective depletions fail to produce the hyperactivity associated with electrolytic raphe lesions. For example, selective neurotoxin lesions that produce forebrain serotonin depletions equivalent to those of electrolytic lesions fail to produce hyperactivity (Hole et al., 1976; Lorens, 1978; Geyer et al., 1980; Lipska et al., 1992). Further, selective neurotoxin-induced depletions of serotonin appear to alter the qualitative features of the behavioral response to amphetamine without necessarily changing the amount of activation produced by amphetamine (Gately et al., 1985; Lipska et al., 1992). Similarly, although the depletion of serotonin with PCPA can increase rat behavioral activity (Fibiger and Campbell, 1971; Jacobs et al., 1975), it also can produce the opposite effect (Callaway et al., 1990). Most importantly, the direction of influence of PCPA appears to depend on environmental and experiential factors (Lorens, 1978). Third, a positive rather than negative correlation exists between raphe firing rates and arousal, both in short-term studies (Trulson and Jacobs, 1979; Steinfels et al., 1983) and across the sleep–wakefulness dimension (Jacobs, 1978). Indeed, the ability of PCPA to disrupt the organization of the sleep–wake cycle (Jacobs et al., 1972) is consistent with the idea that serotonin depletions interfere with the ability to remain quiescent. Fourth, 5-HT$_2$ agonist actions, rather than inhibition of serotonergic activity, have been accepted to be responsible for the perceptual stimulation produced by hallucinogens (Davis et al., 1984; Jacobs, 1984). Thus, the activation of serotonin systems is associated with increases in sensory and cognitive arousal. Fifth, clearly antagonists at dopamine and serotonin receptors have similar effects in the treatment of schizophrenic patients.

One remaining piece of evidence supporting a behaviorally inhibitory effect as the predominant action of global increases in serotonergic tone is the observation that fenfluramine releases presynaptic serotonin (Hekmat-panah and Peroutka, 1990; Sarkissian et al., 1990; Berger et al., 1992) and suppresses motor activity in the rat (Ziance et al., 1972; Linquist and Gotestam, 1977; Aulakh et al., 1988). However, some of the effects of fenfluramine are clearly independent of its ability to release serotonin (Van de Kar et al., 1985a,b). Hence, we examined the effects of fenfluramine in the BPM and explored the relevance of serotonin release to its actions in this paradigm. In concert with the literature, but in contrast with results with MDMA and the other serotonin-releasing drugs, fenfluramine and its metabolite norfenfluramine suppressed activity in the BPM (Callaway et al., 1993). As for MDMA, fenfluramine-induced serotonin release is known to be sensitive to serotonin uptake inhibitors (Hekmatpanah and Peroutka, 1990; Berger et al., 1992). However, the suppression of activity by norfenfluramine was not antagonized by the same fluoxetine and PCPA pretreatments that effectively prevented the locomotor activating effects of MDMA. Thus, these data indicate that the behavioral suppression produced by fenfluramine and norfenfluramine administration resulted from nonserotonergic mechanisms. Interestingly, human psychopharmacology also distinguishes fenfluramine from MDMA-like drugs. In contrast to the effects that characterize the response to MDMA, dysphoria and sedation are the primary subjective effects of fenfluramine (Gotestam and Gunne, 1972; Griffith et al., 1975). Thus, the differences between fenfluramine and the other serotonin releasers in their behavioral profiles provided by unconditioned motor activity paradigms are consistent with their differences in humans. In this case, the locomotor hyperactivity produced by MDMA is more in keeping with human results in literature than is the drug discrimination model, in which fenfluramine generalizes to the discriminative stimulus effects of MBDB (Nichols et al., 1989).

Although we do not yet the mechanism for the fenfluramine-induced suppression of activity, previous studies have reported neuroleptic-like effects of fenfluramine on dopaminergic systems (Bendotti et al., 1980). Fenfluramine-induced suppression of activity may result from the fact that fenfluramine and norfenfluramine differ from MDMA, MBDB, MDA, and PCA by virtue of a substituent in the *ortho* position of the aromatic ring. Supporting this hypothesis, two other phenalkylamine derivatives, MDAI and its congener MMAI, also produce fenfluramine-like and fluoxetine-sensitive suppressions of activity during at least some of their time of activity (Callaway et al., 1993). Formation of the second ring in each of these compounds is accomplished by joining the alkylamine side chain to the *ortho* position of the aromatic ring in the parent phenalkylamine. This *ortho* substitution of phenalkylamines may convey additional pharmacological properties that mask the behaviorally activating effects of drug-induced serotonin release. For example, addition of substituents in the *ortho* position of many phenalkylamines can increase their hallucinogenic effects (Shulgin et al., 1969). Another possibility is suggested by indications from Nova receptor screening that MMAI has high affinity for the α_2-adrenergic receptor (D. Nichols, personal communication). Agonists at this site, such as clonidine, produce dramatic decreases in locomotor activity in rats. Perhaps fenfluramine will be found to have a similar action.

B. Effects of Substituted Amphetamines on Startle Responding

1. Startle Habituation

One of the important behavioral functions of central serotonergic systems is their ability to modulate habituation. For example, LSD and serotonin releasers such as MDMA and PCA impair the habituation of startle responding in rats (Geyer et al., 1978,1990; Braff and Geyer, 1980). Habituation, often considered to represent the simplest form of learning, is defined as the decrement in an unconditioned response when the same stimulus is presented repeatedly at speeds too slow to produce sensory adaptation or receptor fatigue. Indeed, the ability to focus on selected environmental stimuli is a fundamental aspect of sensory information processing that depends on habituation. Unconditioned responses to sensory stimuli have been classified as orienting responses to mild information-laden stimuli and defensive responses elicited by powerful stimuli. The startle response has been considered a prototypical defensive response (Ison and Hoffman, 1983; Davis, 1984) and denotes a pattern of reflexes to sudden intense stimuli. Typically, the startle responses of animals are measured as the whole-body flinch or jump elicited by phasic acoustic or tactile (air puffs) stimuli using stabilimeter chambers. Startle response paradigms have been used frequently to study habituation. The observed habituation of responses is determined theoretically by two behaviorally opposite processes, called habituation and sensitization, that may have discrete neurophysiological and anatomic substrates (Groves and Thompson, 1970). In this chapter, we refer to the observed phenomenon of habituation, not the inferred underlying process of habituation.

2. Effects of Indirect Serotonin Agonists on Startle Habituation

Acute administrations of the indirect serotonin agonists MDMA, MBDB, and PCA have been found to decrease the habituation of both acoustic and tactile startle responses. Kehne et al. (1992) found delayed increases in both tactile and acoustic startle after either MDMA or PCA administration, effects that are consistent with an impairment of habituation. In contrast, fenfluramine decreased acoustic startle and had no significant influence on tactile startle. As were the effects of PCA and MDMA on locomotor hyperactivity and drug-induced tail flicks (see previous discussion), the effects of MDMA on both acoustic and tactile startle habituation were prevented by serotonin uptake blockers, including fluoxetine and MDL 27,777, but not by the norepinephrine uptake blocker desipramine. Moreover, analogous to findings with MDMA-induced hyperactivity, haloperidol blocked the increased reactivity produced by amphetamine without affecting the MDMA-induced alterations in startle responding. Finally, depletion of either ascending or descending serotonergic pathways with 5,7-DHT or 5,6-DHT (dihydroxytryptamines) attenuated the effects of both PCA and MDMA on both acoustic and tactile startle responses. Thus, indirect serotonin agonists such as MDMA impair the habituation of startle independent of the modality of stimulation, and do so via the release of presynaptic serotonin.

3. Effects of Direct Serotonin Agonists and Antagonists on Startle Habituation

Although the serotonin receptor subtype(s) mediating the habituation-decreasing effect of released serotonin remain unknown, some clues to the likely candidates can be obtained by comparing the effects of serotonin agonists and antagonists on startle habituation. The effect of direct serotonin agonists and antagonists on tactile startle reactivity and habituation have been tested in a series of dose–response studies using the same paradigm in which MDMA, MBDB, and PCA were found to decrease startle habituation. In this paradigm, behavioral habituation to 201 presentations of startling air puffs was determined after the administration of various compounds that bind to the major subtypes of putative receptors for serotonin. The mixed 5-HT_{1A} and 5-HT_2 agonist LSD long has been known to produce a dose-dependent impairment of habituation in this and similar paradigms (Key, 1961; Miliaressis and St. Laurent, 1974; Geyer et al., 1978). Similarly, the 5-HT_2 agonist mescaline impairs the habituation of tactile startle (Geyer and Tapson, 1988) and produces a delayed increase in acoustic startle that is consistent with a reduction in habituation (Davis, 1987). The antagonism of the effect of mescaline on acoustic startle by the 5-HT_2 antagonist ritanserin confirms that this effect is mediated by agonist actions at the 5-HT_2 receptor (Davis, 1987). In contrast, 5-HT_{1A} agonists such as 8-OH-DPAT, ipsapirone, and 5-methyl-N, N-dimethyltryptamine (5-Me-ODMT) and the putative 5-HT_{1B} agonist m-trifluoromethylphenylpiperazine (TFMPP) affect startle reactivity but do not have specific effects on habituation (Geyer and Tapson, 1988). Conversely, a variety of 5-HT_2 antagonists, including cyproheptadine, cinanserin, ritanserin, and ketanserin, significantly increases the rate of tactile startle habituation but does not affect initial levels of reactivity. Similarly, nonspecific monoamine antagonists impair the habituation of acoustic as well as tactile startle (Swerdlow et al., 1991,1993). Specifically, several antipsychotic drugs that have affinities for both dopamine and serotonin receptors, including chlorpromazine, perphenazine, clozapine, and haloperidol, all accelerate the habituation of acoustic startle. These effects of both selective 5-HT_2 and mixed antagonists on acoustic and tactile startle habituation are not caused by pharmacokinetic characteristics of these compounds. Moreover, in all these cases, the acceleration of habituation is observed in the absence of any alteration in the initial level of startle reactivity (Geyer and Tapson, 1988; Geyer et al., 1990). Thus, the effects of 5-HT_2 antagonists on startle habituation are found to be diametrically opposed to those found with LSD, mescaline, or MDMA. In general, these results are consistent with the hypothesis (Glennon et al., 1984) that hallucinogenic drugs exert many of their behavioral effects via actions as 5-HT_2 agonists. From these studies we can conclude that (1) 5-HT_2 receptors have important effects on startle habituation and (2) the effects of MDMA on startle habituation are likely to involve the 5-HT_2 receptor system.

4. Effects of Presynaptic Manipulations of Serotonin on Startle Habituation

The findings reviewed earlier indicate that 5-HT_1 and 5-HT_2 receptors have qualitatively different influences on startle behavior (Geyer and Braff,

1987). Whereas 5-HT$_1$ receptors appear to be involved in startle reactivity, 5-HT$_2$ receptors primarily influence startle habituation. To examine the endogenous modulation of the serotonin system via these receptors, presynaptic manipulations of the serotonin system were investigated using serotonin reuptake inhibitors. Fluoxetine, which does not interact directly with either type of receptor, selectively inhibits the reuptake of serotonin (Wong et al., 1983). The direct behavioral effects of this drug are subtle; nevertheless, fluoxetine reduced the rate of tactile startle habituation (Geyer and Tapson, 1988). Presumably, by blocking reuptake, fluoxetine potentiates the actions of the serotonin released in response to startling stimuli (Geyer et al., 1982) and thereby reduces startle habituation.

The effects on startle habituation of the exogenous 5-HT$_2$ antagonists were also consistent with the effects of depletion of endogenous serotonin. Specifically, the serotonin depleting agents PCPA and PCA accelerated startle habituation (Geyer and Tapson, 1988; Geyer et al., 1990). PCPA and PCA deplete presynaptic serotonin by distinctly different mechanisms. PCPA inhibits serotonin synthesis (Koe and Weissman, 1966), whereas PCA stimulates serotonin release and has a poorly understood neurotoxic action (Fuller et al., 1975; Sanders-Bush and Steranka, 1978). Both compounds produce marked and reasonably selective depletion of brain serotonin at the doses and pretreatment times used in these studies. The results indicated that PCPA and PCA produced significant increases in the habituation of tactile startle. The observation that two different kinds of serotonin-depleting agents with presumably different nonspecific effects produced similar increases in startle habituation further supports the hypothesis that endogenous serotonin plays an important role in startle habituation.

5. Indirect Serotonin Agonist Effects on Prepulse Inhibition

Indirect serotonin agonists such as MDMA also disrupt prepulse inhibition, another form of startle plasticity. In prepulse inhibition, weak prestimuli presented at brief intervals prior to a startle-eliciting stimulus reduce, or gate, the amplitude of the startle response. Thus, as with the impairment of habituation produced by LSD and MDMA, the drug-treated animals tested in the prepulse inhibition paradigm exhibit an increased or "unfiltered" responsiveness to sensory stimuli, that is, they fail to exhibit the gating or inhibition of the response normally produced by the prepulse stimulus. As reviewed elsewhere (Ison and Hoffman, 1983; Davis, 1984), this cross-species phenomenon of prepulse inhibition is very robust, unlearned, and ubiquitous. A number of dopamine-releasing agents can disrupt prepulse inhibition, apparently via the indirect activation of D$_2$ receptors (Geyer et al., 1990). MDMA and MDE (Ecstasy and Eve) also disrupt prepulse inhibition. Moreover, the effect of MDE was shown to be stereoselective (Mansbach et al., 1989). To date, no studies have assessed the relevance of serotonin release to these effects of MDMA and MDE, nor have the receptors mediating these effects been examined. However, data from intracranial injections of MDMA suggest that dopamine release can contribute to the behavioral effects of this drug (Callaway and Geyer, 1992a). This amphetamine-like action of MDMA also may contribute to the amphetamine-like disruption of prepulse inhibition.

C. Effects of Substituted Amphetamines on Antinociception

As have other indirect serotonin agonists, MDMA has been reported to produce antinociceptive effects in rats (Crisp et al., 1989). Doses of 3.0–6.0 mg/kg MDMA significantly increased latencies in the hot-plate test without affecting latencies in the tail-flick test. This pattern of results is similar to that observed with PCA or fenfluramine (Ogren and Holm, 1980; Rochat et al., 1982). Further, the effects of MDMA and fenfluramine are attenuated by the mixed serotonin antagonists methysergide and metergoline, respectively (Rochat et al., 1982; Crisp et al., 1989). However, systematic studies using a range of monoamine antagonists have not been done in this paradigm. Similarly, additional research is needed to determine whether the antinociceptive effects of MDMA or fenfluramine are dependent on their ability to release presynaptic serotonin.

III. BEHAVIORAL EFFECTS AFTER REPEATED ADMINISTRATIONS

A. Effects in Conditioned Behavior Paradigms

1. Drug Discrimination Studies

In addition to possessing unique psychopharmacological effects in humans, drugs related to MDMA produce unique behavioral effects in several experimental paradigms using animals. In drug-discrimination studies, for example, MDMA has been reported to produce drug-appropriate responding in rats (Glennon and Young, 1984) and monkeys (Kamien et al., 1986) that have been trained to recognize amphetamine. However, other studies showed that MDMA does not produce drug-appropriate responding in rats trained to recognize amphetamine (Oberlender and Nichols, 1988) or cocaine (Broadbent et al., 1989). In contrast, amphetamine produces only partial drug-appropriate responding in rats trained to recognize MDMA (Schechter, 1989; Glennon and Misenheimer, 1989). Further, animals trained to recognize the more selective serotonin-releasing agent MBDB, which completely substitutes for MDMA (Oberlender and Nichols, 1988), fail to recognize amphetamine (Nichols and Oberlender, 1990). These studies suggest that MDMA produces a complex discriminative stimulus and that a component of this stimulus resembles the cue produced by amphetamine. However, MDMA and its congener MBDB produce an additional cue that differs from that of the psychostimulants. Other data indicate that the nonpsychostimulant MDMA cue is probably not the same as that produced by hallucinogenic amphetamines. For example, rats trained to recognize the hallucinogenic amphetamine derivative DOM, which substitutes for many other hallucinogens (Glennon et al., 1983), do not recognize MDMA (Glennon et al., 1982). Conversely, DOM will not produce completely drug-appropriate responding in rats trained to recognize MDMA (Oberlender and Nichols, 1988) or MBDB (Nichols and Oberlender, 1990). Thus, the stimulus properties of MDMA and several of its congeners in animal drug-discrimination studies differ from those of psychostimulants and hallucinogens.

The discriminative cue that is characteristic of both MDMA and MBDB in drug-discrimination studies may be synonymous with the locomotor-activating effect of MDMA-like drugs in the BPM paradigm. For example, as with locomotor activation, $5-HT_{1B}$ receptor agonists but not $5-HT_{1A}$ or $5-HT_2$ receptor agonists can substitute for MDMA (Schechter, 1989) or MBDB (Nichols and Oberlender, 1990). Further, the MDMA stimulus observed in drug-discrimination studies can be antagonized by serotonin antagonists (Schechter, 1989). Thus, these different paradigms provide converging evidence that indirect agonist actions at $5-HT_1$-like receptors are responsible for the behaviorally salient effects of MDMA-like drugs.

2. Schedule-Controlled Behavior

Relatively few studies have used schedule-controlled behaviors to examine the pharmacological actions of the methylenedioxy-substituted phenalkylamines. Early studies of the optical isomers of MDA and MDMA revealed suppressions of fixed-ratio responding in mice (Glennon et al., 1987). Surprisingly, limited antagonist studies suggested that only the $R-(-)$ isomer of MDA owed its effects to serotonergic actions. In pigeons tested under a multiple 3-min fixed-interval, fixed-ratio 30 schedule, both components of the schedule were disrupted at comparable doses (Nader et al., 1989). This pattern of behavioral change contrasts with the effects of amphetamine-like psychostimulants. The mixed serotonin antagonist metergoline or the $5-HT_2$ antagonist ketanserin restored responding suppressed by MDA, but had no effect on responding suppressed by MDMA. Conversely, the α-adrenergic blocker prazosin attenuated the effects of MDMA but not MDA. These interactions support the idea that MDA and MDMA have dissociable neurochemical and behavioral effects. Since the relevance of serotonin release to the effects of these drugs on schedule-controlled responding has not been examined, and since the available findings are inconsistent with the more extensive studies using other paradigms, additional studies are required before this approach can add to our understanding of these complex drugs.

B. Tolerance Studies

In contrast to the sensitization of the response to amphetamine that is observed after repeated administrations of the drug (Segal and Mandell,

1974; Chapter 4), chronic pretreatment with MDMA generally leads to apparent tolerance to its behavioral effects. This tolerance may be related to the long-term depletions of brain serotonin produced by high doses or repeated administrations of MDMA. For example, the hyperactivity produced by an acute administration of S-MDMA is diminished significantly if the animals are pretreated with R,S-MDMA twice a day for 4 days (Callaway and Geyer, 1992b), as illustrated in Fig. 5.

Studies of cross-tolerance have provided additional evidence for the hypothesis that the specific participation of 5-HT_{1B} receptors contributes to the behavioral profile of MDMA (Callaway and Geyer, 1992b). Previous studies indicated that chronic pretreatments with RU24969, DOI, or 8-OH-DPAT induce tolerance to their respective acute effects (Oberlander *et al.*, 1987; Pranzatelli and Pluchino, 1991; Geyer and Krebs, 1993). Moreover, RU24969-induced tolerance to its locomotor-activating effects has been demonstrated to be receptor specific by showing that chronic RU24969 pretreatment does not affect the behavioral effects of 8-OH-DPAT or amphetamine (Oberlander *et al.*, 1987). In experiments using the BPM paradigm described earlier, pretreatment of animals with the 5-HT_{1B} agonist RU24969 (2.5 mg/kg, twice daily for 3 days) significantly reduced the locomotor-activating effects of MDMA, whereas pretreatment with the 5-HT_2 agonist DOI or the 5-HT_{1A} agonist 8-OH-DPAT (1.0 mg/kg, twice daily for 3 days) failed to alter the effects of MDMA (Callaway and Geyer, 1992b). In a reciprocal experiment, animals pretreated with R,S-MDMA (10 mg/kg, twice daily for 4 days) were tested after acute administration of saline, S-MDMA, RU24969, or *d*-amphetamine. As illustrated in Fig. 5, a chronic R,S-MDMA pretreatment not only produced tolerance to S-MDMA but also reduced the effects of a challenge dose of RU24969 (i.e., cross-tolerance) and potentiated the effects of a *d*-amphetamine challenge. That the development of tolerance to the activating effects of MDMA was associated with reciprocal cross-tolerance to the activating effects of RU24969 suggests that prolonged administration of MDMA stimulates and desensitizes the same neuronal circuitry that mediates the effects of RU24969. Thus, the primary behavioral effects of MDMA are antagonized by 5-HT_{1B} receptor antagonists, are mimicked by 5-HT_{1B} receptor agonists, and are attenuated after selective desensitization of 5-HT_{1B} receptors by chronic drug treatments. The weight of evidence favors the hypothesis that MDMA acts in rats as an indirect agonist at 5-HT_{1B} receptors.

IV. CONCLUSIONS AND FUTURE DIRECTIONS

The studies reviewed in this chapter indicate that serotonergic systems participate extensively in the primary behavioral effects of MDMA and related amphetamine derivatives. Specifically, MDMA-like drugs increase the release of serotonin via direct actions on presynaptic terminals. Although the serotonin released by these drugs probably affects many different serotonin receptors, the predominant behavioral effects of MDMA on unconditioned motor behavior appear to be associated with stimulation of 5-HT_{1B} receptors. The reasons for this apparent selectivity of action remain to be determined in future studies. Also, the alterations in behavioral habituation that are produced by MDMA and related drugs are clearly dependent on the release of presynaptic serotonin. Although the receptors on which this released serotonin acts to produce this effect remain to be determined, the involvement of 5-HT_2 receptors in modulating habituation implicates these sites in the effects of the indirect agonists.

These studies provide a pharmacological basis for the phenomenological distinctions between three classes of substituted amphetamine derivatives suggested by human psychopharmacology, animal drug-discrimination experiments, and tests of exploratory activity. In particular, amphetamine-induced activation appears to depend on the increased release of dopamine in the dorsal and ventral striatum, as reviewed elsewhere (Kelly, 1977; Swerdlow et al., 1986). Other drugs that increase the release of dopamine share the psychostimulant properties of amphetamine; antagonists of dopamine receptors or depletion of presynaptic dopamine can interfere with amphetamine-induced behavior. In contrast, hallucinogenic phenalkylamines such as DOI are inactive as monoamine-releasing agents (McKenna et al., 1991) but are potent agonists at 5-HT_2 receptors (Glennon, 1985). The behavioral effects produced by acute administrations of hallucinogenic phenalkylamines can be blocked by 5-HT_2 receptor antagonists (Wing et al., 1990). Further, structurally unrelated drugs that act at 5-HT_2 receptors exhibit similar behavioral profiles (Adams and Geyer, 1985; Wing et al., 1990). Thus, the indirect agonist action of MDMA-like drugs at 5-HT_1 receptors is mechanistically quite distinct from the actions of both classical dopaminergic psychostimulants and hallucinogenic phenalkylamines (Table I).

The pharmacology of MDMA-like drugs in rats may be relevant to understanding the psychopharmacology of these drugs in humans. Specifically, these results prompt several testable hypotheses about the clinical effects of MDMA-like drugs. First, the interaction studies with serotonin uptake inhibitors and depletors predict that the unique subjective effects of MDMA will be antagonized by preventing presynaptic release of serotonin. Second, the fact that many other serotonin-releasing drugs produce MDMA-like behavioral effects predicts that these drugs will possess MDMA-like subjective effects in humans. This hypothesis has been confirmed for several agents (Shulgin, 1978; Nichols et al., 1986). Third, the studies with receptor agonists and antagonists predict that the subjective effects of MDMA will be antagonized by antagonists of 5-HT_1 receptors. However, the different pharmacology of the human homolog of the rat 5-HT_{1B} receptor, the 5-HT_{1D} receptor, requires that different antagonists be employed for each species.

TABLE I Three Classes of Psychoactive Phenalkylamines

Class	Example	Primary psychopharmacological effect	Effects on unconditioned behavior of animals	Neurotransmitter system implicated in behavioral effects	Locus of action	Receptors implicated
Psychostimulants	Amphetamine	Increased arousal, sympathomimetic	Increased motor activity, stereotypy, increased exploration	Dopamine	Presynaptic	D_1, D_2
"Entactogens"	MDMA[a]	Altered perception of emotions	Increased forward locomotion, decreased exploration	Serotonin	Presynaptic	$5\text{-}HT_{1B}$
Hallucinogens	DOI[b]	Perceptual distortions	Decreased motor activity, neophobia	Serotonin	Postsynaptic	$5\text{-}HT_2$

[a] MDMA, 3,4-Methylenedioxymethamphetamine.
[b] DOI, 2,5-Dimethoxy-4-iodoamphetamine.

Finally, these studies predict that a centrally active agonist at 5-HT$_{1D}$ receptors would produce MDMA-like behavioral effects in humans. Although the toxicity of many serotonin-releasing drugs probably will preclude further psychopharmacological investigations in humans, at least one direct 5-HT$_{1D}$ receptor agonist has been examined already for use in relief of migraine headaches (Cady *et al.*, 1991). Future studies should evaluate the potential psychoactive properties of these compounds.

The demonstrations that the primary behavioral effects of MDMA-like drugs depend on presynaptic serotonin release establishes these compounds as important pharmacological tools for the assessment of the functional roles of serotonergic systems. Just as the indirect agonist amphetamine has been critical to our understanding of the functions of catecholaminergic systems, indirect serotonergic agonists should provide similar information regarding serotonergic systems. To date, most of the literature regarding the behavioral influences of serotonin has been based on the effects of serotonin depletion or drugs acting directly on serotonin receptors. Compounds such as MBDB appear to be selective indirect serotonin agonists, the effects of which reflect the activation of the endogenous transmitter. Given the frequent mismatches between the distributions of the transmitter and the receptors, the effects of direct-acting agonists or antagonists may or may not be related to the endogenous transmitter. In contrast to depleting agents or direct receptor ligands, the behavioral effects of drugs that increase serotonin release may have greater physiological relevance. Presumably, a drug that increases serotonin release exaggerates the normal function of serotonin by releasing this transmitter only from presynaptic sites where it normally is found. By acting acutely, these drugs avoid the long-term adaptations to abnormal levels of transmitters that can confound studies using depleting agents. Hence, further study of these newly indentified indirect serotonin agonists should greatly increase our understanding of the functions of the various serotonergic systems. In this regard, we must remain skeptical about the anomalous effects of fenfluramine, which is the only serotonin-releasing drug that has been studied widely in humans. Although the primary behavioral effect of fenfluramine is the induction of lethargy (Griffith *et al.*, 1975), more recent studies indicate that the suppression of activity by fenfluramine probably is unrelated to its serotonin-releasing properties. In fact, the effects of MDMA are more representative of the actions of many serotonin-releasing drugs. Thus, clinical studies should consider the mood-altering effects of MDMA rather than the sedating properties of fenfluramine as models for serotonin psychopharmacology.

In summary, MDMA is representative of a series of phenalkylamines with psychopharmacological effects that are distinct from both classical phenalkylamine psychostimulants and phenalkylamine hallucinogens, as delineated in Table I. Detailed characterization of drug effects on the exploratory behavior of rats also distinguishes MDMA-like drugs from other phenalkylamines. In rats, the unique behavioral effects of MDMA-like drugs appear to depend on presynaptic serotonin release and activation of 5-HT$_1$ receptors, particularly 5-HT$_{1B}$ receptors. These studies support the hypothesis that MDMA-like drugs constitute a distinct class of phenalkylamine drugs that is phenomenologically and pharmacologically distinct from structurally related psychostimulants and hallucinogens.

REFERENCES

Adams, L. M., and Geyer, M. A. (1982). LSD-induced alterations in locomotor patterns and exploration in rats. *Psychopharmacology* 77, 179–185.

Adams, L. M., and Geyer, M. A. (1985). A proposed animal model for hallucinogens based on LSD's effects on patterns of exploration in rats. *Behav. Neurosci.* 99, 881–900.

Adler, J. (1985). Getting high on 'Ecstasy'. *Newsweek* Apr. 15, 96.

Anderson, G., Braun, G., Braun, U., Nichols, D., and Shulgin, A. (1978). Absolute configuration and psychotomimetic activity. *NIDA Res. Monogr.* 22, 8–15.

Aulakh, C. S., Hill, J. L., Wozniak, K. M., and Murphy, D. L. (1988). Fenfluramine-induced suppression of food intake and locomotor activity is differentially altered by the selective type A monoamine oxidase inhibitor clorgyline. *Psychopharmacology* 95, 313–317.

Battaglia, G., Brooks, B. P., Kulsakdinun, C. and De Souza, E. B. (1988). Pharmacologic profile of MDMA (3,4-methylenedioxymethamphetamine) at various brain recognition sites. *Eur. J. Pharmacol.* 149, 159–163.

Bendotti, C., Borsini, F., Zanini, M. G., Samanin, R., and Garattini, S. (1980). Effect of fenfluramine and norfenfluramine stereoisomers on stimulant effects of *d*-amphetamine and apomorphine in the rat. *Pharmacol. Res. Commun.* 12, 567–574.

Berger, U. V., Gu, X. F., and Azmitia, E. C. (1992). The substituted amphetamines 3,4-methylenedioxymethamphetamine, methamphetamine, *p*-chloroamphetamine, and fenfluramine induce 5-hydroxytryptamine release via a common mechanism blocked by fluoxetine and cocaine. *Eur. J. Pharmacol.* 215, 153–160.

Braff, D. L., and Geyer, M. A. (1980). Acute and chronic LSD effects on rat startle: Data supporting an LSD–rat model of schizophrenia. *Biol. Psychiatry* 15, 909–916.

Broadbent, J., Michael, E. K., and Appel, J. B. (1989). Generalization of cocaine to isomers of 3,4-methylenedioxyamphetamine and 3,4-methylenedioxymethamphetamine: Effects of training dose. *Drug Dev. Res.* 16, 443–450.

Cady, R. K., Wendt, J. K., Kirchner, J. R., Sargent, J. D., Rothrock, J. F., and Skaggs, H. (1991). Treatment of acute migraine with subcutaneous sumatriptan. *J. Am. Med. Assoc.* 265, 2831–2835.

Callaway, C. W., and Geyer, M. A. (1992a). Stimulant effects of 3,4-methylenedioxymethamphetamine in the nucleus accumbens of rat. *Eur. J. Pharmacol.* 214, 45–51.

Callaway, C. W., and Geyer, M. A. (1992b). Tolerance and cross-tolerance to the activating effects of 3,4-methylenedioxymethamphetamine and a serotonin 5HT-1B agonist. *J. Pharmacol. Exp. Ther.* 263, 318–326.

Callaway, C. W., Wing, L., and Geyer, M. A. (1990). Serotonin release contributes to the stimulant effects of 3,4,-methylenedioxymethamphetamine in rats. *J. Pharmacol. Exp. Ther.* 254, 456–464.

Callaway, C. W., Johnson, M. P., Gold, L. H., Nichols, D. E., and Geyer, M. A. (1991a). Amphetamine derivatives produce locomotor hyperactivity by acting as indirect serotonin agonist. *Psychopharmacology* 104, 293–301.

Callaway, C. W., Nichols, D. E., Paulus, M. P., and Geyer, M. A. (1991b). Serotonin release is responsible for the locomotor hyperactivity in rats induced by derivatives of amphetamine related to MDMA. *In* "Serotonin: Molecular Biology, Receptors, and Functional Effects" (J. R. Fozard and P. R. Saxena, eds.), pp. 491–505. Birkhauser, Basel.

Callaway, C. W., Rempel, N., Peng, R. Y., and Geyer, M. A. (1992). Serotonin 5-HT-1-like receptors mediate hyperactivity in rats induced by 3,4-methylenedioxymethamphetamine. *Neuropsychopharmacology,* 7, 113–127.

Callaway, C. W., Wing, L. L., Nichols, D. E., and Geyer, M. A. (1993). Suppression of behavioral activity by norfenfluramine and related drugs in rats is not mediated by serotonin release. *Psychopharmacology* 111, 169–178.

Creese, I., and Iversen, S. D. (1975). The pharmacological and anatomical substrates of the amphetamine response in rats. *Brain Res.* 83, 419–436.

Crisp, T., Stafinsky, J. L., Boja, J. W., and Schechter, M. D. (1989). The antinociceptive effects of 3,4-methylenedioxymethamphetamine (MDMA) in the rat. *Pharmacol. Biochem. Behav.* 34, 497–501.

Davis, M. (1984). The mammalian startle response. *In* "Neural Mechanisms of Startle Behavior" (R. C. Eaton, ed.), pp. 287–342. Plenum Press, New York.

Davis, M. (1987). Mescaline: Excitatory effects on acoustic startle are blocked by serotonin-2 antagonists. *Psychopharmacology* 93, 286–291.

Davis, M., Kehne, J. H., Commissaris, R. L., and Geyer, M. A. (1984). Effects of hallucinogens on unconditioned behaviors in animals. *In* "Hallucinogens: Neurochemical, Behavioral and Clinical Perspectives" (B. L. Jacobs, ed.), pp. 35–76. Raven Press, New York.

Delfs, J. M., Schreiber, L., and Kelley, A. E. (1990). Microinjection of cocaine into the nucleus accumbens elicits locomotor activation in the rat. *J. Neurosci.* 10, 303–310.

Dimpfel, W., Spuler, M., and Nichols, D. E. (1989). Hallucinogenic and stimulatory amphetamine derivatives: Finger-printing DOM, DOI, DOB, MDMA, and MBDB by spectral analysis of brain field potentials in the freely moving rat (Tele-Stereo-EEG). *Psychopharmacology* 98, 297–303.

Fibiger, H. C., and Campbell, B. A. (1971). The effect of parachlorophenylalanine on spontaneous locomotor activity in the rat. *Neuropharmacology* 10, 25–32.

Fuller, R. W., Perry, K. W., and Mollow, B. B. (1975). Reversible and irreversible phases of serotonin depletion by 4-chloroamphetamine. *Eur. J. Pharmacol.* 33, 119–124.

Gately, P. F., Poon, S. L., Segal, D. S., and Geyer, M. A. (1985). Serotonin depletions induced by 5,7-dyhydroxytryptamine alter the response to amphetamine and the habituation of locomotor activity in rats. *Psychopharmacology* 87, 400–405.

Greer, G., and Strassman, R. (1985). Information on "Ecstasy." *Am. J. Psychiatry* 142, 1391.

Gehlert, D., Schmidt, C., Wu, L., and Loveberg, W. (1985). Evidence for specific MDMA (ecstasy) binding sites in the rat brain. *Eur. J. Pharmacol.* 119, 135–136.

Gerson, S. C., and Baldessarini, R. J. (1980). Motor effects of serotonin in the central nervous system. *Life Sci.* 27, 1435–1451.

Geyer, M. A. (1978). Heterogenous functions of discrete serotonergic pathways in brain. *In* "Biochemistry of Mental Disorders" (E. Usdin and A. J. Mandell, eds.), pp. 233–260. Marcel Dekker, New York.

Geyer, M. A. (1982). Variational and probabalistic aspects of exploratory behavior in space: Four stimulant styles. *Psychopharmacol. Bull.* 18, 48–51.

Geyer, M. A. (1990). Approaches to the characterization of drug effects on locomotor activity in rodents. *In* "Modern Methods in Pharmacology" (M. W. Adler and A. Cowan, eds.), Vol. 6, pp. 81–99. Liss, New York.

Geyer, M. A., and Krebs, K. M. (1993). Serotonin receptor involvement in an animal model of the acute effects of hallucinogens. *NIDA Res. Monogr. Ser.,* in press.

Geyer, M. A., and Paulus, M. (1992). Multivariate and non-linear approaches to the characterization of drug effects on the locomotor and investigatory behavior of rats. *NIDA Res. Monogr. Ser.* 124, 203–236.

Geyer, M. A., and Tapson, G. S. (1988). Habituation of tactile startle is altered by drugs acting on serotonin-2 receptors. *Neuropsychopharmacology* **1**, 135–147.

Geyer, M. A., Puerto, A., Menkes, D. B., Segal, D. S., and Mandell, A. J. (1976). Behavioral studies following lesions of the mesolimbic and mesostriatal serotonergic pathways. *Brain Res.* **106**, 256–270.

Geyer, M. A., Petersen, L. R., Rose, G. J., Horwitt, D. D., Light, R. K., Adams, L. M., Zook, J. A., Hawkins, R. L., and Mandell, A. J. (1978). The effects of LSD and mescaline-derived hallucinogens on sensory-integrative function: Tactile startle. *J. Pharmacol. Exp. Ther.* **207**, 837–847.

Geyer, M. A., Petersen, L. R., and Rose, G. J. (1980). Effects of serotonergic lesions on investigatory responding by rats in a holeboard. *Behav. Neural Biol.* **30**, 160–177.

Geyer, M. A., Flicker, C. E., and Lee, E. H. Y. (1982). Effects of tactile startle on serotonin content of midbrain raphe neurons in rats. *Behav. Brain Res.* **4**, 369–376.

Geyer, M. A., Russo, P. V., and Masten, V. L. (1986). Multivariate assessment of locomotor behavior: Pharmacological and behavioral analyses. *Pharmacol. Biochem. Behav.* **25**, 277–288.

Geyer, M. A., and Braff, D. L., (1987). Startle habituation and sensorimotor gating in schizophrenia and related animal models. *Schizophr. Bull.* **13**, 643–668.

Geyer, M. A., Russo, P., Segal, D. S., and Kuczenski, R. (1987). Effects of apomorphine and amphetamine on patterns of locomotor and investigatory behavior in rats. *Pharmacol. Biochem. Behav.* **28**, 393–399.

Geyer, M. A., Sweerdlow, N. R., Mansbach, R. S., and Braff, D. L. (1990). Startle response models of sensorimotor gating and habituation deficits in schizophrenia. *Brain Res. Bull.* **25**, 485–498.

Glennon, R. A. (1985). Involvement of serotonin in the action of hallucinogenic agents. *In* "Neuropharmacology of Serotonin" (A. R. Green, ed.), pp. 253–280. Oxford University Press, Oxford.

Glennon, R. A. (1989). Stimulus properties of hallucinogenic phenalkylamines and related designer drugs: Formulation of structure-activity relationships. *NIDA Res. Monogr.* **94**, 43–67.

Glennon, R. A., and Misenheimer, B. R. (1989). Stimulus effects of N-monethyl-1-(3,4-methylenedioxyphenyl)-2-aminopropane (MDE) and N-hydroxyl-1-(3,4-methylenedioxyphenyl)-2-aminopropane (N-OH MDA) in rats trained to discriminate MDMA from saline. *Pharmacol. Biochem. Behav.* **33**, 909–912.

Glennon, R. A., and Young, R. (1984). Further investigation of the discriminative stimulus properties of MDA. *Pharmacol. Biochem. Behav.* **20**, 501–505.

Glennon, R. A., Young, R., Rosecrans, J. A., and Anderson, G. M. (1982). Discriminative stimulus properties of MDA analogs. *Biol. Psychiatry* **17**, 807–814.

Glennon, R. A., Young, R., and Rosecrans, J. A. (1983). Antagonism of the effects of the hallucinogen DOM and the purported 5-HT agonist quipazine by 5-HT2 antagonists. *Eur. J. Pharmacol.* **91**, 189–194.

Glennon, R. A., Titeler, M., and McKenney, J. D. (1984). Evidence for 5-HT2 involvement in the mechanism of action of hallucinogenic agents. *Life Sci.* **35**, 2505–2511.

Glennon, R. A., Little, P. J., Rosecrans, J. A., and Yousif, M. (1987). The effect of MDMA ("Ecstasy") and its optical isomers on schedule-controlled responding in mice. *Pharmacol. Biochem. Behav.* **26**, 425–426.

Glennon, R. A., Yousif, M., and Patrick, G. (1988). Stimulus properties of 1-(3,4-methylenedioxyphenyl)-2-aminopropane (MDA) analogs. *Pharmacol. Biochem. Behav.* **29**, 443–449.

Gold, L. H., Koob, G. F., and Geyer, M. A. (1988). Stimulant and hallucinogenic behavioral profiles of 3,4-methylenedioxymethamphetamine and N-ethyl-3,4-methylenedioxyamphetamine in rats. *J. Pharmacol. Exp. Ther.* **247**, 547–555.

Gold, L. H., Geyer, M. A., and Koob, G. F. (1989). Neurochemical mechanisms involved in behavioral effects of amphetamines and related designer drugs. *NIDA Res. Monogr. Ser.* **94**, 101–126.

Gotestam, K. G., and Gunne, L. M. (1972). Subjective effects of two anorexigenic drugs fenfluramine and AN448 in amphetamine-dependent subjects. *Br. J. Addict.* **67**, 39–44.

Griffith, J. D., Nutt, J. G., and Jasinski, D. R. (1975). A comparison of fenfluramine and amphetamine in man. *Clin. Pharmacol. Ther.* **18**, 563–570.

Groves, P. M., and Thompson, R. F. (1970). Habituation: A dual process theory. *Psychol. Rev.* **77**, 419–450.

Hekmatpanah, C. R., and Peroutka, S. J. (1990). 5-Hydroxytryptamine uptake blockers attenuate the 5-hydroxytryptamine-releasing effect of 3,4-methylenedioxymethamphetamine and related agents. *Eur. J. Pharmacol.* **177**, 95–98.

Hiramatsu, M., Nabeshima, T., Kameyama, T., Maeda, Y., and Cho, A. K. (1989). The effect of optical isomers of 3,4-methylenedioxymethamphetamine (MDMA) on stereotyped behavior in rats. *Pharmacol. Biochem. Behav.* **33**, 343–347.

Hole, K., Fuxe, K., and Jonsson, G. (1976). Behavioral effects of 5,7-dihydroxytryptamine lesions of ascending 5- hydroxytryptamine pathways. *Brain Res.* **107**, 385–399.

Ison, J. R., and Hoffman, H. S. (1983). Reflex modification in the domain of startle: II. The anomalous history of a robust and ubiquitous phenomenon. *Psychol. Bull.* **94**, 3–17.

Jacobs, B. L. (1978). Dreams and hallucinations: A common mechanism mediating their phenomenological similarities. *Neurosci. Biobehav. Rev.* **2**, 59–69.

Jacobs, B. L. (1984). Postsynaptic serotonergic action of hallucinogens. *In* "Hallucinogens: Neurochemical, Behavioral, and Clinical Perspectives" (B. L. Jacobs, ed.), pp. 183–202. Raven Press, New York.

Jacobs, B. L., Henriksen, S. J., and Dement, W. C. (1972). Neurochemical bases of the PGO wave. *Brain Res.* **48**, 406–411.

Jacobs, B. L., Wise, W. D., and Taylor, K. M. (1974). Differential behavioral and neurochemical effects following lesions of the dorsal or median raphe nuclei in rats. *Brain Res.* **79**, 353–361.

Jacobs, B. L., Trimbach, C., Eubanks, E. E., and Trulson, M. (1975). Hippocampal mediation of raphe lesion- and PCPA-induced hyperactivity in the rat. *Brain Res.* **94**, 253–261.

Johnson, M. P., Hoffman, A. J., and Nichols, D. E. (1986). Effects of enantiomers of MDA, MDMA, and related analogues on [³H]serotonin and [³H]dopamine release from superfused rat brain slices. *Eur. J. Pharmacol.* **132**, 269–276.

Johnson, M. P., Hanson, G. R., and Gibb, J. W. (1987). Effects of N-ethyl-3,4-methylenedioxyamphetamine (MDE) on central serotonergic and dopaminergic systems in the rat. *Biochem. Pharmacol.* **36**, 4085–4093.

Kamien, J. B., Johanson, C. E., Schuster, C. R., and Woolverton, W. L. (1986). The effects of (q)-methylenedioxymethamphetamine and (q)-methylenedioxyamphetamine in monkeys trained to discriminate (+)-amphetamine from saline. *Drug Alcohol Dep.* **18**, 139–147.

Kehne, J. H., McCloskey, T. C., Taylor, V. L., Black, C. K., Fadayel, G. M., and Schmidt, C. J. (1992). Effects of the serotonin releasers 3,4-methylenedioxymethamphetamine (MDMA), 4-chloroamphetamine (PCA) and fenfluramine on acoustic and tactile startle reflexes in rats. *J. Pharmacol. Exp. Ther.* **260**, 78–89.

Kelly, P. H. (1977). Drug induced motor behavior. *In* "Handbook of Psychopharmacology" (L. L. Iversen, S. D. Iversen, and S. H. Snyder, eds.), Vol. 8, pp. 295–331. Raven Press, New York.

Kelly, P., Seviour, P., and Iversen, S. D. (1975). Amphetamine and apomorphine responses in the rat following 6-OHDA lesions of the nucleus accumbens septi and the corpus striatum. *Brain Res.* **94**, 506–522.

Key, B. J. (1961). Effects of chlorpromazine and LSD on the rate of habituation of the arousal response. *Nature (London)* **190**, 275–277.

Koe, B. K., and Weissman, A. (1966). p-Chlorophenylalanine: A specific depletor of brain serotonin. *J. Pharmacol. Exp. Ther.* **154**, 499–516.

Kostowski, W., Giacolone, E., Garattini, S., and Valzelli, L. (1968). Studies on behavioural and biochemical changes in rats after lesions of midbrain raphe. *Eur. J. Pharmacol.* **4**, 371–376.

Linquist, M. P., and Gotestam, K. G. (1977). Open field behavior after intravenous amphetamine analogues in rats. *Psychopharmacology* **55**, 129–133.

Lipska, B. K., Jaskiw, G. E., Arya, A., and Weinberger, D. R. (1992). Serotonin depletion causes long-term reduction of exploration in the rat. *Pharmacol. Biochem. Behav.* **43**, 1247–1252.

Lorens, S. A. (1978). Some behavioral effects of serotonin depletion depend on method: A comparison of 5,7-dihydroxytryptamine, p-chlorophenylalanine, p-chloroamphetamine, and electrolytic raphe lesions. *Ann. N.Y. Acad. Sci.* **305**, 532–555.

Lorens, S. A., and Guldberg, H. C. (1974). Regional 5-hydroxytryptamine following selective midbrain raphe lesions in the rat. *Brain Res.* **78**, 45–56.

Lorens, S. A., Guldberg, H. C., Hole, K., Kohler, C., and Srebro, B. (1976). Activity, avoidance learning and regional 5-hydroxytryptamine following intra-brain stem 5,7-dihydroxytryptamine and electrolytic midbrain raphe lesions in the rat. *Brain Res.* **108**, 97–113.

Lyon, M., and Robbins, T. (1975). The action of central nervous system stimulant drugs: A general theory concerning amphetamine effects. *Curr. Dev. Psychopharmacol.* **2**, 79–163.

Lyon, R., Glennon, R., and Titeler, M. (1986). MDMA: Stereoselective interactions at brain 5-HT$_1$ and 5-HT$_2$ receptors. *Psychopharmacology* **88**, 525–526.

Mansbach, R. S., Braff, D. L., and Geyer, M. A. (1989). Prepulse inhibition of the acoustic startle response is disrupted by N-ethyl-3,4-methylenedioxyamphetamine (MDEA) in the rat. *Eur. J. Pharmacol.* **167**, 49–55.

McCann, U. D., and Ricaurte, G. A. (1991). Major metabolites of (+−)3,4-methylenedioxyamphetamine (MDA) do not mediate its toxic effects in brain serotonin neurons. *Brain Res.* **545**, 279–282.

McKenna, D. J., Guan, X. M., and Shulgin, A. T. (1991). 3,4-Methylenedioxyamphetamine (MDA) analogues exhibit differential effects on synaptosomal release of ^{3}H-dopamine and ^{3}H-5-hydroxytryptamine. *Pharmacol. Biochem. Behav.* **38**, 505–512.

Miliaressis, T., and St. Laurent, J. (1974). Effects de l'amide de l'acid lysergique-25 dur la reaction de sursuat chez le rat. *Can. J. Physiol. Pharmacol.* **52**, 126–129.

Millan, M. J., and Colpaert, F. C. (1991). Methylenedioxymethamphetamine induces spontaneous tail-flicks in the rat via 5-HT$_{1A}$ receptors. *Eur. J. Pharmacol.* **193**, 145–152.

Mittman, S. M., and Geyer, M. A. (1989). Effects of 5HT-1A agonists on locomotor and investigatory behaviors in rats differ from those of hallucinogens. *Psychopharmacology* **98**, 321–329.

Nader, M. A., Hoffmann, S. M., and Barrett, J. E. (1989). Behavioral effects of (±) 3,4-methylenedioxyamphetamine (MDA) and (±) 3,4-methylenedioxymethamphetamine (MDMA) in the pigeon: Interactions with noradrenergic and serotonergic systems. *Psychopharmacology* **98**, 183–188.

Naranjo, C., Shulgin, A. T., and Sargent, T. (1980). Evaluation of 3,4-methylenedioxymethamphetamine (MDA) as an adjunct of psychotherapy. *Med. Pharmacol. Exp.* **17**, 359–364.

Nichols, D. E. (1986). Differences between the action of MDMA, MBDB and the classic hallucinogens. Identification of a new therapeutic class: Entactogens. *J. Psychoact. Drugs* **18**, 305–313.

Nichols, D. E., and Oberlender, R. (1990). Structure-activity relationships of MDMA and related compounds: A new class of psychoactive drugs? *Ann. N.Y. Acad. Sci.* **600**, 613–625.

Nichols, D., Lloyd, D., Hoffman, A., Nichols, M., and Yim, G. (1982). Effects of certain hallucinogenic amphetamine analogues in the release of [^{3}H]serotonin from rat brain synaptosomes. *J. Med. Chem.* **25**, 530–535.

Nichols, D. E., Hoffman, A. J., Oberlender, R. A., Jacob, P., and Shulgin, A. T. (1986). Derivatives of 1-(1,3-benzodiox0l-5-yl)-2-butanamine: Representatives of a novel therapeutic class. *J. Med. Chem.* **29**, 2009–2015.

Nichols, D. E., Oberlender, R., Burris, K., Hoffman, A. J., and Johnson, M. P. (1989). Studies of dioxole ring substituted 3,4-methylenedioxymethamphetamine (MDA) analogues. *Pharmacol. Biochem. Behav.* **34**, 571–576.

Nisbet, A. R., and Marsden, C. A. (1984). Increased behavioral response to 5-methoxy-*N,N*-dimethyltryptamine but not to RU24969 after intraventricular 5,7-dihydroxytryptamine administration. *Eur. J. Pharmacol.* **104**, 177–180.

Nozaki, M., Vaupel, D. B., and Martin, W. R. (1977). A pharmacological comparison of 3,4-methylendioxyamphetamine and LSD in the chronic spinal dog. *Eur. J. Pharmacol.* **46**, 339–349.

Oberlander, C., Blaquiere, B., and Pujol, J. F. (1986). Distinct functions for dopamine and serotonin in locomotor behavior: Evidence using the $5HT_1$ agonist RU24969 in globus pallidus-lesioned rats. *Neurosci. Lett.* **67**, 113–118.

Oberlander, C., Demassey, Y., Verdu, A., Van de Velde, D., and Bardelay, C. (1987). Tolerance to the serotonin $5\text{-}HT_1$ agonist RU24969 and effects on dopaminergic behavior. *Eur. J. Pharmacol.* **139**, 205–214.

Oberlender, R., and Nichols, D. E. (1988). Drug discrimination studies with MDMA and amphetamine. *Psychopharmacology* **95**, 71–76.

Ogren, S. O., and Holm, A. C. (1980). Test-specific effects of the 5-HT reuptake inhibitors alaproclate and zimelidine on pain sensitivity and morphine analgesia. *J. Neural Transm.* **47**, 253–271.

Paulus, M., and Geyer, M. A. (1991). A temporal and spatial scaling hypothesis for the behavioral effects of psychostimulants. *Psychopharmacology* **104**, 6–16.

Paulus, M. P., and Geyer, M. A. (1992). The effects of MDMA and other methylenedioxy substituted phenylalkylamines on the structure of rat locomotor activity. *Neuropsychopharmacology* **7**, 15–31.

Paulus, M. P., Geyer, M. A., Gold, L. H., and Mandell, A. J. (1990). Application of entropy measures derived from the ergodic theory of dynamical systems to rat locomotor behavior. *Proc. Nat. Acad. Sci. U.S.A.* **87**, 723–727.

Paulus, M. P., Geyer, M. A., and Mandell, A. J. (1991). Statistical mechanics of a neurobiological dynamical system: The spectrum of local entropies ($S(\alpha)$) applied to cocaine-perturbed behavior. *Physica A* **174**, 567–577.

Pijnenburg, A. J. J., Honig, W. M. M., Van der Heyden, J. A. M., and Van Rossum, J. M. (1976). Effect of chemical stimulation of the mesolimbic dopamine system upon locomotor activity. *Eur. J. Pharmacol.* **35**, 45–58.

Pranzatelli, M. R., and Pluchino, R. S. (1991). The relation of central $5\text{-}HT_{1A}$ and $5\text{-}HT_2$ receptors: Low dose agonist-induced selective tolerance in the rat. *Pharmacol. Biochem. Behav.* **39**, 407–413.

Rempel, N., Callaway, C. W., and Geyer, M. A. (1993). Serotonin 5-HT-1B receptor activation mimics behavioral effects of presynaptic serotonin release. *Neuropsychopharmacology* **8**, 201–212.

Rochat, C., Cervo, L., Romandini, S., and Saminin, R. (1982). Differences in the effects of *d*-fenfluramine and morphine on various responses of rats to painful stimuli. *Psychopharmacology* **76**, 188–192.

Sanders-Bush, E., and Steranka, L. R. (1978). Immediate and long-term effects of *p*-chloroamphetamine on brain amines. *Ann. N.Y. Acad. Sci.* **305**, 208–221.

Sarkissian, C. F., Wurtman, R. J., Morse, A. N., and Gleason, R. (1990). Effects of fluoxetine or *d*-fenfuramine on serotonin release from, and levels in, rat frontal cortex. *Brain Res.* **529**, 294–301.

Schecter, M. D. (1989). Serotonergic-dopaminergic mediation of 3,4-methylenedioxymethamphetamine (MDMA, "Ecstasy"). *Pharmacol. Biochem. Behav.* **31**, 817–824.

Schmidt, C. (1987). Neurotoxicity of the psychedelic amphetamine, MDMA. *J. Pharmacol. Exp. Ther.* **240**, 1–7.

Schmidt, C. J., Levin, J. A., and Lovenberg, W. (1987). In vitro and in vivo neurochemical effects of methylenedioxymethamphetamine on striatal monoaminergic systems in the rat brain. *Biochem. Pharmacol.* **36**, 747–755.

Segal, D. S. (1975). Behavioral and neurochemical correlates of repeated d-amphetamine administration. *In* "Neurobiological Mechanisms of Adaptation and Behavior" (A. J. Mandell, ed.), pp. 247–262. Raven Press, New York.

Segal, D. S., and Mandell, A. J. (1974). Long-term administration of d-amphetamine: Progressive augmentation of motor activity and stereotypy. *Pharmacol. Biochem. Behav.* **2**, 249–255.

Shulgin, A. T. (1978). Psychotomimetic drugs: Structure–activity relationships. *In* "Handbook of Psychopharmacology" (L. L. Iversen, S. D. Iversen, and S. H. Snyder, eds.), Vol. 11, pp. 243–333. Raven Press, New York.

Shulgin, A. (1986). The background and chemistry of MDMA *J. Psychoact. Drugs* **18**, 291–304.

Shulgin, A. T., Sargent, T., and Naranjo, C. (1969). Structure–activity relationships of one-ring psychotomimetics. *Nature (London)* **221**, 537–541.

Slikker, W., Holson, R. R., Ali, S. F., Kolta, M. G., Paule, M. G., Scallet, A. C., McMillan, D. E., Bailey, J. R., Hong, J. S., and Scalzo, F. M. (1989). Behavioral and neurochemical effects of orally administered MDMA in the rodent and nonhuman primate. *Neurotoxicology* **10**, 529–542.

Soubrie, P. (1986). Reconciling the role of central serotonin neurons in human and animal behavior. *Behav. Brain Sci.* **9**, 319–364.

Spanos, L. J., and Yamamoto, B. K. (1989). Acute and subchronic effects of methylenedioxymethamphetamine (qMDMA) on locomotion and serotonin syndrome behavior in the rat. *Pharmacol. Biochem. Behav.* **32**, 835–840.

Steinfels, G. F., Heym, J., Strecker, R. E., and Jacobs, B. L. (1983). Raphe unit activity in freely moving cats is altered by manipulations of central but not peripheral motor systems. *Brain Res.* **279**, 77–84.

Stone, D., Stahl, D., Hanson, G., and Gibb, J. (1986). The effects of MDMA and MDA on monoaminergic systems in the rat brain. *Eur. J. Pharmacol.* **128**, 41–48.

Swerdlow, N. R., Vaccarino, F. J., Amalric, M., and Koob, G. F. (1986). The neural substrates for the motor activating properties of psychostimulants: A review of recent findings. *Pharmacol. Biochem. Behav.* **25**, 233–248.

Swerdlow, N. R., Braff, D. L., Masten, V. L., and Geyer, M. A. (1990). Schizophrenic-like sensorimotor gating abnormalities in rats following dopamine infusions into the nucleus acumbens. *Psychopharmacology* **101**, 414–420.

Swerdlow, N. R., Keith, V. A., Braff, D. L., and Geyer, M. A. (1991). The effects of spiperone, SCH 23390 and clozapine on apomorphine-inhibition of sensorimotor gating of the startle response in the rat. *J. Pharmacol. Exp. Ther.* **256**, 530–536.

Swerdlow, N. R., Braff, D. L., Taaid, N., and Geyer, M. A. (1993). Assessing the validity of an animal model of deficient sensorimotor gating in schizophrenic patients. *Arch. Gen. Psychiatry,* in press.

Teitler, M., Leonhardt, S., Appel, N. M., De Souza, E. B., and Glennon, R. A. (1990). Receptor pharmacology of MDMA and related hallucinogens. *Ann. N.Y. Acad. Sci.* **600**, 626–639.

Tricklebank, M. D., Middlemiss, D. N., and Neill, J. (1986). Pharmacological analysis of the behavioral and thermoregulatory effects of the putative 5-HT-1B receptor agonist, RU 24969, in the rat. *Neuropharmacology* **25**, 877–886.

Trulson, M. E., and Jacobs, B. L. (1979). Raphe unit activity in freely moving cats: Correlation with level of behavioral arousal. *Brain Res.* **163**, 135–150.

Van De Kar, L. D., Richardson, K. D., and Urban, J. H. (1985a). Serotonin and norepinephrine-dependent effects of fenfluramine on plasma renin activity in conscious male rats. *Neuropharmacology* **24,** 487–494.

Van De Kar, L. D., Urban, J. H., Richardson, K. D., and Bethea, C. L. (1985b). Pharmacological studies on the serotonergic and nonserotonin-mediated stimulation of prolactin and corticosterone secretion by fenfluramine. *Neuroendocrinology* **41,** 283–288.

Wing, L. L., Tapson, G. S., and Geyer, M. A. (1990). 5-HT$_2$ mediation of acute behavioral effects of hallucinogens in rats. *Psychopharmacology* **100,** 417–425.

Wong, D. T., Bymaster, F. P., Reid, L. R., and Threlkeld, P. G. (1983). Fluoxetine and two other serotonin uptake inhibitors without affinity for neuronal receptors. *Biochem. Pharmacol.* **32,** 1287–1293.

Ziance, R. J., Sipes, I. G., Kinnard, W. J., and Buckley, J. P. (1972). Central nervous system effects of fenfluramine hydrochloride. *J. Pharmacol. Exp. Ther.* **180,** 110–117.

Ray W. Fuller
Mark G. Henderson

7
Neurochemistry of Halogenated Amphetamines

I. INTRODUCTION AND BACKGROUND

Initially, the halogenated derivatives of amphetamine were synthesized primarily in the search for anorexigenic drugs with less central stimulant action than amphetamine (Owen, 1963; Kaergaard Nielsen et al., 1967). The discovery that p-chloromethamphetamine selectively depleted brain serotonin (Pletscher et al., 1963, 1964) triggered interest in this most widely known neurochemical effect of halogenated amphetamines. In early structure–activity relationship studies, p-chloroamphetamine (PCA) was found to be the most potent serotonin-depleting member of the halogenated amphetamine family (Fuller et al., 1965). PCA and fenfluramine (N-ethyl-m-trifluoromethylamphetamine) have become the two most extensively studied halogenated amphetamines. Although some halogenated amphetamines retain some of the varied neurochemical effects of amphetamine, for example, influencing catecholamines and serotonin, the prominent characteristic associated with halogenated amphetamines is their relatively selective effects on serotonin neurons. PCA and several related compounds release serotonin acutely and cause prolonged effects on serotonin neurons. These effects are the major focus of this chapter.

II. ACUTE NEUROCHEMICAL EFFECTS OF p-CHLOROAMPHETAMINE

Within 1 hr of PCA injection into rats, neurochemical changes can be detected. Figure 1 (*top*) shows brain concentrations of serotonin, dopamine, norepinephrine, and their metabolites 4 hr after PCA injection into rats. Concentrations of serotonin and of its major metabolite 5-hydroxyindoleacetic acid (5-HIAA) are decreased. This result distinguishes PCA from other serotonin-releasing drugs such as reserpine, tetrabenazine, Ro4-1284, and high doses of L-DOPA, which increase 5-HIAA concentration while decreasing serotonin concentration in rat brain (Fuller and Perry, 1975; Saner and Pletscher, 1978; Tachiki *et al.*, 1978). A rapid decrease in tryptophan hydroxylase (TPH) activity also occurs after PCA administration (Sanders-Bush *et al.*, 1972a). PCA is a reasonably potent, reversible, and competitive inhibitor of monoamine oxidase (MAO) (Fuller, 1966), particularly MAO-A (Fuller and Hemrick-Luecke, 1982). The concentration of PCA in brain is adequate to inhibit MAO based on its *in vitro* potency (Fuller, 1966), and diminished conversion of radioactive serotonin to radioactive 5-HIAA has been shown after PCA administration (Fuller and Hines, 1970). Inhibition of MAO would not cause the lowering of brain serotonin concentration but might contribute to the lowering of 5-HIAA concentration. The decrease in 3,4-dihydroxyphenylacetic acid (DOPAC) concentration 4 hr after PCA injection (Fig. 1, *top*) probably is the result of inhibition of MAO-A.

At early times after injection of PCA in rats, an increase in extracellular concentrations of serotonin is seen, as measured by intracerebral microdialysis (Sharp *et al.*, 1986; Adell *et al.*, 1989; Hutson and Curzon, 1989). This increase in extracellular serotonin, due to release of intraneuronal granular stores of serotonin, is the cause of several acute effects of PCA such as suppression of food intake (Kaergaard Nielsen *et al.*, 1967), the serotonin behavioral syndrome (Trulson and Jacobs, 1976; Growdon, 1977; Lucki *et al.*, 1984), and increases in serum corticosterone and prolactin (Fuller and Snoddy, 1980; Fuller and Clemens, 1981).

III. LONG-TERM DEPLETION OF BRAIN SEROTONIN BY p-CHLOROAMPHETAMINE

Not until the studies by Sanders-Bush *et al.* (1972b) showed that brain serotonin was depleted for as long as 4 mo after a single dose of PCA in rats

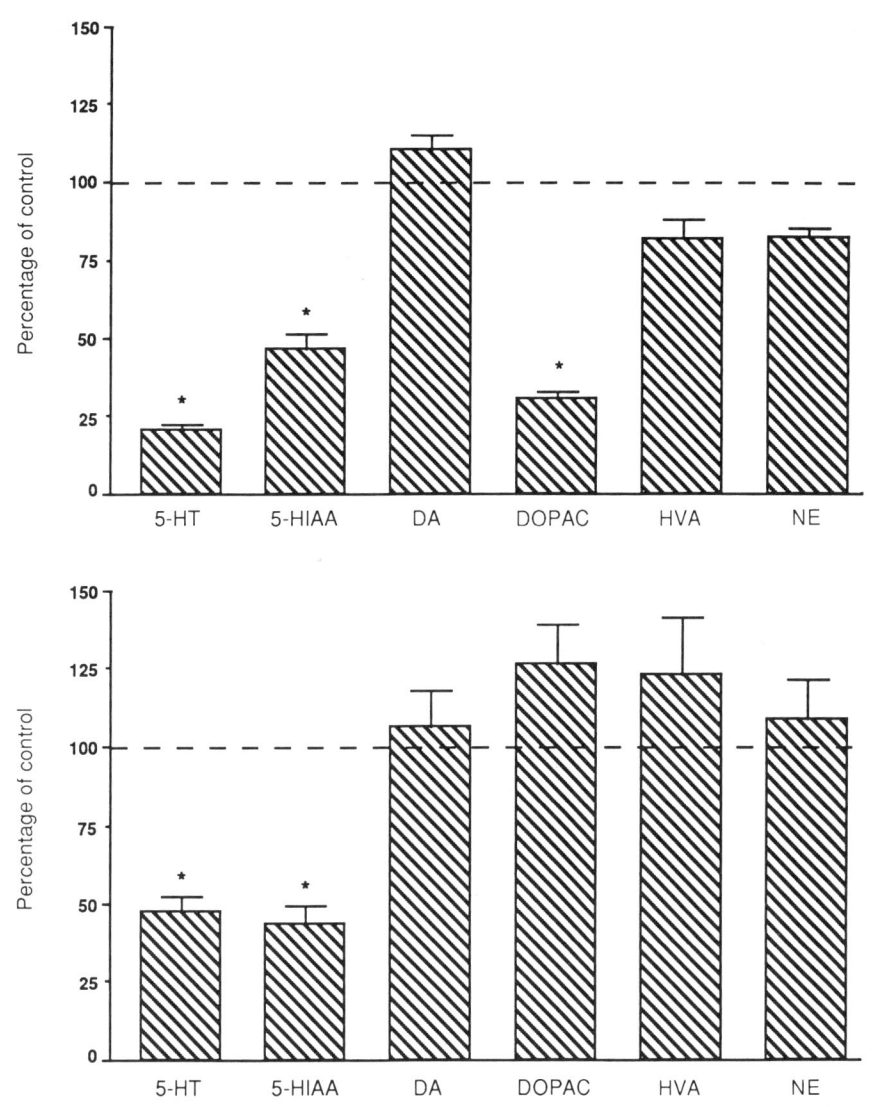

FIGURE 1 Brain monoamines and their metabolites at 4 hr (*top*) and 1 wk (*bottom*) after injection of 0.1 mmol/kg (20.6 mg/kg) *p*-chloroamphetamine (PCA) in rats (ip). Serotonin and other constituents were measured by high performance liquid chromatography (HPLC) with electrochemical detection in these and subsequent experiments. Abbreviations: 5-HT, serotonin; 5-HIAA, 5-hydroxyindoleacetic acid; DA, dopamine; DOPAC, 3,4-dihydroxyphenylacetic acid; HVA, homovanillic acid; NE, norepinephrine. Mean values ± standard errors for 5 rats are shown. Asterisks indicate significant difference from control group ($P < .05$).

did researchers widely appreciate that PCA affected brain serotonin neurons by a mechanism additional to reversible depletion of serotonin by release of stores from intraneuronal vesicles. Actually, Frey (1970) noted earlier that repeated doses of PCA for as short a time as 5 days led to depletion of brain serotonin that persisted up to 4 wk in rats, but that observation, published in a large book about amphetamines, was not widely noticed by investigators in the field. Figure 1 (*bottom*) shows the neurochemical effects of PCA 1 wk after injection of a single dose into rats. Once the long-lasting nature of the PCA-induced serotonin depletion became the subject of research attention, several characteristics of the depletion were found that indicated the effect to be one of neurotoxicity.

A. Characteristics of Long-Term Effects

Accompanying the prolonged decrease in brain serotonin content after PCA injection are changes in all the parameters that have been measured to be specifically associated with brain serotonin neurons. Steady state concentrations of the serotonin metabolite 5-HIAA are decreased, as is the rate of 5-HIAA accumulation after probenecid is administered to block its efflux from brain (Sanders-Bush *et al.*, 1972a; Fuller and Snoddy, 1974). TPH activity as measured by *in vitro* assay is decreased (Sanders-Bush *et al.*, 1972a), as is tryptophan hydroxylation measured *in vivo* by the rate of accumulation of 5-hydroxytryptophan (5-HTP) after decarboxylase inhibition by NSD 1015 (*m*-hydroxybenzylhydrazine) (Fuller and Perry, 1983) or the conversion of radioactive tryptophan to radioactive serotonin (Sanders-Bush and Sulser, 1970). The mechanism of the decrease in TPH is not fully known, since PCA does not inhibit the enzyme when added directly to it *in vitro* (Sanders-Bush *et al.*, 1972b). Ross and Froden (1977) suggested that the acute decrease in TPH activity after PCA injection into rats is secondary to the release of serotonin and is mediated by serotonin receptor stimulation. These investigators found that metergoline (a serotonin receptor antagonist) and reserpine (a serotonin depletor) antagonized the PCA-induced decrease in TPH activity in rat brain. Such a mechanism would not explain the long-term loss of TPH, however.

Serotonin uptake capacity is decreased for prolonged periods of time after PCA injection (Massari and Sanders-Bush, 1975; Sanders-Bush *et al.*, 1975; Sanders-Bush and Steranka, 1978), as is binding of numerous radioligands to the serotonin uptake carrier, including norzimelidine (Hall, 1984), imipramine, citalopram (D'Amato *et al.*, 1987), 6-nitroquipazine (Hashimoto and Goromaru, 1990), sertraline (Koe *et al.*, 1990), and paroxetine (Hrdina *et al.*, 1990; Newman *et al.*, 1991; Dewar *et al.*, 1992). Such a decrease in radioligand binding to the uptake carrier was not seen after serotonin depletion by p-chlorophenylalanine (Dewar *et al.*, 1992), suggesting that the decreases after PCA administration relate to neurotoxic effects on serotonin nerve terminals, not simply to depletion of serotonin stores. All these changes point to permanent damage of serotonin nerve terminals or axons. As further documentation of such damage, Mamounas and Molliver (1988) and Berger *et al.* (1989) reported loss of serotonin-containing nerve fibers in serotonin projection areas of rat brain at long times after PCA injection. Also, Mamounas and Molliver (1988) and Fritschy *et al.* (1988) described loss of serotonin-positive neurons in the dorsal raphe nucleus that were labeled retrogradely by injection of tracer into rat forebrain areas at long times after PCA injection. Commins *et al.* (1987) presented histological evidence for degenerating axon terminals at shorter times after PCA injection.

Researchers long have recognized that brain serotonin depletion by PCA is incomplete, even after repeated doses (Pletscher *et al.*, 1964; Fuller *et al.*, 1973b). Meek (1978) attempted to increase the extent of serotonin depletion by giving doses of PCA higher than those that ordinarily would be tolerated in rats. He found that chlorpromazine prevented the hyperthermia and hyperactivity produced by PCA and allowed a dose of 80 mg/kg of PCA (usually lethal) to be given to rats. However, brain serotonin depletion was still not complete, even when a second dose of PCA was given 2 days after the first. These findings suggested that certain brain serotonin neurons are resistant to PCA.

Now investigators know that not all serotonin neurons in brain are susceptible to the neurotoxic actions of PCA. Fritschy *et al.* (1987) showed that axon terminals from the dorsal raphe nucleus were more susceptible than axon terminals from other serotonergic nuclei to PCA. Blier *et al.* (1990) found that serotonin nerve terminals in the dentate gyrus, a region innervated exclusively by the median raphe nucleus, were not affected by PCA. Mamounas *et al.* (1991; Mamounas and Molliver, 1988) reported that fine axons with minute varicosities in rat cerebral cortex degenerated selectively after PCA injection in rats, whereas beaded axons with large spherical varicosities were spared. The axons with small fusiform varicosities are known to come from the dorsal raphe nucleus, whereas the axons with large round varicosities are projections from the median raphe nucleus (Kosofsky and Molliver, 1987). The two types of axon terminals are known to have different ontogenic development (Vu and Tork, 1992), although the full physiological significance of these discrete types of terminals is unknown. The fact that these axons are differentially vulnerable to PCA neurotoxicity may reflect some differences in their fundamental properties and eventually may aid in elucidation of their functional differences.

B. Use of *p*-Chloroamphetamine for Chemical Lesioning of Serotonin Pathways

The ability of PCA to cause neurotoxic lesions of some brain serotonin neurons has resulted in its use as a chemical neurotoxin to study the projections and physiological functions of serotonin neurons in brain. In several ways, PCA and agents related to it differ from a second class of chemical neurotoxins that lesion serotonin neurons, namely the dihydroxytryptamines 5,6- and 5,7 dihydroxytryptamine (5,6-DHT and 5,7-DHT).

C. Comparison of *p*-Chloroamphetamine with Hydroxylated Indoleamine Neurotoxins

1. Route of Administration

Direct intracerebroventricular (icv), intracisternal, or intracerebral injections are needed with 5,6- or 5,7-DHT because these compounds, like serotonin, do not cross the blood–brain barrier. Not only is this direct injection not needed with PCA, but Sherman et al. (1975) found that, when PCA is injected intracerebroventricularly, doses must be almost as high to deplete brain serotonin as when PCA is injected intraperitoneally (ip). That effect apparently is due to the fact that PCA crosses the blood–brain barrier easily in both directions and leaves the brain quickly after icv injection. Berger et al. (1990) reported that direct intracerebral injection, even infusion for up to 48 hr, does not result in any neurotoxicity of PCA in rat cerebral cortex.

2. Regions of CNS Affected

The magnitude of serotonin depletion induced by PCA in the central nervous system (CNS) varies by region. The pattern of serotonin depletion by PCA is different from that seen with 5,6- or 5,7-DHT.

Acutely, PCA administration causes a marked decrease in serotonin levels 1 hr after injection in rat cortex, hippocampus, and striatum. The midbrain, where the serotonin cell bodies are found, shows no decrease, nor are serotonin levels reduced in the hypothalamus or spinal cord (Ogren, 1982). By 4 hr after PCA administration, large decreases in serotonin and TPH are seen in the hypothalamus, midbrain, cortex, hippocampus, and striatum, whereas somewhat smaller decreases are noted in the medulla–pons area and the spinal cord (Sanders-Bush et al., 1975). A similar trend to that seen at 4 hr also is seen at 14 days (Sanders-Bush et al., 1975) and 30 days (Harvey, 1978). Figure 2 shows the decrease in serotonin levels as a percentage of control 3 days after a 10 mg/kg ip dose of PCA. Thus, changes produced by PCA result in a long-lasting decrease in serotonin and TPH activity in the areas examined in the mesencephalon (midbrain, raphe), diencephalon (hypothalamus), and telencephalon (hippocampus, frontal cortex), while less dramatic serotonin decreases are seen in the spinal cord and areas surveyed in the metencephalon (pons) and myelencephalon (medulla).

The distribution of serotonin depletion induced by 5,6- or 5,7-DHT differs from that seen with PCA, perhaps because of the difference in the method of administration of the drug. PCA is given systemically, but the dihydroxytryptamines must be administered directly into the brain, thus

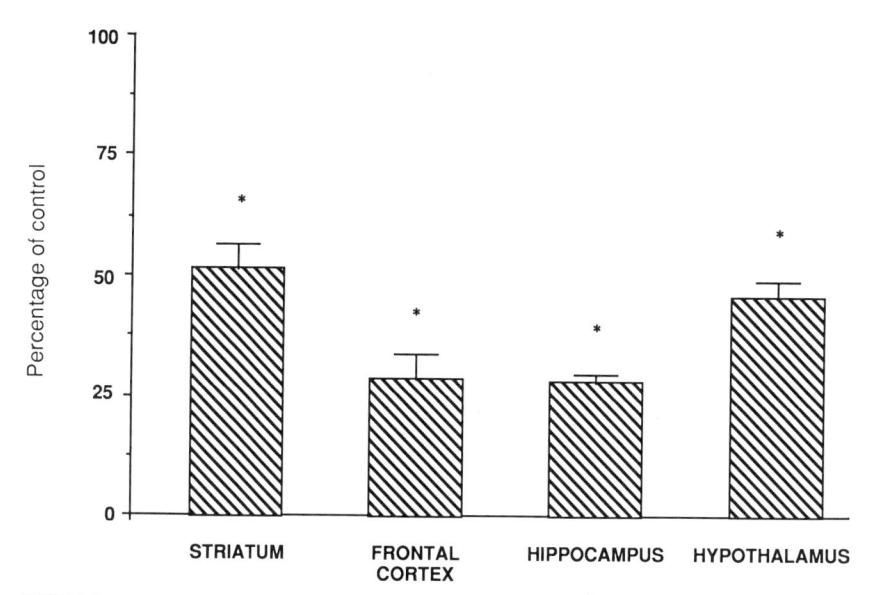

FIGURE 2 Serotonin concentration in rat brain regions 3 days after a 10 mg/kg ip dose of p-chloroamphetamine (PCA). Mean values ± standard errors for 5 rats per group are shown. Asterisks indicate significant difference from control group ($P < .05$).

limiting the regions that are affected by the diffusion of the drug from the site of injection. Unlike PCA, 5,7-DHT causes a marked decrease in spinal cord serotonin concentration. The forebrain and brain stem regions also were found to have decreased serotonin concentrations (Bjorklund et al., 1975). Similarly, serotonin uptake was decreased by 5,6-DHT 10–17 days after injection into the hypothalamus, medulla oblongata, and spinal cord. By 3 mo, only the cortex and spinal cord exhibited decreased serotonin uptake (Bjorklund et al., 1973).

Although PCA and the dihydroxytryptamines cause long-term depletion of serotonin, a difference exists in their regional specificity. The most notable difference is that PCA does not affect spinal cord serotonin concentration whereas the dihydroxytryptamines severely deplete serotonin in spinal cord.

3. Ontogeny of Susceptibility

Clemens et al. (1978) found that PCA, even at high doses, did not cause either short-term or long-term depletion of brain serotonin in newborn (3- to 5-day-old) rats but was effective in 7- to 20-day-old rats just as it is in older rats. Figure 3 shows the lack of serotonin depletion in rats given PCA at 3 days of age, whereas rats given PCA at 20 days of age show serotonin depletion similar to that found in adult rats. Sherman and Gal (1979) found

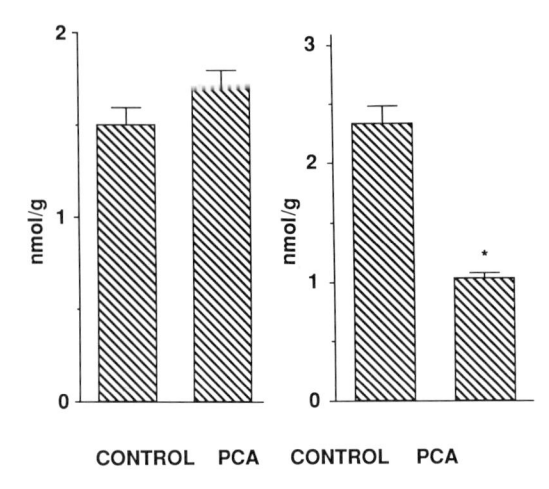

FIGURE 3 Depletion of brain serotonin 6 hr after administration of p-chloroamphetamine (PCA, 20 mg/kg ip) to 20-day-old rats (*right*). Administration of same dose did not affect 3-day-old rats (*left*). Mean values ± standard errors for 5 rats per group are shown. Asterisk indicates significant difference from control group ($P < .05$). Data are from Clemens et al. (1978).

that PCA and its metabolites appeared rapidly in the brains of 3-day-old rats, suggesting that their resistance to serotonin depletion was not the result of an inability of PCA to reach the brain or to be metabolized.

In contrast to their resistance to PCA neurotoxicity, serotonin neurons in neonatal rats are destroyed readily by 5,6- or 5,7-DHT. Administration of dihydroxytryptamines to neonatal rats 1–5 days old has been used routinely to lesion serotonin neurons (Krieger, 1975; Hamon et al., 1981; Hard et al., 1983; Mueller et al., 1985; Pranzatelli, 1990). Because the dihydroxytryptamines, as well as PCA, require the uptake carrier on serotonin neurons, the insensitivity to PCA could not be attributed to lack of development of the uptake carrier on serotonin neurons. Some other factor(s) must account for the resistance of these neurons to PCA neurotoxicity in neonatal rats.

4. Time Course and Reversibility of Effects

Within a few hours of the injection of PCA or 5,6- or 5,7-DHT into rats, marked depletion of brain serotonin is noted (Fuller et al., 1973b; Bjorklund et al., 1974). This depletion of serotonin persists for weeks or months. In the case of PCA and related halogenated amphetamines, the initial depletion of brain serotonin is not the result of neurotoxicity, however, and actually can be reversed by subsequent administration of an inhibitor of the serotonin uptake carrier (Meek et al., 1971; Fuller et al., 1975b; Fuller and Perry, 1978). To our knowledge, no reports have been made of reversal of the serotonin depletion after 5,6- or 5,7-DHT injection by subsequent administration of an uptake inhibitor or other pharmacological agent. Indeed, although the depletion of brain serotonin by PCA and related halogenated amphetamines is prevented readily and completely by prior administration of inhibitors of the serotonin uptake carrier, the serotonin depletion by dihydroxytryptamines is difficult to prevent with uptake inhibitors (Breese and Mueller, 1978).

5. Susceptibility across Species

Most of the studies on serotonin depletion by PCA have been in rats, but data in other species are available as well. In the original study of *p*-chloromethamphetamine, Pletscher *et al.* (1964) stated that this compound did not cause major changes in brain serotonin content in mice or rabbits. PCA *does* deplete brain serotonin in mice (Fuller *et al.*, 1974); we have used the antagonism of that depletion as a means of comparing uptake-inhibiting antidepressant drugs for many years (Fuller *et al.*, 1974,1975c). However, the depletion of brain serotonin in mice after injection of a single dose of PCA is short term (Steranka and Sanders-Bush, 1978a; Burke *et al.*, 1979). Steranka and Sanders-Bush (1978a) showed that this short duration of effect was largely the result of the comparatively short half-life of PCA in mouse brain. These researchers administered PCA subcutaneously (sc) via osmotic minipumps for a period of 3 days and observed depletion of brain serotonin that lasted at least 4 wk in mice.

Alesci and Bagnoli (1988) showed that PCA is useful for lesioning serotonin neurons in pigeon brain. Oral administration of PCA to dogs over a period of 90 days caused a dose-dependent depletion of brain serotonin that persisted, in part, 2 wk after treatment was discontinued (Fuller *et al.*, 1979). Long-lasting depletion of brain serotonin by PCA in guinea pigs has been reported (Fuller and Perry, 1974). PCA at doses tested by Sherman and Gal (1976b) did not deplete brain serotonin in chickens. Thus PCA depletes brain serotonin in several species, but not in all species.

IV. STRUCTURE–ACTIVITY RELATIONSHIPS

Both enantiomers of PCA and of some related chlorinated amphetamines deplete brain serotonin initially (Fuller *et al.*, 1965; Sekerke *et al.*, 1975), but the prolonged depletion of brain 5-hydroxyindoles and TPH activity is produced to a much greater extent by (+)-PCA than by (−)-PCA (Sekerke *et al.*, 1975). The different neurotoxicity of the two enantiomers appears not to be the result of any difference in their persistence in brain (Sekerke *et al.*, 1975), but may be the result of different pharmacological interactions. Rudnick and Wall (1992) have observed that (+)-PCA has higher affinity than (−)-PCA for the serotonin transporter. Figure 4 shows chemical structures of many of the analogs of PCA that have been studied, as discussed in the following sections.

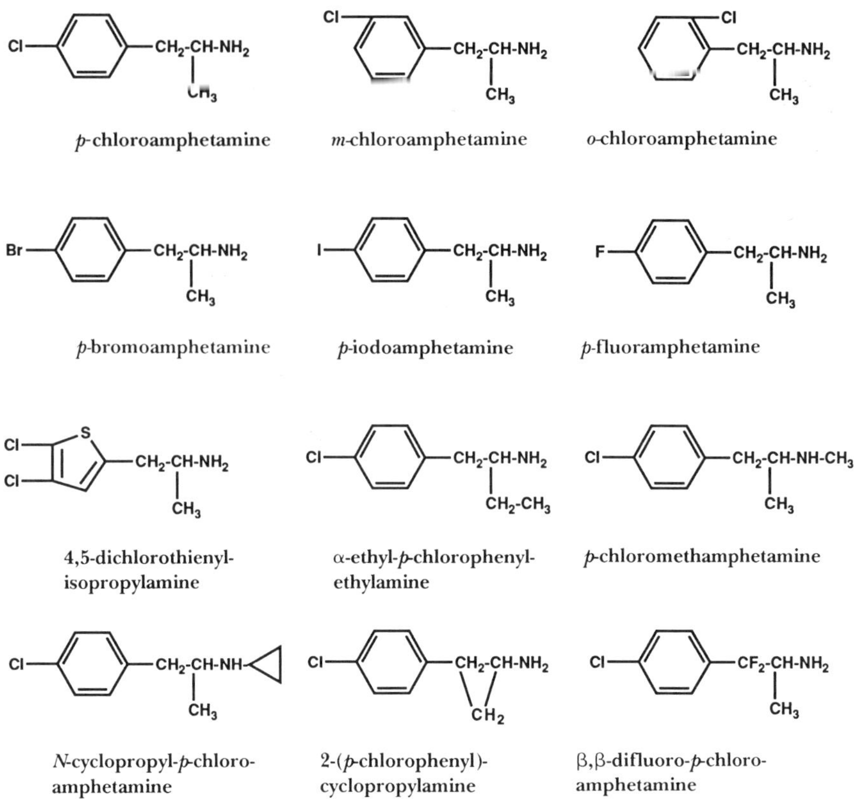

FIGURE 4 Chemical structures of *p*-chloroamphetamine and some of its analogs that are studied as serotonin depletors.

A. Ring-Substituted Amphetamines

Numerous amphetamines with substituents other than *p*-chloro on the phenyl ring have been studied in comparison with PCA. In rats, *o*-chloroamphetamine does not deplete brain serotonin and *m*-chloroamphetamine causes a much smaller effect than PCA (Fuller *et al.*, 1972). Brain concentrations of the *o*-chloro and *m*-chloro compounds decline much more rapidly than those of PCA, because the site of metabolic hydroxylation—the *para* position of the phenyl ring—is accessible in the *o*-chloro and *m*-chloro compounds but is occupied in PCA. Amphetamine is metabolized in rats predominantly by hydroxylation at the *para* position of the phenyl ring (Axelrod, 1954). This ring hydroxylation is blocked by drugs such as desipramine (Dolfini *et al.*, 1969) and iprindole (Freeman and Sulser, 1972). Pretreatment with desipramine or iprindole slows the disappearance of *o*-chloro- and *m*-chloroamphetamines from brain so brain concentrations are similar to those of PCA (Fuller et al., 1972; Fuller and Baker, 1974). When metabolic differences among the chloroamphetamines are eliminated in this way, m-chloroamphetamine causes brain serotonin depletion similar to that caused by PCA, but o-chloroamphetamine causes a slight increase in brain serotonin concentration, apparently by MAO inhibition (Fuller et al., 1972). In a species in which ring hydroxylation is not a metabolic pathway for amphetamines—the guinea pig—m-chloroamphetamine and PCA cause similar depletion of brain serotonin (Fuller, 1972).

p-Bromoamphetamine and p-iodoamphetamine cause depletion of brain serotonin similar to that caused by PCA, but p-fluoroamphetamine has a smaller effect (Fuller et al., 1975a). p-Methylamphetamine, p-methoxyamphetamine, p-phenoxyamphetamine, and p-trifluoromethylamphetamine cause little or no depletion of brain serotonin (Fuller et al., 1973b).

Conde et al. (1978) synthesized some ring-chlorinated thienylisopropylamines, thiophene analogs of chloroamphetamines, and found that 4,5-dichlorothienylisopropylamine caused PCA-like depletion of brain serotonin that persisted at 7 days in rats. Monochloro analogs were less potent.

B. Side-Chain-Substituted Amphetamines

Several N-alkyl analogs of PCA can deplete brain serotonin (Fuller and Baker, 1977). Indeed, N-methyl PCA was the first compound in this series to be found to deplete serotonin (Pletscher et al., 1963, 1964). These N-alkyl compounds are metabolized by N dealkylation to PCA. Discerning whether the serotonin-depleting effects are accounted for entirely by PCA formation or whether the parent compounds are able to deplete serotonin has not been possible (Fuller and Baker, 1977).

Johnson et al. (1990) synthesized the α-ethyl homolog of PCA and found it to be effective but less potent than PCA in causing long-lasting depletion of serotonin and 5-HIAA levels and of tritiated paroxetine binding. The α-ethyl homolog was also less potent than PCA in blocking serotonin uptake in vitro.

N-Cyclopropyl PCA was reported to be an irreversible inhibitor of MAO that increased brain serotonin concentration transiently before causing a PCA-like long-lasting depletion of serotonin (Fuller and Molloy, 1973; Fuller and Perry, 1977). The inactivation of MAO by N-cyclopropyl PCA, dependent on MAO activity, could be prevented by harmaline (a reversible short-acting inhibitor of MAO), resulting in an earlier decline in brain serotonin concentration (Fuller et al., 1979). Conversely, the serotonin-depleting action of N-cyclopropyl PCA, being dependent on the serotonin uptake carrier, could be prevented by fluoxetine pretreatment; the result was an exaggerated increase in serotonin concentration as a consequence of the MAO inhibition by N-cyclopropyl PCA (Fuller and Perry, 1977). N-Cyclopropyl PCA was proposed to have PCA-like properties with the additional action of inhibiting MAO irreversibly. However, Fagervall and Ross (1986) later reported an inability to detect any transport of N-cyclopropyl PCA by the uptake carrier on serotonin neurons. These investigators suggested that metabolic N-dealkylation to form PCA might account for the serotonin-depleting action of N-cyclopropyl PCA. This suggestion was correct, as shown by Fuller et al. (1987), who measured brain concentrations of PCA sufficient to account for the serotonin depletion that occurred after N-cyclopropyl PCA administration to rats.

Another PCA analog that is an irreversible inhibitor of MAO is 2-(p-chlorophenyl)cyclopropylamine (Fuller and Kaiser, 1980). This compound inhibits MAO-A and MAO-B irreversibly but does not have the serotonin-depleting activity of PCA.

A particularly interesting analog is β,β-difluoro PCA (Fuller et al., 1973a). The two fluorines withdraw electrons so the nitrogen is much less basic, that is, much less willing to accept a proton. The result is a lowering of the pK_a value from ~9.3 for PCA to 6.8, below physiological pH. As a result, β,β-difluoro PCA exists mainly as a neutral nonprotonated molecule, whereas PCA is virtually entirely in the protonated cationic form. Therefore, β,β-difluoro PCA distributes much differently in tissues than does PCA and is metabolized much more rapidly than PCA. β,β-Difluoro PCA localizes mainly in fat (the tissue containing lowest concentrations of PCA) at the expense of other tissues, including brain. β,β-Difluoro PCA is metabolized much more rapidly and extensively than PCA (Fuller et al., 1973a). For β,β-difluoro PCA to deplete brain serotonin, it must be injected at 4 times the doses of PCA to produce equivalent brain concentrations of the two drugs; then the depletion of serotonin is transient, unlike the long-lasting depletion caused by PCA. If the metabolism of β,β-difluoro PCA could be inhibited, prolonged brain concentrations of the drug *might* cause longer lasting depletion of brain serotonin.

Owen et al. (1991) studied a PCA analog with a monofluoro substituent on the α-methyl carbon atom. This analog attained lower brain concentrations than PCA at comparable doses and disappeared from brain more rapidly than PCA in rats. The α-methyl substituted PCA depleted brain serotonin less than did PCA; that difference was not accounted for fully by differences in drug concentrations in brain.

Studies with PCA analogs have demonstrated the importance of persistence of PCA in the brain for the long-term neurotoxic effects to be produced. m-Chloroamphetamine is an analog that does not cause long-term depletion of brain serotonin unless it persists in brain with a half-life similar to that of PCA, as in rats treated with desipramine or iprindole to block aromatic hydroxylation or in guinea pigs, which do not metabolize amphetamines by aromatic hydroxylation. Steranka and Sanders-Bush (1978b) showed that induction of liver microsomal enzymes by 3-methylcholanthrene markedly accelerated the disappearance of PCA from rat brain and antagonized its neurotoxic effects on brain serotonin neurons.

V. ARE p-CHLOROAMPHETAMINE AND ITS ANALOGS SUBSTRATES FOR THE SEROTONIN TRANSPORTER?

The ability of inhibitors of the serotonin uptake carrier to block serotonin depletion of PCA (discussed subsequently), coupled with the affinity of PCA for the serotonin uptake carrier (Carlsson, 1970; Wong et al., 1973), has led to the idea that PCA is accumulated in serotonin neurons by the membrane uptake carrier and that this accumulation is essential to the short-term and long-term depletion of serotonin (Meek et al., 1971; Fuller, 1980). However, attempts to measure the transport of PCA by the serotonin transporter directly have failed (Ross, 1976; Sanders-Bush and Steranka, 1978; Ask et al., 1989). Some evidence is consistent with the idea that PCA is a substrate for the serotonin transporter, however. After the injection of PCA into rats, much of the drug in the brain is associated with synaptosomes after differential centrifugation (Wong et al., 1972). Fagervall and Ross (1986) said that PCA preferentially inhibits MAO inside serotonergic nerve terminals, suggesting that the drug might be accumulated by the membrane transporter. Rudnick and Wall (1992) provided indirect evidence that PCA is a substrate for the serotonin transporter on platelet membranes. Taddei and Mennini (1989) observed that a related halogenated amphetamine, d-fenfluramine, is accumulated by rat brain synaptosomes *in vitro*, possibly by the serotonin transporter.

VI. IS A METABOLITE OF p-CHLOROAMPHETAMINE INVOLVED IN ITS NEUROCHEMICAL EFFECTS?

The idea that a cytotoxic metabolite of PCA may be involved in its neurotoxic effects on brain serotonin neurons has been attractive, in part because the PCA molecule itself does not easily suggest a mechanism for neurotoxicity in the same way that autoxidation of 5,6- or 5,7-DHT does for those serotonin neurotoxins. Some studies of PCA metabolism have been reported, although our understanding of its metabolism is far from complete.

Parli (1976) found that much of the administered PCA is excreted unchanged in the urine of rats. Four metabolites were identified in urine: p-chlorophenylacetone, p-chlorophenylisopropanol, p-chlorobenzoic acid, and 4-hydroxy-3-chloroamphetamine; other metabolites were found but not identified. Silverman and Ho (1979) reported a brain metabolite of PCA. Sherman and Gal (1975, 1976a) and Sherman et al. (1977) reported that PCA can be metabolized in brain, showed the presence of metabolites in brain after administering tritiated PCA, and identified 3,4-dimethoxyamphetamine and p-chloronorephedrine as two of the metabolites. Ames et al. (1977) reported that PCA was metabolized to a chemically reactive substance. Miller et al. (1986) showed that incubation of tritiated PCA with rat liver microsomes resulted in covalent binding of radioactivity to microsomal protein, suggesting that reactive and toxic metabolites might be formed from PCA that might be related to the neurotoxic effects of PCA on brain serotonin neurons.

One approach to investigating whether an active metabolite of PCA is involved in its depletion of serotonin has been determining whether inhibitors of drug metabolism antagonize that depletion or whether inducers of drug metabolism facilitate the depletion. Steranka and Sanders-Bush (1978b) reported that 3-methylcholanthrene and phenobarbital, inducers of hepatic microsomal drug-metabolizing enzymes, did not enhance PCA neurotoxicity, nor did piperonyl butoxide, an inhibitor of drug metabolism, decrease it. In rats treated with iprindole at doses effective in blocking ring hydroxylation of amphetamine and 3-chloroamphetamine, PCA caused the usual long-lasting depletion of brain serotonin (Fuller and Baker, 1974). Fuller and Perry (1974) found long-lasting depletion of brain serotonin by PCA in guinea pigs, a species that does not metabolize amphetamines by ring hydroxylation. These findings argue against a ring-hydroxylated metabolite or intermediate being involved in the neurotoxic action of PCA.

Another approach has been administering postulated or possible metabolites of PCA to see whether they are more potent than PCA in depleting brain serotonin. Fuller et al. (1974) found that the oxime of PCA had only weak serotonin-depleting properties; the hydroxylamine was as effective as PCA but apparently acted via metabolism to PCA (instead of vice versa). Sherman and Gal (1976b) examined 5-chloroindole, 6-chloro-2-methylindole, 5-chloroindazole, and 5-chloro-2-methylindole and concluded that none of these compounds represented or was converted to a metabolite possibly responsible for the neurotoxic effects of PCA. On the basis of the finding by Parli and Schmidt (1975) that 3-chloro-4-hydroxyamphetamine is a metabolite of PCA in rats, that substance was administered intraperitoneally and intraventricularly to rats, but did not mimic PCA in causing short-term or long-term depletion of brain serotonin (Sherman et al., 1975).

Currently, no compelling evidence exists that a metabolite of PCA is involved in either its short-term or its long-term serotonin-depleting effects.

VII. POSSIBLE MECHANISMS IN ACUTE AND LONG-TERM SEROTONIN DEPLETION

A. Prevention of p-Chloroamphetamine Effects by Uptake Inhibitors

Inhibitors of the serotonin transporter completely prevent the acute effects of PCA and related halogenated amphetamines that are caused by the released serotonin acting on synaptic receptors, the acute depletion of brain serotonin, and the long-term depletion of brain serotonin, as discussed in the

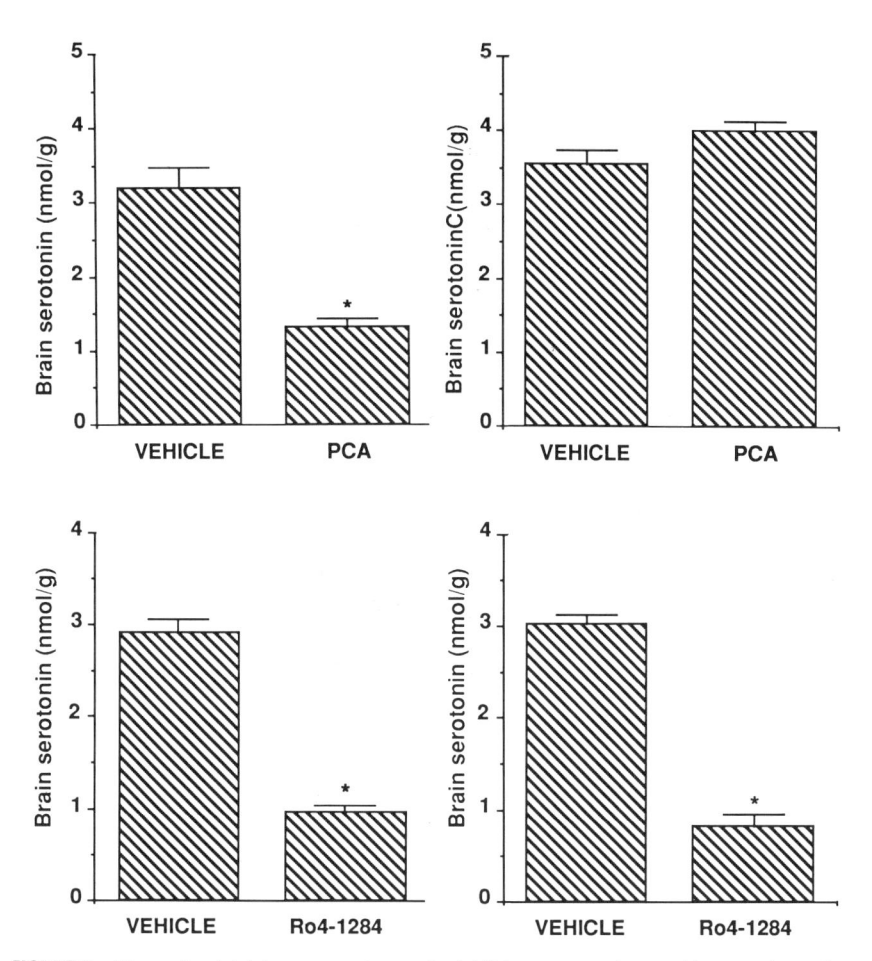

FIGURE 5 Fluoxetine (*right*), a serotonin uptake inhibitor, antagonizes *p*-chloroamphetamine (PCA)-induced (*top*) but not Ro4-1284-induced (*bottom*) depletion of brain serotonin in rats. Fluoxetine was injected (ip) at a dose of 10 mg/kg, PCA (ip) at 10 mg/kg, and Ro4-1284 (ip) at 2.5 mg/kg. PCA was injected 1 hr after fluoxetine and 5 hr before rats were killed. Ro4-1284 was injected 1 hr after fluoxetine and 1 hr before rats were killed. Mean values ± standard errors for 5 rats per group are shown. Asterisks indicate significant difference from control group ($P < .05$) (*left*).

following sections. Figure 5 illustrates that pretreatment with fluoxetine, a selective inhibitor of the serotonin uptake carrier, prevented serotonin depletion by PCA but not serotonin depletion by Ro4-1284, a tetrabenazine-like agent. Thus, the depletion of brain serotonin by PCA (and other halogenated amphetamines) is carrier dependent, whereas the depletion of brain serotonin by some other agents is not.

1. Prevention of Acute Effects Caused by Serotonin Release

Growdon (1977) reported postural abnormalities, tremor, myoclonus, and autonomic signs in rats within minutes of PCA injection. These effects were suppressed by *p*-chlorophenylalanine, which depletes serotonin by inhibiting its synthesis, or by serotonin receptor antagonists (methiothepin and metergoline). Growdon (1977) found that fluoxetine pretreatment prevented these effects of PCA, apparently by preventing the carrier-dependent release of serotonin by PCA. Buus Lassen (1978) reported that hypermotility induced by PCA in rats was antagonized by pretreatment with uptake inhibitors. Ogren *et al.* (1977) found that a serotonin uptake inhibitor prevented the acute effect of PCA on avoidance performance in rats, apparently by blocking the acute release of serotonin.

Similar effects of other halogenated amphetamines that are believed to result from serotonin release also are antagonized or prevented by pretreatment with uptake inhibitors. For instance, many effects of fenfluramine have been shown to be blocked by pretreatment with fluoxetine or other serotonin uptake inhibitors: anorectic effects in rats (Jespersen and Scheel-Kruger, 1973; Garattini *et al.*, 1975); discriminative stimulus properties in rats (McElroy and Feldman, 1984); hyperthermia in rats housed at 27–28°C (Sugrue, 1984); elevation of serum corticosterone concentration in rats (McElroy *et al.*, 1984); elevation of serum prolactin concentration in rats (Van de Kar *et al.*, 1985); increase in striatal acetylcholine in rats (Consolo *et al.*, 1979); hyperthermia and behavioral stimulation in rabbits (Quock and Beal, 1976); and increase in twitch frequency of the suprahyoideal muscles in anesthetized rats (Clineschmidt and McGuffin, 1978).

2. Prevention of Acute Depletion of Brain Serotonin

Meek and colleagues (1971) first reported that relatively nonselective inhibitors of the serotonin uptake carrier—chlorimipramine, chlorpheniramine, and meperidine—blocked serotonin depletion by *p*-chloromethamphetamine in rats. Antagonism by uptake inhibitors of serotonin depletion subsequently has been shown for PCA (Fuller *et al.*, 1974, 1975a,b; Ross, 1976) and for many of its different analogs (Harvey *et al.*, 1977; Fuller *et al.*, 1980). Indeed, blockade of PCA-induced depletion of brain serotonin has become a standard test for *in vivo* efficacy of inhibitors of the serotonin uptake carrier (Vaatstra *et al.*, 1981; Thomas *et al.*, 1987; Wentland *et al.*, 1987; Perrone *et al.*, 1990), useful in comparing the relative potency of various uptake inhibitors and in defining their duration of action. Antagonism of fenfluramine depletion of brain serotonin has been used in a similar way to study uptake inhibitors (Ghezzi *et al.*, 1973; Garattini *et al.*, 1976; Pugsley and Lippmann, 1977).

3. Prevention of Long-Term Depletion of Brain Serotonin

Not only does pretreatment with uptake inhibitors prevent the acute depletion of serotonin by PCA, but treatment with an uptake inhibitor *after* the initial depletion of serotonin has occurred can reverse the effects of PCA, preventing the long-term depletion of serotonin (Meek *et al.*, 1971; Fuller *et al.*, 1975b). The ability of an uptake inhibitor to reverse serotonin depletion by PCA lasts for several hours but not indefinitely. By 24 hr after PCA injection, the depletion of serotonin has become irreversible (Fuller *et al.*, 1975b). The rate at which serotonin depletion becomes irreversible is the same for *p*-bromoamphetamine as for PCA (Fuller and Perry, 1978). When a long-acting serotonin uptake inhibitor is given prior to PCA injection, depletion of brain serotonin at any time is prevented. However, a short-acting serotonin uptake inhibitor can prevent the initial depletion of serotonin without affecting the long-term depletion. For example, clomipramine, given at a 40 mg/kg ip dose 30 min before PCA injection, prevents brain serotonin depletion at 3 hr, but clomipramine is metabolized rapidly whereas PCA persists longer in brain, so depletion of brain serotonin 24 hr after PCA injection is scarcely affected by this dose of clomipramine as a pretreatment (Fuller *et al.*, 1978).

The ability of uptake inhibitors to prevent both acute and long-term effects of PCA demonstrates the importance of the transporter in serotonin neurons to these actions of PCA. One interpretation is that PCA is accumulated into serotonin neurons via the uptake carrier and that this accumulation is necessary for the short-term release and depletion of serotonin as well as for the long-term neurotoxic effects. The acute release of serotonin may be mediated by the transporter operating in reverse, transporting serotonin out rather than transporting PCA in. This action would explain why uptake inhibitors block the acute effects of PCA, but would not explain their block of the neurotoxic effects as well. PCA is a lipophilic molecule and, no doubt, can enter cells readily. However, PCA may, in addition, be accumulated inside serotonin nerve terminals because of its accumulation via the membrane transporter; accumulation to concentrations higher than those in other cells may account for the selective actions of PCA on serotonin neurons and for the ability of uptake inhibitors to block all these effects of PCA. Another possibility is that PCA leaks out of the serotonin nerve terminal as rapidly as it is transported in and is transported continually via the uptake carrier. The prolonged overactivity of the transporter might lead to the neurotoxicity; hence, inhibitors of the transporter would block that neurotoxicity. Additional experiments are necessary to clarify the mechanism by which uptake inhibitors prevent the acute and long-term effects of PCA.

B. Other Protective Measures

Steranka and Rhind (1987) reported that injection of L-cysteine prevented the persistent decreases of brain serotonin and 5-HIAA 1 wk after the administration of PCA to rats, and interpreted that some neurotoxic electrophilic intermediates probably were inactivated by the cysteine. Invernizzi *et al.* (1989) confirmed that L-cysteine antagonized brain serotonin effects 1

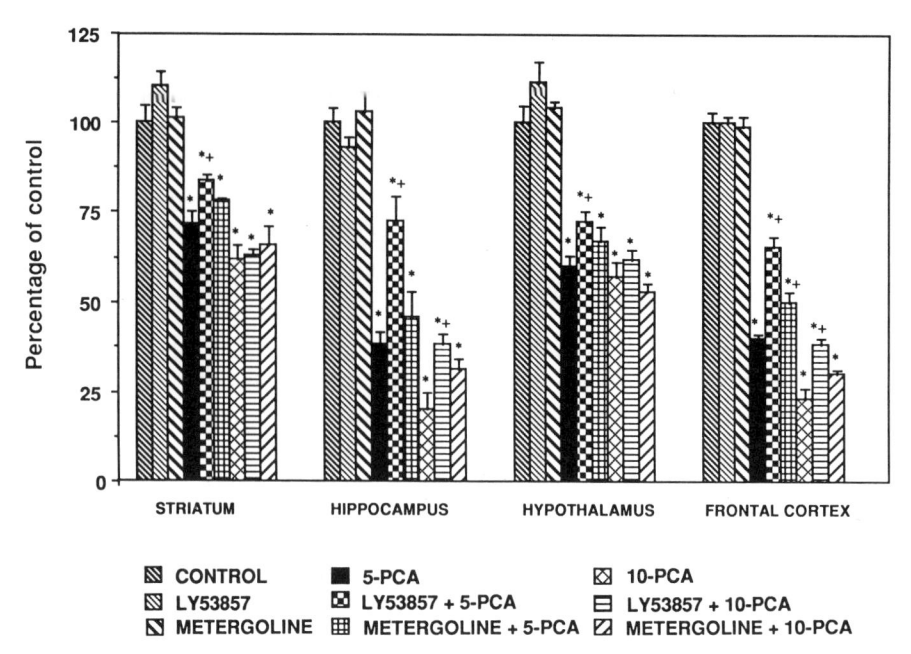

FIGURE 6 Antagonism by 5-HT$_2$ receptor antagonists (LY53857 and metergoline) of the p-chloroamphetamine (PCA)-induced depletion of brain serotonin in brain regions. LY53857 (10 mg/kg sc) or metergoline (3 mg/kg sc) was given as a 1-hr pretreatment to PCA (5 or 10 mg/kg, ip). Rats were killed 3 days later; the striatum, hippocampus, hypothalamus, and frontal cortex were dissected. Mean values ± standard errors for 5 rats per group are shown. Asterisks indicate significant difference from the respective control group ($P < .05$). Plus symbols (+) indicate significant difference from the respective group treated with PCA alone ($P < .05$).

wk after treatment with PCA or with d-fenfluramine and showed similar protective effects at earlier times (16–24 hr) after injection of PCA or d-fenfluramine. However, these authors suggested that the effect of L-cysteine was related to its ability to increase the rate of elimination of the substituted amphetamines, as evident by reduced concentrations of the drugs in brain and blood.

Serotonin receptor antagonists have been shown to protect against the neurotoxicity produced by methylenedioxymethamphetamine (MDMA) as measured by changes in serotonin and TPH activity (Schmidt et al., 1990, 1991b). Figure 6 shows the effects of the 5-HT$_2$ receptor antagonists LY53857 and metergoline on the depletion of serotonin in various brain areas by PCA. LY53857 attenuated the depletion of serotonin by the lower dose of PCA in all the brain areas analyzed, whereas metergoline only antagonized serotonin depletion in the frontal cortex. At the higher dose of PCA, the protective effect of LY53857 was seen only in the hippocampus and frontal cortex; no protection was afforded by metergoline. Although the mechanism by which 5-HT$_2$ receptor antagonists block serotonin depletion by PCA or MDMA is not known, these compounds have been suggested to act by blocking the facilitatory influence of the 5-HT$_2$ receptor on dopamine synthesis (Schmidt *et al.*, 1991b). Dopamine, as discussed in a later section, may play a role in the neurotoxicity of MDMA and PCA for serotonin neurons (Schmidt *et al.*, 1991a).

The possibility that a neurotoxic metabolite of PCA is involved in its neurotoxic action has been discussed already. Another possibility, that a neurotoxic metabolite of serotonin is formed as a result of PCA administration, also has been considered. Berger *et al.* (1989) have postulated this idea based on the finding that PCA does not cause serotonergic neurotoxicity in rats pretreated with both *p*-chlorophenylalanine and reserpine to deplete serotonin. Because *p*-chlorophenylalanine alone was not protective, despite greater than 90% depletion of brain serotonin, the authors believed peripheral sources of serotonin were required and suggested that source to be blood platelets. These researchers postulated that PCA-induced release of serotonin from platelets might result in the formation of a neurotoxic metabolite of serotonin which was then responsible for the prolonged depletion of brain serotonin after PCA injection. Berger *et al.* (1992) demonstrated that daily injections of *p*-chlorophenylalanine alone for 6 days attenuated the degeneration of serotonin axons in the anterior parietal cortex after a 10 mg/kg dose of PCA. Even a single dose of *p*-chlorophenylalanine partially protected against the serotonergic neurotoxicity caused by PCA, in contrast to earlier findings (Fuller *et al.*, 1975b; Ross, 1976). Berger *et al.* (1992) suggested that PCA might elevate plasma serotonin levels by stimulating release from peripheral storage sites, leading to formation of 5,6- or 5,7-DHT, which causes the neurotoxic effect in brain. Ordinarily, these neurotoxic dihydroxytryptamines would not cross the blood–brain barrier (see previous discussion), but Berger *et al.* (1992) postulated that transient disruption of the blood–brain barrier might occur after PCA injection because of hypertension or other mechanisms. Commins *et al.* (1987) suggested earlier that endogenously (but locally) produced 5,6-DHT may mediate the neurotoxic effects of PCA. They measured 5,6-DHT levels in rat hippocampus after a single injection of PCA.

The neurotoxic effects of amphetamine and methamphetamine on nigrostriatal dopamine neurons in mice (Sonsalla *et al.*, 1989) and rats (Henderson *et al.*, 1992) are prevented by pretreatment with MK-801, an antagonist of the *N*-methyl-D-aspartate (NMDA) subclass of glutamate receptors. Excitotoxic actions of amino acid neurotransmitters may be involved in that dopaminergic neurotoxicity, or glutamate receptors may be involved in the neurotoxic mechanism in some other way. To determine whether NMDA receptors are involved in the serotonergic neurotoxic effects of PCA, we investigated whether similar pretreatment with MK-801 would

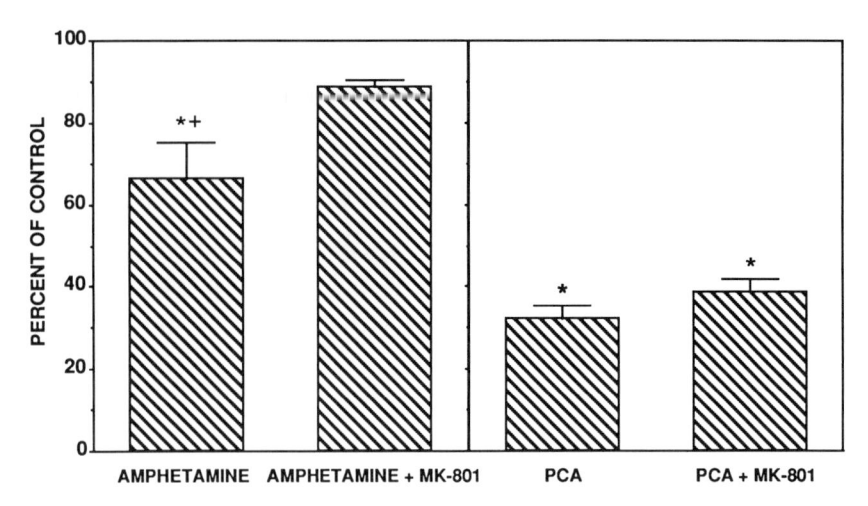

FIGURE 7 MK-801, an N-methyl-D-aspartate (NMDA) receptor antagonist, antagonizes amphetamine-induced depletion of striatal dopamine (*left*) but not p-chloroamphetamine (PCA)-induced depletion of brain serotonin (*right*) in rats. (±)-Amphetamine (0.1 mmol/kg ip) was injected into rats pretreated with iprindole (10 mg/kg ip, 10 min earlier) alone or 15 min after MK-801 (2 mg/kg ip). PCA (0.1 mmol/kg ip) was injected into rats 1 hr after MK-801 (2 mg/kg ip). In both cases, rats were killed 3 days later. Dopamine was measured in striatum; serotonin was measured in whole brain. Mean values ± standard errors for 5 rats per group are shown. Asterisks indicate significant difference from the respective control group ($P < .05$). Plus symbol (+) indicates significant difference from respective group treated with amphetamine + MK-801 ($P < .05$).

prevent the long-term depletion of brain serotonin by PCA. Figure 7 shows that MK-801, at a dose effective in preventing dopamine depletion by amphetamine, did not prevent serotonin depletion by PCA. This finding suggests that NMDA receptors are not involved in the neurotoxic actions of PCA on serotonin neurons.

Finnegan *et al.* (1991) reported that dextromethorphan protected against the neurotoxic effects of PCA on brain serotonin neurons in rats. These researchers interpreted their findings as probable consequences of the reported ability of dextromethorphan to block calcium channels *in vitro* (Carpenter *et al.*, 1988) and to antagonize glutamine-induced neurotoxicity in neuronal cell culture (Choi, 1987). These investigators proposed a role for intracellular calcium and possibly NMDA receptors in the neurotoxicity of PCA. Finnegan *et al.* (1991) did not consider what we judged to be a more likely explanation for the ability of dextromethorphan to block PCA effects, namely, its previously reported potency as an inhibitor of the serotonin uptake carrier *in vitro* (Ahtee, 1975; Moffat and Jhamandas, 1976; Sinclair and Lo, 1977) and *in vivo* (Ahtee, 1975; Sinclair and Lo, 1977). We have found that dextromethorphan blocks the acute depletion of brain serotonin

2 hr after PCA injection (Henderson and Fuller, 1992), which is not a neurotoxic effect, and also blocks serotonin depletion by H75/12, a nonhalogenated amphetamine analog that is not neurotoxic but causes transient depletion of brain serotonin via a carrier-dependent mechanism (Fuller *et al.*, 1976). Additionally, dextromethorphan decreased brain concentrations of 5-HIAA without affecting serotonin (Henderson and Fuller, 1992), as uptake inhibitors are known to do. All these data suggest that dextromethorphan is blocking the serotonin uptake carrier at doses that protect against PCA neurotoxicity; hence, block of the uptake carrier is adequate to explain the protective effect without invoking an involvement of excitatory amino acid receptors or of calcium.

C. Role for Dopamine in *p*-Chloroamphetamine Neurotoxicity

A role for dopamine in the neurotoxic effects of amphetamine on dopamine neurons has been postulated since the early 1980s (Fuller and Hemrick-Luecke, 1982). The possibility that dopamine might be involved in the neurotoxic effects of substituted amphetamines on serotonin neurons was considered soon thereafter, first with methamphetamine (Schmidt and Lovenberg, 1985), then with MDMA (Schmidt *et al.*, 1987; Stone *et al.*, 1988; see Chapter 5), and most recently with PCA.

Axt and Seiden (1990) reported that α-methyl-*p*-tyrosine pretreatment attenuated the depletion of brain serotonin by PCA and suggested that newly synthesized dopamine may be required for the maximum depletion of brain serotonin by PCA. Schmidt *et al.* (1991a) found that L-DOPA administration potentiated the depletion of serotonin in three different brain regions 1 wk after the injection of a single high dose of PCA. These findings suggest that release of dopamine by PCA (Sharp *et al.*, 1986) may contribute to its neurotoxic actions on serotonin neurons. Johnson *et al.* (1990) inferred from their studies with the α-ethyl homolog of PCA, because it was less potent than PCA in releasing dopamine acutely and in causing neurotoxic effects on serotonin neurons, that dopamine release might be involved in the serotonin neurotoxicity of PCA.

VIII. FUNCTIONAL EFFECTS ASSOCIATED WITH LONG-TERM NEUROCHEMICAL DEFICITS INDUCED BY *p*-CHLOROAMPHETAMINE

After several hours, rats recover from the initial effects of PCA that are caused by the release of serotonin and/or dopamine, and are grossly normal in appearance. At 1 wk and later, when brain serotonin remains depleted, no marked behavioral or other effects are detected by simple observation, as is true for other halogenated amphetamines such as fenfluramine or *d*-fenfluramine and for MDMA. Some workers have suggested that the neurotoxic actions on brain serotonin neurons have little functional consequence (see discussion following Molliver et al., 1990).

Although PCA and p-chloromethamphetamine have received limited clinical trials in humans (van Praag et al., 1971; van Praag and Korf, 1973), the compounds have not been used in humans in recent years. Concerns have been expressed about possible serotonergic neurotoxic effects in humans of fenfluramine and d-fenfluramine, which are used as medicinal agents, and of MDMA, which is a widely abused drug. Certainly no catastrophic events have been associated with the use of these drugs in humans that would be in any way analogous to the severe neurological symptoms that have resulted from neurotoxic effects of 1-methyl-4-phenyl-1,2,3,6-tetrahydropyridine (MPTP) on nigrostriatal dopamine neurons (Langston et al., 1986). On the other hand, investigators have pointed out that subtle changes resulting from destruction of serotonin neurons might have gone undetected. Events such as increased frequency of headache, decreased ability to cope, intensified anxiety, or minor mood disorders might occur without suspicion of their having resulted from a prior period of drug use. Indeed, animal studies suggest that long-term serotonin depletion after PCA administration does have functional consequences of a relatively subtle nature.

Table I lists many such changes that have been described in rats, not acutely after PCA treatment, but at times at least a few days later and, in most cases, at least 1 wk later. The changes are listed in the chronological order in which they were described. Some of the changes represent reduced acute responses to PCA that are very likely the result of diminished stores of serotonin, so less is released by PCA. Other changes may involve responses to a drug stimulus or other stimulus that are mediated by serotonin neurons; hence the responses are attenuated in rats with reduced serotonin content. Endocrine, behavioral, cognitive, neurological, and other functions can be altered. Thus, concluding that functional consequences do result from serotonin neurotoxicity caused by halogenated amphetamines in animals seems appropriate, but these consequences generally are subtle rather than catastrophic.

IX. SUMMARY AND CONCLUSIONS

PCA and certain other substituted amphetamines cause the acute release of serotonin and dopamine from intraneuronal storage granules and increase extracellular concentrations of these neurotransmitters in brain, leading to functional effects caused by increased activation of postsynaptic receptors. The release of serotonin by PCA, and additional actions including reduction in TPH activity, lead to acute depletion of serotonin in brain. In addition to these acute effects, PCA causes long-term depletion of brain serotonin even after a single dose. The long-term depletion is accompanied

TABLE I Functional Changes Associated with the Long-Term Depletion of Brain Serotonin by *p*-Chloroamphetamine

Effects	References
Intensified apomorphine-induced stimulation of locomotion	Grabowska and Michaluk (1974)
Hypoactivity in open field and increased defecation; facilitated acquisition in a shock avoidance Y-maze task	Vorhees et al. (1975)
Impaired acquisition of a two-way conditioned avoidance task	Ogren et al. (1976); Kohler et al. (1978)
Decreased open-field exploratory behavior; decreased reaction to novel object introduced into home cage	Kohler et al. (1978)
Reduced suckling-induced increase in prolactin	Rowland et al. (1978)
Stimulated sexual behavior in male rats; increased behavioral sensitivity to testosterone	Sodersten et al. (1978)
Abolished daily surge of luteinizing hormone in estrogen-treated ovariectomized rats	Coen and MacKinnon (1979)
Attenuated acute sterotypic behavior response to PCA	Kutscher and Yamamoto (1979)
Attenuated serotonin release and behavior response to acute PCA	Marsden (1979)
Delayed tolerance development to phenobarbital depressant effects	Lyness and Mycek (1980)
Blocked fear retention impairment due to acute serotonin release by PCA or fenfluramine	Archer et al. (1982)
Attenuation of morphine analgesia	Berge et al. (1983)
Prolonged duration of postdecapitation convulsions; no change in acute response to N,N-dimethyl-5-methoxytryptamine	Archer and Tandberg (1984)
Accentuated down-regulation of $5-HT_2$ receptors in rat brain by desipramine	Hall et al. (1984)
Attenuated acute antinociceptive response to PCA	Hunskaar et al. (1986)
Potentiated apomorphine-, methamphetamine-, and phencyclidine-induced dopaminergic behaviors (sniffing, licking, gnawing, biting)	Yamaguchi et al. (1986)
Enhanced ambulatory and exploratory activity and grooming	Gob et al. (1987)
Abolished sex differences in passive avoidance	Heinsbroek et al. (1988)
Impaired learning of a self-shaped avoidance behavior in a peripheral field avoidance test	Kollner et al. (1988)
Potentiation of phencyclidine-induced behaviors; antagonism of serotonin-mediated behaviors induced by acute PCA injection	Nabeshima et al. (1989)
Enhanced development of neuroleptic-induced orofacial dyskinesias	Sandyk and Fisher (1989)
Improved acquisition in 14-unit Stone T-maze test; no effect in radial arm maze	Altman et al. (1989); Normile et al. (1990)
Inhibition of MDMA[a]-elicited, serotonin-mediated decrease in extracellular dopamine	Gazzara et al. (1989)

(*continued*)

TABLE I (*Continued*)

Effects	References
Abolished behavioral effects of *m*-chlorophenylpiperazine; attenuated behavioral effects of *m*-trifluoromethylphenylpiperazine	Klodzinska *et al.* (1989)
Attenuated scopolamine impairment of non-matching-to-sample operant task	Sakurai and Wenk (1990)
Altered cerebral metabolic response to serotonin agonist in brain regions	Freo *et al.* (1991)
Potentiated behavioral deficit induced by cholinergic lesion	Markowska and Wenk (1991)
Attenuated serum corticosterone elevation due to acute serotonin release by PCA	Fuller (1992)

a MDMA, 3,4-Methylenedioxymethamphetamine.

by loss of several parameters associated with serotonin neurons, for example, 5-HIAA, TPH, serotonin uptake capacity, and radioligand binding to uptake sites. These changes, as well as histological evidence, support the idea that the long-term effect is one of selective neurotoxicity. The long-term neurotoxic depletion of serotonin by PCA has numerous subtle functional consequences that have been demonstrated in rats, often as altered behavioral, neuroendocrine, or other responses to physiological or pharmacological stimuli. All the effects of PCA on serotonin neurons are dependent on the membrane uptake carrier and are blocked by inhibitors of that uptake carrier. The precise mechanisms of the neurotoxic action of PCA, beyond dependence on the membrane transporter, are not well understood. Based on some evidence, suggestions have been made that serotonin itself (via a metabolite) or dopamine might be involved, or that accumulation of PCA into serotonin nerve terminals occurs, but additional studies are required before the mechanisms of these actions can be established. Despite the need for a more complete understanding of its mechanisms of action, PCA is a useful pharmacological tool for the acute or long-term manipulation of serotonergic function.

REFERENCES

Adell, A., Sarna, G. S., Hutson, P. H., and Curzon, G. (1989). An in vivo dialysis and behavioural study of the release of 5-HT by *p*-chloroamphetamine in reserpine-treated rats. *Br. J. Pharmacol.* 97, 206–212.

Ahtee, L. (1975). Dextromethorphan inhibits 5-hydroxytryptamine uptake by human blood platelets and decreases 5-hydroxyindoleacetic acid content in rat brain. *J. Pharm. Pharmacol.* 27, 177–180.

Alesci, R., and Bagnoli, P. (1988). Endogenous levels of serotonin and 5-hydroxyindoleacetic acid in specific areas of the pigeon CNS: Effects of serotonin neurotoxins. *Brain Res.* **450**, 259–271.

Altman, H. J., Ogren, S. O., Berman, R. F., and Normile, H. J. (1989). The effects of *p*-chloroamphetamine, a depletor of brain serotonin, on the performance of rats in two types of positively reinforced complex spatial discrimination tasks. *Behav. Neural Biol.* **52**, 131–144.

Ames, M. M., Nelson, S. D., Lovenberg, W., and Sasame, H. A. (1977). Metabolic activation of *para*-chloroamphetamine to a chemically reactive metabolite. *Commun. Psychopharmacol.* **1**, 455–460.

Archer, T., and Tandberg, B. (1984). Effects of acute administration of 5-methoxy-N,N-dimethyltryptamine upon the latency and duration of post-decapitation convulsions. *Acta Pharmacol. Toxicol. (Copenhagen)* **55**, 224–230.

Archer, T., Ogren, S. O., and Ross, S. B. (1982). Serotonin involvement in aversive conditioning: Reversal of the fear retention deficit by long-term *p*-chloroamphetamine but not *p*-chlorophenylalanine. *Neurosci. Lett.* **34**, 75–82.

Ask, A.-L., Fagervall, I., Huang, R.-B., and Ross, S. B. (1989). Release of ^{3}H-5-hydroxytryptamine by amiflamine and related phenylalkylamines from rat occipital cortex slices. *Naunyn-Schmiedeberg's Arch. Pharmacol.* **339**, 684–689.

Axelrod, J. (1954). Studies on sympathomimetic amines. II. The biotransformation and physiological disposition of *d*-amphetamine, *d-p*-hydroxyamphetamine and *d*-methamphetamine. *J. Pharmacol. Exp. Ther.* **110**, 315–326.

Axt, K. J., and Seiden, L. S. (1990). α-Methyl-*p*-tyrosine partially attenuates *p*-chloroamphetamine-induced 5-hydroxytryptamine depletions in the rat brain. *Pharmacol. Biochem. Behav.* **35**, 995–997.

Berge, O.-G., Hole, K., and Ogren, S.-O. (1983). Attenuation of morphine-induced analgesia by *p*-chlorophenylalanine and *p*-chloroamphetamine: Test-dependent effects and evidence for brainstem 5-hydroxytryptamine involvement. *Brain Res.* **271**, 51–64.

Berger, U. V., Grzanna, R., and Molliver, M. E. (1989). Depletion of serotonin using *p*-chlorophenylalanine (PCPA) and reserpine protects against the neurotoxic effects of *p*-chloroamphetamine (PCA) in the brain. *Exp. Neurol.* **103**, 111–115.

Berger, U. V., Molliver, M. E., and Grzanna, R. (1990). Unlike systemic administration of *p*-chloroamphetamine, direct intracerebral injection does not produce degeneration of 5-HT axons. *Exp. Neurol.* **109**, 257–268.

Berger, U. V., Grzanna, R., and Molliver, M. E. (1992). The neurotoxic effects of *p*-chloroamphetamine in rat brain are blocked by prior depletion of serotonin. *Brain Res.* **578**, 177–185.

Bjorklund, A., Nobin, A., and Stenevi, U. (1973). Regeneration of central serotonin neurons after axonal degeneration induced by 5,6-dihydroxytryptamine. *Brain Res.* **50**, 214–220.

Bjorklund, A., Baumgarten, H.-G., and Nobin, A. (1974). Chemical lesioning of central monoamine axons by means of 5,6-dihydroxytryptamine and 5,7-dihydroxytryptamine. *Adv. Biochem. Psychopharmacol.* **10**, 13–33.

Bjorklund, A., Baumgarten, H. G., and Rensch, A. (1975). 5,7-Dihydroxytryptamine: Improvement of its selectivity for serotonin neurons in the CNS by pretreatment with desipramine. *J. Neurochem.* **24**, 833–835.

Blier, P., Serrano, A., and Scatton, B. (1990). Differential responsiveness of the rat dorsal and median raphe 5-HT systems to 5-HT$_1$ receptor agonists and *p*-chloroamphetamine. *Synapse* **5**, 120–133.

Breese, G. R., and Mueller, R. A. (1978). Alterations in the neurocytotoxicity of 5,7-dihydroxytryptamine by pharmacologic agents in adult and developing rats. *Ann. N.Y. Acad. Sci.* **305**, 160–174.

Burke, D. H., Brooks, J. C., Ryan, R. P., and Treml, S. B. (1979). *p*-Chloroamphetamine antagonism of cobaltous chloride-induced hypothermia in mice. *Eur. J. Pharmacol.* **60,** 241–243.

Buus Lassen, J. (1978). Influence of the new 5-HT-uptake inhibitor paroxetine on hypermotility in rats produced by *p*-chloroamphetamine (PCA) and 4,α-dimethyl-*m*-tyramine (H77/77). *Psychopharmacology* **57,** 151–153.

Carlsson, A. (1970). Structural specificity for inhibition of [^{14}C]-5-hydroxytryptamine uptake by cerebral slices. *J. Pharm. Pharmacol.* **22,** 729–732.

Carpenter, C. L., Marks, S. S., Watson, D. L., and Greenberg, D. A. (1988). Dextromethorphan and dextrorphan as calcium channel antagonists. *Brain Res.* **439,** 372–375.

Choi, D. W. (1987). Dextrorphan and dextromethorphan attenuate glutamate neurotoxicity. *Brain Res.* **403,** 333–336.

Clemens, J. A., Fuller, R. W., Perry, K. W., and Sawyer, B. D. (1978). Effects of *p*-chloroamphetamine on brain serotonin in immature rats. *Commun. Psychopharmacol.* **2,** 11–16.

Clineschmidt, B. V., and McGuffin, J. C. (1978). Pharmacological differentiation of the central 5-hydroxytryptamine-like actions of MK-212 (6-chloro-2-[1-piperazinyl]-pyrazine), *p*-methoxyamphetamine and fenfluramine in an in vivo model system. *Eur. J. Pharmacol.* **50,** 369–375.

Coen, C. W., and MacKinnon, P. C. (1979). Serotonin involvement in the control of phasic luteinizing hormone release in the rat: Evidence for a critical period. *J. Endocrinol.* **82,** 105–113.

Commins, D. L., Axt, K. J., Vosmer, G., and Seiden, L. S. (1987). Endogenously produced 5,6-dihydroxytryptamine may mediate the neurotoxic effects of *para*-chloroamphetamine. *Brain Res.* **419,** 253–261.

Conde, S., Madronero, R., Fernandez-Tome, M. P., and del Rio, J. (1978). Effects of thiophene analogues of chloroamphetamines on central serotonergic mechanisms. *J. Med. Chem.* **21,** 978–981.

Consolo, S., Ladinsky, H., Tirelli, A. S., Crunelli, V., Samanin, R., and Garattini, S. (1979). Increase in rat striatal acetylcholine content by *d*-fenfluramine, a serotonin releaser. *Life Sci.* **25,** 1975–1981.

D'Amato, R. J., Largent, B. L., Snowman, A. M., and Snyder, S. H. (1987). Selective labeling of serotonin uptake sites in rat brain by [^{3}H]citalopram contrasted to labeling of multiple sites by [^{3}H]imipramine. *J. Pharmacol. Exp. Ther.* **242,** 364–371.

Dewar, K. M., Grondin, L., Carli, M., Lima, L., and Reader, T. A. (1992). [^{3}H]Paroxetine binding and serotonin content of rat cortical areas, hippocampus, neostriatum, ventral mesencephalic tegmentum, and midbrain raphe nuclei region following *p*-chlorophenylalanine and *p*-chloroamphetamine treatment. *J. Neurochem.* **58,** 250–257.

Dolfini, E., Tansella, M., Valzelli, L., and Garattini, S. (1969). Further studies on the interaction between desipramine and amphetamine. *Eur. J. Pharmacol.* **5,** 185–190.

Fagervall, I., and Ross, S. B. (1986). Failure to detect any transport of the irreversible monoamine oxidase inhibitor N-cyclopropyl-4-chloroamphetamine by the carrier of 5-hydroxytryptamine in the brain of the rat in vivo. *Neuropharmacology* **25,** 911–913.

Finnegan, K. T., Kerr, J. T., and Langston, J. W. (1991). Dextromethorphan protects against the neurotoxic effects of *p*-chloroamphetamine in rats. *Brain Res.* **558,** 109–111.

Freeman, J. J., and Sulser, F. (1972). Iprindole-amphetamine interactions in the rat: The role of aromatic hydroxylation of amphetamine in its mode of action. *J. Pharmacol. Exp. Ther.* **183,** 307–315.

Freo, U., Larson, D. M., Tolliver, T., Rapoport, S. I., and Soncrant, T. T. (1991). Para-chloroamphetamine selectively alters regional cerebral metabolic responses to the serotonergic agonist metachlorophenylpiperazine in rats. *Brain Res.* **544,** 17–25.

Frey, H.-H. (1970). *p*-Chloroamphetamine—Similarities and dissimilarities to amphetamine. *In* "Amphetamines and Related Compounds: Proceedings of the Mario Negri Institute for

Pharmacological Research" (E. Costa and S. Garattini, ed.), pp. 343–355. Raven Press, New York.

Fritschy, J. M., Lyons, W. E., Molliver, M. E., and Grzanna, R. (1987). Serotonergic (5-HT) projections from dorsal raphe, raphe obscurus, and raphe pallidus to the motor nucleus of the trigeminal nerve: Differential vulnerability of their axon terminals to p-chloroamphetamine (PCA). *Soc. Neurosci. Abstr.* **13**, 907.

Fritschy, J. M., Lyons, W. E., Molliver, M. E., and Grzanna, R. (1988). Neurotoxic effects of p-chloroamphetamine on the serotoninergic innervation of the trigeminal motor nucleus: a retrograde transport study. *Brain Res.* **473**, 261–270.

Fuller, R. W. (1966). Serotonin oxidation by rat brain monoamine oxidase: Inhibition by 4-chloroamphetamine. *Life Sci.* **5**, 2247–2252.

Fuller, R. W. (1972). Species difference in the lowering of brain 5-hydroxytryptamine by m-chloroamphetamine. *J. Pharm. Pharmacol.* **24**, 88.

Fuller, R. W. (1980). Mechanism by which uptake inhibitors antagonize p-chloroamphetamine-induced depletion of brain serotonin. *Neurochem. Res.* **5**, 241–245.

Fuller, R. W. (1992). Effects of p-chloroamphetamine on brain serotonin neurons. *Neurochem. Res.* **17**, 449–456.

Fuller, R. W., and Baker, J. C. (1974). Long-lasting reduction of brain 5-hydroxytryptamine concentration by 3-chloroamphetamine and 4-chloroamphetamine in iprindole-treated rats. *J. Pharm. Pharmacol.* **26**, 912–914.

Fuller, R. W., and Baker, J. C. (1977). The role of metabolic N-dealkylation in the action of p-chloromethamphetamine and related drugs on brain 5-hydroxytryptamine. *J. Pharm. Pharmacol.* **29**, 561–562.

Fuller, R. W., and Clemens, J. A. (1981). Role of serotonin in the hypothalamic regulation of pituitary function. *In* "Serotonin: Current Aspects of Neurochemistry and Function" (B. Haber, S. Gabay, M. R. Issidorides, and S. G. A. Alivisatos, eds.), pp. 431–444. Plenum Publishing, New York.

Fuller, R. W., and Hemrick-Luecke, S. K. (1982). Influence of ring and side chain substituents on the selectivity of amphetamine as a monoamine oxidase inhibitor. *Res. Commun. Subst. Abuse* **3**, 159–164.

Fuller, R. W., and Hines, C. W. (1970). Inhibition by p-chloroamphetamine of the conversion of 5-hydroxytryptamine to 5-hydroxyindoleacetic acid in rat brain. *J. Pharm. Pharmacol.* **22**, 634–635.

Fuller, R. W., and Kaiser, C. (1980). Effect of 2-(p-chlorophenyl)cyclopropylamine on 5-hydroxyindole concentration and monoamine oxidase activity in rat brain. *Biochem. Pharmacol.* **29**, 3328–3330.

Fuller, R. W., and Molloy, B. B. (1973). Effects of N-cyclopropyl-4-chloroamphetamine on brain serotonin metabolism in rats. *Res. Commun. Chem. Pathol. Pharmacol.* **6**, 407–418.

Fuller, R. W., and Perry, K. W. (1974). Long-lasting depletion of brain serotonin by 4-chloroamphetamine in guinea pigs. *Brain Res.* **82**, 383–385.

Fuller, R. W., and Perry, K. W. (1975). Inability of an inhibitor of amine uptake (Lilly 110140) to block depletion of brain 5-hydroxytryptamine by L-DOPA. *J. Pharm. Pharmacol.* **27**, 618–620.

Fuller, R. W., and Perry, K. W. (1977). Further studies on the effects of N-cyclopropyl-p-chloroamphetamine in rat brain. *Neuropharmacology* **16**, 495–497.

Fuller, R. W., and Perry, K. W. (1978). Fluoxetine reversal of rat brain serotonin depletion by p-bromoamphetamine. *IRCS Med. Sci.* **6**, 117.

Fuller, R. W., and Perry, K. W. (1983). Decreased accumulation of brain 5-hydroxytryptophan after decarboxylase inhibition in rats treated with fenfluramine, norfenfluramine or p-chloroamphetamine. *J. Pharm. Pharmacol.* **35**, 597–598.

Fuller, R. W., and Snoddy, H. D. (1974). Long-term effects of 4-chloroamphetamine on brain 5-hydroxyindole metabolism in rats. *Neuropharmacology* **13**, 85–90.

Fuller, R. W., and Snoddy, H. D. (1980). Effect of serotonin releasing drugs on serum corticosterone concentration in rats. *Neuroendocrinology* **31**, 96–100.

Fuller, R. W., Hines, C. W., and Mills, J. (1965). Lowering of brain serotonin level by chloroamphetamines. *Biochem. Pharmacol.* **14**, 483–488.

Fuller, R. W., Schaffer, R. J., Roush, B. W., and Molloy, B. B. (1972). Drug disposition as a factor in the lowering of brain serotonin by chloroamphetamines in the rat. *Biochem. Pharmacol.* **21**, 1413–1417.

Fuller, R. W., Snoddy, H. D., and Molloy, B. B. (1973a). Effect of β,β-difluoro substitution on the disposition and pharmacological effects of 4-chloroamphetamine in rats. *J. Pharmacol. Exp. Ther.* **184**, 278–284.

Fuller, R. W., Snoddy, H. D., Roush, B. W., and Molloy, B. B. (1973b). Further structure-activity studies on the lowering of brain 5-hydroxyindoles by 4-chloroamphetamine. *Neuropharmacology* **12**, 33–42.

Fuller, R. W., Perry, K. W., Snoddy, H. D., and Molloy, B. B. (1974). Comparison of the specificity of 3-(p-trifluoromethylphenoxy)-N-methyl-3-phenylpropylamine and chlorimipramine as amine uptake inhibitors in mice. *Eur. J. Pharmacol.* **28**, 233–236.

Fuller, R. W., Baker, J. C., Perry, K. W., and Molloy, B. B. (1975a). Comparison of 4-chloro-, 4-bromo- and 4-fluoroamphetamine in rats: Drug levels in brain and effects on brain serotonin metabolism. *Neuropharmacology* **14**, 739–746.

Fuller, R. W., Perry, K. W., and Molloy, B. B. (1975b). Reversible and irreversible phases of serotonin depletion by 4-chloroamphetamine. *Eur. J. Pharmacol.* **33**, 119–124.

Fuller, R. W., Snoddy, H. D., and Molloy, B. B. (1975c). Blockade of amine depletion by nisoxetine in comparison to other uptake inhibitors. *Psychopharmacol. Commun.* **1**, 455–464.

Fuller, R. W., Perry, K. W., and Baker, J. C. (1976). Duration of the effects of H75/12, α-ethyl-4-methyl-meta-tyramine, on brain 5-hydroxyindole levels in rats. *J. Pharm. Pharmacol.* **28**, 649–650.

Fuller, R. W., Snoddy, H. D., Perry, K. W., Bymaster, F. P., and Wong, D. T. (1978). Importance of duration of drug action in the antagonism of p-chloroamphetamine depletion of brain serotonin—Comparison of fluoxetine and chlorimipramine. *Biochem. Pharmacol.* **27**, 193–198.

Fuller, R. W., Meyers, D. B., Gibson, W. R., and Snoddy, H. D. (1979). Depletion of brain serotonin by chronic administration of p-chloroamphetamine orally to rats and dogs. *Toxicol. Appl. Pharmacol.* **48**, 369–374.

Fuller, R. W., Snoddy, H. D., Snoddy, A. M., Hemrick, S. K., Wong, D. T., and Molloy, B. B. (1980). p-Iodoamphetamine as a serotonin depletor in rats. *J. Pharmacol. Exp. Ther.* **212**, 115–119.

Fuller, R. W., Snoddy, H. D., and Perry, K. W. (1987). p-Chloroamphetamine formation responsible for long-term depletion of brain serotonin after N-cyclopropyl-p-chloroamphetamine injection in rats. *Life Sci.* **40**, 1921–1927.

Garattini, W., Buczko, W., Jori, A., and Samanin, R. (1975). The mechanism of action of fenfluramine. *Postgrad. Med. J. (Suppl. 1)*, **51**, 27–35.

Garattini, S., de Gaetano, G., Samanin, R., Bernasconi, S., and Roncaglioni, M. C. (1976). Effects of trazodone on serotonin in the brain and platelets of the rat. *Biochem. Pharmacol.* **25**, 13–16.

Gazzara, R. A., Takeda, H., Cho, A. K., and Howard, S. G. (1989). Inhibition of dopamine release by methylenedioxymethamphetamine is mediated by serotonin. *Eur. J. Pharmacol.* **168**, 209–217.

Ghezzi, D., Samanin, R., Bernasconi, S., Tognoni, G., Gerna, M., and Garattini, S. (1973). Effect of thymoleptics on fenfluramine-induced depletion of brain serotonin in rats. *Eur. J. Pharmacol.* **24**, 205–210.

Gob, R., Kollner, U., Kollner, O., and Klingberg, F. (1987). Early postnatal development of open field behaviour is changed by single doses of fenfluramine or *p*-chloroamphetamine. *Biomed. Biochim. Acta* **46**, 189–198.

Grabowska, M., and Michaluk, J. (1974). On the role of serotonin in apomorphine-induced locomotor stimulation in rats. *Pharmacol. Biochem. Behav.* **2**, 263–266.

Growdon, J. H. (1977). Postural changes, tremor, and myoclonus in the rat immediately following injections of *p*-chloroamphetamine. *Neurology* **27**, 1074–1077.

Hall, H. (1984). Characterization of the binding of ^{3}H-norzimelidine, a 5-HT uptake inhibitor, to rat brain homogenates. *Acta Pharmacol. Toxicol.* **55**, 33–40.

Hall, H., Ross, S. B., and Sallemark, M. (1984). Effect of destruction of central noradrenergic and serotonergic nerve terminals by systemic neurotoxins on the long-term effects of antidepressants on beta-adrenoceptors and 5-HT$_2$ binding sites in the rat cerebral cortex. *J. Neural Transm.* **59**, 9–23.

Hamon, M., Nelson, D. L., Mallat, M., and Bourgoin, S. (1981). Are 5-HT receptors involved in the sprouting of serotoninergic terminals following neonatal 5,7-dihydroxytryptamine treatment in the rat? *Neurochem. Int.* **3**, 69–79.

Hard, E., Ahlenius, S., and Engel, J. (1983). Effects of neonatal treatment with 5,7-dihydroxytryptamine or 6-hydroxydopamine on the ontogenetic development of the audiogenic immobility reaction in the rat. *Psychopharmacology* **80**, 269–274.

Harvey, J. A. (1978). Neurotoxic action of halogenated amphetamines. *Ann. N.Y. Acad. Sci.* **305**, 289–304.

Harvey, J. A., McMaster, S. E., and Fuller, R. W. (1977). Comparison between the neurotoxic and serotonin-depleting effects of various halogenated derivatives of amphetamine in the rat. *J. Pharmacol. Exp. Ther.* **202**, 581–589.

Hashimoto, K., and Goromaru, T. (1990). High-affinity [^{3}H]6-nitroquipazine binding sites in rat brain. *Eur. J. Pharmacol.* **180**, 273–281.

Heinsbroek, R. P., Feenstra, M. G., Boon, P., Van Haaren, F. and Van de Poll, N. E. (1988). Sex differences in passive avoidance depend on the integrity of the central serotonergic system. *Pharmacol. Biochem. Behav.* **31**, 499–503.

Henderson, M. G., and Fuller, R. W. (1992). Dextromethorphan antagonizes the acute depletion of brain serotonin by *p*-chloroamphetamine and H75/12 in rats. *Brain Res.* **594**, 323–326.

Henderson, M. G., Hemrick-Luecke, S. K., and Fuller, R. W. (1992). MK-801 protects against amphetamine-induced striatal dopamine depletion in iprindole-treated rats, but not against brain serotonin depletion after *p*-chloroamphetamine. *Ann. N.Y. Acad. Sci.* **648**, 286–288.

Hrdina, P. D., Foy, B., Hepner, A., and Summers, R. J. (1990). Antidepressant binding sites in brain: Autoradiographic comparison of [^{3}H]paroxetine and [^{3}H]imipramine localization and relationship to serotonin transporter. *J. Pharmacol. Exp. Ther.* **252**, 410–418.

Hunskaar, S., Berge, O. G., Broch, O. J., and Hole, K. (1986). Lesions of the ascending serotonergic pathways and antinociceptive effects after systemic administration of *p*-chloroamphetamine in mice. *Pharmacol. Biochem. Behav.* **24**, 709–714.

Hutson, P. H., and Curzon, G. (1989). Concurrent determination of effects of *p*-chloroamphetamine on central extracellular 5-hydroxytryptamine concentration and behaviour. *Br. J.Pharmacol.* **96**, 801–806.

Invernizzi, R., Fracasso, C., Caccia, S., Di Clemente, A., Garattini, S., and Samanin, R. (1989). Effect of L-cysteine on the long-term depletion of brain indoles caused by *p*-chloroamphetamine and *d*-fenfluramine in rats. Relation to brain drug concentrations. *Eur. J. Pharmacol.* **163**, 77–83.

Jespersen, S., and Scheel-Kruger, J. (1973). Evidence for a difference in mechanism of action between fenfluramine- and amphetamine-induced anorexia. *J. Pharm. Pharmacol.* **25**, 49–54.

Johnson, M. P., Huang, X. M., Oberlender, R., Nash, J. F., and Nichols, D. E. (1990). Behav-

ioral, biochemical and neurotoxicological actions of the alpha-ethyl homologue of p-chloroamphetamine. *Eur. J. Pharmacol.* **191**, 1–10.
Kaergaard Nielsen, C., Magnussen, M. P., Kampmann, E., and Frey, H.-H. (1967). Pharmacological properties of racemic and optically active p-chloroamphetamine. *Arch. Int. Pharmacodyn.* **170**, 428–444.
Klodzinska, A., Jaros, T., Chojnacka-Wojcik, E., and Maj, J. (1989). Exploratory hypoactivity induced by *m*-trifluoromethylphenylpiperazine (TFMPP) and *m*-chlorophenylpiperazine (m-CPP). *J. Neural Transm.* **1**, 207–218.
Koe, B. K., Lebel, L. A., and Welch, W. M. (1990). [^{3}H]Sertraline binding to rat brain membranes. *Psychopharmacology* **100**, 470–476.
Kohler, C., Ross, S. B., Srebro, B., and Ogren, S.-O. (1978). Long-term biochemical and behavioral effects of p-chloroamphetamine in the rat. *Ann. N.Y. Acad. Sci.* **305**, 645–663.
Kollner, O., Kollner, U., Gob, R., and Klingberg, F. (1988). Postnatal application of p-chloroamphetamine or fenfluramine reduces response selection during early ontogenetic development of rat avoidance behavior. *Biomed. Biochim. Acta* **47**, 997–1005.
Kosofsky, B. E., and Molliver, M. E. (1987). The serotoninergic innervation of cerebral cortex: Different classes of axon terminals arise from dorsal and median raphe nuclei. *Synapse* **1**, 153–168.
Krieger, D. T. (1975). Effect of intraventricular neonatal 6-OH dopamine or 5,6-dihydroxytryptamine administration on the circadian periodicity of plasma corticosteroid levels in the rat. *Neuroendocrinology* **17**, 62–75.
Kutscher, C. L., and Yamamoto, B. K. (1979). A frequency analysis of behavior components of the serotonin syndrome produced by p-chloroamphetamine. *Pharmacol. Biochem. Behav.* **11**, 611–616.
Langston, J. W., Irwin, I., Langston, E. B., DeLanney, L. E., and Ricaurte, G. A. (1986). MPTP-induced parkinsonism in humans: A review of the syndrome and observations relating to the phenomenon of tardive toxicity. *In* "MPTP: A Neurotoxin Producing a Parkinsonian Syndrome" (S. P. Markey, N. Castagnoli, Jr., A. J. Trevor, and I. J. Kopin, eds.), pp. 9–21. Academic Press, Orlando, Florida.
Lucki, I., Nobler, M. S., and Frazer, A. (1984). Differential actions of serotonin antagonists on two behavioural models of serotonin receptor activation in the rat. *J. Pharmacol. Exp. Ther.* **228**, 133–139.
Lyness, W. H., and Mycek, M. J. (1980). The role of cerebral serotonin in the development of tolerance to centrally administered phenobarbital. *Brain Res.* **187**, 443–456.
Mamounas, L. A., and Molliver, M. E. (1988). Evidence for dual serotonergic projections to neocortex: Axons from the dorsal and median raphe nuclei are differentially vulnerable to the neurotoxin p-chloroamphetamine (PCA). *Exp. Neurol.* **102**, 23–36.
Mamounas, L. A., Mullen, C. A., O'Hearn, E., and Molliver, M. E. (1991). Dual serotoninergic projections to forebrain in the rat: Morphologically distinct 5-HT axon terminals exhibit differential vulnerability to neurotoxic amphetamine derivatives. *J. Comp. Neurol.* **314**, 558–586.
Markowska, A. L., and Wenk, G. L. (1991). Serotonin influences the behavioral recovery of rats following nucleus basalis lesions. *Pharmacol. Biochem. Behav.* **38**, 731–737.
Marsden, C. A. (1979). Long term effects of p-chloroamphetamine on hippocampal 5-hydroxytryptamine release. *Br. J. Pharmacol.* **66**, 120P.
Massari, V. J., and Sanders-Bush, E. (1975). Synaptosomal uptake and levels of serotonin in rat brain areas after p-chloroamphetamine or B-9 lesions. *Eur. J. Pharmacol.* **33**, 419–422.
McElroy, J. F., and Feldman, R. S. (1984). Discriminative stimulus properties of fenfluramine: Evidence for serotonergic involvement. *Psychopharmacology* **83**, 172–178.
McElroy, J. F., Miller, J. M., and Meyer, J. S. (1984). Fenfluramine, p-chloroamphetamine and p-fluoroamphetamine stimulation of pituitary-adrenocortical activity in rat: Evidence for differences in site and mechanism of action. *J. Pharmacol. Exp. Ther.* **228**, 593–599.

Meek, J. L. (1978). Studies of serotonin neurotoxins in discrete brain nuclei. *Ann. N.Y. Acad. Sci.* **305**, 190–197.

Meek, J. L., Fuxe, K., and Carlsson, A. (1971). Blockade of p-chloromethamphetamine induced 5-hydroxytryptamine depletion by chlorimipramine, chlorpheniramine and meperidine. *Biochem. Pharmacol.* **20**, 707–709.

Miller, K. J., Anderholm, D. C., and Ames, M. M. (1986). Metabolic activation of the serotonergic neurotoxin para-chloroamphetamine to chemically reactive intermediates by hepatic and brain microsomal preparations. *Biochem. Pharmacol.* **35**, 1737–1742.

Moffat, J. A., and Jhamandas, K. (1976). Effects of acute and chronic methadone treatment of the uptake of ^{3}H-5-hydroxytryptamine in rat hypothalamic slices. *Eur. J. Pharmacol.* **36**, 289–297.

Molliver, M. E., Berger, U. V., Mamounas, L. A., Molliver, D. C., O'Hearn, E., and Wilson, M. A. (1990). Neurotoxicity of MDMA and related compounds: Anatomic studies. *Ann. N.Y. Acad. Sci.* **600**, 640–664.

Mueller, R. A., Towle, A., and Breese, G. R. (1985). Serotonin turnover and supersensitivity after neonatal 5,7-dihydroxytryptamine. *Pharmacol. Biochem. Behav.* **22**, 221–225.

Nabeshima, T., Yamaguchi, K., Ishikawa, K., Furukawa, H., and Kameyama, T. (1989). Potentiation of phencyclidine and serotonin agonist-induced behaviors after administration of p-chloroamphetamine in rats. *Res. Commun. Subst. Abuse* **10**, 37–61.

Newman, M. E., Ben-Zeev, A., and Lerer, B. (1991). Chloroamphetamine did not prevent the effects of chronic antidepressants on 5-hydroxytryptamine inhibition of forskolin-stimulated adenylate cyclase in rat hippocampus. *Eur. J. Pharmacol.* **207**, 209–213.

Normile, H. G., Jenden, D. J., Kuhn, D. M., Wolf, W. A., and Altman, H. J. (1990). Effects of combined serotonin depletion and lesions of the nucleus basalis magnocellularis on acquisition of a complex spatial discrimination task in the rat. *Brain Res.* **536**, 245–250.

Ogren, S. O. (1982). Forebrain serotonin and avoidance learning: Behavioural and biochemical studies on the acute effect of p-chloroamphetamine one one-way active avoidance learning in the male rat. *Pharmacol. Biochem. Behav.* **16**, 881–895.

Ogren, S. O., Kohler, C., Ross, S. B., and Srebro, B. (1976). 5-Hydroxytryptamine depletion and avoidance acquisition in the rat. Antagonism of the long-term effects of p-chloroamphetamine with a selective inhibitor of 5-hydroxytryptamine uptake. *Neurosci. Lett.* **3**, 341–347.

Ogren, S.-O., Ross, S. B., Holman, A.-C., and Baumann, L. (1977). 5-Hydroxytryptamine and avoidance performance in the rat: Antagonism of the acute effect of p-chloroamphetamine by zimelidine, an inhibitor of 5-hydroxytryptamine uptake. *Neurosci. Lett.* **7**, 331–336.

Owen, J. E., Jr. (1963). Psychopharmacological studies of some 1-(chlorophenyl)-2-aminopropanes. I. Effects on appetite-controlled behavior. *J. Pharm. Sci.* **52**, 679–683.

Owen, M. L., Baker, G. B., Coutts, R. T., and Dewhurst, W. G. (1991). Analysis of p-chloroamphetamine and a side-chain monofluorinated analogue in rat brain. *J. Pharmacol. Meth.* **25**, 147–155.

Parli, C. J. (1976). Metabolism of para-chloroamphetamine in rats. *Psychopharmacol. Bull.* **12**, 54–55.

Parli, C. J., and Schmidt, B. (1975). Metabolism of 4-chloroamphetamine to 3-chloro-4-hydroxyamphetamine in rat: Evidence for an *in vivo* "NIH shift" of chlorine. *Res. Commun. Chem. Pathol. Pharmacol.* **10**, 601–604.

Perrone, M. H., Luttinger, D., Hamel, L. T., Fritz, P. M., Ferraino, R., and Haubrich, D. R. (1990). In vivo assessment of napamezole, an alpha-2 adrenoceptor antagonist and monoamine re-uptake inhibitor. *J. Pharmacol. Exp. Ther.* **254**, 476–483.

Pletscher, A., Burkard, W. P., Bruderer, H., and Gey, K. F. (1963). Decrease of cerebral 5-hydroxytryptamine and 5-hydroxyindoleacetic acid by an arylalkylamine. *Life Sci.* **11**, 828–833.

Pletscher, A., Bartholini, G., Bruderer, H., Burkard, W. P., and Gey, K. F. (1964). Chlorinated arylalkylamines affecting the cerebral metabolism of 5-hydroxytryptamine. *J. Pharmacol. Exp. Ther.* **145**, 344–350.

Pranzatelli, M. R. (1990). Neonatal 5,7-DHT lesions upregulate [³H]mesulergine labelled spinal 5-HT$_{1C}$ binding sites in the rat. *Brain Res. Bull.* **25**, 151–153.

Pugsley, T. A., and Lippmann, W. (1977). 1-[2-Amino-1-[1-(naphthyloxy)methyl]ethyl]piperidine hydrochloride: Specific inhibition of brain serotonin uptake and related activities. *Arch. Int. Pharmacodyn.* **228**, 322–338.

Quock, R. M., and Beal, G. A. (1976). Fenfluramine-induced hyperthermia and stimulation in the rabbit. *Res. Commun. Chem. Pathol. Pharmacol.* **13**, 401–409.

Ross, S. B. (1976). Antagonism of the acute and long-term biochemical effects of 4-chloroamphetamine on the 5-HT neurones in the rat brain by inhibitors of the 5-hydroxytryptamine uptake. *Acta Pharmacol. Toxicol.* **39**, 456–476.

Ross, S. B., and Froden, O. (1977). On the mechanism of the acute decrease of rat brain tryptophan hydroxylase activity by 4-chloroamphetamine. *Neurosci. Lett.* **5**, 215–220.

Rowland, D., Steele, M., and Moltz, H. (1978). Serotonergic mediation of the suckling-induced release of prolactin in the lactating rat. *Neuroendocrinology* **26**, 8–14.

Rudnick, G., and Wall, S. C. (1992). *p*-Chloroamphetamine induces serotonin release through serotonin transporters. *Biochemistry* **31**, 6710–6718.

Sakurai, Y., and Wenk, G. L. (1990). The interaction of acetylcholinergic and serotonergic neural systems on performance in a continuous nonmatching to sample task. *Brain Res.* **519**, 118–121.

Sanders-Bush, E., and Steranka, L. R. (1978). Immediate and long-term effects of *p*-chloroamphetamine on brain amines. *Ann. N.Y. Acad. Sci.* **305**, 208–221.

Sanders-Bush, E., and Sulser, F. (1970). *p*-Chloroamphetamine: *In vivo* investigations on the mechanism of action of the selective depletion of cerebral serotonin. *J. Pharmacol. Exp. Ther.* **175**, 419–426.

Sanders-Bush, E., Bushing, J. A., and Sulser, F. (1972a). *p*-Chloroamphetamine—Inhibition of cerebral tryptophan hydroxylase. *Biochem. Pharmacol.* **21**, 1501–1510.

Sanders-Bush, E., Bushing, J. A., and Sulser, F. (1972b). Long-term effects of *p*-chloroamphetamine on tryptophan hydroxylase activity and on the levels of 5-hydroxytryptamine and 5-hydroxyindoleacetic acid in brain. *Eur. J. Pharmacol.* **20**, 385–388.

Sanders-Bush, E., Bushing, J. A., and Sulser, F. (1975). Long-term effects of *p*-chloroamphetamine and related drugs on central serotonergic mechanisms. *J. Pharmacol. Exp. Ther.* **192**, 33–41.

Sandyk, R., and Fisher, H. (1989). Treatment with *p*-chloroamphetamine enhances the development of neuroleptic-induced orofacial dyskinesias in the rat. *Int. J. Neurosci.* **48**, 129–131.

Saner, A., and Pletscher, A. (1978). Enhancement of 5-hydroxytryptamine synthesis in brain by monoamine-depleting drugs. *J. Pharm. Pharmacol.* **30**, 115–117.

Schmidt, C. J., and Lovenberg, W. (1985). *In vitro* demonstration of dopamine uptake by neostriatal serotonergic neurons of the rat. *Neurosci. Lett.* **59**, 9–14.

Schmidt, C. J., Levin, J. A., and Lovenberg, W. (1987). *In vitro* and *in vivo* neurochemical effects of methylenedioxymethamphetamine on striatal monoaminergic systems in the rat brain. *Biochem. Pharmacol.* **36**, 747–755.

Schmidt, C. J., Abbate, G. M., Black, C. K., Taylor, V. L. (1990). Selective 5-hydroxytryptamine$_2$ receptor antagonists protect against the neurotoxicity of methylenedioxymethamphetamine in rats. *J. Pharmacol. Exp. Ther.* **255**, 478–483.

Schmidt, C. J., Black, C. K., and Taylor, V. L. (1991a). L-DOPA potentiation of the serotonergic deficits due to a single administration of 3,4-methylenedioxymethamphetamine, *p*-chloroamphetamine or methamphetamine in rats. *Eur. J. Pharmacol.* **203**, 41–49.

Schmidt, C. J., Taylor, V. L., Abbate, G. M., and Niedzuak, T. R. (1991b). 5-HT$_2$ antagonists stereoselectively prevent the neurotoxicity of 3,4-methylenedioxymethamphetamine by blocking the acute stimulation of dopamine synthesis: Reversal by L-DOPA. *J. Pharmacol. Exp. Ther.* **256,** 230–235.

Sekerke, H. J., Smith, H. E., Bushing, J. A., and Sanders-Bush, E. (1975). Correlation between brain levels and biochemical effects of the optical isomers of *p*-chloroamphetamine. *J. Pharmacol. Exp. Ther.* **193,** 835–844.

Sharp, T., Zetterstrom, T., Christmanson, L., and Ungerstedt, U. (1986). *p*-Chloroamphetamine releases both serotonin and dopamine into rat brain dialysates in vivo. *Neurosci. Lett.* **72,** 320–324.

Sherman, A. D., and Gal, E. M. (1975). ³H-*p*-Chloroamph tamine: Cerebral levels and distribution. *Psychopharmacol. Commun.* **1,** 261–273.

Sherman, A. D., and Gal, E. M. (1976a). Mass-spectrographic evidence of the conversion of *p*-chloroamphetamine to 3,4-dimethoxyamphetamine. *Psychopharmacol. Commun.* **2,** 421–427.

Sherman, A. D., and Gal, E. M. (1976b). Studies on the metabolism of 5-hydroxytryptamine (serotonin). VII. Effects of haloindoles on cerebral 5-HT in various species. *Psychopharmacol. Commun.* **2,** 285–293.

Sherman, A. D., and Gal, E. M. (1979). Levels of *p*-chloroamphetamine and its metabolites in brains of immature rats. *Commun. Psychopharmacol.* **3,** 31–34.

Sherman, A., Gal, E. M., Fuller, R. W., and Molloy, B. B. (1975). Effects of intraventricular *p*-chloroamphetamine and its analogues on cerebral 5-HT. *Neuropharmacology* **14,** 733–737.

Sherman, A. D., Hsiao, W.-C., and Gal, E. M. (1977). Cerebral metabolism of [³H]-*p*-chloroamphetamine. *Neuropharmacology* **16,** 17–24.

Silverman, P. B., and Ho, B. T. (1979). *p*-Chloroamphetamine: Evaluation of a brain metabolite. *Commun. Psychopharmacol.* **3,** 291–294.

Sinclair, J. G., and Lo, G. F. (1977). The blockade of serotonin uptake into synaptosomes: Relationship to an interaction with monoamine oxidase inhibitors. *Can. J. Physiol. Pharmacol.* **55,** 180–187.

Sodersten, P., Berge, O. G., and Hole, K. (1978). Effects of *p*-chloroamphetamine and 5,7-dihydroxytryptamine on the sexual behavior of gonadectomized male and female rats. *Pharmacol. Biochem. Behav.* **9,** 499–508.

Sonsalla, P. K., Nicklas, W. J., and Heikkila, R. E. (1989). Role for excitatory amino acids in methamphetamine-induced nigrostriatal dopaminergic toxicity. *Science* **243,** 398–400.

Steranka, L. R., and Rhind, A. W. (1987). Effect of cysteine on the persistent depletion of brain monoamines by amphetamine, *p*-chloroamphetamine and MPTP. *Eur. J. Pharmacol.* **133,** 191–197.

Steranka, L. R., and Sanders-Bush, E. (1978a). Long-term effects of continuous exposure to *p*-chloroamphetamine on central serotonergic mechanisms in mice. *Biochem. Pharmacol.* **27,** 2033–2037.

Steranka, L. R., and Sanders-Bush, E. (1978b). Long-term reduction of brain serotonin by *p*-chloroamphetamine: Effects of inducers and inhibitors of drug metabolism. *J. Pharmacol. Exp. Ther.* **206,** 460–467.

Stone, D. M., Johnson, M., Hanson, G. R., and Gibb, J. W. (1988). Role of endogenous dopamine in the central serotonergic deficits induced by 3,4-methylenedioxymethamphetamine. *J. Pharmacol. Exp. Ther.* **247,** 79–87.

Sugrue, M. F. (1984). Antagonism of fenfluramine-induced hyperthermia in rats by some, but not all, selective inhibitors of 5-hydroxytryptamine uptake. *Brit. J. Pharmacol.* **81,** 651–657.

Tachiki, K. H., Takagi, A., Tateishi, T., Kido, A., Nishiwaki, K., Nakamura, E., Nagayama, H., and Takahashi, R. (1978). Animal model of depression. III. Mechanism of action of tetrabenazine. *Biol. Psychiatry* **13,** 429–443.

Taddei, C., and Mennini, T. (1989). High affinity ³H-*d*-fenfluramine uptake by rat brain synaptosomes. *In* "Abstracts of the International Symposium on Serotonin from Cell Biology to Pharmacology and Therapeutics," Florence, Italy, p. 167.

Thomas, D. R., Nelson, D. R., and Johnson, A. M. (1987). Biochemical effects of the antidepressant paroxetine, a specific 5-hydroxytryptamine uptake inhibitor. *Psychopharmacology (Berlin)* **93,** 193–200.

Trulson, M. E., and Jacobs, B. L. (1976). Behavioural evidence for the rapid release of CNS serotonin by PCA and fenfluramine. *Eur. J. Pharmacol.* **36,** 149–154.

Vaatstra, W. J., Deiman-Van Aalst, W. M., and Eigeman, L. (1981). DU 24565, a quipazine derivative, a potent selective serotonin uptake inhibitor. *Eur. J. Pharmacol.* **70,** 195–202.

Van de Kar, L. D., Richardson, K. D., and Urban, J. H. (1985). Serotonin and norepinephrine-dependent effects of fenfluramine on plasma renin activity in conscious male rats. *Neuropharmacology* **24,** 487–494.

van Praag, H. M., and Korf, J. (1973). 4-Chloroamphetamines. Chance and trend in the development of new antidepressants. *J. Clin. Pharmacol.* **13,** 3–14.

van Praag, H. M., Schut, T., Bosma, E., and van den Bergh, R. (1971). A comparative study of the therapeutic effects of some 4-chlorinated amphetamine derivatives in depressive patients. *Psychopharmacologia (Berlin)* **20,** 66–76.

Vorhees, C. V., Schafer, G. J., and Barrett, R. J. (1975). p-Chloroamphetamine: Behavioral effects of reduced cerebral serotonin in rats. *Pharmacol. Biochem. Behav.* **3,** 279–284.

Vu, D. H., and Tork, I. (1992). Differential development of the dual serotoninergic fiber system in the cerebral cortex of the cat. *J. Comp. Neurol.* **317,** 156–174.

Wentland, M. P., Bailey, D. M., Alexander, E. J., Castaldi, M. J., Ferrari, R. A., Haubrich, D. R., Luttinger, D. A., and Perrone, M. H. (1987). Synthesis and antidepressant properties of novel 2-substituted 4,5-dihydro-¹H-imidazole derivatives. *J. Med. Chem.* **30,** 1482–1489.

Wong, D. T., Van Frank, R. M., Horng, J.-S., and Fuller, R. W. (1972). Accumulation of amphetamine and p-chloroamphetamine into synaptosomes of rat brain. *J. Pharm. Pharmacol.* **24,** 171–173.

Wong, D. T., Horng, J.-S., and Fuller, R. W. (1973). Kinetics of serotonin accumulation into synaptosomes of rat brain–Effects of amphetamine and chloroamphetamines. *Biochem. Pharmacol.* **22,** 311–322.

Yamaguchi, K., Nabeshima, T., and Kameyama, T. (1986). Potentiation of phencyclidine-induced dopamine-dependent behaviors in rats after pretreatments with serotonin-depletors. *J. Pharmacobio-Dyn.* **9,** 179–189.

Alison McGregor
David C. S. Roberts

8
Mechanisms of Abuse

I. INTRODUCTION

This chapter reviews evidence concerning the brain mechanisms involved in amphetamine abuse. Amphetamine and its active analogs are potent central nervous system (CNS) stimulants. With short-term use, these compounds induce feelings of intense euphoria and increased strength and mental capacity. Long-term abuse causes severe anxiety, agitation, and insomnia, paranoid thinking, and even psychosis. To overcome these effects of chronic amphetamine intake, addicts often take other drugs of abuse and become multiple-drug users. Therefore, amphetamine addiction obviously has extremely deleterious effects on the family life and social relationships of the user (see Part IV).

Understanding the brain processes involved in amphetamine addiction is a daunting challenge. Suggesting that we have more than an inkling about the neural mechanisms involved would be a mistake. Just as most behaviors are a manifestation of the integrated activity of many brain systems, so must addictive behaviors have a multitude of neural mechanisms that contribute in some way to the final response pattern. We are far from understanding how these components interrelate. Nonetheless, some progress has been made. Monoamine terminals appear to be the locus at which amphetamines gain access to the neural circuitry (see Chapter 3). We review the evidence that addresses where and to what degree these actions are associated with the addictive process.

We concentrate, in this chapter, on studies that have addressed the neural mechanisms responsible for amphetamine reinforcement. Most of these studies have used self-administration techniques that require laboratory animals, usually rats, to meet specific response requirements before receiving an intravenous injection of amphetamine. Self-administration techniques measure response-contingent reinforcement mechanisms.

Response-contingent reinforcement sometimes can be obscured by other, often powerful, drug-induced behavioral effects. Moderate doses of amphetamine induce intense locomotor stimulation, inhibit sleep processes, suppress food intake, and, at higher doses, produce stereotyped behavior (see Chapter 4). Generally these behavioral effects are not studied with respect to understanding amphetamine reinforcement, although these other behaviors may compete with normal behaviors or may become incorporated into the behavioral repertoire simply by their association with the drug. Addictive behavior patterns, therefore, may contain a variety of these drug-induced effects. Further, amphetamine may produce reinforcing effects even when delivered independent of an explicit response contingency. The conditioned place preference (CPP) paradigm is a good example of such independent reinforcement. An animal need not perform a specific response for the drug to reinforce particular associations. We review those studies that have used the CPP paradigm to investigate the site of action of amphetamine and we also discuss other ways in which amphetamine might affect the behavioral repertoire directly in the absence of formal drug-response contingencies.

II. DOPAMINE

A. Dopamine Antagonists and Amphetamine Self-Administration

Of the possible neurochemical systems, the evidence is strongest for involvement of dopamine (DA) in the control of amphetamine self-administration behavior. Treatment with DA antagonists is well established to interfere with the reinforcing qualities of amphetamine during self-administration. In an early study, Wilson and Schuster (1972) showed that pretreatment with chlorpromazine caused an increase in amphetamine intake in the monkey. Similarly, in the dog, pretreatment with the DA antagonists pimozide or chlorpromazine caused a dose-dependent increase in amphetamine intake (Risner and Jones, 1976). This effect also has been produced in the rat (e.g., Davis and Smith, 1974; Yokel and Wise, 1975, 1976). The phenomenon is extremely robust, although interpretation of the data is less straightforward. Yokel and Wise (1975, 1976) argued that the increase in drug intake represents a compensatory response to a decrease in drug potency. Rate of drug intake on simple fixed ratio (FR) schedules is related inversely to unit drug dose (Pickens and Thompson, 1968); animals will self-inject at a faster rate on low unit injection doses. Animals are capable of maintaining a relatively constant level of drug intake across a large dose range simply by adjusting the interinfusion interval. The fact the animals increase their drug intake after DA receptor blockade has been interpreted as an analogous response (Wilson and Schuster, 1972; Yokel and Wise, 1975).

Figure 1 shows the effect of various doses of pimozide on the pattern of amphetamine self-administration in the rat. At low doses of pimozide, the animal displays a dose-dependent increase in responding whereas, at very high doses, the animal initially shows very high levels of responding followed by cessation of responding. This behavioral pattern mimics the extinction behavior produced when saline is substituted for amphetamine, suggesting that at high doses the effect of DA antagonists is blocking all reinforcing action of amphetamine.

Investigation of the contribution made by DA receptor subtypes to the reinforcing action of amphetamine has been extremely limited. One study

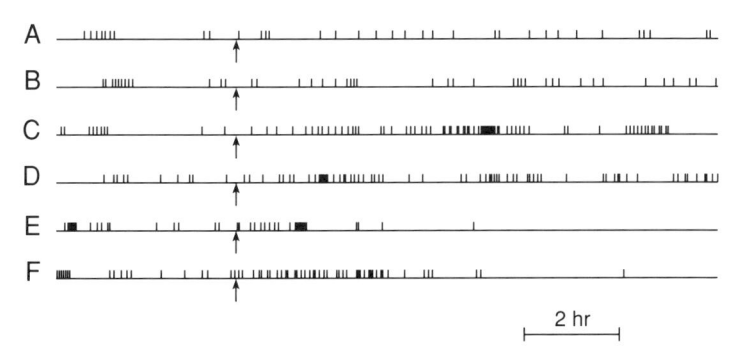

FIGURE 1 Representative event records from an animal self-administering intravenous amphetamine (0.25 mg/kg/injection). Each vertical mark represents an amphetamine injection. Vehicle (saline) or one of various doses of the dopaminergic antagonist pimozide were administered to the animal (marked by the arrows). (A) Vehicle treatment; (B) 0.0625 mg/kg pimozide; (C) 0.125 mg/kg pimozide; (D) 0.25 mg/kg pimozide; (E) 0.5 mg/kg pimozide; (F) intravenous saline replaced for amphetamine. Rate of intake increased with the dose of antagonist until, at the highest dose (E), extinction-like responding was produced (see F). Reproduced with permission from Yokel and Wise (1975). Copyright © 1975 by the AAAS.

has looked at the effect of the specific D_2 antagonist remoxipride on amphetamine self-administration. The results show a small increase in self-administration at lower doses and a decrease in responding at the highest dose (Amit and Smith, 1991). This result concurs with the previous effects of DA antagonism on amphetamine self-administration produced with pimozide, haloperidol, or butaclamol (see previous discussion)—all relatively but not exclusively selective for the D_2 receptor. Some drugs show a greater selectivity for this receptor subtype (e.g., sulpiride or spiperone) but have not been investigated with respect to amphetamine self-administration behavior. Further, no study to date has examined the effect of specific D_1 receptor subtype antagonism on amphetamine self-administration.

More study has been done of the involvement of different receptor subtypes in cocaine self-administration behavior. Note that, although certain parallels can be drawn between these two drugs of abuse and their mechanisms of action, significant differences also exist (Spyraki *et al.*, 1982a,b; Goeders and Smith, 1983; Hoebel *et al.*, 1983; Mackey and van der Kooy, 1985; Hemby *et al.*, 1992; Steketee and Kalivas, 1992). However, the results from cocaine studies that have focused on specific D_2 antagonists reveal a profile of behavioral effects similar to those reported for amphetamine self-administration and treatment with the less selective D_2 antagonists (Britton *et al.*, 1991; Corrigal and Coen, 1991; Hubner and Moreton, 1991). Similarly, treatment with the selective D_1 antagonist SCH 23390 caused a significant increase in responding for cocaine at low doses and cessation of responding at higher doses (Koob *et al.*, 1987; Britton *et al.*, 1991; Corrigal and Coen, 1991; Hubner and Moreton, 1991). If these results are applicable to amphetamine, then both the D_1 and the D_2 receptor will prove to be involved in mediating the reinforcing effects of amphetamine within the self-administration paradigm. However, the nature of the contribution made by these receptor subtypes to these effects remains unclear (see Beninger, 1992, for discussion).

B. Dopamine Projection Areas and Amphetamine Self-Administration

Systemic injections of antagonists can yield valuable information concerning the involvement of various receptors, although such studies do not contribute to our understanding of the relative importance of specific brain areas. Addressing this issue requires intracerebral injections of antagonists or neurotoxins. Given the pivotal role of DA, several studies have used the neurotoxin 6-hydroxydopamine (6-OHDA) to destroy the DA-containing terminals within a particular neural site, and thus determine the contribution made by DA in that neural site to amphetamine mechanisms of reinforcement. Figure 2 illustrates the major projection areas of the ascending DA systems in the rat. Given the importance of the A10 dopaminergic

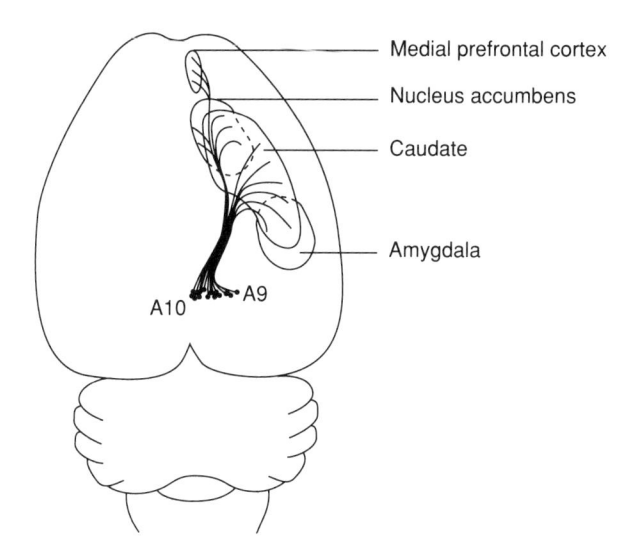

FIGURE 2 Horizontal rat brain section showing the main projection areas of the ventral tegmental area (A10) dopaminergic neurons. Innervation of the medial prefrontal cortex, nucleus accumbens, and amygdala is heavy. In contrast, the A9 neurons of the substantia nigra heavily innervate the caudate. Modified with permission from Ungerstedt (1971).

neurons in reward mechanisms, neural sites receiving innervation by these neurons may make contributions to amphetamine reinforcement mechanisms.

1. Nucleus Accumbens

The nucleus accumbens (NACC) is perhaps the most heavily implicated structure in the reinforcing mechanisms of drug action. This area receives very heavy DA innervation from the DA-containing cells of the ventral tegmental area (VTA) in the brainstem. Lyness and colleagues (1979) placed 6-OHDA lesions in the NACC and abolished not only the intake of amphetamine in self-administering rats but also the acquisition of the behavioral response in naive rats. The lack of postlesion responding in the experienced animals is perhaps surprising, since this result suggests an effect other than rendering the amphetamine nonreinforcing. Such an effect would be expected to induce initially high levels of postlesion responding, that is, extinction behavior, which has been reported in rats trained to self-administer cocaine prior to 6-OHDA lesions of the NACC (Roberts *et al.*, 1980). These authors report a similar lack of responding if the animals are given drug access immediately after lesioning; in contrast, extinction-like behavior is produced if the animals are allowed 5 days to recover after lesioning. This result suggests a postlesion time interaction with drug action may affect the interference in amphetamine reinforcement mechanisms induced by the lesion.

Whatever the time course of the effects of lesion, the study shows that DA modulation of the NACC plays a pivotal role in the control of amphetamine intake. Further, rats have been found to self-administer amphetamine directly into the NACC (Hoebel et al., 1983), again suggesting that the action of the drug in the NACC is powerfully reinforcing.

2. Medial Prefrontal Cortex

Like the NACC, the medial prefrontal cortex (mPFC) is a recipient of heavy DA innervation from the VTA. However, this structure has received much less attention with respect to its role in amphetamine reinforcement mechanisms.

Leccese and Lyness (1987) reported that 6-OHDA lesions of the mPFC had no effect on amphetamine self-administration behavior. Similarly, injections of amphetamine into the mPFC did not induce a CPP (Carr and White, 1986; see Fig. 3). Thus, the mPFC appears not to be involved in mediating the reinforcing action of amphetamine.

However, this result is perhaps surprising given the implication of this area in the reinforcement mechanisms of cocaine. Although 6-OHDA lesions of the mPFC have been found not to affect cocaine self-administration (Martin-Iverson et al., 1986), rats have been reported to self-administer both cocaine (Goeders and Smith, 1983, 1986) and stimulation (e.g., Mora and Myers, 1977) directly into this brain region in a DA-dependent fashion. Similarly, 6-OHDA lesions of the mPFC did prevent a cocaine-induced CPP (Isaac et al., 1989).

These discrepancies between the neural substrates of cocaine and amphetamine mechanisms of reinforcement are interesting and obviously require further investigation. The apparent involvement of the mPFC in cocaine reinforcement mechanisms suggests that this neural site should not be ignored as a possible neuroanatomical substrate of amphetamine action.

3. Amygdaloid Complex

The amygdaloid complex (AMY) also receives DA innervation from the VTA, but generally has been overlooked by individuals studying the mesolimbic DA system. For this reason, this region has received even less investigation than the mPFC as a possible mediator of drug action.

However, Deminiere et al. (1988) showed that 6-OHDA lesions of the AMY led to a greater sensitivity in rats acquiring the self-administration of amphetamine. That is, lesioned rats would acquire the response for a dose of amphetamine below one for which the controls would respond. However, these authors did not study the maintenance of responding in experienced rats after such a lesion.

C. Microdialysis and Amphetamine Self-Administration

In vitro data have suggested that amphetamine releases newly synthesized monoamines from nerve terminals and blocks their reuptake. The net result is presumed to be an increased concentration of neurotransmitter in the synaptic cleft. With the advent of *in vivo* microdialysis, some of these assumptions could be tested in anesthetized and, in some cases, behaving animals.

Many laboratories now have confirmed that amphetamine increases the concentration of extracellular DA in the striatum and the nucleus accumbens (Zetterstrom *et al.*, 1988; Hernandez *et al.*, 1989; Kuczenski and Segal, 1989; Robinson and Camp, 1990; Yamamoto and Pehek, 1990). Since the pharmacological and lesion data already discussed have indicated that the mesolimbic DA system is involved critically in psychostimulant self-administration, many of the microdialysis studies have focused on the ventral striatum and nucleus accumbens. For example, DiChiara and Inoki (1988) argued that "preferential" release of mesolimbic DA, compared with striatal DA, is a defining feature of many drugs of abuse. Whether, in fact, "preferential" release occurs is a matter of some debate (Robinson and Camp, 1991; Di Chiara, 1991), although researchers generally agree that extracellular DA is increased by other reinforcing drugs such as cocaine (Hurd and Ungerstedt, 1989; Pettit *et al.*, 1990), methylenedioxyamphetamine (MDA), methylenedioxymethamphetamine (MDMA) (Nash and Nichols, 1991), phencyclidine (Hernandez *et al.*, 1989), and nicotine (Mifsud *et al.*, 1989; Brazell *et al.*, 1990).

This new technology has provided a new means of correlating neurochemical events with behavioral data. For example, microdialysis studies in behaving animals have shown that increased extracellular DA levels are associated with consummatory responses such as feeding (Hernandez *et al.*, 1989) and sexual behavior (Damsma *et al.*, 1992). Another observation is that decreases in extracellular DA are correlated with withdrawal from drugs of abuse (Imperato *et al.*, 1992; Rossetti *et al.*, 1992; Weiss *et al.*, 1992). An interesting speculation is whether such neurochemical measures could define a motivational state such as "craving."

Several studies have used *in vivo* microdialysis techniques to study cocaine self-administration behavior (Hurd *et al.*, 1989; Pettit and Justice, 1991; Weiss *et al.*, 1992). The results confirm that extracellular DA levels rise following each injection. No studies to date have examined amphetamine self-administration.

One behavioral aspect of amphetamine that has been investigated through microdialysis is behavioral sensitization. Sensitization describes the finding that animals will show increased locomotor activity and/or stereotypy with each successive injection of a psychomotor stimulant (see Chapter

4). Attempts have been made to link the behavioral response to an increased neurochemical event. Several groups have confirmed that an increase in extracellular DA correlates with the augmented behavioral response to repeated injections of either cocaine (Akimoto et al., 1989; Kalivas and Duffy, 1990; Pettit et al., 1990) or amphetamine (Robinson et al., 1988; Kazahaya et al., 1989; Akimoto et al., 1990; Patrick et al., 1991). Other studies, however, have shown that behavioral and neurochemical sensitization are not necessarily linked, casting doubt on whether increased release of DA can explain fully the changes in the locomotor response (Hurd et al., 1989; Kuczenski and Segal, 1990; Segal and Kuczenski, 1992; Kalivas and Duffy, 1993). Whether DA release is increased or decreased appears to depend on an interaction of dose and time following acute or chronic exposure (see Chapter 4). These data emphasize that, although the *in vivo* microdialysis technology allows neurochemical events to be monitored, the approach is nonetheless correlational, so conclusions regarding causality must be considered speculative.

D. Conditioned Place Preference

Much of the data from CPP studies parallel findings from self-administration studies, although important differences do exist. Pharmacological manipulations and lesion studies support the idea that DA mechanisms mediate amphetamine-induced CPP. Thus, that amphetamine-induced CPP can be blocked by pretreatment with DA antagonists is well established (Spyraki et al., 1982b; Mackey and Van der Kooy, 1985; Mithani et al., 1986). Moreover, researchers have shown that both the D_1 and the D_2 receptor subtypes appear to be involved in the acquisition of amphetamine CPP. Both the D_1 receptor antagonist SCH 23390 and the D_2 receptor antagonist sulpiride blocked the acquisition of amphetamine CPP (Leone and DiChiara, 1987; Hoffman and Beninger, 1989; Hiroi and White, 1991a).

However, a distinction must be made between the acquisition and ex-

---

FIGURE 3 Results from amphetamine-induced conditioned place preference (CPP) experiments demonstrating the differential contributions made by the amygdala and nucleus accumbens to this measure of drug action. The mean ($\pm$ SEM) time (sec) spent in the amphetamine- (■) and saline-paired (□) environments is shown. (A) Bilateral injections of amphetamine (10 μg/0.5 μl/side) were made into the nucleus accumbens, amygdala, or medial prefrontal cortex *prior* to placing the animal in a specific environment. Only amphetamine injection into the nucleus accumbens supported a CPP. (B) Bilateral lesions, either electrolytic (E-Lesion) or excitotoxic (N-methyl-D-aspartate, NMDA), of the lateral amygdala were placed *after* the animal had received all the conditioning trials. Both types of lesion significantly attenuated the amphetamine-induced CPP produced in the control animals (sham), suggesting that the amygdala plays a role in the expression of CPP. Modified with permission from Carr and White (1986) and Hiroi and White (1991b).

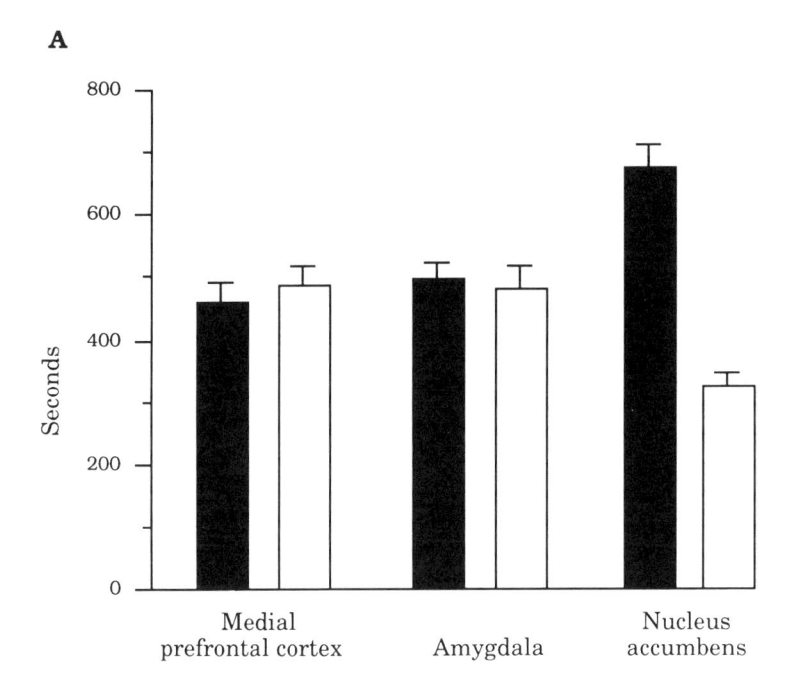

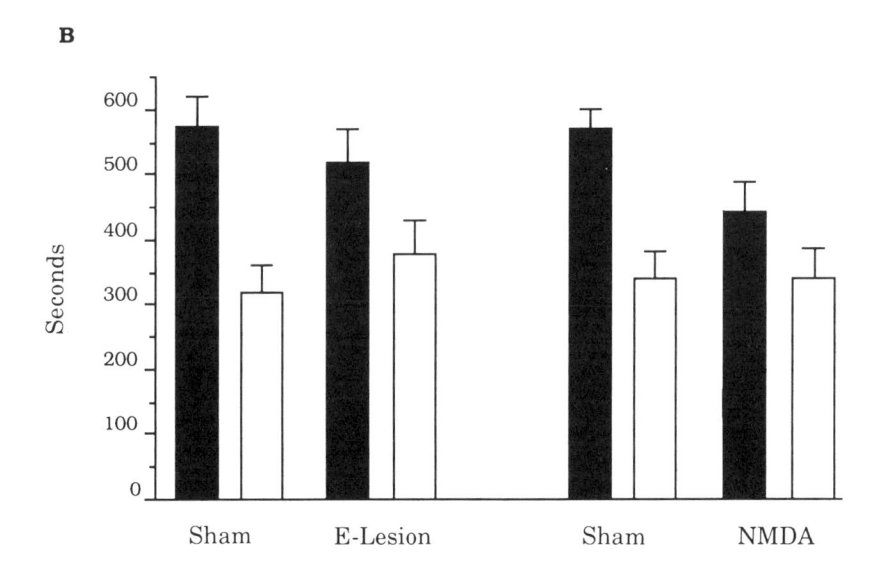

A

B

pression of CPP. Hiroi and White (1991a) demonstrated that, although the D_1 and D_2 receptor subtypes both contributed to the acquisition of the CPP, only the D_1 receptor appeared to be involved in the expression of this behavior, that is, D_2 antagonists only blocked CPP expression at doses capable of D_1 receptor binding.

In parallel with such self-administration results, amphetamine-induced CPP also appears to rely on dopaminergic NACC mechanisms. Spyraki *et al.* (1982a) demonstrated that 6-OHDA lesions of the NACC prevented amphetamine-induced CPP; that is, animals with the greatest DA depletion in the NACC showed the greatest resistance to CPP. In agreement with this result, Carr and White (1986) showed that injections of amphetamine directly into the NACC induced a strong CPP. Figure 3A shows how intra-NACC injections of amphetamine made prior to placing the animal in a particular environment increase the time spent in that environment when the animal has free access to both the amphetamine- and the saline-paired environments. This effect of amphetamine action within the NACC appears to be relatively anatomically specific, since preconditioning amphetamine injections into the mPFC or AMY have no effect on CPP.

With respect to amphetamine-induced CPP, Carr and White (1986) showed that amphetamine injected into the AMY did not induce a CPP. However, Hiroi and White (1991b) reported that excitotoxic lesions of the lateral AMY did interfere with an amphetamine-induced CPP. The lesions were placed either before or after the conditioning trials to define whether the interference involved the acquisition or the expression of the CPP. Both sets of lesions impaired the CPP, indicating that the lesion disrupted the expression of the CPP. Figure 3 illustrates this effect of postconditioning lesions on amphetamine-induced CPP. Electrolytic and excitotoxic lesions were used; both types of lesion attenuated expression of the CPP. Thus, although the available data are limited, this limbic site does appear to make some as yet undefined contribution to amphetamine neural mechanisms of action.

III. SEROTONIN

A significant amount of work has addressed the contribution made by serotonin (5-HT) to amphetamine self-administration behavior. However, the nature of this contribution is still open to debate.

Lyness and co-workers (1980) administered the 5-HT toxin 5,7-dihydroxytryptamine (5,7-DHT) intracerebroventrically (icv) and showed a significant increase in amphetamine intake, both during the acquisition of the behavioral response and during the maintenance of the already acquired response.

In agreement with this lesion study, researchers have shown that pretreatment with 5-HT-enhancing substances caused decrements in amphetamine self-administration. This result has been found repeatedly following administration of the 5-HT reuptake inhibitor fluoxetine (Leccese and Lyness, 1984; Porrino *et al.*, 1989; Yu *et al.*, 1990) and administration of L-tryptophan (the precursor amino acid of 5-HT), which increases brain levels of 5-HT and causes an increase in 5-HT release (Lyness, 1983; Leccese and Lyness, 1984).

Various interpretations of these results have been put forward. Initially, investigators proposed that 5-HT might contribute to the positively reinforcing action of amphetamine (Lyness *et al.*, 1980; Leccese and Lyness, 1984). More specifically, the increase following lesioning and the decrease following 5-HT augmenting treatments were suggested to represent the same changes produced when the unit dose of the amphetamine injection was decreased or increased, respectively. However, this argument is less convincing when more recent findings are also considered. The problems with interpretation of rate data can be avoided using a progressive ratio (PR) schedule of reinforcement (Roberts *et al.*, 1989). Thus, researchers have shown that 5,7-DHT lesions of the medial forebrain bundle (MFB) increase the reinforcing efficacy of cocaine under a PR schedule (Loh and Roberts, 1990) whereas pretreatment with fluoxetine (Richardson and Roberts, 1991) and L-tryptophan (McGregor and Roberts, 1993) decreases the reinforcing efficacy of cocaine. Although caution should be exercised in making assumptions about the mechanism of amphetamine action from cocaine-derived data, the effects on cocaine rate of self-administration after fluoxetine and L-tryptophan pretreatment are similar to those on amphetamine self-administration behavior (Carroll *et al.*, 1989,1990).

These PR results lend support to an alternative interpretation. Investigators have shown that amphetamine does have some aversive action in addition to its positively reinforcing effects (Wise et al., 1976; Van Haaren and Hughes, 1990; Ferrari *et al.*, 1990). Thus, 5-HT mechanisms may mediate some aspect of this action. After lesioning, any "aversive tone" in the action of amphetamine is lost and the drug becomes more reinforcing, whereas increasing 5-HT function would increase the "aversive tone" of amphetamine, making it less reinforcing (Smith *et al.*, 1986; Loh and Roberts, 1990). However, this interpretation is somewhat complicated by the action of 5-HT antagonists on amphetamine self-administration. If 5-HT action is aversive, then antagonism of this effect might be expected to increase the reinforcing efficacy of amphetamine and, thus, produce a change in self-administration behavior. This expectation is, indeed, borne out; however, the change in direction of intake is the same as that following agonist pretreatments. Pretreatment with the 5-HT antagonists methysergide, cyproheptadine (Leccese and Lyness, 1984), and cinanserin (Porrino *et al.*, 1989) all decrease amphetamine self-administration—the behavioral response produced after fluoxetine and L-tryptophan treatment. Thus, although a decrease in rate of self-administration can be interpreted as an increase in drug reinforcing efficacy, when both the agonist and antagonist treatments induce the same response, this result underlines the problems with interpretation of such data. An examination of the effects of these treatments on amphetamine self-administration under a PR schedule of reinforcement is required.

The problem of comparing cocaine self-administration results with those of amphetamine must be emphasized here. Although the PR data from agonist treatments suggest an inhibitory role for 5-HT in stimulant reinforcement mechanisms, pretreatment with 5-HT antagonists produces no effect on cocaine self-administration, either under FR (Porrino et al., 1989; Peltier and Schenck, 1991) or PR (Lacosta and Roberts, 1993) reinforcement. Thus, this discrepancy suggests that, in fact, significantly different 5-HT mechanisms may influence the control of these two stimulants over behavior.

In summary, 5-HT evidently has a significant role in amphetamine self-administration behavior. However, since only the rate of amphetamine intake has been used as the dependent variable, defining that role as contributing to the reinforcing aspect of the drug or, alternatively, to an aversive action of the drug is still not possible.

In contrast, less work has examined the role of 5-HT in the CPP paradigm. Spyraki et al. (1988) did show that 5,7-DHT lesions of the NACC did not affect acquisition of amphetamine CPP; however, no icv 5-HT lesions have been administered. Thus, determining whether 5-HT plays a significant role in CPP is difficult, although 5-HT_3 receptor antagonism with MDL 72222 and ICS 205-930 also has been demonstrated not to affect the acquisition of amphetamine CPP (Carboni et al., 1989).

IV. NORADRENALINE

Noradrenaline (NA) generally has been assumed not to play a significant role in the reinforcement mechanisms underlying amphetamine self-administration behavior. Risner and Jones (1976) reported that pretreatment with NA agonists or antagonists did not alter amphetamine self-administration in dogs. Moreover, the NA agonist methoxamine was not self-administered when substituted for amphetamine. Concurring with these results, Woolverton (1987) reported similar results from monkeys self-administering amphetamine, showing no effect after pretreatment with an α-adrenergic receptor antagonist.

However, Davis et al. (1975) reported that manipulation of NA systems in the rat altered amphetamine self-administration behavior. Pretreatment with the NA-depleting agents diethyldithiocarbamate (DDC) and U-14,624 prevented amphetamine intake and also blocked the establishment of conditioned reinforcers. Note that DDC has been reported to induce a strong conditioned taste aversion (Roberts and Fibiger, 1976), suggesting that this agent may reduce amphetamine intake because of illness. Still, Davis and Smith (1977) found that self-administration of clonidine, a NA agonist, was acquired by rats and also served to establish a neutral stimulus successfully as a conditioned reinforcer. Both these effects were blocked by the adrenergic receptor antagonist phenoxybenzamine.

No work to date has studied the role of NA within the CPP paradigm.

V. DISCUSSION

Much of the literature that has explored the neural substrates of amphetamine reinforcement has described studies using the operant approach. As indicated in the introduction and just reviewed, self-administration techniques have been used to examine response-contingent reinforcement mechanisms. In this section, we explore possible other aspects of amphetamine-induced effects related to drug abuse.

A. Drug-Induced Perseveration

Stereotyped behavior is a prominent feature of a number of neurological disorders including autism, obsessive compulsive disorder, and schizophrenia (Randrup and Munkvad, 1967; Robbins, 1976). Perseverative behavior patterns also are produced by psychomotor stimulant drugs, prompting the speculation that amphetamine-induced stereotypy might be a useful animal model for investigating pathological behavior (Randrup and Munkvad, 1967; Segal and Schuckit, 1983). Depending on the dose, amphetamine elicits a variety of species-typical behaviors; at high dosages, the pattern of activity breaks down into short sequences that are emitted at high frequency (Lyon and Robbins, 1975). A vast literature has explored the topography and neural substrates of behaviors elicited by high doses of amphetamine (Kelly *et al.*, 1975; Szechtman *et al.*, 1988). At lower doses, however, the behavioral repertoire of animals is affected in much more subtle ways. Researchers have suggested that the changes brought about by the lower doses of stimulant drugs may provide a better model for neurological disorders (Ridley *et al.*, 1988).

High doses of amphetamines tend to elicit behaviors such as licking, chewing, and gnawing in the rat (Randrup and Munkvad, 1967); however, lower doses do not seem to elicit behavior as much as modify existing responses. For example, Ellinwood and colleagues (1972; Ellinwood and Kilbey, 1975) chronicled the effects of amphetamines on the behavior of a number of species. This work elegantly established that amphetamine promotes perseverative behaviors that are determined largely by the behavior being displayed at the time of drug onset. This idea has been confirmed in a number of experimental situations (Randrup and Munkvad, 1967; Szechtman, 1983; Beck et al., 1986). Thus, perseveration appears to be a primary consequence of psychostimulant administration. Descriptions of the effects of low doses of amphetamine appear to offer important insights into how one class of reinforcing drug might affect ongoing behavior.

A distinction should be made here between the responses elicited by high doses of amphetamine and the more subtle changes produced by lower doses. The former will be referred to as "stereotypy." Stereotyped behavior might be thought of as an elicited response—an unconditioned response—such as licking or chewing. The lower dose effects are qualitatively different and may have very different neural substrates. The lower dose effects are referred to as perseverative behavior.

The usefulness of the concept of perseveration hinges largely on whether the more subtle drug-induced changes can be quantified. Amphetamine-induced stereotypy has been measured using categorical scales (Fray et al., 1980) or choreographic observations (Szechtman, 1983). We have developed a method of assessing perseverative behavior by examining the way in which rats explore an environment for food. The apparatus is an 8-arm radial maze but, unlike researchers that use this maze to study learning and memory, we rebait the goal box each time the animal leaves the arm. In this situation, food is available in every arm and the animal is required simply to explore the maze and collect the food. Perseverative behavior can be assessed by examining whether the animal shows repetitive patterns of arm entry during its explorations.

We have investigated how various drugs of abuse affect the exploratory pattern of rats in this radial arm maze. During baseline testing, animals display a slight bias in the way they explore the maze. Each animal is likely to repeat certain "angles of turn" over the course of the session. After amphetamine administration, the foraging pattern becomes much more perseverative. This robust effect is seen at doses as low as 0.25 mg/kg (see Fig. 4). Beck and Loh (1990) previously reported that ethanol and diazepam increase perseveration. These findings have been extended to include amphetamine, heroin, and nicotine (Loh et al., 1993). Nonreinforcing drugs such as scopolamine and haloperidol have no effect (or even decrease perseveration).

Reinforcing drugs, by tradition, have been defined by their ability to increase the probability of a particular response. Of course, an increase in the rate of one particular response, such as lever responding, would be at the expense of other response categories, such as exploration. Clearly, reinforcing stimuli usually cause a focusing of the behavioral repertoire, sometimes to the point that one particular response prevails over all others. Conversely, withholding reinforcement (extinction) usually causes behavior to become more variable and diversified (Beck and Loh, 1990). We suggest that the

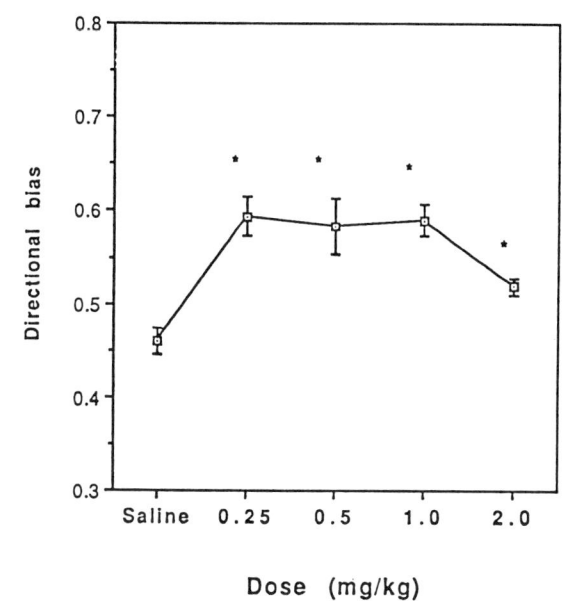

FIGURE 4 Perseverative responding induced by various doses of amphetamine in an 8-arm radial maze. Points represent means (± SEM) of 8 rats pretreated with amphetamine or saline 30 min prior to testing. Each animal was permitted to explore the maze until it had completed 25 arm entries. Once a rat had obtained the food reward from a particular arm and had re-entered the central hub of the maze, the food cup was rebaited. This testing procedure placed no restrictions on the route of exploration. The directional bias score can vary from a theoretical minimum of 0.354 (random pattern) to a maximum of 1.0 (perseverative re-entry into a single arm). See Loh et al. (1993) for details.

behavioral repertoire might be described along a continuum. At one end, an organism perseverates in a single response category (highly reinforced responding) whereas, at the other extreme, behavior would be highly variable. If this formulation is correct, one might predict that the behavioral repertoire might be shifted in the perseverative direction by reinforcing drugs.

We hypothesize that reinforcing drugs predictably affect behavior despite the absence of formal response contingencies. Psychomotor stimulants enhance those categories of behavior that happen at the time of drug delivery (Broekkamp and Van Rossum, 1974; Ellinwood and Kilbey, 1975; Ridley et al., 1981; Carr and White, 1987). Predominant behaviors would be expected to increase in frequency, forcing less frequent behaviors out of the repertoire. In other words, the result may be "an increased frequency in a decreased number of categories." This description is precisely that used by Lyon and Robbins (1975) to describe stimulant stereotypy. Other reinforcing drugs may produce similar effects, although the degree to which ongoing behavior would be influenced by such factors as speed of drug onset, clearance rate, and other competing responses would vary.

The reinforcing effects that influence ongoing behavior should be considered an important aspect of the addictive process and may account for "habits" and "superstitious" response patterns associated with drug taking. Exploratory behavior in the radial arm maze appears to be effective in detecting and quantifying drug-induced perseveration, which may provide a useful tool in investigating how reinforcing and other centrally acting drugs influence ongoing behavior.

Much of the literature that has explored amphetamine-induced changes in behavior has concentrated on stereotyped responses (high dose effects). For the most part, striatal DA mechanisms have been implicated (Kelly *et al.*, 1975; Kuczenski and Segal, 1989; Florin *et al.*, 1992). What brain areas are involved in the perseverative response (low dose effects) is not yet clear, although perseverativity has been noted as a hallmark of some hippocampal lesions (Devenport *et al.*, 1981). Lesions of the hippocampal–accumbens pathway have been reported to block methamphetamine-induced sensitization (Yoshikawa *et al.*, 1991). Given the role of forebrain DA in amphetamine self-administration and CPP, investigations of the role of the DA innervation of the various limbic structures in perseveration might be fruitful.

B. Multiple Paradigms to Measure Reinforcement

Does a single paradigm exist that best assesses drug reinforcement? We suggest not. Some controversy exists over whether CPP and self-administration techniques measure the same phenomenon, that is, drug reinforcement (see Wise, 1989). Attempting to equate the two may be futile. Perhaps a better way to view them is as measuring overlapping but different aspects of a broader concept.

The similarity between the results from the two paradigms, revealing DA as an essential mediator of amphetamine reinforcement mechanisms, is very striking. Moreover, the involvement of the NACC in this DA function is also strongly suggested by both paradigms. Therefore, whatever aspect of reinforcement is represented in the NACC by its DA function is common to both paradigms and is equally important in supporting the different types of behavior measured in them.

In contrast, the differences in results between the two paradigms show how amphetamine reinforcement mechanisms might differ. For example, the dissociation of the acquisition and expression of amphetamine-induced CPP, with respect to both DA receptor subtype and neuroanatomical site, indicates a neurochemical and functional action of amphetamine reinforcement that can be revealed only by one of these paradigms.

The overlap between the two sets of results suggests the validity in viewing both paradigms as measuring the same aspect of drug action, that is, reinforcement. However, these results also indicate that reinforcement can be viewed as a multifaceted phenomenon, of which these two paradigms measure different, but probably equally fundamental aspects of drug abuse. Whether future investigations reveal that amphetamine-induced perseveration uses the same or additional neural mechanisms remains to be seen.

C. Neural Mechanisms of Reinforcement

Three neuroanatomical substrates of amphetamine reinforcement mechanisms have been considered here. However, the role or even involvement of the mPFC in amphetamine abuse remains so unclear that it will not be discussed.

The NACC plays a vital and well-discussed role in mediating the reinforcing action of amphetamine. Without this neural structure, animals will neither self-administer drug nor show any drug-induced conditioning effects, suggesting that the NACC is in some way responsible for mediating or processing some primary action of the drug. Without this action, no other reinforcement-related effects can be produced.

Although the AMY has been relatively ignored with respect to drug reinforcement mechanisms, it has been implicated more generally in mechanisms of reward, especially in stimulus–reward associations (e.g., Gaffan and Harrison, 1987; Cador *et al.*, 1989; Everitt *et al.*, 1989,1991). The neuroanatomy of the AMY supports such a role; the structure is a recipient of a DA projection from the VTA, a recipient of highly processed sensory information from all modalities, and a recipient via the globus pallidus of NACC information. In return, the AMY heavily innervates the NACC, association cortex, and PFC (see DeOlmos *et al.*, 1985; Price *et al.*, 1987). Thus, a role in mechanisms of drug action is perhaps not unexpected.

However, to date this role is undetermined. Hiroi and White (1991b) demonstrated that the lateral nucleus of the AMY is involved in the expression (and possibly the acquisition) of an amphetamine-induced CPP, whereas Carr and White (1986) had shown that injections of amphetamine into the AMY did not induce a CPP. Together these results suggest that the AMY is not involved in mediating the primary effects of amphetamine reinforcement, but is involved in the formation or utilization of the associations formed between the primary effects of the drug and stimuli such as those required for a CPP (see Fig. 3).

The self-administration paradigm generally is thought to measure the intrinsic reinforcing value of the drug; as such, involvement of the AMY might not be expected. However, the paradigm does in fact include some elements of drug-induced conditioning. In most self-administration experiments, a stimulus light accompanies the drug injection. Such a cue may serve as a conditioned stimulus through its predictive power for the reinforcement that will follow. Similarly, the response lever in these experiments must acquire incentive value and thus have an increased ability to elicit an approach response (see Deininger, 1992). Such considerations suggest that the AMY could make a contribution to amphetamine reinforcement mechanisms involved in controlling self-administration behavior as well as CPP behavior.

D. Mechanisms of Abuse

The results of the self-administration and CPP studies demonstrate that the NACC and DA are involved fundamentally in amphetamine mechanisms of action. If a single neural component is responsible for mediating the primary reinforcement or rewarding effect of amphetamine, then the DA innervation of the NACC is a prime candidate for this role.

The impact of such action will, however, have various secondary effects on neural mechanisms. The stimulus–reward associations formed by the reinforcing impact of amphetamine action constitute one such effect. This phenomenon was discussed earlier, and a role for the AMY in this effect appears likely. However, undoubtedly other effects may require other paradigms to reveal them behaviorally. For example, the stimulus–response association formed between the stimulus lever and the appropriate response presumably contributes to the control of the drug over the behavioral response, but dissociating this effect from the primary effects of amphetamine within the self-administration paradigm would be difficult. Similarly, stimulus–stimulus associations also may be set up by amphetamine reinforcement mechanisms that contribute to the behavioral output of an animal in relatively subtle ways that remain undetected.

Mechanisms of drug abuse clearly are related to a wider issue than the primary reinforcing action of drugs. Understanding the various aspects of drug action that lead to abuse—including, for example, craving and relapse—may require looking beyond a single phenomenon or neural site. A reinforcing drug will have many effects on learning, memory, affect, and behavioral repertoire of an animal that may or may not be related to reinforcing properties of the drug.

Interestingly, systemic posttraining amphetamine injections can enhance memory function (e.g., White, 1988). For example, having paired a neutral stimulus with an aversive stimulus, the animal shows greater recall and learning about this association if amphetamine is administered after training. This effect also can be produced using other reinforcing stimuli such as brain stimulation (Coulombe and White, 1980, 1982) and intake of sucrose and saccharin solutions (Messier and White, 1984). Further, researchers long have recognized that "noncontingent" injections of psychomotor stimulants can induce perseverative behavior. Such repetitive response patterns may be intensified by the reinforcing action of the drugs (Loh et al., 1993).

Thus, reinforcing stimuli appear to be able to alter associational processes even when not performing as reinforcers in the conventional sense.

Collectively, such a wide spectrum of effects could have profound consequences on the abuse profile and abuse symptoms of a drug. Therefore, gaining a greater understanding of amphetamine abuse rather than amphetamine reinforcement alone calls for using a broader selection of behavioral paradigms that address these issues.

REFERENCES

Akimoto, K., Hamamura, T., and Otsuki, S. (1989). Subchronic cocaine treatment enhances cocaine-induced dopamine efflux: Studies by *in vivo* intracerebral dialysis. *Brain Res.* **490**, 339–344.

Akimoto, K., Hamamura, T., Kazahaya, Y., Akiyama, K., and Otsuki, S. (1990). Enhanced extracellular dopamine level may be the fundamental neuropharmacological basis of cross-behavioral sensitization between methamphetamine and cocaine—An *in vivo* dialysis study in freely moving rats. *Brain Res.* **507**, 344–346.

Amit, Z., and Smith, B. R. (1991). Remoxipride, a specific D_2 dopamine antagonist: An examination of its self-administration liability and its effects on d-amphetamine self-administration. *Pharmacol. Biochem. Behav.* **41**, 259–261.

Beck, C. H. M., and Loh, E. A. (1990). Reduced behavioral variability in extinction: Effect of chronic treatment with the benzodiazepine diazepam or with ethanol. *Psychopharmacology* **100**, 328–333.

Beck, C. H. M., Chow, H. L., and Cooper, S. J. (1986). Initial environmental influences amphetamine-induced stereotypy: Subsequent environmental change has little effect. *Behav. Neural. Biol.* **46**, 383–397.

Beninger, R. J. (1992). D-1 receptor involvement in reward-related learning. *J. Psychopharmacol.* **6**, 34–42.

Brazell, M. P., Mitchell, S. N., Joseph, M. H., and Gray, J. A. (1990). Acute administration of nicotine increases the *in vivo* extracellular levels of dopamine, 3,4-dihydroxyphenylacetic acid and ascorbic acid preferentially in the nucleus accumbens of the rat: Comparison with caudate-putamen. *Neuropharmacology* **29**, 1177–1185.

Britton, D. R., Curzon, P., Mackenzie, R. G., Kebabian, J. W., Williams, J. E. G., and Kerkman, D. (1991). Evidence for involvement of both D_1 and D_2 receptors in maintaining cocaine self-administration. *Pharmacol. Biochem. Behav.* **39**, 911–915.

Broekkamp, C. L. E., and Van Rossum, J. M. (1974). Effects of apomorphine on self-stimulation behavior. *Psychopharmacology* **34**, 71–80.

Cador, M., Robbins, T. W., and Everitt, B. J. (1989). Involvement of the amygdala in stimulus–reward associations: Interactions with the ventral striatum. *Neuroscience* **30**, 77–86.

Carboni, E., Leone, A. P., and Di Chiara, G. (1989). $5HT_3$ receptor antagonists block morphine- and nicotine- but not amphetamine-induced reward. *Psychopharmacology* **97**, 175–178.

Carr, G. D., and White, N. M. (1986). Anatomical disassociation of amphetamine's rewarding and aversive effects: An intracranial microinjection study. *Psychopharmacology* **89**, 340–346.

Carr, G. D., and White, N. M. (1987). Effects of systemic and intracranial amphetamine injections on behavior in the open field: A detailed analysis. *Pharmacol. Biochem. Behav.* **27**, 113–123.

Carroll, M. E., Lac; S. T., Asencio, M., and Kragh, R. (1989). Fluoxetine reduces intravenous cocaine self-administration in rats. *Pharmacol. Biochem. Behav.* **35**, 237–244.

Carroll, M. E., Lac, S. T., Asencio, M., and Kragh, R. (1990). Intravenous cocaine self-administration is reduced by dietary L-tryptophan. *Psychopharmacology* **100**, 293–300.

Corrigall, W. A., and Coen, K. M. (1991). Cocaine self-administration is increased by both D$_1$ and D$_2$ dopamine antagonists. *Pharmacol. Biochem. Behav.* **39**, 799–802.

Coulombe, D., and White, N. M. (1980). The effects of post-training lateral hypothalamic self-stimulation on aversive and appetitive classical conditioning. *Physiol. Behav.* **25**, 267–272.

Coulombe, D., and White, N. M. (1982). The effects of post-training lateral hypothalamic self-stimulation on sensory preconditioning in rats. *Can. J. Psychol.* **36**, 57–66.

Damsma, G., Pfaus, J. G., Wenkstern, D., Phillips, A. G., and Fibiger, H. C. (1992). Sexual behavior increases dopamine transmission in the nucleus accumbens and striatum of male rats: Comparison with novelty and locomotion. *Behav. Neurosci.* **106**, 181–191.

Davis, W. M., and Smith, S. G. (1974). Positive reinforcing effects of apomorphine, d-amphetamine and morphine: Interaction with haloperidol. *Pharmacologist* **16**, 193.

Davis, W. M., and Smith, S. G. (1977). Catecholaminergic mechanisms of reinforcement: Assessment by drug self-administration. *Life Sci.* **20**, 483–492.

Davis, W. M., Smith, S. G., and Khalsa, J. H. (1975). Noradrenergic role in the self-administration of morphine or amphetamine. *Pharmacol. Biochem. Behav.* **3**, 477–484.

Deminiere, J. M., Taghzouti, K., Tassin, J. P., LeMoal, M., and Simon, H. (1988). Increased sensitivity to amphetamine and facilitation of amphetamine self-administration after 6-hydroxydopamine lesions of the amygdala. *Psychopharmacology* **94**, 232–236.

DeOlmos, J., Alheid, G. F., and Beltramino, C. A. (1985). Amygdala. *In* "The Rat Nervous System" (G. Paxinos, ed.), Vol. 1, pp. 233–334. Academic Press, Orlando, Florida.

Devenport, L. D., Devenport, J. A., and Holloway, F. A. (1981). Reward-induced stereotypy: Modulation by the hippocampus. *Science* **212**, 1288–1289.

Di Chiara, G. (1991). On the preferential release of mesolimbic dopamine by amphetamine. *Neuropsychopharmacology* **5**, 243–244.

DiChiara, G., and Inoki, R. (1988). Drugs abused by humans preferentially increase dopamine concentrations in the mesolimbic system of freely moving rats. *Proc. Natl. Acad. Sci. U.S.A.* **85**, 5274–5278.

Ellinwood, E. H., Jr., and Kilbey, M. M. (1975). Amphetamine stereotypy: The influence of environmental factors and prepotent behavioral patterns on its topography and development. *Biol. Psychiatry* **10**, 3–16.

Ellinwood, E. H., Sudilovski, A., and Nelson, L. (1972). Behavioral analysis of chronic amphetamine intoxication. *Biol. Psychiatry* **4**, 215.

Everitt, B. J., Cador, M., and Robbins, T. W. (1989). Interactions between the amygdala and ventral striatum in stimulus–reward associations: Studies using a second-order schedule of sexual reinforcement. *Neuroscience* **30**, 63–75.

Everitt, B. J., Morris, K. A., O'Brien, A., and Robbins, T. W. (1991). The basolateral amygdala–ventral striatal system and conditioned place preference: Further evidence of limbic–striatal interactions underlying reward-related processes. *Neuroscience* **42**, 1–18.

Ferrari, C. M., O'Conner, D. A., and Riley, A. L. (1990). Cocaine-induced taste aversions: Effect of route of administration. *Pharmacol. Biochem. Behav.* **38**, 267–271.

Florin, S. M., Kuczenski, R., and Segal, D. S. (1992). Amphetamine-induced changes in behavior and caudate extracellular acetylcholine. *Brain Res.* **581**, 53–58.

Fray, P. J., Sahakian, B. J., Robbins, T. W., Koob, G. F., and Iversen, S. D. (1980). An observational method for quantifying the behavioral effects of dopamine agonists: Contrasting effects of d-amphetamine and apomorphine. *Psychopharmacology* **69**, 253–259.

Gaffan, D., and Harrison, S. (1987). Amygdalaectomy and disconnection in visual learning for auditory secondary reinforcement by monkeys. *J. Neurosci.* **7**, 2285–2292.

Goeders, N. E., and Smith, J. E. (1983). Cortical dopaminergic involvement in cocaine reinforcement. *Science* **221**, 773–775.

Goeders, N. E., and Smith, J. E. (1986). Reinforcing properties of cocaine in the medial prefrontal cortex: Primary action on presynaptic dopaminergic terminals. *Pharmacol. Biochem. Behav.* **25,** 191–199.

Hemby, S. E., Jones, G. H., Justice, J. B., and Neill, D. B. (1992). Conditioned locomotor activity but not conditioned place preference following intra-accumbens infusions of cocaine. *Psychopharmacology* **106,** 330–336.

Hernandez, L., Lee, F., and Hoebel, B. G. (1989). Microdialysis in the nucleus accumbens during feeding or drugs of abuse: Amphetamine, cocaine and phencyclidine. *Ann. N.Y. Acad. Sci.* **575,** 508–511.

Hiroi, N., and White, N. M. (1991a). The amphetamine conditioned place preference: Differential involvement of dopamine receptor subtypes and two dopaminergic terminal areas. *Brain Res.* **552,** 141–152.

Hiroi, N., and White, N. M. (1991b). The lateral nucleus of the amygdala mediates expression of the amphetamine-produced conditioned place preference. *J. Neurosci.* **11,** 2107–2116.

Hoebel, B. G., Monaco, A. P., Hernandez, L., Aulisi, E. P., Stanley, B. G., and Lenard, L. (1983). Self-injection of amphetamine directly into the brain. *Psychopharmacology* **81,** 158–163.

Hoffman, D. C., and Beninger, R. J. (1989). The effects of selective dopamine D_1 or D_2 antagonists on the establishment of agonist-induced place conditioning in rats. *Pharmacol. Biochem. Behav.* **33,** 273–279.

Hubner, C. B., and Moreton, J. E. (1991). Effects of selective D_1 and D_2 dopamine antagonists on cocaine self-administration. *Psychopharmacology* **105,** 151–156.

Hurd, Y. L., and Ungerstedt, U. (1989). Cocaine: An in vivo microdialysis evaluation of its acute action on dopamine transmission in rat striatum. *Synapse* **3,** 48–54.

Hurd, Y. L., Weiss, F., Koob, G. F., And, N. E., and Ungerstedt, U. (1989). Cocaine reinforcement and extracellular dopamine overflow in rat nucleus accumbens: An in vivo microdialysis study. *Brain Res.* **498,** 199–203.

Imperato, A., Mele, A., Scrocco, M. G., and Puglisi-Allegra, S. (1992). Chronic cocaine alters limbic extracellular dopamine. Neurochemical basis for addiction. *Eur. J. Pharmacol.* **212,** 299–300.

Isaac, W. L., Nonneman, A. J., Neiswander, A. J., Landers, T., and Bardo, M. T. (1989). Prefrontal cortex lesions differentially disrupt cocaine-reinforced conditioned place preference but not conditioned taste aversion. *Behav. Neurosci.* **103,** 345–356.

Kalivas, P. W., and Duffy, P. (1990). Effect of acute and daily neurotensin and enkephalin treatments on extracellular dopamine in the nucleus accumbens. *J. Neurosci.* **10,** 2940–2949.

Kalivas, P. W., and Duffy, P. (1993). Time course of extracellular dopamine and behavioural sensitization to cocaine: I. Dopamine axon terminals. *J. Neurosci.* **13,** 266–275.

Kazahaya, Y., Akimoto, K., and Saburo, O. (1989). Subchronic methamphetamine treatment enhances methamphetamine- or cocaine-induced dopamine efflux in vivo. *Biol. Psychiatry* **25,** 903–912.

Kelly, P. H., Seviour, P. W., and Iversen, S. D. (1975). Amphetamine and apomorphine responses in the rat following 6-OHDA lesions of the nucleus accumbens septi and corpus striatum. *Brain Res.* **94,** 507–522.

Koob, G. F., Le, H. T., and Creese, I. (1987). The D_1 dopamine receptor antagonist SCH 23390 increases cocaine self-administration in the rat. *Neurosci. Lett.* **79,** 315–320.

Kuczenski, R., and Segal, D. (1989). Concomitant characterization of behavioral and striatal neurotransmitter response to amphetamine using in vivo microdialysis. *J. Neurosci.* **9,** 2051–2065.

Lacosta, S., and Roberts, D. C. S. (1993). MDL 72222, ketanserin and methysergide pretreatments fail to alter breaking points on a progressive ratio schedule reinforced by intravenous cocaine. *Pharmacol. Biochem. Behav.* **44,** 161–166.

Leccese, A. P., and Lyness, W. H. (1984). The effects of putative 5-hydroxytryptamine receptor active agents on d-amphetamine self-administration in controls and rats with 5,7-dihydroxytryptamine median forebrain bundle lesions. *Brain Res.* **303**, 153–162.

Leccese, A. P., and Lyness, W. H. (1987). Lesions of dopamine neurons in the medial prefrontal cortex: Effects on self-administration of amphetamine and dopamine synthesis in the brain of the rat. *Neuropharmacology* **26**, 1295–1302.

Leone, P., and Di Chiara, G. (1987). Blockade of D_1 receptors by SCH 23390 antagonises morphine- and amphetamine-induced place preference conditioning. *Eur. J. Pharmacol.* **135**, 251–254.

Loh, E. A., and Roberts, D. C. S. (1990). Break points on a progressive ratio schedule reinforced by intravenous cocaine increase following depletion of forebrain serotonin. *Psychopharmacology* **101**, 262–266.

Loh, E. A., Smith, A. M., and Roberts, D. C. S. (1993). Evaluation of response perseveration in the radial arm maze following reinforcing and non-reinforcing drugs. *Pharmacol. Biochem. Behav.* **44**, 735–740.

Lyness, W. H. (1983). Effect of L-tryptophan pretreatment on d-amphetamine self-administration. *Substance Alcohol Actions/Misuse* **4**, 305–312.

Lyness, W. H., Friedle, N. M., and Moore, K. E. (1979). Destruction of dopaminergic nerve terminals in nucleus accumbens: Effects of d-amphetamine self-administration. *Pharmacol. Biochem. Behav.* **11**, 553–556.

Lyness, W. H., Friedle, N. M., and Moore, K. E. (1980). Increased self-administration of d-amphetamine after destruction of 5-hydroxytryptaminergic neurons. *Pharmacol. Biochem. Behav.* **12**, 937–941.

Lyon, M., and Robbins, T. W. (1975). The action of central nervous system stimulant drugs: A general theory concerning amphetamine effects. *Curr. Dev. Psychopharmacol.* **2**, 81–163.

Mackey, W. B., and van der Kooy, D. (1985). Neuropleptics block the positive reinforcing effects of amphetamine but not of morphine as measured by place conditioning. *Pharmacol. Biochem. Behav.* **22**, 101–105.

Martin-Iverson, M. T., Szostak, C., and Fibiger, H. C. (1986). 6-Hydroxydopamine lesions of the medial prefrontal cortex fail to influence intravenous self-administration of cocaine. *Psychopharmacology* **88**, 310–314.

McGregor, A., and Roberts, D. C. S. (1993). L-Tryptophan decreases the breaking point under a progressive ratio of intravenous cocaine reinforcement in the rat. *Pharmacol. Biochem. Behav.* **44**, 651–655.

Messier, C., and White, N. M. (1984). Contingent and non-contingent actions of sucrose and saccharin reinforcers: Effects on taste preference and memory. *Pharmacol. Biochem. Behav.* **32**, 195–203.

Mifsud, J. C., Hernandez, L., and Hoebel, B. G. (1989). Nicotine infused into the nucleus accumbens increases synaptic dopamine as measured by *in vivo* microdialysis. *Brain Res.* **478**, 365–367.

Mithani, S., Martin-Iverson, M. T., Phillips, A. G., and Fibiger, H. C. (1986). The effects of haloperidol on amphetamine- and methylphenidate-induced conditioned place preferences and locomotor activity. *Psychopharmacology* **90**, 247–252.

Mora, F., and Myers, R. D. (1977). Brain self-stimulation: Direct evidence for the involvement of dopamine in the prefrontal cortex. *Science* **197**, 1387–1389.

Nash, J. F., and Nichols, D. E. (1991). Microdialysis studies on 3,4-methylenedioxyamphetamine and structurally related analogues. *Eur. J. Pharmacol.* **200**, 53–58.

Patrick, S. L., Thompson, T. L., Walker, J. M., and Patrick, R. L. (1991). Concomitant sensitization of amphetamine-induced behavioral stimulation and *in vivo* dopamine release from rat caudate nucleus. *Brain Res.* **538**, 343–346.

Peltier, R., and Schenk, S. (1991). GR38032F, a serotonin $5HT_3$ antagonist, fails to alter cocaine self-administration in rats. *Pharmacol. Biochem. Behav.* **39**, 133–136.

Pettit, H. O., and Justice, J. B., Jr. (1991). Effect of dose on cocaine self-administration behavior and dopamine levels in the nucleus accumbens. *Brain Res.* **539**, 94–102.
Pettit, H. O., Pan, H.-T., Parsons, L. H., and Justice, J. B., Jr. (1990). Extracellular concentrations of cocaine and dopamine are enhanced during chronic cocaine administration. *J. Neurochem.* **55**, 798–804.
Pickens, R., and Thompson, T. (1968). Cocaine-reinforced behaviour in rats: Effects of reinforcement magnitude and fixed-ratio size. *J. Pharmacol. Exper. Therapeutics* **161**, 122–129.
Porrino, L. J., Ritz, M. C., Goodman, N. L., Sharpe, L. G., Kuhar, M. J., and Goldberg, S. R. (1989). Differential effects of the pharmacological manipulation of the serotonin system on cocaine and amphetamine self-administration. *Life Sci.* **45**, 1529–1536.
Price, J. L., Russchen, F. T., and Amaral, D. G. (1987). The amygdaloid complex. In "Integrated Systems of the CNS" (A. Bjorklund, T. Hokfelt, and L. W. Swanson, eds.), Part 1, Handbook of Chemical Neuroanatomy, Vol. 5, pp. 279–388. Elsevier, Amsterdam.
Randrup, A., and Munkvad, I. (1967). Stereotyped behavior produced by amphetamine in several animal species and man. *Psychopharmacology* **11**, 300–310.
Richardson, N. R., and Roberts, D. C. S. (1991). Fluoxetine pretreatment reduces breaking points on a PR schedule reinforced by intravenous cocaine self-administration in the rat. *Life Sci.* **49**, 833–840.
Ridley, R. M., Haystead, T. A. J., and Baker, H. F. (1981). An involvement of dopamine in higher order choice mechanisms in the monkey. *Psychopharmacology* **72**, 173–177.
Ridley, R. M., Baker, H. F., Frith, C. D., Dowdy, J., and Crow, T. J. (1988). Stereotyped responding on a two-choice guessing task by marmosets and humans treated with amphetamine. *Psychopharmacology* **95**, 560–564.
Risner, M. E., and Jones, B. E. (1976). Role of noradrenergic and dopaminergic processes in amphetamine self-administration. *Pharmacol. Biochem. Behav.* **5**, 477–482.
Roberts, D. C. S., and Fibiger, H. C. (1976). Conditioned taste aversion induced by diethyldithiocarbamate (DDC). *Neurosci. Lett.* **2**, 339–342.
Roberts, D. C. S., Koob, G. F., and Fibiger, H. C. (1980). Extinction and recovery of cocaine self-administration following 6-hydroxydopamine lesions of the nucleus accumbens. *Pharmacol. Biochem. Behav.* **12**, 781–787.
Roberts, D. C. S., Loh, E. A., and Vickers, G. (1989). Self-administration of cocaine on a progressive ratio schedule in rats: Dose–response relationship and effect of haloperidol pretreatment. *Psychopharmacology* **97**, 535–538.
Robbins, T. W. (1976). Relation between reward-enhancing and stereotypical effects of psychomotor stimulant drugs. *Nature (London)* **264**, 57–59.
Robinson, T. E., and Camp, D. M. (1990). Does amphetamine *preferentially* increase the extracellular concentration of dopamine in the mesolimbic system of freely moving rats. *Neuropsychopharmacology* **3**, 163–173.
Robinson, T. E., and Camp, D. M. (1991). On the preferential release of mesolimbic dopamine by amphetamine. Reply. *Neuropsychopharmacology* **5**, 245–247.
Robinson, T. E., Jurson, P. A., Bennett, J. A., and Bentgen, K. M. (1988). Persistent sensitization of dopamine neurotransmission in ventral striatum (nucleus accumbens) produced by prior experience with (+)-amphetamine: A microdialysis study in freely moving rats. *Brain Res.* **462**, 211–222.
Rossetti, Z. L., Melis, F., Carboni, S., and Gessa, G. L. (1992). Dramatic depletion of mesolimbic extracellular dopamine after withdrawal from morphine, alcohol or cocaine: A common neurochemical substrate for drug dependence. *Ann. N.Y. Acad. Sci.* **654**, 513–516.
Segal, D. S., and Kuczenski, R. (1992). *In vivo* microdialysis reveals a diminished amphetamine-induced dopamine response corresponding to behavioural sensitization produced by repeated amphetamine pretreatment. *Brain Res.* **571**, 330–337.
Segal, D. S., and Schuckit, M. A. (1983). Animal models of stimulant-induced psychosis. *In*

"Stimulants: Neurochemical, Behavioral and Clinical Perspectives" (I. Creese, ed.), pp. 131–167. Raven Press, New York.

Smith, F. L., Yu, D. S., Smith, D. G., Leccese, A. P., and Lyness, W. H. (1986). Dietary tryptophan supplements attenuate amphetamine self-administration in the rat. *Pharmacol. Biochem. Behav.* 25, 849–855.

Spyraki, C., Fibiger, H. C., and Phillips, A. G. (1982a). Dopaminergic substrates of amphetamine-induced place preference conditioning. *Brain Res.* 253, 185–192.

Spyraki, C., Fibiger, H. C., and Phillips, A. G. (1982b). Cocaine-induced place-preference conditioning: Lack of effects of neuroleptics and 6-hydroxydopamine lesions. *Brain Res.* 253, 195–203.

Spyraki, C., Nomikos, G. G., Galanopuolou, P., and Daifotis, Z. (1988). Drug-induced place preference in rats with 5,7-dihydroxytryptamine lesions of the nucleus accumbens. *Behav. Brain Res.* 29, 127–134.

Steketee, J. D., and Kalivas, P. W. (1992). Microinjection of the D_2 agonist quinpirole into the A10 dopamine region blocks amphetamine- but not cocaine-stimulated motor activity. *J. Pharmacol. Exp. Ther.* 261, 811–818.

Szechtman, H. (1983). Peripheral sensory input directs apomorphine-induced circling in rats. *Brain Res.* 264, 332–335.

Szechtman, H., Eilam, D., Teitelbaum, P., and Golani, I. (1988). A different look at measurement and interpretation of drug-induced stereotyped behavior. *Psychobiology* 16, 164–173.

Understedt, U. (1971). Stereotaxic mapping of the monamine pathways in the rat brain. *Acta Physiol. Scand.* 367, 1–48.

Van Haaren, F., and Hughes, C. E. (1990). Cocaine-induced conditioned taste aversions in male and female Wistar rats. *Pharmacol. Biochem. Behav.* 37, 693–696.

Weiss, F., Markou, A., Lorang, M. T., and Koob, G. F. (1992). Basal extracellular dopamine levels in the nucleus accumbens are decreased during cocaine withdrawal after unlimited-access self-administration. *Brain Res.* 593, 314–318.

White, N. M. (1988). Effect of nigrostriatal dopamine depletion on the post-training memory-improving action of amphetamine. *Life Sci.* 43, 7–12.

Wilson, M. C., and Schuster, C. R. (1972). The effects of chlorpromazine on psychomotor stimulant self-administration in the rhesus monkey. *Psychopharmacology* 26, 115–126.

Wise, R. A. (1989). The brain and reward. In "The Neuropharmacological Basis of Reward" (J. M. Liebmen and S. J. Cooper, eds.), pp. 377–424. Oxford University Press, Oxford.

Wise, R. A., Yokel, R. A., and de Wit, H. (1976). Both positive reinforcement and conditioned aversion from amphetamine and apomorphine in rats. *Science* 191, 1273–1274.

Woolverton, W. L. (1987). Evaluation of the role of norepinephrine in the reinforcing effects of psychomotor stimulants in rhesus monkeys. *Pharmacol. Biochem. Behav.* 26, 835–839.

Yamamoto, B. K., and Pehek, E. A. (1990). A neurochemical heterogeneity of the rat striatum as measured by in vivo electrochemistry and microdialysis. *Brain Res.* 506, 236–242.

Yokel, R. A., and Wise, R. A. (1975). Increased lever pressing for amphetamine after pimozide in rats: Implication for a dopamine theory of reward. *Science* 187, 547–549.

Yokel, R. A., and Wise, R. A. (1976). Attenuation of intravenous amphetamine reinforcement by central dopamine blockade in rats. *Psychopharmacology* 48, 311–318.

Yoshikawa, T., Shibuya, H., Kaneno, S., and Toru, M. (1991). Blockade of behavioral sensitization to methamphetamine by lesion of hippocampo-accumbal pathway. *Life Sci.* 48, 1325–1332.

Yu, D. S. L., Smith, F. L., Smith, D. G., and Lyness, W. H. (1990). Fluoxetine-induced attenuation of amphetamine self-administration in rats. *Life Sci.* 39, 1383–1388.

Zetterstrom, T., Sharp, T., Collin, A. K., and Ungerstedt, U. (1988). In vivo measurement of extracellular dopamine and DOPAC in rat striatum after various dopamine-releasing drugs: Implications for the origin of extracellular DOPAC. *Eur. J. Pharmacol.* 148, 327–334.

III
TOXICOLOGY

James W. Gibb
Glen R. Hanson
Michel Johnson

9
Neurochemical Mechanisms of Toxicity

I. HISTORY OF NEUROTOXICITY BY AMPHETAMINE AND ITS ANALOGS

Although the first reports of amphetamine misuse appeared in 1936 (Grinspoon and Hedblom, 1975), the initial indication that amphetamine and its analogs are neurotoxic was reported by Pletscher *et al.* (1963,1964), who observed that *p*-chloroamphetamine (PCA) caused a sustained depletion of 5-hydroxytryptamine (5-HT) and its metabolite 5-hydroxyindoleacetic acid (5-HIAA). Over subsequent years, Fuller and colleagues (1965,1973, 1980; Fuller and Perry, 1977) correlated the structure–activity relationship of amphetamine analogs with their neurotoxicity; as judged by decreases in brain 5-HT content, these investigators concluded that PCA was the most toxic of the analogs studied.

Further neurochemical evidence that PCA was neurotoxic was obtained by Sanders-Bush and Sulser (1970), who observed that, in addition to depleting 5-HT, PCA also inhibited the conversion of radiolabeled tryptophan to 5-HT. These investigators later found that tryptophan hydroxylase (TPH) activity was impaired for as long as 4 mo after a single dose of PCA; 5-HT uptake was compromised as well (Sanders-Bush and Sulser, 1970; Sanders-Bush et al., 1975). Anatomic evidence for PCA neurotoxicity was provided by Harvey et al. (1975), who observed histological alterations within 1 day of administering a single dose (20 mg/kg) of the drug; these morphological changes were still evident 30 days later.

II. RESPONSE OF DOPAMINERGIC SYSTEM TO METHAMPHETAMINE

An extensive study of the neurotoxicity of methamphetamine (METH) was begun after the reports by Koda and Gibb (1971,1973) and Fibiger and McGeer (1971) that prolonged exposure to METH compromised the dopaminergic system. After administering large doses of METH [10–15 mg/kg, subcutaneously (sc), every 6 hr for 5 doses], tyrosine hydroxylase (TH) activity in the rat neostriatum declined to approximately 60% of control (Koda and Gibb, 1973; Hotchkiss and Gibb, 1980a). The decline in enzyme activity persisted for extended periods of time after discontinuing the drug (Hotchkiss et al., 1979; Fig. 1). Concurrent with the compromised TH activity was a depletion of dopamine (DA) and its metabolites dihydroxyphenylacetic acid (DOPAC) and homovanillic acid (HVA) (Koda and Gibb; Schmidt and Gibb, 1985a). A similar decrease in rat striatal DA concentration occurred after a single administration of amphetamine and iprindole, an inhibitor of amphetamine para-hydroxylation that increases amphetamine half-life (Fuller and Hemrick-Luecke, 1980).

In 1975, Seiden et al. (1975/76) reported a decrease in DA concentrations in the brain of monkeys that received large doses of METH; suggesting that the human brain may respond in a similar fashion. Moreover, these investigators later observed that the decrease in DA concentration induced by METH was accompanied by a decrease in the number of DA uptake sites, consistent with the destruction of nerve terminals (Wagner et al., 1979, 1980).

III. SEROTONERGIC RESPONSE TO METHAMPHETAMINE

In an effort to determine whether other neurotransmitter systems also were altered by METH, Hotchkiss and Gibb (1980a) determined the response of the serotonergic system to the large doses of METH just described. Compared with the 60% loss in neostriatal TH activity, this treatment induced more than a 90% decrease in the activity of TPH with a concurrent decrease in the concentration of 5-HT and its metabolite 5-HIAA. Although the enzymatic activity recovered slightly over time, the decrement persisted

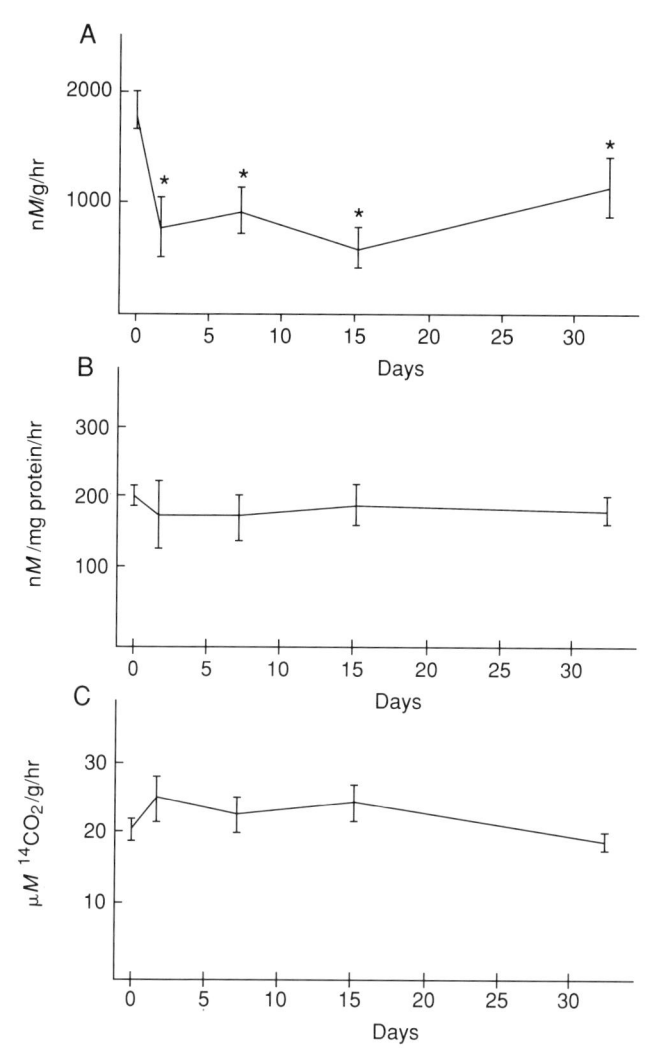

FIGURE 1 Effect of multiple administrations of methampethamine (METH) on neostriatal dopaminergic, cholinergic, and GABA-ergic systems in rats. METH was give at 0, 6, 12, 18, and 24 hr (15 mg/kg, sc). Animals were killed after varying periods of time. Enzymatic activity was evaluated for (A) tyrosine hydroxylase, (B) choline acetyltransferase, and (C) glutamate acid decarboxylase. Results are expressed as means ± SEM; 5–12 rats per group were studied. Asterisks indicate $p < 0.05$ relative to control by Student's *t*-test. Adapted from *Life Sci.* **25**, A. Hotchkiss, M. E. Morgan, and J. W. Gibb. The long-term effects of multiple doses of methamphetamine on neostriatal tryptophan hydroxylase, tyrosine hydroxylase, choline acetyltransferase and glutamate decarboxylase activities. Pp. 1373–1378 (1979), with permission from Pergamon Press Ltd., Headington Hills Hall, Oxford OX3 OBW, UK.

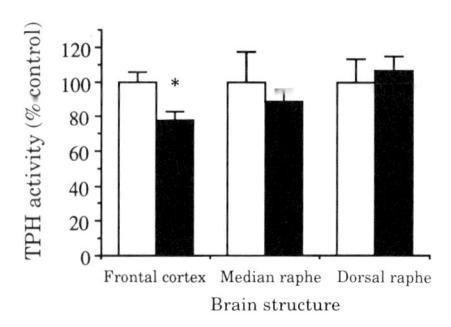

FIGURE 2 Acute effect of methamphetamine (METH) on tryptophan hydroxylase (TPH) activity (■). Male Sprague–Dawley rats (180–220 g) were administered METH (15 mg/kg, sc) and killed 1 hr later. TPH activity was measured by high performance liquid chromatography (HPLC) as described by Johnson et al. (1992). Means ± SEM are expressed as percentage of control (□) measured in each brain structure. Control TPH activity in the frontal cortex was 123 ± 7 nmol hydroxylated tryptophan/hr/g tissue. Control TPH activity in the median raphe and in the dorsal raphe were 7.3 ± 1.3 and 12.8 ± 1.7 nmol hydroxylated tryptophan/hr/mg protein, respectively. Each group contained 5 rats. Statistical analysis was performed using Student's t-test; asterisk indicates $P < 0.05$ relative to control.

for extended periods of time (see Section IV). Additional significant differences were noted between the effects of METH on the serotonergic system and the response in the dopaminergic system. Whereas in the dopaminergic system, little change was observed in TH activity, DA, or its metabolites until 12–18 hr after initiating METH treatment, in the serotonergic system marked alterations were apparent within minutes. More specifically, neostriatal TPH activity was diminished significantly within 20 min after rats received a single dose of METH (10 mg/kg, sc); Bakhit and Gibb, 1981. The decrement was observed in all serotonergic terminal areas of the brain, but not in the cell bodies of the dorsal or median raphe (Fig. 2). This rapid decrease in serotonin synthesis also was reported after a single dose of amphetamine or PCA (Knapp et al., 1974). This initial acute decrease in TPH activity induced by METH and other amphetamine analogs differed from the decrease in TPH activity observed after multiple METH administrations since it was reversed by incubating the enzyme under reducing conditions (Stone et al., 1989a,b; see Section X).

IV. NEUROCHEMICAL ALTERATIONS PERSIST FOR EXTENDED PERIODS OF TIME

In view of the widespread abuse of amphetamine and its congeners, the possible irreversible or prolonged effects of these compounds are of particular concern. Ellison et al. (1978) reported a long-lasting depression of TH activity in the neostriatum 115 days after discontinuing the drug. Lorez (1981) observed a decrease in striatal dopaminergic nerve terminals after multiple doses of METH, as measured by fluorescent histochemistry. These results are in agreement with the neurochemical data and support the concept that METH is neurotoxic to the central dopaminergic system.

Bakhit et al. (1981) investigated the regional response of the serotonergic system to the effects of multiple doses of METH. Rats received METH (15 mg/kg, sc) every 6 hr for 5 doses. The animals were sacrificed 36 hr or 110 days after the 5th and final dose of METH. As reported earlier, TPH activity was depressed markedly 36 hr after the final dose of METH was administered. At 110 days after the last dose of METH, TPH activity returned to normal in the hypothalamus, olfactory tubercle, and spinal cord; however, enzyme activity remained suppressed in the hippocampus, cerebral cortex, neostriatum, and nucleus accumbens. The decrease in concentrations of 5-HT and 5-HIAA observed at 36 hr also persisted for 110 days in all brain areas that were investigated. Ricaurte and collaborators (1980) also reported that the number of 5-HT uptake sites was reduced in the rat brain 6 wk after multiple administrations of METH. The central serotonergic system of monkeys was still impaired 4 yr after the administration of large doses of METH (Woolverton et al., 1989).

These long-term persisting effects of METH on both the dopaminergic and the serotonergic systems, coupled with morphological studies (Lorez, 1981; Axt and Molliver, 1991), provide significant evidence that METH does, indeed, cause long-lasting neurotoxicity when given in large repeated doses.

V. DOES METHAMPHETAMINE CAUSE GENERALIZED NEUROTOXICITY?

The question arises of whether METH, when administered in large repeated doses, causes generalized neurochemical deficits. Hotchkiss et al. (1979) monitored enzyme activity in the dopaminergic, serotonergic, GABA-ergic, and cholinergic systems simultaneously. Choline acetyltransferase (ChAT) and glutamate decarboxylase (GAD) were used as enzyme markers to determine the integrity of the cholinergic and GABA-ergic systems. As reported earlier, neostriatal TPH and TH activities, which were used as enzyme markers for the catecholaminergic and serotonergic systems, respectively—were decreased by METH. Since ChAT and GAD activities were not altered after METH administration (Fig. 1), we concluded that the neurotoxicity associated with multiple doses of METH is limited to the dopaminergic and serotonergic systems and that other neurotransmitter systems remain intact. However, continuous amphetamine treatment for 45 days does increase GAD activity in the substantia nigra (Pérez-de la Mora et al., 1990), again suggesting an absence of neurotoxic damage to this system.

Hanson and co-workers (1992) conducted extensive studies on the effects of METH on the brain content of various neuropeptides. METH increased neurotensin concentration by as much as 300% in the neostriatum and nucleus accumbens; significant changes also were observed in the substance P (Ritter et al., 1984), neurokinin A, and dynorphin A (Hanson et al., 1987) concentrations. However, in all cases, these alterations in neuropeptide content were transient and returned to normal within 2–3 days of administering METH. Apparently, then, persisting alterations of neurotransmitter/neuromodulator systems observed after multiple doses of METH are limited to the dopaminergic and serotonergic systems, except for some unidentified neurons in the somatosensory cortex. Commins and Seiden (1986) observed that METH may destroy nondopaminergic or nonserotonergic neurons in that structure 2 days after the administration of a single dose of METH (100 mg/kg). The identity of these cortical neurons has not been established, but this brain area does not contain dopaminergic or serotonergic cell bodies (Emson and Lindvall, 1979; Steinbusch, 1984).

VI. NEUROCHEMICAL ALTERATIONS BY METHYLENEDIOXY CONGENERS OF AMPHETAMINE

Based on the evidence just described, determining whether other amphetamine analogs used by the public can produce long-term alterations in the central nervous system is of interest. Extensive research was conducted during the 1980s on methylenedioxy congeners of amphetamine since they were being used in psychotherapy (Shulgin, 1990) and for hedonistic purposes by the public (Peroutka, 1987). Initial research established that these amphetamine analogs shared some of the characteristics of amphetamine, METH, and PCA.

Ricaurte et al. (1985) reported that 3,4-methylenedioxyamphetamine (MDA) caused long-lasting decreases in concentrations of 5-HT and 5-HIAA in the hippocampus and neostriatum. 5-HT uptake sites were impaired also. Using the Fink–Heimer technique, these researchers also provided morphological evidence of tissue damage. This initial work indicated that the methylenedioxy analog of METH also may produce long-lasting alterations in the serotonergic system.

Stone et al. (1986,1987b,c) investigated the effects of MDA and its analog 3,4-methylenedioxymethamphetamine (MDMA) on TH and TPH activity in the neostriatum. Neostriatal, hippocampal, and cortical TPH activity was decreased markedly by either methylenedioxy analog. This decrease was accompanied by a comparable decrease in the concentration of 5-HT and 5-HIAA. In contrast to METH, MDA or MDMA failed to alter neostriatal TH activity or deplete DA (Stone et al., 1986,1987c). Although the decrease in hypothalamic TPH activity induced by MDMA was not as dramatic as in other brain structures and eventually returned to normal, the decrease in TPH activity measured in the neostriatum, cortex, and hippocampus was still observed after 110 days (Stone et al., 1987b). These long-term changes suggest that MDMA, like MDA, is neurotoxic to the serotonergic system but not to the dopaminergic system. The decrease in the number of 5-HT reuptake sites, but not DA sites, supports the relative selectivity of MDMA for the 5-HT system (Commins et al., 1987c; Schmidt, 1987). Histological studies verified the neurotoxic response to MDMA (Commins et al., 1987c; O'Hearn et al., 1988). As reported for METH, some neurons in the neocortex that do not contain 5-HT or DA can be affected adversely by MDMA (Commins et al., 1987c).

Not all animal species respond to MDMA in a quantitatively similar manner. The activity of TPH, the content of 5-HT and 5-HIAA, and the number of 5-HT reuptake sites all are compromised in the rat brain after administration of MDMA. These deficits were not observed in mice treated with similar doses (Stone et al., 1987b; Logan et al., 1988; Peroutka, 1988), but when larger doses of MDMA were administered, similar changes in serotonergic parameters were observed (Stone et al., 1987b).

VII. ROLE OF DOPAMINE IN METHAMPHETAMINE-INDUCED CHEMICAL CHANGES IN DOPAMINERGIC SYSTEMS

Buening and Gibb (1974) provided the first evidence of DA involvement in the neurochemical response to large doses of METH. These researchers found that when either of the DA receptor antagonists, chlorpromazine or haloperidol, was administered concurrently with METH, the drug-induced decrease in TH activity was prevented in a dose-dependent fashion. These investigators concluded that DA released by METH is involved in the changes in the dopaminergic system. This laboratory extended the study and found that the D_1 receptor antagonist, SCH 23390, blocked the METH-induced decrease in TH activity (Sonsalla et al., 1986a). Blockade of the D_2 receptors by sulpiride also prevented the METH effects on the dopaminergic system, suggesting that the neurotoxic effects of METH on the dopaminergic system are D_1- and D_2-mediated events.

Confirmation of the role of DA in the response of the dopaminergic system to METH was provided by Gibb and his collaborators. Gibb and Kogan (1975,1979) determined that the METH-induced decrease in TH activity was blocked completely by inhibiting DA synthesis at the rate-limiting step. When an inhibitor of TH, α-methyl-p-tyrosine (α-MT), was administered concurrently with METH, the usual decrease in TH activity was prevented. DA synthesis was restored in an α-MT-treated animal by administering L-DOPA, which circumvents the rate-limiting TH step. To confirm the requirement for DA in METH neurotoxicity, METH-treated animals were administered both α-MT and L-DOPA (Fig. 3). In this study, multiple METH administrations reduced TH activity. When α-MT was ad-

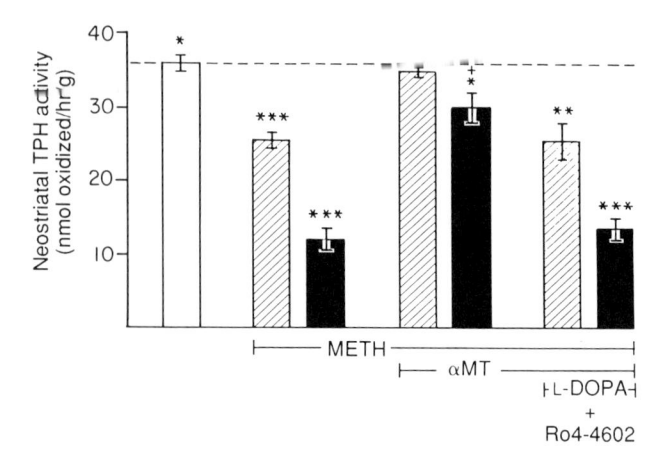

FIGURE 3 Effect of L-DOPA and R04-4602 on the ability of α-methyl-p-tyrosine (α-MT) to antagonize the depression of neostriatal tryptophan hydroxylase (TPH) activity by multiple injections of methamphetamine (METH). Male Sprague–Dawley rats (180–225 g) were given 5 doses of METH [10 (□) or 15 (■) mg/kg, sc], one every 6 hr, and killed 18 hr after the last drug injection. Each value represents the mean ± SEM for 6 or more animals. Significant differences of $P < 0.05$ (*), $P < 0.01$ (**), $P < 0.001$ (***) vs control (□) and $P < 0.001$ (+) vs METH alone by Student's t-test (two tailed). Reprinted with permission from C. J. Schmidt et al. role of dopamine in the neurotoxic effects of methamphetamine *J. Pharmacol., Exp. Ther.* 233, 539–544, (1985), © by Am. Soc. for Pharmacology and Experimental Therapeutics.

ministered concurrently with METH, enzyme activity remained normal. However, when L-DOPA (in addition to a peripheral DOPA aromatic L-amino acid decarboxylase inhibitor, R04-4602) was administered with the α-MT, the deficit in TH activity was observed again. Seiden and co-workers also demonstrated that α-MT prevented the damage in cell bodies that occurred in the somatosensory cortex (Commins and Seiden, 1986). Researchers also showed that α-MT prevented the persistent depletion in DA induced by amphetamine (Fuller and Hemrick-Luecke, 1982).

The role of DA in METH-induced neurotoxicity also was supported by the ability of DA reuptake blockers, such as amfonelic acid, to block the decrease in TH activity after multiple doses of METH (Schmidt and Gibb, 1985a). Inhibitors of DA reuptake prevented the release of DA induced by amphetamine or METH (Raiteri *et al.*, 1979; Liang and Rutledge, 1982; Steranka, 1982; Schmidt and Gibb, 1985b) without altering the induced release of 5-HT (Schmidt and Gibb, 1985b). The protective effect of DA reuptake blockers could be explained by the ability of these compounds to prevent amphetamine entry into dopaminergic neurons, but several reports indicated that amphetamine and other amphetamine analogs did not require the transporter to enter the intracellular compartment (Masuoka *et al.*, 1975; Fuller and Snoddy, 1979; Steranka, 1982). However, the dopaminergic reuptake carrier plays an essential role in the exchange diffusion mechanism by which amphetamine analogs release DA, by acting as a reverse transporter (Fisher and Cho, 1979; Liang and Rutledge, 1982). Thus, these observations suggest that toxicity to dopaminergic neurons is not mediated by the presence of amphetamine in the nerve terminals alone, but that DA also must be released through the transporter system.

VIII. ROLE OF DOPAMINE IN METHAMPHETAMINE-INDUCED ALTERATIONS IN THE SEROTONERGIC SYSTEM

The METH-induced changes in the dopaminergic and serotonergic systems are prevented by DA receptor antagonists, indicating that the decreases in TPH activity and in the concentration of 5-HT are mediated by DA (Hotchkiss and Gibb, 1980a; Sonsalla et al., 1986a). These workers found that the D_1 receptor antagonist, SCH 23390, attenuated the effects of METH on the serotonergic system. Although blockade of D_2 receptors by sulpiride prevented the METH effects on the dopaminergic system, it failed to block the changes in the serotonergic system. This result suggests that the neurotoxic effects of METH on the serotonergic system are D_1-mediated events.

The role of DA as a mediator of the decrease in TPH activity that is induced by multiple administrations of METH was confirmed by preventing the changes in the serotonergic system with α-MT or by destroying DA neurons with 6-hydroxydopamine (Hotchkiss and Gibb, 1980a; Schmidt et al., 1985a; Sonsalla et al., 1986b). Administration of L-DOPA to animals treated with α-MT restored the ability of METH to decrease TPH activity (Schmidt et al., 1985a). Although a low concentration of DA is measured in the hippocampus, α-MT prevented the METH-induced decrease in hippocampal TPH activity, suggesting that hippocampal DA also may mediate the response to METH (Hotchkiss and Gibb, 1980a). This possibility is supported by the absence of protection when hippocampal norepinephrine concentration is depleted with N-(2-chloroethyl)-N-ethyl-2-bromobenzylamine (DSP4) (Johnson et al., 1991).

The mechanism by which DA mediates the alterations in the serotonergic system after multiple administrations of METH remains unknown. In addition to the interaction between DA and its receptors, researchers have suggested that DA mediates the detrimental effects of METH directly inside serotonergic nerve terminals after entering via the serotonergic reuptake carrier (Schmidt and Lovenberg, 1985). This hypothesis is based on the observation that 5-HT reuptake blockers prevent neurotoxic damage induced by amphetamine analogs in the serotonergic system but not in the dopaminergic system (Fuller et al., 1975; Hotchkiss and Gibb, 1980a; Fuller and Hemrick-Luecke, 1982; Schmidt and Gibb, 1985b), and that DA is taken up into serotonergic neurons by the serotonergic uptake carrier

(Schmidt and Lovenberg, 1985; Waldmeier, 1985). Once in the nerve terminal, DA could be converted nonenzymatically into 6-hydroxydopamine (Seiden and Vosmer, 1984) or autoxidized (Graham et al., 1978) to form toxic free radicals and quinones. Since 5-HT uptake blockers prevent the METH-induced release of 5-HT (Schmidt and Gibb, 1985a), this transmitter could be required for the destruction of serotonergic nerve terminals. However, in contrast to the inhibition of DA synthesis, inhibition of 5-HT synthesis with p-chlorophenylalanine fails to protect the serotonergic system from the changes induced by PCA or METH (Fuller et al., 1975; Schmidt and Gibb, 1983).

Interestingly, DA also mediates the rapid decrease in TPH activity measured shortly after a single METH or MDMA injection (see Section III). This rapid decrease in TPH activity is reversible and does not reflect neuronal damage (see Section X). Depleting the striatal DA content by destroying the nigrostriatal dopaminergic pathway with 6-hydroxydopamine prevented the METH- and MDMA-induced decrease in striatal TPH activity (Johnson et al., 1987a, Stone et al., 1988; Fig. 4). The brain areas that remained intact after 6-hydroxydopamine lesion still displayed the METH-induced reduction in TPH activity, indicating that local DA released by METH in each brain structure is responsible for the loss in enzyme activity. Although the decrease in TPH activity induced shortly after a single dose of METH and the decrease induced by multiple administrations of METH are mediated by DA, the mechanisms mediating these decreases appear to differ since the

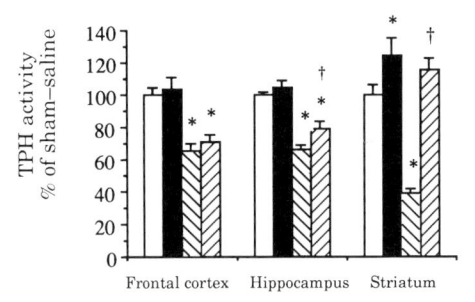

FIGURE 4 Effects of lesioning the nigrostriatal dopaminergic pathway with 6-hydroxydopamine (6-OHDA) on methamphetamine (METH)-induced decreases in tryptophan hydroxylase (TPH) activity measured 3 hr after drug administration. Rats received 6-OHDA (8 μg/4 μl 0.1% ascorbate saline; ■, ▨) or vehicle (0.1% ascorbate; □, ▧) bilaterally into the substantia nigra 11 days before METH (10 mg/kg, sc; ▨, ▧) or saline (□, ■). Animals were killed 3 hr after receiving a single dose of METH. Means ± SEM of TPH activity are expressed as percentage of the sham–saline control group. The enzymatic activities of the sham–saline groups expressed in nmol hydroxylated/hr/g tissue were 103 ± 4.2 in the frontal cortex ($n = 10$); 62.3 ± 1.1 in the hippocampus ($n = 10$); and 39.8 ± 2.4 in the striatum ($n = 10$). Statistical analysis of enzymatic activities performed with a Student's t-test yielded $P < 0.05$ (*) vs sham–saline and $P < 0.05$ (†) vs sham–METH. Adapted with permission from Johnson et al. (1987a).

rapid decrease in activity is not blocked by dopaminergic receptor antagonists (Johnson *et al.,* 1988).

IX. ROLE OF TOXIC METABOLITES IN NEUROTOXICITY

The nonenzymatic transformation of DA to 6-hydroxydopamine or the autoxidation of DA to toxic quinones has been proposed to explain the damaging effect of METH on the serotonergic and dopaminergic systems. Seiden and Vosmer (1984) reported the formation of 6-hydroxydopamine in the rat caudate nucleus shortly after a single administration of METH (100 mg/kg). These investigators also reported the formation of 5,6-dihydroxytryptamine, a serotonergic neurotoxin, in the hippocampus of rats treated with METH or PCA (Commins *et al.,* 1987a,b). METH may favor the nonenzymatic formation of these two neurotoxins by hydroxylation of DA and 5-HT, but the role of these proposed metabolites remains to be established since only small amounts of these toxins were detected (less than 1 ng/mg tissue). Moreover, the identity of these compounds has not been confirmed by mass spectrometry, and other researchers have failed to detect the formation of 6-hydroxydopamine in METH-treated animals (Rollema *et al.,* 1986). The lability of these compounds may explain the difficulty in measuring these reactive metabolites.

Since direct injection of MDMA or PCA into the ventricle or into a brain structure (Molliver *et al.,* 1986; Berger *et al.,* 1990; Paris and Cunningham, 1990) does not elicit neurotoxicity, the parent amphetamine analogs themselves are unlikely to induce the neurotoxic effect. The possibility that metabolites of amphetamine analogs are responsible for the neurotoxicity has, therefore, been explored (Gál *et al.,* 1975; Sherman *et al.,* 1975; Schmidt, 1987). Lim and Foltz (1988,1991) demonstrated that MDMA is metabolized to 17 different metabolites by N-demethylation, O-dealkylation, aromatic hydroxylation, deamination, and conjugation. Intracerebroventricular administration of 3,4-dihydroxymethamphetamine, 3-O-methyl-α-methyldopamine, 2-hydroxy-4,5-methylenedioxymethamphetamine, or 2-hydroxy-4,5-methylenedioxyamphetamine, four metabolites of MDMA, failed to induce neurotoxicity in the central monoaminergic systems (McCann and Ricaurte, 1991; Steele *et al.,* 1991; Elayan *et al.,* 1992; Johnson *et al.,* 1992; Zhao *et al.,* 1992). Lim *et al.* have suggested that 2,4,5-trihydroxymethamphetamine (THM) and 2,4,5-trihydroxyamphetamine (THA) are generated after MDMA administration since these metabolites are formed in a liver preparation from 2-hydroxy-4,5-methylenedioxymethamphetamine and 2-hydroxy-4,5-methylenedioxyamphetamine, two products of MDMA metabolism (Lim and Foltz, 1988,1990,1991; Lim *et al.,* 1991). However, THM and THA have not yet been detected in the brain of MDMA-treated rats. Five to 7 days after intracerebroventricular administration of THM or THA, a decrease in TPH activity in the rat cortex, striatum, and hippocampus was observed (Elayan et al., 1992; Johnson et al., 1992). Moreover, striatal TH activity also was reduced in a dose-dependent manner by both toxins. This result is in contrast with the effects of MDMA, which alters only the serotonergic system (see Section VI).

In an earlier study, Matsuda et al. (1989) reported the effects of the amphetamine metabolites, p-hydroxyamphetamine (POHA) and p-hydroxynorephedrine (POHNOR), administered intrastriatally, on dopaminergic and serotonergic parameters. POHA decreased concentrations of DA more than those of 5-HT. POHNOR decreased both neostriatal DA and 5-HT content. Neither compound altered neostriatal TH activity, but TPH activity was increased by both metabolites. Systemic administration of POHA also decreased DA and its metabolites in a dose-dependent manner. Larger doses of POHA were required to decrease 5-HT content in the hippocampus. Note that administration of amfonelic acid, a DA uptake inhibitor, attenuated the effects of systemically administered POHA on the content of DA, DOPAC, and HVA. POHA and POHNOR are not likely to constitute the principal cause of amphetamine-induced neurotoxicity since (1) these metabolites fail to reduce TH and TPH activity and (2) the inhibition of para-hydroxylation with iprindole fails to prevent toxicity (Steranka, 1982).

Thus, the role of amphetamine metabolites in mediating central neurotoxicity is unclear. MDMA metabolites measured in the brain are not neurotoxic. THM and THA lack the selectivity of the parent drug for the serotonergic system. Moreover, amphetamine metabolites do not penetrate the central nervous system easily because of their hydrophilic nature. Whether the brain metabolism of amphetamine analogs generates a sufficient amount of the neurotoxins to cause neuronal damage is not certain. Finally, the role of DA in response to amphetamine-derived neurotoxins in mediating amphetamine neurotoxicity remains to be reconciled.

X. ROLE OF OXIDATIVE PROCESSES IN THE NEUROTOXIC RESPONSE

Within minutes of a single subcutaneous administration of amphetamine, METH, PCA, fenfluramine, or MDMA, TPH activity is decreased significantly (Fig. 5; Knapp et al., 1974; Stone et al., 1987d, 1989a,b). In a kinetic analysis, Schmidt and Taylor (1987) observed that MDMA decreased the V_{max} of TPH, but did not alter the enzyme affinity for the substrate tryptophan or the synthetic pterin cofactor 6-methyl-5,6,7,8-tetrahydropterin (6-MPH$_4$). These changes in enzyme kinetics are similar to those induced by molecular oxygen, which inactivates TPH by oxidizing its sulfhydryl sites (Kuhn et al., 1980). This inactivation induced by oxygen can be reversed by incubating the inactive enzyme in an anaerobic atmosphere in the presence of a sulfhydryl reducing agent such as dithiothreitol (Kuhn et

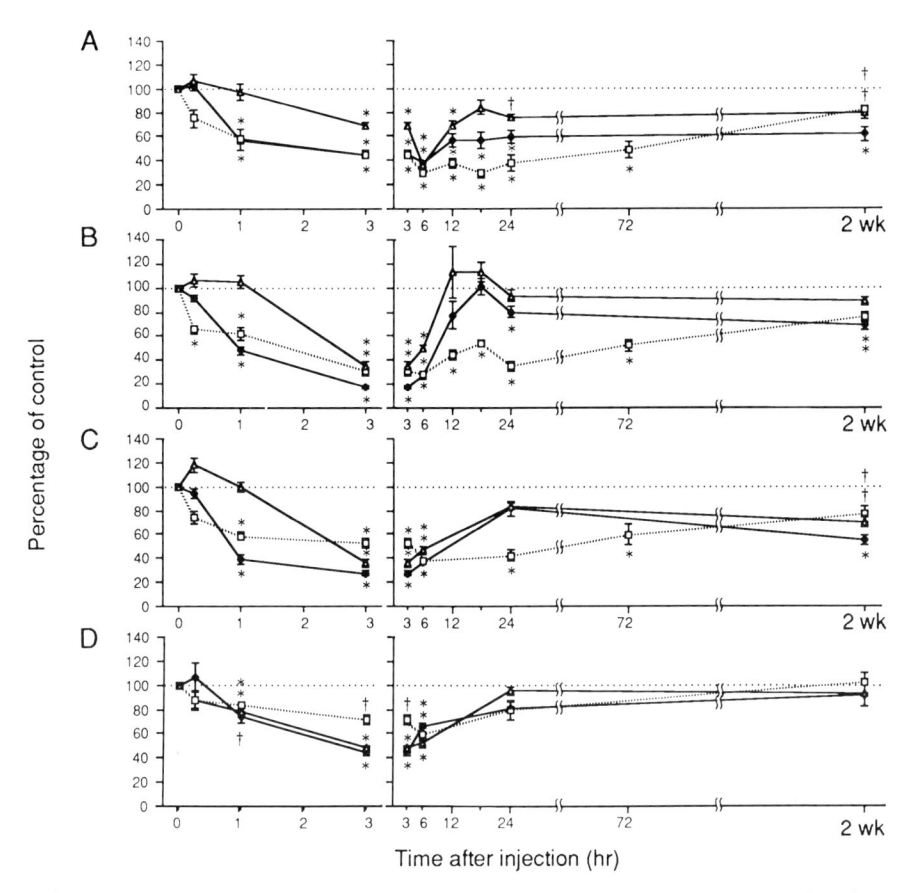

FIGURE 5 Time course of the regional serontonergic effects of acute administration of methylenedioxymethamphetamine (MDMA). A single dose of MDMA (10 mg/kg, sc) or saline (control) was injected; rats were killed at specific times thereafter. Each point represents the mean ± SEM from 4–6 rats, expressed as a percentage of the corresponding control. Immediate effects (up to 3 hr after injection) are represented on the left; long-term regional responses (from 3 hr to 2 wk after injection) are diagrammed on the right. Control values ± SEM (at 1 hr) for neostriata (A), frontal cortex (B), hippocampus (C), and hypothalamus (D) were as follows: activity of tryptophan hydroxylase (nmol/g tissue/hr)—39.9 ± 3.6, 80.1 ± 4.9, 63.0 ± 2.6, and 249.3 ± 7.0, respectively; concentrations of 5-hydroxytryptamine and 5-hydroxyindoleacetic acid (μg/g tissue)—0.533 ± 0.019 and 0.557 ± 0.023, 0.518 ± 0.018 and 0.218 ± 0.013, 0.359 ± 0.040 and 0.338 ± 0.009, and 0.905 ± 0.064 and 0.396 ± 0.018, respectively. Control values at other times did not vary significantly from those listed above. By the two-tailed Student's t-test, $P < 0.05$ (†) and $P < 0.005$ (*) vs corresponding control. Reprinted from *Neuropharmacol.* **26**, D. M. Stone, K. M. Merchant, G. R. Hanson, and J. W. Gibb. Immediate and long-term effects of 3,4-methylenedioxymethamphetamine on serotonin pathways in the brain of rat. Pp. 1677–1683, 1987, with permission from Pergamon Press Ltd., Headington Hill Hall, Oxford OX3 OBW, U.K.

al., 1980). Thus, Stone *et al.* (1989a) explored the possibility that the early decrease in TPH activity observed after one dose of MDMA might be restored by exposure to *in vitro* reducing conditions.

Rats were administered a single dose of MDMA and killed 3 hr later after drug administration. TPH activity in the cerebral cortex was assessed, with or without prior incubation with the reducing agents dithiothreitol and Fe^{2+}, for varying periods lasting up to 48 hr. Cortical TPH activity from saline-treated animals remained normal when incubated at 25°C under nitrogen for 24 hr with reducing agents. When dithiothreitol and Fe^{2+} were removed from the incubation mixture, cortical TPH activity in saline-treated animals declined during the course of incubation. The enzyme activity from MDMA-treated animals was depressed markedly. However, the enzyme activity gradually was restored to control activity during the anaerobic incubation with dithiothreitol and Fe^{2+}. Dithiothreitol but not ascorbate (M. Johnson, unpublished results) reversed the inactivation of the enzyme, which suggests that oxidation of sulfhydryl sites on TPH caused the loss of activity induced by amphetamine analogs. The decline in TPH activity measured 3 hr after administering METH, PCA, or fenfluramine also was reversed by the reducing conditions (Stone *et al.*, 1989a,b). These results suggest that TPH is present in active and inactive forms in serotonergic nerve terminals shortly after administration of amphetamine analogs.

Interestingly, reducing conditions failed to return TPH activity to control levels when measured 1 wk after a single MDMA administration, or 18 hr after the last of 5 doses (at 6-hr intervals) of METH or MDMA (Stone *et al.*, 1989a,b; M. Johnson, unpublished results). This result suggests that TPH is inactivated permanently after a certain period of time following drug administration. Since neuronal damage occurs after a certain period of time, a likely explanation for the irreversible decline in enzyme activity is that serotonergic nerve terminals have been destroyed.

Of significant interest is the fact that TPH, inactivated soon after a single dose of several amphetamine congeners, could be restored *in vitro*. These results suggest that a similar mechanism may underlie TPH inactivation caused by all these amphetamine compounds. The oxidation of TPH sulfhydryl sites may reflect the occurrence of oxidative stress within the serotonergic terminals. Interestingly, the possibility that oxidative stress and free radicals play a major role in the induction of serotonergic neuron neurotoxicity is under consideration (Kappus and Sies, 1981; De Vito and Wagner, 1989). Stone *et al.* (1989a) suggested previously that quinone species, formed by oxidative metabolism of either the parent amphetamine compound or of drug-liberated catecholamines, represent potential mediators of oxidative stress by virtue of their ability to undergo one-electron oxidation reaction. Superoxide anion, hydrogen peroxide, and other reactive oxygen species formed by an intracellular redox cycling might inactivate TPH directly, or indirectly by elevating cellular oxidized glutathione levels, since detoxification of hydrogen peroxide by glutathione peroxidase involves the oxidation of glutathione. Elevated levels of oxidized glutathione have been associated with increased mixed disulfide formations.

The condition causing the rapid inactivation of TPH alone does not necessarily lead to neuronal damage, as demonstrated with N-ethyl-3,4-methylenedioxyamphetamine (MDE). A single administration of MDE (10 mg/kg) induces a rapid decrease in TPH activity similar to rapid decreases induced by other amphetamine analogs (Johnson *et al.*, 1987b,1989b), but multiple administrations of MDE fail to induce toxicity in the serotonergic system 2 wk after treatment (Stone *et al.*, 1987c). A factor other than the condition inducing the rapid inactivation of TPH may also be required to induce neurotoxicity.

XI. ROLE OF GLUTAMINERGIC AND GABA-ERGIC SYSTEMS IN NEUROTOXICITY INDUCED BY AMPHETAMINE CONGENERS

The possible role of glutamate in METH-induced toxicity has been explored. Excitatory amino acids (EAA) injected into the brain cause neurodegeneration; this response has been attributed to excessive stimulation of EAA receptors. Since a significant glutamatergic neuronal input from the cortex to the neostriatum exists, glutamate may contribute to the METH-induced neurotoxicity. These glutamatergic neurons have presynaptic dopaminergic D_2 receptors that decrease glutamatergic uptake (Nieoullon *et al.*, 1983; Kerkerian *et al.*, 1987) and inhibit glutamate release (Crowder and Bradford, 1987; Maura *et al.*, 1988; Yamamoto and Davy, 1992).

To determine whether glutamate contributes to METH-induced toxicity, Sonsalla *et al.* (1989) evaluated the effects of noncompetitive N-methyl-D-aspartate (NMDA) antagonists on changes in monoaminergic systems after METH treatment. In their experiments, mice were administered MK-801 concomitantly with METH; MK-801 attenuated the METH-induced deficits in neostriatal TH activity and DA content. Since phencyclidine and ketamine similarly antagonize NMDA receptors, these agents were tested also. Each agent attenuated the METH-induced decrease in DA content and TH activity. Johnson *et al.* (1989a; Fig. 6, Table I) confirmed the observation of Sonsalla *et al.* (1989) that MK-801 prevented the METH-induced decrease in dopaminergic parameters, and also found that this glutamate antagonist similarly protected against serotonergic changes. However, these investigators observed no protection by MK-801 against MDMA-induced neurotoxicity.

In a preliminary report, Farfel and Seiden (1992) suggested a hypothermic component in the protection afforded by MK-801. A similar mechanism of action was proposed by Buchan and Pulsinelli (1990) to explain the protective effect of MK-801 against neuronal damage induced by ischemia.

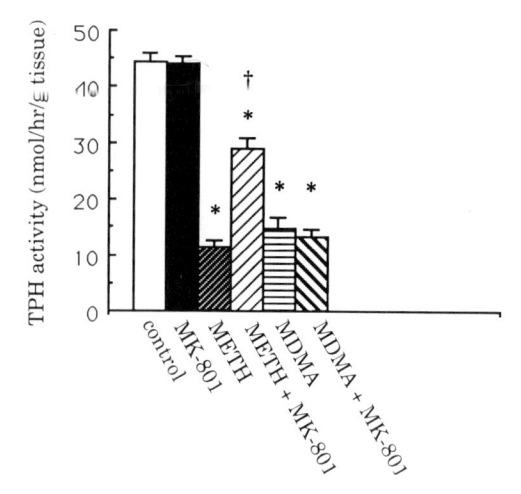

FIGURE 6 Effects of MK-801 on the decrease in neostriatal tryptophan hydroxylase (TPH) activity induced by methamphetamine (METH) or methylenedioxymethamphetamine (MDMA). Rats received 4 doses of METH (15 mg/kg, sc), MDMA (10 mg/kg, sc), or vehicle (0.9% NaCl) at 6-hr intervals. The MK-801-treated animals received a dose of 2.5 mg/kg intraperitoneally 15 min prior to the injection of METH, MDMA, or vehicle. The animals were killed by decapitation 18–20 hr after the last administration. TPH activity (means ± SEM expressed as nmol hydroxylated tryptophan/hr/g tissue) shown in this figure is the result of two separate experiments. Each group contained 10–18 animals. Statistical analysis of enzymatic activities was performed with a one-way analysis of variance followed by a Scheffe multiple comparison test. Statistical significance between groups was defined at $P < 0.05$.*, significantly different from control group; †, significantly different from METH-treated group. Adapted with permission from Johnson et al. (1989a).

TABLE I Effects of MK-801 on the Decrease in Neostriatal Tyrosine Hydroxylase Activity Induced by Methamphetamine

Treatment[a]	Dose (mg/kg)	TH activity[b] (nmol/hr/g tissue)
Control (5)	—	2196 ± 116
MK-801 (3)	2.5	2369 ± 263
METH (7)	15	1408 ± 112[c]
METH + MK-801 (5)	15 + 2.5	1878 + 64[d]

[a] Rats were treated as described in Fig. 6. The number of rats per group is indicated in parentheses.
[b] Statistical analysis of enzymatic activities was performed with a one-way analysis of variance followed by a Fisher test.
[c] $P < 0.05$ vs control.
[d] $P < 0.05$ vs METH.

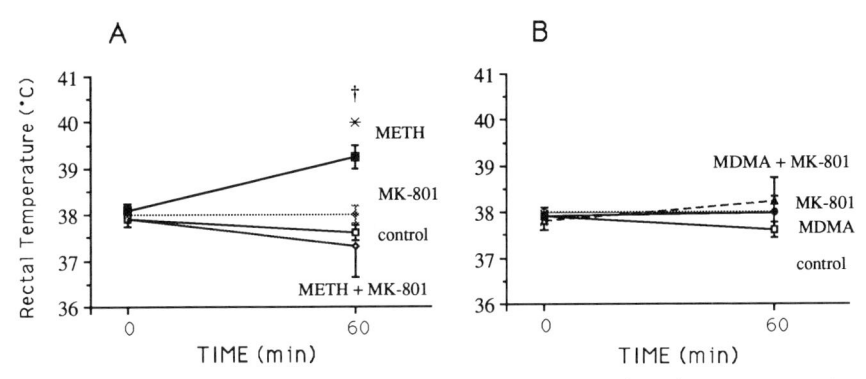

FIGURE 7 Effects of methamphetamine (METH), mehtylenedioxymethamphetamine (MDMA), and MK-801 on rectal temperature measured in rats. Rectal temperature was measured in male Sprague–Dawley rats (260–300 g) before and 1 hr after administration of (A) METH (15 mg/kg, sc) or (B) MDMA (10 mg/kg, sc) with or without MK-801 (2.5 mg/kg, ip). Control animals were injected with 0.9% saline. Statistical analysis was performed with an ANOVA test followed with a Scheffe multiple comparison test. $P < 0.05$ vs control (at 60 min) (*); $P < 0.05$ vs METH + MK-801 (at 60 min) (†) ($n = 4-8$).

Hyperthermia has been shown to increase amphetamine-induced mortality and to cause the amphetamine-induced decrease in protein synthesis (Craig and Kupferberg, 1972; Nowak, 1988). Farfel and Seiden (1992) reported that MK-801 in combination with METH or MDMA decreased body temperature. Figure 7 shows the effect of MK-801 on rectal temperature measured in rats 1 hr after they received METH (15 mg/kg, sc) or MDMA (10 mg/kg, sc). METH induced a significant increase in body temperature that was blocked by MK-801. In contrast to the report by Farfel and Seiden (1992), MDMA failed to alter body temperature, probably because of the lower dose used. Since METH and MDMA reduced TPH activity after multiple administrations of these drugs (Fig. 6) but MDMA failed to raise body temperature, hyperthermia may not be the only factor involved in the neurochemical changes. However, interestingly MK-801 blocked the increase in body temperature and attenuated the decrease in TPH induced by METH (Figs. 6,7). Of significant importance is establishing whether the effect of MK-801 on body temperature plays a role in protecting against METH toxicity or is only a coincidental response.

We have demonstrated that NMDA, when administered concurrently with subthreshold doses of METH, decreased TPH activity in the prefrontal cortex but had no effect in other brain regions. Neither NMDA nor METH alone (in the subthreshold doses administered) decreased TPH activity (Table II), suggesting a synergism between METH and NMDA. However, the decrease in cortical TPH activity was not observed after 1 wk (results not shown). More studies are required to identify the synergistic interaction between METH and NMDA.

TABLE II Effects of N-Methyl-D-aspartate on Cortical Tryptophan Hydroxylase Activity when Given with Subthreshold Doses of Methamphetamine

Treatment[a]	Dose (mg/kg)	TPH activity[b] (nmol/hr/g tissue)	
		Frontal cortex	Striatum
Control	—	53 ± 3	14 ± 1
NMDA	110	50 ± 4	16 ± 2
METH	5	45 ± 2	14 ± 1
METH + NMDA	5 + 110	29 + 3[c,d]	13 ± 1

[a] Male Sprague–Dawley rats (200–230 g) were administered 5 injections of METH (5 mg/kg, sc) or saline with or without NMDA (110 mg/kg, ip) at 6-hr intervals and were killed 18 hr after the last drug administration.
[b] TPH activity is expressed in nmol hydroxylated tryptophan/hr/g tissue ± SEM ($n = 6-9$). Statistical analysis was performed with an ANOVA analysis followed by a Scheffe multiple comparison test.
[c] $P < 0.05$ vs control.
[d] $P < 0.05$ vs METH.

A. Effect of Amphetamine and Methamphetamine on Glutamate Release

The possibility that glutamate plays a role in METH-induced neurotoxicity has been strengthened by experiments in which glutamate release is monitored in awake rats that receive METH. Nash and Yamamoto (1992a) monitored extracellular concentrations of glutamate, DA, and DOPAC by *in vivo* dialysis. METH or MDMA was administered at 2-hr intervals; extracellular concentrations of the respective compounds were measured in the freely moving animals. Glutamate concentrations remained normal until approximately the third dose of METH was administered, at which time the glutamate content rose to approximately 10 times normal. However, similar treatment with MDMA did not elevate glutamate concentrations. Both METH and MDMA increased striatal DA concentrations but decreased DOPAC concentrations. The authors hypothesized a causal relationship between the elevated glutamate concentration and the neurotoxicity resulting from METH but not MDMA. This hypothesis is consistent with the observations that glutamate antagonists attenuate the METH-induced dopaminergic changes (Sonsalla *et al.*, 1989) but MDMA-induced neurochemical deficits are not attenuated by glutamate antagonists (Johnson *et al.*, 1989a).

A number of explanations are possible for the proposed METH-induced glutamate-mediated neurotoxicity. One hypothesis is based on the observation that, under normal conditions, released neostriatal DA inhibits activated corticostriatal glutamate neurons (Maura et al., 1988; Yamamoto and Davy, 1992). After the initial large doses of METH, the increase in DA release inhibits glutamate release and reuptake but, as DA release returns to normal during the following hours, a disinhibition of the glutamatergic corticostriatal neurons occurs; the resulting increase in glutamate release causes the neurotoxicity. Left unresolved, however, is the fact that DA release is caused also by MDMA but does not appear to alter glutamate activity. Interestingly, Nash and Yamamoto (1992b) also observed an increase in glutamate release in iprindole-treated rats receiving amphetamine. This increase could be blocked by the D_2 receptor antagonist haloperidol. Haloperidol failed to block the amphetamine-induced DA release but prevented the long-term depletion of DA induced by the treatment. These results support the concept that both the dopaminergic and the glutamatergic system are involved in the mechanism causing the amphetamine-induced neurotoxicity in the dopaminergic system.

B. Role of γ-Aminobutyric Acid in Amphetamine Toxicity

γ-Aminobutyric acid (GABA) also may participate in mediating METH- and MDMA-induced toxicity. GABA agonists depress the activity of the nigrostriatal dopaminergic system (Wood, 1982). Administration of aminooxyacetic acid, γ-acetylenic GABA, and other inhibitors of GABA transaminase, the catabolic enzyme for GABA, prevents the METH-induced decrease in central TH and TPH activity as well as the MDMA-induced decrease in TPH activity (Hotchkiss and Gibb, 1980a,b; Stone et al., 1987a). This result suggests that an increase in GABA-ergic transmission protects the dopaminergic and serotonergic systems. Moreover, the report that GABA release is reduced by repeated low doses of amphetamine (Lindefors et al., 1992) provides evidence that a reduction in GABA release may facilitate the induction of long-term changes in the dopaminergic and serotonergic systems.

XII. DEVELOPMENT OF TOLERANCE TO METHAMPHETAMINE-INDUCED TOXICITY

It is well established that repeated exposure to the same dose of amphetamines causes a diminished response to the drug. The possibility that rats become tolerant to the neurotoxic effects of large doses of amphetamine was explored by Schmidt et al. (1985b). In these studies, rats were injected with METH or saline on Days 1, 3, or 5. Rats received 2.5 mg/kg every 6 hr for 5 doses on Day 1. The dose was increased to 5.0 mg/kg on Day 3 and to 7.5 mg/kg on Day 5. A 24-hr drug-free period was interspersed between each increase in dose. On Day 7, the saline group was divided into a saline control and a naive METH challenge group; the pretreated METH group was divided into a METH-pretreated–saline and a METH-pretreated–METH challenge group. Naive and pretreated rats then were challenged with a high-dose METH regimen of 15 mg/kg every 6 hr for 5 doses. In the naive animals challenged with the high-dose regimen of METH, the usual marked decrease in activities of TH and TPH was observed. Similarly, the concentrations of DA and 5-HT and their respective metabolites were compromised severely. However, in the animals pretreated with the gradually increasing doses of METH, then challenged with the high-dose METH, the activities of TH and TPH, as well as the concentrations of DA, 5-HT, and their metabolites, were affected significantly less than in naive animals.

In a subsequent study, Schmidt and collaborators (1985c) reported that [^{3}H] sulpiride binding in the neostriatum and nucleus accumbens of naive rats challenged with high-dose METH was reduced. However, in rats pretreated with gradually increasing doses of METH, as just described, the decrease in D_2 binding was attenuated. Forebrain concentrations of METH and its metabolite amphetamine also were compared in the naive and pretreated rats that received the challenge high dose of METH. The brain concentrations of both METH and amphetamine were decreased markedly in the pretreated group. According to the findings just described, tolerance appears to develop to the neurotoxicity induced by large doses of METH as a result of a decrease in drug concentration in the brain.

In an effort to clarify the response of the tolerant animals, Alburges *et al.* (1990) compared concentrations of METH, amphetamine, *p*-hydroxymethamphetamine, and *p*-hydroxyamphetamine over time in the brain, liver, and blood of rats treated according to the protocol described earlier. Surprisingly, whereas the brain levels of METH and its metabolites decreased in pretreated animals, the levels of these compounds were actually substantially higher in the liver and blood of the pretreated groups. Providing an explanation of these unexpected findings will require considerable experimentation. Whether these results can be explained by an alteration of the metabolism of METH remains to be elucidated.

XIII. CONCLUSION

In toxic repeated doses, METH decreases brain dopaminergic (TH activity, concentrations of DA and its metabolites) and serotonergic (TPH activity, concentrations of 5-HT and its metabolite) parameters. Although the response is reversible after a single lower dose of METH, the neurotoxic effects persist for extended periods of time when the drug is administered in large multiple doses. The methylenedioxy analogs of amphetamine induce similar responses in the serotonergic system but have no lasting effect in the dopaminergic system. Since interruption of dopaminergic function attenuates these effects, DA and its reactive metabolites are concluded to play a significant role in this neurotoxic response. Glutamate and GABA also appear to be involved.

ACKNOWLEDGMENTS

The research performed by Drs. Gibb, Hanson, and Johnson was supported by the United States Public Health Service Grants DA-00869 and DA-04221.

REFERENCES

Alburges, M. E., Hanson, G. R., and Gibb, J. W. (1990). Role of methamphetamine metabolism in the development of CNS tolerance to the drug. *Inves. Clin.* 31, 165–176.
Axt, K. J., and Molliver, M. E. (1991). Immunocytochemical evidence for methamhetamine-induced serotonergic axon loss in the rat brain. *Synapse* 9. 302–313.
Bakhit, C., and Gibb, J. W. (1981). Methamphetamine-induced depression of tryptophan hydroxylase: Recovery following acute treatment. *Eur. J. Pharmacol.* 76, 229–233.
Bakhit, C., Morgan, M. E., Peat, M. A., and Gibb, J. W. (1981). Long-term effects of methamphetamine on synthesis and metabolism of 5-hydroxytryptamine in various regions of the rat brain. *Neuropharmacology* 20, 1135–1140.
Berger, U. V., Molliver, M. E., and Grzanna, R. (1990). Unlike systemic administration of p-chloroamphetamine, direct intracerebral injection does not produce degeneration of 5-HT axons. *Exp. Neurol.* 109, 257–268.
Buchan, A., and Pulsinelli, W. A. (1990). Hypothermia but not the N-methyl-D-aspartate antagonist, MK-801, attenuates neuronal damage in gerbils subjected to transient global ischemia. *J. Neurosci.* 10, 311–316.
Buening, M. K., and Gibb, J. W. (1974). Influence of methamphetamine and neuroleptic drugs on tyrosine hydroxylase activity. *Eur. J. Pharmacol.* 26, 30–34.
Commins, D. L., and Seiden, L. S. (1986). α-Methyltyrosine blocks methamphetamine-induced degeneration in the rat somatosensory cortex. *Brain Res.* 365, 15–20.
Commins, D. L., Axt, K. J., Vosmer, G., and Seiden, L. S. (1987a). 5,6-Dihydroxytryptamine, a serotonergic neurotoxin, is formed endogenously in the rat brain. *Brain Res.* 403, 7–14.
Commins, D. L., Axt, K. J., Vosmer, G., and Seiden, L. S. (1987b). Endogenously produced 5,6-dihydroxytryptamine may mediate the neurotoxic effects of *para*-chloroamphetamine. *Brain Res.* 419, 253–261.
Commins, D. L., Vosmer, G., Virus, R. M., Woolverton, W. L., Schuster, C. R., and Seiden, L. S. (1987c). Biochemical and histological evidence that methylenedioxymethamphetamine (MDMA) is toxic to neurons in the rat brain. *J. Pharmacol. Exp. Ther.* 241, 338–345.
Craig, A. L., and Kupferberg, H. J. (1972). Hyperthermia in *d*-amphetamine toxicity in aggregated mice of different strains. *J. Pharmacol. Exp. Ther.* 180, 616–624.
Crowder, J. M., and Bradford, H. F. (1987). Inhibitory effects of noradrenaline and dopamine on calcium influx and neurotransmitter glutamate release in mammalian brain slices. *Eur. J. Pharmacol.* 143, 343–352.
De Vito, M. J., and Wagner, G. C. (1989). Methamphetamine-induced neuronal damage: A possible role for free radicals. *Neuropharmacology* 28, 1145–1150.
Elayan, I., Gibb, J. W., Hanson, G. R., Foltz, R. L., Lim, H. K., and Johnson, M. (1992). Long-term alteration in the central monoaminergic systems of the rat by 2,4,5-trihydroxyamphetamine but not by 2-hydroxy-4,5-methylenedioxymethamphetamine or 2-hydroxy-4,5-methylenedioxyamphetamine. *Eur. J. Pharmacol.* 221, 281–88.

Ellison, G., Eison, M. S., Huberman, H. S., and Daniel, F. (1978). Long-term changes in dopaminergic innervation of caudate nucleus after continuous amphetamine administration. *Science (Washington, D.C.)* **201**, 276–278.

Emson, P. C., and Lindvall, O. (1979). Distribution of putative neurotransmitters in the neocortex. *Neuroscience* **4**, 1–30.

Farfel, G. M., and Seiden, L. S. (1992). Temperature decrease may mediate protection by MK-801 against serotonergic neurotoxicity. *Soc. Neurosci. Abstr.* **18**, 1602.

Fibiger, H. C., and McGeer, E. G. (1971). Effect of acute and chronic methamphetamine treatment on tyrosine hydroxylase activity in brain and adrenal medulla. *Eur. J. Pharmacol.* **16**, 176–180.

Fisher, J. F., and Cho, A. K. (1979). Chemical release of dopamine from striatal homogenates: Evidence for an exchange diffusion model. *J. Pharmacol. Exp. Ther.* **208**, 203–209.

Fuller, R. W., and Hemrick-Luecke, S. K. (1980). Long-lasting depletion of striatal dopamine by a single injection of amphetamine in iprindole-treated rats. *Science (Washington, D.C.)* **209**, 305–307.

Fuller, R. W., and Hemrick-Luecke, S. K. (1982). Further studies on the long-term depletion of striatal dopamine in iprindole-treated rats by amphetamine. *Neuropharmacology*, **21**, 433–438.

Fuller, R. W., and Perry, K. W. (1977). Further studies on the effects of N-cyclopropyl-p-chloroamphetamine in rat brain. *Neuropharmacology* **16**, 495–497.

Fuller, R. W., and Snoddy, H. D. (1979). Inability of methylphenidate or mazindol to prevent the lowering of 3,4-dihydroxyphenylacetic acid in rat brain by amphetamine. *J. Pharm. Pharmacol.* **31**, 183–184.

Fuller, R. W., Hines, C. W., and Mills, J. (1965). Lowering of brain serotonin level by chloroamphetamines. *Biochem. Pharmacol.* **14**, 483–488.

Fuller, R. W., Snoddy, H. D., Roush, B. W., and Molloy, B. B. (1973). Further structure-activity studies on the lowering of brain 5-hydroxyinoles by 4-chloroamphetamines. *Neuropharmacology* **12**, 33–42.

Fuller, R. W., Perry, K. W., and Molloy, B. B. (1975). Reversible and irreversible phases of serotonin depletion by 4-chloroamphetamine. *Eur. J. Pharmacol.* **33**, 119–124.

Fuller, R. W., Snoddy, H. D., Snoddy, A. M., Hemrick, S. K., Wong, D. T., and Molloy, B. B. (1980). p-Iodoamphetamine as a serotonin depletor in rats. *J. Pharmacol. Exp. Ther.* **212**, 115–119.

Gál, E. M., Christiansen, P. A., and Yunger, L. M. (1975). Effect of p-chloroamphetamine on cerebral tryptophan-5-hydroxylase in vivo: A reexamination. *Neuropharmacology* **14**, 31–39.

Gibb, J. W., and Kogan, F. J. (1975). The effect of α-methyl-p-tyrosine on methamphetamine-induced alterations of tyrosine hydroxylase activity. *Int. Congr. Pharmacol.* **6**, 1300.

Gibb, J. W., and Kogan, F. J. (1979). Influence of dopamine synthesis on methamphetamine-induced changes in striatal and adrenal tyrosine hydroxylase activity. *Naunyn-Schmiedeberg's Arch. Pharmacol.* **310**, 185–187.

Graham, D. G., Tiffany, S. M., Bell, W. R., Jr., and Gutknecht, W. F. (1978). Autoxidation versus covalent binding of quinones as the mechanism of toxicity of dopamine, 6-hydroxydopamine, and related compounds toward C1300 neuroblasma cells in vitro. *Mol. Pharmacol.* **14**, 644–653.

Grinspoon, L., and Hedblom, P. (1975). "The Speed Culture. Amphetamine Use and Abuse in America." Harvard University Press, Cambridge, Massachusetts.

Hanson, G. R., Merchant, K. M., Letter, A. A., Bush, L., and Gibb, J. W. (1987). Methamphetamine-induced changes in the striatal-nigral dynorphin system: Role of D_1 and D_2 receptors. *Eur. J. Pharmacol.* **144**, 245–246.

Hanson, G. R., Singh, N., Merchant, K., Johnson, M., Bush, L., and Gibb, J. W. (1992). Response of limbic and extrapyramidal neurotensin systems to stimulants of abuse. *Ann. N.Y. Acad. Sci.* **668**, 165–172.

Harvey, J. A., McMaster, S. E., and Yunger, L. M. (1975). p-Chloroamphetamine: Selective neurotoxic action in brain. *Science (Washington, D.C.)* **187**, 841–843.
Hotchkiss, A., and Gibb, J. W. (1980a). Long-term effects of multiple doses of methamphetamine on tryptophan hydroxylase and tyrosine hydroxylase activity in rat brain. *J. Pharmacol. Exp. Ther.* **214**, 257–262.
Hotchkiss, A., and Gibb, J. W. (1980b). Blockade of methamphetamine-induced depression of tyrosine hydroxylase by GABA transaminase inhibitors. *Eur. J. Pharmacol.* **66**, 201–205.
Hotchkiss, A., Morgan, M. E., and Gibb, J. W. (1979). The long-term effects of multiple doses of methamphetamine on neostriatal tryptophan hydroxylase, tyrosine hydroxylase, choline acetyltransferase and glutamate decarboxylase activities. *Life Sci.* **25**, 1373–1378.
Johnson, M., Stone, D. M., Hanson, G. R., and Gibb, J. W. (1987a). Role of the dopaminergic nigrostriatal pathway in methamphetamine-induced depression of the neostriatal serotonergic system. *Eur. J. Pharmacol.* **135**, 231–234.
Johnson, M., Stone, D. M., Hanson, G. R., and Gibb, J. W. (1987b). Effects of N-ethyl-3,4-methylenedioxyamphetamine (MDE) on central serotonergic and dopaminergic systems of the rat. *Biochem. Pharmacol.* **36**, 4085–4093.
Johnson, M., Hanson, G. R., and Gibb, J. W. (1988). Effects of dopaminergic and serotonergic receptor blockade on neurochemical changes induced by acute administration of methamphetamine and 3,4-methylenedioxymethamphetamine. *Neuropharmacology* **27**, 1089–1096.
Johnson, M., Hanson, G. R., and Gibb, J. W. (1989a). Effect of MK-801 on the decrease in tryptophan hydroxylase induced by methamphetamine and its methylenedioxy analog. *Eur. J. Pharmacol.* **165**, 315–318.
Johnson, M., Hanson, G. R., and Gibb, J. W. (1989b). Characterization of acute N-ethyl-3,4-methylenedioxyamphetamine (MDE) action on the central serotonergic system. *Biochem. Pharmacol.* **38**, 4333–4338.
Johnson, M., Hanson, G. R., and Gibb, J. W. (1991). Norepinephrine does not contribute to methamphetamine-induced changes in hippocampal serotonergic system. *Neuropharmacology* **30**, 617–622.
Johnson, M., Elayan, I., Hanson, G. R., Foltz, R. L., Gibb, J. W., and Lim, H. K. (1992). Effects of 3,4-dihydroxymethamphetamine and 2,4,5-trihydroxymethamphetamine, two metabolites of 3,4-methylenedioxymethamphetamine, on central serotonergic and dopaminergic systems. *J. Pharmacol. Exp. Ther.* **261**, 447–253.
Kappus, H., and Sies, H. (1981). Toxic drug effects associated with oxygen metabolism: Redox cycling and lipid peroxidation. *Experientia* **37**, 1233–1258.
Kerkerian, L., Dusticier, N, and Nieoullon, A. (1987). Modulatory effect of dopamine on high-affinity glutamate uptake in the rat striatum. *J. Neurochem.* **48**, 1301–1306.
Knapp, S., Mandell, A. J., and Geyer, M. A. (1974). Effects of amphetamines on regional tryptophan hydroxylase activity and synaptosomal conversion of tryptophan to 5-hydroxytryptamine in rat brain. *J. Pharmacol. Exp. Ther.* **189**, 676–689.
Koda, L. Y., and Gibb, J. W. (1971). The effect of repeated large doses of methamphetamine on adrenal and brain tyrosine hydroxylase. *Pharmacologist* **13**, 253.
Koda, L. Y., and Gibb, J. W. (1973). Adrenal and striatal tyrosine hydroxylase activity after methamphetamine. *J. Pharmacol. Exp. Ther.* **185**, 42–48.
Kuhn, D. M., Ruskin, B., and Lovenberg, W. (1980). Tryptophan hydroxylase: The role of oxygen, iron and sulfhydryl groups as determinants of stability and catalytic activity. *J. Biol. Chem.* **255**, 4137–4143.
Liang, N. Y., and Rutledge, C. O. (1982). Comparison of the release of [^{3}H]dopamine from isolated corpus striatum by amphetamine, fenfluramine, and unlabelled dopamine. *Biochem. Pharmacol.* **31**, 983–992.
Lim, H. K., and Foltz, R. L. (1988). In vivo and in vitro metabolism of 3,4-(methylenedioxy)methamphetamine in the rat: Identification of metabolites using an ion trap detector. *Chem. Res. Toxicol.* **1**, 370–378.

Lim, H. K., and Foltz, R. L. (1990). Ion trap MS/MS techniques: Structural elucidation of potentially neurotoxic aromatic hydroxylated metabolites of 3,4-(methylenedioxy)methamphetamine (MDMA). *Proc. 38th ASMS Conf. Mass Spec. Allied Topics* **38**, 1002–1003.

Lim, H. K., and Foltz, R. L. (1991). Ion trap tandem mass spectrometric evidence for the metabolism of 3,4-(methylenedioxy)methamphetamine to the potent neurotoxins 2,4,5-trihydroxymethamphetamine and 2,4,5-trihydroxyamphetamine. *Chem. Res. Toxicol.* **4**, 626–632.

Lim, H. K., Stevens, W., and Foltz, R. L. (1991). In vitro metabolism of 6-hydroxy-(3,4-methylenedioxy)methamphetamine to the neurotoxin, 2,4,5-trihydroxymethamphetamine. *Soc. Neurosci. Abstr.* **17**, 1248.

Lindefors, N., Hurd, Y. L., O'Connor, W. T., Brené, S., Persson, H., and Ungerstedt, U. (1992). Amphetamine regulation of acetylcholine and γ-aminobutyric acid in nucleus accumbens. *Neuroscience* **48**, 439–448.

Logan, B. J., Laverty, R., Sanderson, W. D., and Yee, Y. B. (1988). Differences between rats and mice in MDMA (methylenedioxymethamphetamine) neurotoxicity. *Eur. J. Pharmacol.* **152**, 227–234.

Lorez, H. (1981). Fluorescence histochemistry indicates damage of striatal dopamine nerve terminals in rats after multiple doses of methamphetamine. *Life Sci.* **28**, 911–916.

Masuoka, D. T., Kokka, N., and Earle, R. W. (1975). Effect of 6-hydroxydopamine on the distribution and action of amphetamine. In "Chemical Tools in Catecholamine Research" (G. Jonsson, T. Malmfors, and C. Sachs, eds), Vol. 1, pp. 291–301. North Holland, Amsterdam.

Matsuda, L. A., Hanson, G. R., and Gibb, J. W. (1989). Neurochemical effects of amphetamine metabolites on central dopaminergic and serotonergic systems. *J. Pharmacol. Exp. Ther.* **251**, 901–908.

Maura, G., Giardi, A., and Raiteri, M. (1988). Release-regulating D-2 dopamine receptors are located on striatal glutamatergic nerve terminals. *J. Pharmacol. Exp. Ther.* **247**, 680–684.

McCann, U. D,. and Ricaurte, G. A. (1991). Major metabolites of (±) 3,4-methylenedioxyamphetamine (MDA) do not mediate its toxic effects on brain serotonin neurons. *Brain Res.* **545**, 279–282.

Molliver, M. E., O'Hearn, E., Battaglia, G., and De Souza, E. B. (1986). Direct intracerebral administration of MDA and MDMA does not produce serotonin neurotoxicity. *Soc. Neurosci. Abstr.* **12**, 1234.

Nash, J. F., and Yamamoto, B. K. (1992a). Methamphetamine neurotoxicity and striatal glutamate release: Comparison to 3,4-methylenedioxymethamphetamine. *Brain Res.* **581**, 237–243.

Nash, J. F., and Yamamoto, B. K. (1992b). Effect of d-amphetamine on the extracellular concentrations of dopamine and glutamate in iprindole treated rats. *Soc. Neurosci. Abstr.* **18**, 363.

Nieoullon, A., Kerkerian, L., and Dusticier, N. (1983). Presynaptic dopaminergic control of high affinity glutamate uptake in the striatum. *Neurosci. Lett.* **43**, 191–196.

Nowak, T. S., Jr. (1988). Effects of amphetamine on protein synthesis and energy metabolism in mouse brain: Role of drug-induced hyperthermia. *J. Neurochem.* **50**, 285–294.

O'Hearn, E., Battaglia, G., De Souza, E. B., Kuhar, M. J., and Molliver, M. E. (1988). Methylenedioxyamphetamine (MDA) and methylenedioxymethamphetamine (MDMA) cause selective ablation of serotonergic axon terminals in forebrain: Immunocytochemical evidence for neurotoxicity. *J. Neurosci.* **8**, 2788–2803.

Paris, J. M., and Cunningham, K. A. (1990). Lack of neurotoxicity after intra-raphe microinjection of MDMA ("Ecstasy"). *NIDA Res. Monogr.* **105**, 333.

Pérez-de la Mora, M., López-Quiroz, D., Méndez-Franco, J., and Drucker-Colín, R. (1990). Chronic administration of amphetamine increases glutamic acid decarboxylase activity in the rat substantia nigra. *Neurosci. Res.* **109**, 315–320.

Peroutka, S. J. (1987). Incidence of recreational use of 3,4-methylenedioxymethamphetamine (MDMA, "Ecstasy") on an undergraduate campus. *N. Engl. J. Med.* **317,** 1542–1543.

Peroutka, S. J. (1988). Relative insensitivity of mice to 3,4-methylenedioxymethamphetamine (MDMA) neurotoxicity. *Res. Commun. Subs. Abuse* **9,** 193–206.

Pletscher, A., Burkard, W. P., Bruderer, H., and Gey, K. F. (1963). Decrease of cerebral 5-hydroxytryptamine and 5-hydroxyindolacetic acid by arylalklamine. *Life Sci.* **11,** 828–833.

Pletscher, A., Bartholini, G., Bruderer, H., Burkard, W. P., and Gey, K. F. (1964). Chlorinated arylalkylamines affecting the cerebral metabolism of 5-hydroxytryptamine. *J. Pharmacol. Exp. Ther.* **145,** 344–350.

Raiteri, M., Cerrito, F., Cervoni, A. M., and Levi, G. (1979). Dopamine can be released by two mechanisms differentially affected by the dopamine transport inhibitor nomifensine, *J. Pharmacol. Exp. Ther.* **208,** 195–202.

Ricaurte, G., Schuster, C. R., and Seiden, L. S. (1980). Long-term effects of repeated methylamphetamine administration on dopamine and serotonin neurons in the rat brain: a regional study. *Brain Res.* **193,** 153–163.

Ricaurte, G., Bryan, G., Strauss, L., Seiden, L., and Schuster, C. (1985). Hallucinogenic amphetamine selectively destroys brain serotonin nerve terminals. *Science (Washington, D.C.)* **229,** 986–988.

Ritter, J. K., Schmidt, C. J., Gibb, J. W., and Hanson, G. R. (1984). Increases of substance P-like immunoreactivity within striatal–nigral structures after subacute methamphetamine treatment. *J. Pharmacol. Exp. Ther.* **229,** 487–492.

Rollema, H., DeVries, J. B., Westerink, B. H. C., Van Putten, F. M. S., and Horn, A. S. (1986). Failure to detect 6-hydroxydopamine in rat striatum after the dopamine releasing drugs dexamphetamine, methylamphetamine and MPTP. *Eur. J. Pharmacol.* **132,** 65–69.

Sanders-Bush, E., and Sulser, F. (1970). *p*-Chloroamphetamine: in vivo investigations on the mechanism of action of the selective depletion of cerebral serotonin. *J. Pharmacol. Exp. Ther.* **175,** 419–426.

Sanders-Bush, E., Bushing, J. A., and Sulser, F. (1975). Long-term effects of *p*-chloroamphetamine and related drugs on central serotonergic mechanisms. *J. Pharmacol. Exp. Ther.* **192,** 33–41.

Schmidt, C. J. (1987). Neurotoxicity of psychedelic amphetamine, methylenedioxymethamphetamine. *J. Pharmacol. Exp. Ther.* **240,** 1–7.

Schmidt, C. J., and Gibb, J. W. (1983). Effect of 5-HT synthesis inhibition by *p*-chlorophenylalanine on the neurotoxic response of the dopaminergic and serotonergic system to methamphetamine. *Fed. Proc.* **42,** 879.

Schmidt, C. J., and Gibb, J. W. (1985a). Role of the dopamine uptake carrier in the neurochemical response to methamphetamine: Effects of amfonelic acid. *Eur. J. Pharmacol.* **109,** 73–80.

Schmidt, C. J., and Gibb, J. W. (1985b). Role of the serotonin uptake carrier in the neurochemnical response to methamphetamine. *Neurochem. Res.* **10,** 637–648.

Schmidt, C. J., and Lovenberg, W. (1985). In vitro demonstration of dopamine uptake by neostriatal serotonergic neurons of the rat. *Neurosci. Lett.* **59,** 9–14.

Schmidt, C. J., and Taylor, V. L. (1987). Depression of rat brain tryptophan hydroxylase activity following the acute administration of methylenedioxymethamphetamine. *Biochem. Pharmacol.* **36,** 4095–4102.

Schmidt, C. J., Ritter, J. K., Sonsalla, P. K., Hanson, G. R., and Gibb, J. W. (1985a). Role of dopamine in the neurotoxic effects of methamphetamine. *J. Pharmacol. Exp. Ther.* **233,** 539–544.

Schmidt, C. J., Sonsalla, P. K., Hanson, G. R., Peat, M. A., and Gibb, J. W. (1985b). Methamphetamine-induced depression of monoamine synthesis in the rat: Development of tolerance. *J. Neurochem.* **44,** 852–855.

Schmidt, C. J., Gehlert, D. R., Peat, M. A., Sonsalla, P. K., Hanson, G. R., Wamsley, J. K., and

Gibb, J. W. (1985c). Studies on the mechanism of tolerance to methamphetamine. *Brain Res.* **343**, 305–313.

Seiden, L. S., and Vosmer, G. (1984). Formation of 6-hydroxydopamine in caudate nucleus of the rat brain after a single large dose of methylamphetamine. *Pharmacol. Biochem. Behav.* **21**, 29–31.

Seiden, L. S., Fischman, M. W., and Schuster, C. R. (1975/76). Long-term methamphetamine induced changes in brain catecholamines in tolerant rhesus monkeys. *Drug Alcohol Dep.* **1**, 215–219.

Sherman, A., Gál, E. M., Fuller, R. W., and Molloy, B. B. (1975). Effects of intraventricular *p*-chloroamphetamine and its analogues on cerebral 5-HT. *Neuropharmacology* **14**, 733–737.

Shulgin, A. T. (1990). History of MDMA. In "Ecstasy: The Clinical, Pharmacological and Neurotoxicological Effects of the Drug MDMA" (S. J. Peroutka, ed.), pp. 1–20. Kluwer Academic Publishers, Boston.

Sonsalla, P. K., Gibb, J. W., and Hanson, G. R. (1986a). Roles of D_1 and D_2 dopamine receptor subtypes in mediating the methamphetamine-induced changes in monoamine systems. *J. Pharmacol. Exp. Ther.* **238**, 932–937.

Sonsalla, P. K., Gibb, J. W., and Hanson, G. R. (1986b). Nigrostriatal dopamine actions on the D_2 receptors mediate methamphetamine effects on the striatonigral substance P system. *Neuropharmacology* **25**, 1221–1230.

Sonsalla, P. K., Nicklas, W. J., and Heikkila, R. E. (1989). Roles for excitatory amino acids in methamphetamine-induced nigrostriatal dopaminergic toxicity. *Science (Washington, D.C.)* **243**, 398–400.

Steele, T. D., Brewster, W. K., Johnson, M. P., Nichols, D. E, and Yim, G. K. W. (1991). Assessment of the role of α-methylepinine in the neurotoxicity of MDMA. *Pharmacol. Biochem. Behav.* **38**, 345–351.

Steinbusch, H. W. M. (1984). Serotonin-immunoreactive neurons and their projections in the CNS. In "Handbook of Chemical Neuroanatomy" (A. Björkfelt, T. Hökfelt, and M. J. Kuhar, eds.), Vol. 3, Part II, pp. 68–125. Elsevier Science, Amsterdam.

Steranka, L. R. (1982). Long-term decreases in striatal dopamine, 3,4-dihydroxyphenylacetic acid, and homovanillic acid after a single injection of amphetamine in iprindole-treated rats: Time course and time-dependent interactions with amfonelic acid. *Brain Res.* **234**, 123–136.

Stone, D. M., Stahl, D. M., Hanson, G. R., and Gibb, J. W. (1986). The effects of 3,4-methylenedioxymethamphetamine (MDMA) and 3,4-methylenedioxyamphetamine (MDA) on monoaminergic systems in the rat brain. *Eur. J. Pharmacol.* **128**, 41–48.

Stone, D. M., Hanson, G. R., and Gibb, J. W. (1987a). GABA transaminase inhibitor protects against methylenedioxymethamphetamine (MDMA)-induced neurotoxicity. *Soc. Neurosci. Abstr.* **13**, 904.

Stone, D. M., Hanson, G. R., and Gibb, J. W. (1987b). Differences in the central serotonergic effects of methylenedioxymethamphetamine (MDMA) in mice and rats. *Neuropharmacology* **26**, 1657–1661.

Stone, D. M., Johnson, M., Hanson, G. R., and Gibb, J. W. (1987c). A comparison of the neurotoxic potential of methylenedioxyamphetamine (MDA) and its N-methylated and N-ethylated derivatives. *Eur. J. Pharmacol.* **134**, 245–248.

Stone, D. M., Merchant, K. M., Hanson, G. R., and Gibb, J. W. (1987d). Immediate and long-term effects of 3,4-methylenedioxymethamphetamine on serotonin pathways in brain of rat. *Neuropharmacology* **26**, 1677–1683.

Stone, D. M., Merchant, K. M., Hanson, G. R., and Gibb, J. W. (1988). Role of endogenous dopamine in the central serotonergic deficits induced by 3,4-methylenedioxymethamphetamine. *J. Pharmacol. Exp. Ther.* **247**, 79–87.

Stone, D. M., Hanson, G. R., and Gibb, J. W. (1989a). In vitro reactivation of rat cortical tryptophan hydroxylase following in vivo inactivation by methylenedioxymethamphetamine. *J. Neurochem.* **53,** 572–581.

Stone, D. M., Johnson, M., Hanson, G. R., and Gibb, J. W. (1989b). Acute inactivation of tryptophan hydroxylase by amphetamine analogs involves the oxidation of sulfhydryl sites. *Eur. J. Pharmacol.* **172,** 93–97.

Wagner, G. C., Seiden, L. S., and Schuster, C. R. (1979). Methamphetamine induced changes in brain catecholamines in rats and guinea pigs. *Drug Alcohol Dep.* **4,** 135–138.

Wagner, G. C., Ricaurte, G. A., Seiden, L. S., Schuster, C. R., Miller, R. J., and Westley, J. (1980). Long-lasting depletions of striatal dopamine and loss of dopamine uptake sites following repeated aministration of methamphetamine. *Brain Res.* **181,** 151–160.

Waldmeier, P. C. (1985). Displacement of striatal 5-hydroxytryptamine by dopamine released from endogenous stores. *J. Pharm. Pharmacol.* **37,** 58–60.

Wood, P. L. (1982). Actions of GABAergic agents on dopamine metabolism in the nigrostriatal pathway of the rat. *J. Pharmacol. Exp. Ther.* **222,** 674–679.

Woolverton, W. L., Ricaurte, G. A., Forno, L. S., and Seiden, L. S. (1989). Long-term effects of chronic methamphetamine administration in rhesus monkeys. *Brain Res.* **486,** 73–78.

Yamamoto, B. K., and Davy, S. (1992). Dopaminergic modulation of glutamate release in striatum as measured by microdialysis. *J. Neurochem.* **58,** 1736–1742.

Zhao, Z., Castagnoli, N., Jr., Ricaurte, G. A., Steele, T., and Martello, M. (1992). Synthesis and neurotoxicological evaluation of putative metabolites of the serotonergic neurotoxin 2-(methylamino)-1-[3,4-(methylenedioxy)phenyl]propane [(methylenedioxy)methamphetamine]. *Chem. Res. Toxicol.* **5,** 89–94.

George A. Ricaurte
Karen E. Sabol
Lewis S. Seiden

10
Functional Consequences of Neurotoxic Amphetamine Exposure

I. INTRODUCTION

The neurotoxic effects of amphetamine analogs on central dopamine and serotonin neurons are now well documented. In other chapters, Gibb and colleagues (Chapter 9) summarized the neurochemical evidence and Molliver *et al.* (Chapter 11) detailed the neuroanatomic data. Despite these compelling lines of evidence that amphetamine analogs can damage brain dopamine and serotonin neurons, few behavioral correlates of amphetamine neurotoxicity have been identified, perhaps because the primary research emphasis to date has been on the neurochemical, morphological, and acute behavioral effects of toxic amphetamine analogs. However, the scarcity of known functional correlates of amphetamine neurotoxicity may relate to the fact that few such correlates exist, or, if they do, that they are subtle and difficult to detect using standard procedures. This may be especially true for serotonin, for which the type of functions that are influenced (e.g., mood, anxiety states, aggression, impulse control, perception of pain) is difficult to assess in animals. In this chapter, we address the behavioral consequences of amphetamine neurotoxicity and discuss both positive and pertinent negative findings. We also discuss possible factors that may underlie the lack of known long-lasting behavioral changes following toxic amphetamine exposure. Since a search for functional correlates of amphetamine neurotoxicity requires a working knowledge of the functional role of dopamine and serotonin in the mammalian central nervous system (CNS), we begin with a brief discussion of these topics.

II. FUNCTIONAL ROLE OF BRAIN DOPAMINE AND SEROTONIN

A. Dopamine

A large body of biochemical, pharmacological, and electrophysiological evidence indicates that dopamine in the corpus striatum (i.e., caudate nucleus, putamen, and globus pallidus) plays a key role in extrapyramidal motor function (Hornykiewicz, 1966; Moore and Bloom, 1978). These data, coupled with more recent observations with the dopamine neurotoxin 1-methyl-4-phenyltetrahydropyridine (MPTP) (Langston et al., 1984; Bergman et al., 1990), clearly establish that dopamine in the corpus striatum is important for the proper initiation and control of movement. Clinical and preclinical observations demonstrate that interference with nigrostriatal dopamine function, if sufficiently severe, can lead to the development of parkinsonism, an extrapyramidal movement disorder characterized by slow movement, muscular rigidity, rest tremor, and postural instability. For a parkinsonian syndrome to develop in humans, nigrostriatal dopamine must be depleted by an estimated 80% or more (Koller, 1992). Interestingly, in animals (nonhuman primates) an even larger striatal dopamine deficit may be necessary (Schneider et al., 1992).

In addition to its role in extrapyramidal motor function, brain dopamine, specifically dopamine in the limbic forebrain, has been implicated in the control of mood and affect in humans (Roth et al., 1987) and locomotion in rodents (Creese and Iversen, 1975). In the hypothalamic tuberoinfundibular system, dopamine is known to be the factor that inhibits prolactin secretion (Fuxe and Hokfelt, 1969); hypothalamic dopamine may influence the secretion of other anterior pituitary hormones as well (see Malven, 1993). Additionally, brain dopamine neurons mediate many of the effects of psychostimulant drugs (Creese and Iversen, 1972, 1975), including their effects on locomotion, food intake, and schedule-controlled behavior (see Seiden et al., 1993, for review). In the nucleus accumbens, dopamine is involved in drug reward mechanisms (Wise, 1987; Koob and Bloom, 1988).

The major clinical neurological disorder with which dopamine is associated is Parkinson's disease. Dopamine dysfunction is thought to be an important feature of schizophrenia, but the nature of the dysfunction is not known. However, since symptoms of schizophrenia often are treated effectively with dopamine receptor blockers, mesolimbic dopamine overactivity has been hypothesized to be central to the pathogenesis of this disorder (Losonczy et al., 1987). Alterations in dopamine function also have been postulated in narcolepsy, a sleep disorder characterized by excessive daytime sleepiness, cataplexy, and sleep paralysis (Mefford et al., 1983). As in schizophrenia, the exact nature of the dopamine dysfunction in narcolepsy remains to be defined.

B. Serotonin

Compared with the neurobiology of dopamine, the biobehavioral role of serotonin is less well defined, perhaps because the functions in which serotonin has been implicated (e.g., mood regulation, impulse control) are studied less easily in animals. On the basis of anatomic and electrophysiological data, Jacobs and Azmitia (1992) and others (Moore et al., 1978) proposed that an important role of brain serotonin may be modulating the effects of other neurotransmitter systems. Although the precise role of serotonin in the CNS is not entirely clear, considerable preclinical (see Seiden and Dykstra, 1977) and clinical (see Coccaro and Murphy, 1990; Siever et al., 1991) evidence suggests that serotonin is involved directly or indirectly in mood, sleep, anxiety states, food intake, sexual behavior, pain sensitivity, cognition, aggression, and neuroendocrine function. Major neuropsychiatric disorders that have a prominent serotonergic component include depression (Meltzer and Lowy, 1987), generalized anxiety, panic disorder (Charney, et al., 1990), obsessive compulsive disorder (Insel et al., 1990), and certain personality disorders characterized by impulsivity (Coccaro et al., 1989).

III. NEUROTOXIC AMPHETAMINE EXPOSURE IN ANIMALS

In this section, we focus on behavioral findings in animals previously treated with one of four amphetamine derivatives with known neurotoxic activity: p-chloroamphetamine (PCA), methamphetamine, 3,4-methylenedioxymethamphetamine (MDMA), and fenfluramine. These amphetamine analogs are emphasized because they have been studied most extensively from the standpoint of probing for possible lasting functional consequences. The neurotoxic profile of each of these analogs is not identical: PCA, MDMA, and fenfluramine, when tested in moderate doses, tend to damage brain serotonin neurons selectively, whereas methamphetamine is toxic to both dopamine and serotonin neurons (see subsequent discussion). For the sake of clarity, positive findings with each of these analogs are considered first; pertinent negative findings are emphasized in the ensuing section to highlight areas in which a lack of consensus is noted. With the exception of fenfluramine, these analogs are discussed in the order in which their neurotoxic effects were discovered and their behavioral effects tested.

A. Positive Findings

1. p-Chloroamphetamine

Sanders-Bush and colleagues (1972) were the first to note that PCA, a halogenated amphetamine derivative, selectively damaged brain serotonin axons and axon terminals. Soon after this neurotoxic effect was confirmed and extended by others (Harvey et al., 1975), Sheard and Davis (1976) found that rats treated with PCA initially (2 hr after drug) displayed decreased shock-elicited aggression but that, 4 wk later, they exhibited increased aggression. These authors recognized that, although the initial behavioral effects of PCA may have been pharmacological (i.e., related to the continued presence of drug or a metabolite), the long-term effects could be related to the toxic action of PCA on serotonin neurons. Gianutsus and Lal (1975) also noted increased aggression in PCA-treated animals, but did not observe the animals longer than 3 days after PCA treatment and, therefore, could not distinguish between a pharmacological and a neurotoxic action of PCA. Moreover, these investigators did not confirm the serotonin-depleting effectiveness of the PCA regimen they employed in mice [20 mg/kg, intraperitoneally (ip)], so ascertaining whether the increased aggression observed was, in fact, related to brain serotonin depletion is difficult.

Studying an acoustic startle response, Davis and Sheard (1976) observed that, within the first 2 hr of PCA treatment, rats exhibited a decreased startle response but that, 4 wk later, when brain serotonin was depleted by 52%, the startle response was augmented. The time course of these changes paralleled the changes in aggressive behavior described earlier (Sheard and Davis, 1976).

Another behavioral change noted in PCA-treated animals was reduced responsiveness to morphine analgesia. Mice treated with PCA (5 mg/kg, ip) showed decreased sensitivity to morphine analgesia 24 hr after PCA treatment (Takemori et al., 1975; Tulunay, 1976). Unfortunately, the duration of this change was not investigated, nor was the level of brain serotonin at the time of morphine administration measured to insure that the altered responsiveness to morphine analgesia was related to a depletion of brain serotonin. Since the dose of PCA used was only 5 mg/kg, and since the experiment employed mice (which tend to be resistant to PCA neurotoxicity), this consideration may be important.

Other functional changes noted in PCA-treated animals include reduced open field activity (Vorhees *et al.*, 1975), decreased exploratory behavior in a novel environment (Kohler *et al.*, 1978), and impaired acquisition of a two-way active avoidance (Ogren *et al.*, 1975). All these changes were observed at least 2 wk after PCA exposure, and were shown to be associated with a depletion of brain serotonin. Notably, however, some of these changes have proved difficult to reproduce in other laboratories (see subsequent discussion).

2. Methamphetamine

Methamphetamine is an amphetamine derivative that has the potential to damage both brain dopamine and serotonin neurons (Hotchkiss and Gibb, 1980; Ricaurte *et al.*, 1980). Soon after the neurotoxicity of methamphetamine came to light (Kogan *et al.*, 1976; Seiden *et al.*, 1976), Fischman and Schuster (1977) observed that rhesus monkeys treated with repeated high doses of methamphetamine were less sensitive to the suppressant action of methamphetamine on lever pressing maintained by a particular schedule of reinforcement (Differential Reinforcement of Low Rate, 40 sec, DRL-40). Finnegan and colleagues (1982) extended this finding by demonstrating that monkeys given neurotoxic doses of methamphetamine had enduring alterations in sensitivity to two other drugs known to influence dopaminergic neurotransmission: apomorphine and haloperidol. These early studies were the first to demonstrate that pharmacological challenge was an effective technique for unmasking clinically silent depletions of caudate dopamine.

Because of the involvement of dopamine in motor function (see preceding discussion), Ando *et al.*, (1985) treated rhesus monkeys with neurotoxic doses of methamphetamine and examined their performance on a fine motor control task. These researchers found that, 4 wk later, baseline ability of the animals to perform the task was unaltered. However, when the same animals were challenged with drugs known to interact with dopamine systems, differences between methamphetamine-treated and control animals were unmasked. Specifically, methamphetamine-treated monkeys were less sensitive to the disruptive effects of methamphetamine and more sensitive to the disruptive effects of haloperidol. (Sensitivity to apomorphine was not altered in a consistent manner.) Subsequently, Ando and colleagues (1986) made similar observations using eye-tracking ability as a functional end point. As before, baseline performance on the eye-tracking task was not altered by prior methamphetamine treatment.

Walsh and Wagner (1992) demonstrated that a high-dose methamphetamine regimen resulted in an increased number of footslips on a balance beam task and an increased latency on an active avoidance task; no methamphetamine-induced deficits were observed on rotorod or passive avoidance performance. Rats in this experiment were trained before the methamphetamine regimen, and tested for performance deficits after lesioning. The balance beam deficit was observed up to 1 mo and the shock avoidance latency deficit up to 2 mo after methamphetamine treatment. When administered a fenfluramine challenge, the methamphetamine-treated rats were less affected on rotorod and active avoidance performance than were control rats receiving fenfluramine.

Another long-lasting functional change reported in methamphetamine-treated rats is a lasting impairment in the acquisition and performance of a reaction-time task (Baggot et al., 1992; Richards et al., 1993). For 3 mo post-regimen, rats were tested on their ability to acquire a reaction-time task. Control rats exhibited a time-dependent decrease in reaction time, whereas methamphetamine-treated rats failed to improve over the 3-mo test period.

Finally, Myers and Wagner (1992) presented a preliminary evidence that force lever performance is impaired in rats treated with a high-dose regimen of methamphetamine, and that the impairment is exacerbated by the cholinergic agent oxotremorine. Collectively, these reports indicate that pharmacological challenge is an effective means of unmasking lasting functional consequences of partial dopamine or serotonin depletions, particularly when sensitive behavioral measures are used to evaluate methamphetamine-treated animals.

3. Methylenedioxymethamphetamine

Like PCA, MDMA selectively damages brain serotonin neurons (Stone et al., 1986; Commins et al., 1987; Schmidt, 1987; Insel et al., 1989). Functional correlates of MDMA neurotoxicity have been sought in both animals and humans, since the neurotoxic dose of MDMA in nonhuman primates closely approaches that used by humans (Ricaurte et al., 1988). In MDMA-treated animals, only two lasting behavioral changes have been identified. In rat pups, Winslow and Insel (1992) found that MDMA caused a long-lasting deficit in the rate of isolation calling, a behavior critical for mother–infant affiliation. In adult rats, Nencini and colleagues (1988) observed that MDMA produced a lasting increase in responsiveness to morphine analgesia. This finding is seemingly at odds with the observation of reduced responsiveness to morphine analgesia in PCA-treated animals (Takemori et al., 1975; Tulunay, 1976). However, the different findings may stem from a number of procedural differences in the two studies, including use of different species, drugs, and times and methods of testing, as well as differences in degree of serotonin depletion.

To the best of our knowledge, the behavior of MDMA-treated monkeys has yet to be investigated systematically. However, in our laboratories, no gross behavioral changes have been observed in either squirrel or rhesus monkeys with large brain serotonin deficits.

In humans, evidence is currently insufficient to determine whether

MDMA produces serotonergic neurotoxicity. However, several neuropsychiatric disturbances have been noted in MDMA users (McCann and Ricaurte, 1991; Chapter 12), some of which could be related to impaired serotonin function. These disturbances and other functional alterations in human MDMA users are discussed in detail in Chapter 12. Although concluding that the behavioral changes noted in some MDMA users are related to serotonin dysfunction would be premature, intriguingly, some of the changes noted involve behavioral spheres in which serotonin has been implicated (mood, anxiety, and memory). As suggested by Seiden (1990), retrospective analysis of human clinical data may help identify changes in subtle behavioral functions such as those in which serotonin putatively is involved.

4. Fenfluramine

Fenfluramine is also selectively toxic to brain serotonin neurons in animals (Harvey and McMaster, 1975; Clineschmidt et al., 1976). Unlike MDMA, however, fenfluramine has an accepted medical use and is marketed as an appetite suppressant in the United States and Europe. Although fenfluramine neurotoxicity in humans remains to be established, long-term toxic effects of fenfluramine on central serotonin neurons in animals are well documented (Schuster et al., 1986; Kleven et al., 1988; Ricaurte et al., 1991).

A clear and striking functional change that occurs in animals as a consequence of chronic fenfluramine exposure is the development of tolerance to the anorectic action of the drug (Kleven et al., 1988). In rats, this form of tolerance does not dissipate for as long as 8 wk after discontinuing fenfluramine, suggesting that the effect is related to a lasting depletion of brain serotonin. This notion is consistent with data demonstrating that animals treated with 5,7-dihydroxytryptamine (5,7-DHT), another documented serotonin neurotoxin (Baumgarten et al., 1978), are also tolerant to the anorectic action of fenfluramine (Fuxe et al., 1975).

An important but as yet unanswered question is whether *persistent* tolerance develops toward the anorectic action of fenfluramine in humans. Unfortunately, clinical data bearing on the issue of tolerance are not always consistent. Rowland and Carlton (1986) suggest that true tolerance to the anorectic action of fenfluramine may not develop. However, Woodward (1970), in reviewing the clinical experience with fenfluramine in the United States, observed that when fenfluramine was given chronically, its dose needed to be increased every 2–3 mo to maintain anorectic efficacy. These observations, suggestive of tolerance, are in accord with those of Lewis and colleagues (1971), who noted dissipation of the effects of fenfluramine with repeated dosing, and with the conclusion by Pinder et al. (1975) that tolerance to the anorectic action of fenfluramine does develop. If, as these reports suggest, tolerance to the anorectic action of fenfluramine develops in humans, one question remains. Is tolerance to the anorectic action of fenfluramine persistent or transient? If persistent, humans, like animals, may incur long-lasting neurotoxic effects of fenfluramine.

In humans, two adverse effects of fenfluramine that have been noted to last beyond the period of drug administration are depression (Oswald et al., 1971; Harding, 1972; Steel and Briggs, 1972) and worsening of psychosis (Marshall et al., 1989). Depression was most profound 4 days after discontinuation of fenfluramine and dissipated thereafter. Aggravated psychosis was still apparent 4 wk after discontinuation of fenfluramine, but resolved 6–18 mo later.

Schecter (1990) reported that rats treated with 6.25 mg/kg fenfluramine twice a day for 4 days show increased ability to discriminate a fenfluramine cue when tested 2 wk later. Since the aforementioned regimen of fenfluramine is known to produce a large depletion of brain serotonin (Kleven et al., 1988), Schecter postulated that this lasting functional effect (improved discrimination) is related to supersensitivity of serotonin receptors.

B. Negative Findings

In this section, we direct the reader to studies that either failed to replicate an effect reported in one of the aforementioned studies or probed for a given behavioral deficit but did not find one. In some instances, we draw attention to studies that have yielded positive findings in the past but have not been confirmed. These studies are mentioned only to highlight research approaches that previous results suggest may hold promise.

1. p-Chloroamphetamine

To the best of our knowledge, only Sheard and Davis (1976) and Gianutsus and Lal (1975) have documented alterations in aggressive behavior in animals previously depleted of brain serotonin with PCA (i.e., no further reports of similar findings have been published by other investigators). Similarly, the persistent enhancement of the startle response in PCA-treated rats (Davis and Sheard, 1976) has not been reported by other laboratories. As the search for functional correlates of partial brain serotonin depletion continues, these paradigms deserve further study.

The reduction in responsiveness to morphine analgesia found in PCA-treated mice (Takemori et al., 1975; Tulunay, 1976) is consistent with the observation that, in rats with lesions of the midbrain raphe nuclei, the antinociceptive action of morphine is reduced persistently (Samanin et al., 1970). Findings in PCA-treated mice, however, are not in accord with the observation that MDMA-treated rats are more responsive to morphine analgesia (Nencini et al., 1988; but see subsequent discussion). As mentioned earlier, procedural differences may account for the different findings.

The lasting decrease in open field activity following PCA administration

(Vorhees et al., 1975) has not been observed consistently by others. Kohler et al. (1978) found reductions in open field activity but noted that the finding was dependent on the procedure that was used to measure activity. In his review, Lorens (1978) commented that he observed increased rather than decreased activity in rats given PCA 18 days previously. Finally, Messing found no change in the locomotor activity of rats treated with neurotoxic doses of PCA 2 wk previously. As emphasized by various investigators (Messing et al., 1976); Lorens, 1978), changes in locomotor activity may be influenced by the method of testing, the testing conditions, the time after PCA injection, and the dosage schedule used, as well as by effects of PCA on sensory processes (Davis and Sheard, 1976).

As noted previously, one of the more striking lasting behavioral deficits reported in PCA-treated rats is impaired acquisition of a two-way conditioned avoidance response (CAR) (Ogren et al., 1975, 1976; Kohler et al., 1978). However, Lorens (1978) reported that rats treated with PCA (10 mg/kg, ip) 18 days prior to testing had normal one-way CAR acquisition. Whether the different findings are related to differences in the parameters of the avoidance conditioning procedure (one- versus two-way), escape training and/or shock exposure prior to conditioning, or differences in drug treatment (10 versus 20 mg/kg PCA) has not been determined.

2. Methamphetamine

Seiden and co-workers (1993b) have carried out an extensive series of studies in which rats were treated with various doses of methamphetamine [12.5, 25, or 50 mg/kg, subcutaneously (sc)] twice daily (12 hr apart) for 4 days, then allowed a 2-wk drug-free period before undergoing a number of behavioral tests. Behavioral tests included measures of locomotor activity, food intake, open field behavior, one-way avoidance, two-way avoidance, swimming behavior, radial maze performance, morphine analgesia, home cage intrusion, and schedule-controlled behavior. In all these paradigms, the performance of methamphetamine-treated rats did not differ significantly from that of controls. However, note that in short-term studies (2 wk), monoamine depletions in methamphetamine-treated animals were smaller than anticipated (Seiden et al., 1993b).

3. Methylenedioxymethamphetamine

Several investigative teams have evaluated MDMA-treated animals for evidence of functional impairment and have obtained negative findings. Robinson and colleagues (1993), for example, tested the effects of MDMA neurotoxicity on several behavioral tasks known to be sensitive to neocortical and hippocampal damage. These investigators also attempted to unmask the behavioral effects of serotonergic deficits through cholinergic challenge. In a number of behavioral paradigms including a place-navigation learning set task, a skilled forelimb use task, and a stimulus discrimination task,

MDMA-treated animals performed as well as controls. In addition, MDMA-treated animals were not more sensitive to cholinergic challenge than saline-treated controls.

Ricaurte et al. (1993) also evaluated MDMA-treated rats for possible memory impairment by testing their performance in a T-maze alternation task. In agreement with the findings of Robinson and colleagues (1993), MDMA-treated rats were found to perform normally on the task. The lack of impairment did not appear to be related to insufficiently large regional brain serotonin depletions.

Seiden and colleagues (1993b) evaluated MDMA-treated rats with the same test battery used to evaluate methamphetamine-treated rats (see previous discussion). MDMA-treated rats performed normally in all tasks, including the task that assessed their sensitivity to morphine analgesia. The latter finding is at odds with the previous report that MDMA-treated rats have increased sensitivity to morphine analgesia (Nencini et al., 1988). The basis for the different findings remains to be identified, but could be related to differences in serotonin depletion extent or time of testing.

Dornan et al. (1991) tested the effects of repeated MDMA administration on sexual behavior in male rats. Although MDMA produced a transient disruption of the expression of male copulatory behavior, its disruptive effects dissipated within 1 wk of cessation of MDMA administration.

4. Fenfluramine

The persistent tolerance to the anorectic action of fenfluramine reported by Kleven et al. (1988) has not been investigated further in other laboratories. However, this phenomenon is reminiscent of the enduring tolerance to the effects of methamphetamine observed in methamphetamine-treated rhesus monkeys (Fischman and Schuster, 1977; Ando et al., 1985, 1986). More studies on this effect of fenfluramine are needed in animals and in humans.

In general, the small number of side-effects that have been associated with fenfluramine use or withdrawal appear to dissipate within a few weeks or months. For example, depression after fenfluramine withdrawal is reported to be most acute 4 days after the drug is discontinued, and to disappear in the ensuing week (Oswald et al., 1971; Steel and Briggs, 1972). The same result appears to occur for polysomnographically characterized sleep disturbances documented in subjects taking fenfluramine (Lewis et al., 1971). Loss of libido, anxiety, drowsiness, and gastrointestinal disturbances are other side-effects of fenfluramine that either wear off within days of continued treatment or disappear once the drug is discontinued (Pinder et al., 1975).

Therefore, aside from persistent tolerance to the anorectic action of fenfluramine, few, if any, lasting behavioral effects of fenfluramine have been documented in animals or in humans. However, whether the paucity of findings is the result of their nonexistence or of the use of insufficiently sensitive or specific methods of testing is not clear.

Given the apparent absence of lasting untoward effects of fenfluramine, some individuals have maintained that fenfluramine neurotoxicity is unlikely to occur in humans (Derome-Trembley and Nathan, 1989). Although this assertion may be correct, the absence of systematic prospective studies specifically probing for evidence of serotonergic dysfunction in individuals previously treated with repeated high doses of fenfluramine should be recognized. Such studies are needed to insure that mood, sleep, and sexual disturbances, which often are observed in obese patients (regardless of whether they have been treated with fenfluramine), are in fact comorbid conditions rather than lasting untoward effects of fenfluramine.

IV. CONSIDERATIONS

Several factors may account for the relative absence of enduring functional changes identified to date in amphetamine-treated animals. The largest effects of the substituted amphetamines with respect to their neurotoxicity has been on dopamine- or serotonin-containing neurons. Judging from the role that these transmitters play in normal brain function and in the expression of drug effects, the result seems paradoxical that long-lasting or permanent depletions of the transmitters would not be expressed more robustly in behavior. Several factors should be considered before reaching the conclusion that the depletion of these transmitters is of little importance to brain function.

First, the size of the dopamine or serotonin depletion induced by amphetamine analogs may be insufficiently large to produce behavioral deficits. Lending credence to this view is the finding that, for a parkinsonian syndrome to become apparent in MPTP-treated primates, a near-total depletion of striatal dopamine is needed (Schneider *et al.*, 1992). Whether lesion size is as important in the case of serotonin neurons is not known.

Second, compensatory changes in the injured neuronal systems or in other systems may decrease the likelihood that partial depletions of dopamine or serotonin have a functional consequence. The development of increased dopamine metabolism (Agid *et al.*, 1973; Zigmund and Stricker, 1974) and increased dopamine receptor number (Creese *et al.*, 1977) following 6-hydroxydopamine (6-OHDA) lesions is well known. Increased dopamine metabolism after methamphetamine lesions also has been demonstrated (Ricaurte *et al.*, 1983). Also, redundancy exists in the CNS, so the destruction or compromise of a single system may not show any behavioral change until the redundant system is compromised as well.

Third, the absence of enduring behavioral deficits in amphetamine-treated animals may, at least in part, be related to regeneration of damaged axons and axon terminals. Since amphetamines typically damage axons and axon terminals but spare the nerve cell body, behavioral deficits may develop transiently but may disappear as axonal regeneration takes place. Thus, the timing of the behavioral testing may be a critical factor.

Fourth, as alluded to earlier, the method used to study behavior may not always be appropriate to or optimal for detecting the functional consequences of the transmitter depletion in question. For instance, in the case of dopamine depletion, clearly a method that assesses extrapyramidal motor function is optimal, whereas one that probes cerebellar or pyramidal motor function would not be suitable. In the case of serotonin neurons the issue is more complex, since the functional role of serotonin is less well defined and knowing whether the behavior being measured is under direct serotonergic control is difficult. Moreover, the behavioral paradigms that have been used most often are ones that can detect the effects of drugs useful in the treatment of specific diagnostic categories (e.g., antipsychotics, antidepressants); the effects of these drugs are often dependent on more than one neurotransmitter. Thus, the methods used may not be optimal for detecting the effects of depleting a single neurotransmitter.

Fifth, the development of functional consequences of amphetamine neurotoxicity may be influenced, at least in part, by the species being tested. To date, most studies of amphetamine neurotoxicity have been performed in rodents. Rodents may be less likely to express behavioral deficits that primates, as shown for MPTP (Chiueh et al., 1983; Langston, 1985).

Sixth, in human studies, the frequent occurrence of comorbid conditions (e.g., obesity with sleep disturbance or depression) complicates determining whether a particular functional change is part of a pre-existing or comorbid disorder, or is related to possible drug-induced neurotoxicity. For instance, in an obese patient treated with fenfluramine, is a coincident sleep disturbance related to obesity or to serotonin depletion induced by fenfluramine?

Finally, assuming an isomorphism between a transmitter system and behavior would be illogical. Evidence suggests that dopamine and serotonin systems play a role in normal brain function and drug-induced modifications of behavior. Therefore, before the admixture of positive and negative results summarized in this paper can be interpreted properly, we must await the application of systematic testing approaches in which variables such as dose, route of administration, number of injections, time of testing after drug exposure, and depletion size are explored and held constant systematically as needed.

V. SUMMARY AND CONCLUSIONS

Regardless of whether one considers data derived from studies with PCA, methamphetamine, MDMA, or fenfluramine, few long-lasting behavioral correlates of amphetamine neurotoxicity can be identified. In animals, one of the only changes that is observed consistently is the altered sensitivity to specific pharmacological challenge: methamphetamine-treated animals have altered sensitivity to dopaminergic agents; fenfluramine-treated animals have altered sensitivity to serotonergic agents. This apparent trend deserves further scrutiny, since pharmacological challenge may be an effective way to detect otherwise functionally silent brain dopamine or serotonin depletions. In humans, several neuropsychiatric sequelae have been associated with neurotoxic amphetamine use but, as discussed earlier, any causal relationship is unclear. Indeed, whether humans, like animals, develop neurotoxic changes and whether these are of functional consequence remains to be determined. Numerous factors may account for the apparent paucity of enduring behavioral changes following toxic amphetamine exposure. Lesion size, neuronal compensation, neuronal recovery, method of testing, functional end point selected, species studies, and intercurrent conditions all are factors that could influence the expression and/or detection of functional consequences of amphetamine neurotoxicity. Controlled studies designed to determine whether humans exposed to neurotoxic amphetamines develop lasting neuropsychiatric sequelae are needed. Such research will help define the public health consequences of toxic amphetamine exposure and could enhance our understanding of the role of dopamine and serotonin in human brain function.

ACKNOWLEDGMENTS

We thank U. McCann for reviewing the manuscript and for her many helpful suggestions. Preparation of this manuscript was supported in part by NIDA grants DA-06275, DA-05707, and DA-05938 to G. A. Ricaurte.

REFERENCES

Agid, Y., Javoy, F., and Glowinski, J. (1973). Hyperactivity of the remaining dopaminergic neurons after partial destruction of the nigrostriatal dopaminergic system in the rat. *Nature New Biol.* **245,** 150–151.

Ando, K., Johanson, C. E., Seiden, L. S., and Schuster, C. R. (1985). Sensitivity changes to dopaminergic agents in fine motor control of rhesus monkeys after repeated methamphetamine administration. *Pharmacol. Biochem. Behav.* **22,** 737–743.

Ando, K., Johanson, C. E., and Schuster, C. R. (1986). Effects of dopaminergic agents on eye tracking before and after repeated methamphetamine. *Pharmacol. Biochem. Behav.* **24,** 693–699.

Baggott, M. J., Richards, J. B., Sabol, K. E., and Seiden, L. S. (1992). Large doses regimen of methamphetamine produces long lasting deficits in the acquisition and performance of a reaction time task. *Soc. Neurosci.* **18**(1), 914.

Barranger, D., Marshall, B. D., Jr., Glynn, S. M., Midha, K. K., Hubbard, J. W., Bowen, L. L., Banzett, M. S. W., Mintz, J., and Liberman, R. P. (1989). Adverse effects of fenfluramine in treatment refractor schizophrenia. *J. Clin. Psychopharmacol.* **9**(2), 110–115.

Baumgarten, H. G., Klemm, H. P., and Lachenmayer, L. (1978). Mode and mechanism of action of neurotoxic indoleamines: A review and a progress report. *Ann. N.Y. Acad. Sci.* **305,** 1–7.

Bergman, H., Whichmann, F., and Delong, M. (1990). Reversal of experimental parkinsonism by lesions of the subthalamic nucleus. *Science* **249**, 1436–1438.

Charney, D. S., Woods, S. W., Krystal, J. H., and Heninger, G. R. (1990). Serotonin function and human anxiety disorders. *Ann. N.Y. Acad. Sci.* **600**, 558–573.

Chiueh, C. C., Markey, S. P., Burns, R. S., Johannessen, J., Jacobwitz, D. M., and Kopin, I. J. (1983). N-Methyl-4-phenyl-1,2,3,6-tetrahydropyridine, a parkinsonian syndrome causing agent in man and monkey, produces different effects in guinea pig and rat. *Pharmacologist* **25**, 131.

Clineschmidt, B. V., Totaro, J. A., McGuffin, J. C., and Pflueger, A. B. (1976). Fenfluramine: Long-term reduction in brain serotonin (5-hydroxytryptamine). *Eur. J. Pharmacol.* **35**, 211–214.

Coccaro, E. F., and Murphy, D. L. (1990). "Serotonin in Major Psychiatric Disorders," American Psychiatric Press, Washington, D.C.

Coccaro, E. F., Siever, L. J., Klar, H. M., Maurer, G., Cochrane, K., Cooper, T., Mohs, R., and Davis, K. L. (1989). Serotonergic studies in patients with affective and personality disorders. *Arch. Gen. Psychiatry* **46**, 587–599.

Commins, D. L., Vosmer, G., Virus, R. M., Woolverton, W. L., Schuster, C. R., and Seiden, L. S. (1987). Biochemical and histological evidence that methylenedioxymethamphetamine (MDMA) is toxic to neurons in the rat brain. *J. Pharmacol. Exp. Ther.* **242**(1), 338–345.

Creese, I., and Iversen, S. D. (1972). Amphetamine response in rats after dopamine neuron destruction. *Nature New Biol.* **238**, 247–248.

Creese, I., and Iversen, S. D. (1975). The pharmacological and anatomical substrates of amphetamine response in the rat. *Brain Res.* **83**, 419–436.

Creese, I., Burt, D. R., and Snyder, S. H. (1977). Dopamine receptor binding enhancement accompanies lesion-induced behavioral supersensitivity. *Science* **197**, 596–598.

Davis, M., and Sheard, M. H. (1976). p-Chloroamphetamine (PCA): Acute and chronic effects on habituation and sensitization of the acoustic startle response in rats. *Eur. J. Pharmacol.* **35**, 261–273.

Derome-Trembley, M., and Nathan, C. (1989). Fenfluramine studies. *Science* **243**, 991.

Dornan, W. A., Katz, J. L., and Ricaurte, G. A. (1991). The effects of repeated administration of MDMA on the expression of sexual behavior in the male rat. *Pharmacol. Biochem. Behav.* **39**, 813–816.

Finnegan, K. T., Ricaurte, G. A., Seiden, L. S., and Schuster, C. R. (1982). Altered sensitivity to d-methylamphetamine, apomorphine, and haloperidol in rhesus monkeys depleted of caudate dopamine by repeated administration of d-methylamphetamine. *Psychopharmacology* **77**, 43–52.

Fischman, M. W., and Schuster, C. R. (1977). Long-term behavioral changes in the rhesus monkey after multiple daily injections of d-methylamphetamine. *J. Pharmacol. Exp. Ther.* **201**, 593–605.

Fuxe, K., and Hokfelt, T. (1969). Catecholamines in the hypothalamus and the pituitary gland. *In* "Frontiers in Neuroendocrinology" (W. F. Ganong and L. Martini, eds.), pp. 47–96. Oxford University Press, New York.

Fuxe, K., Hamberger, B., Farnebo, L. O., and Ogren, S. O. (1975). On the *in vivo* and *in vitro* actions of fenfluramine and its derivatives on central monoamine neurons, especially 5-hydroxytryptamine neurons, and their relation to the anorectic activity of fenfluramine. *Postgrad. Med. J.* **51**, 35–45.

Gianutsos, G., and Lal, H. (1975). Aggression in mice after p-Chloroamphetamine. *Res. Commun. Chem. Pathol. Pharmacol.* **10**(2), 379–382.

Harvey, J. A., and McMaster, S. E. (1975). Fenfluramine: Evidence for a neurotoxic action on midbrain and a long-term depletion of serotonin. *Psychopharmacol. Commun.* **1**, 217–228.

Harvey, J. A., McMaster, S. E., and Yunger, L. M. (1975). p-Chloroamphetamine: Selective neurotoxic action in brain. *Science* **187**, 841–843.

Hornykiewicz, O. (1966). Dopamine (3-hydroxytyramine) and brain function. *Pharmacol. Rev.* **18**(2), 925–964.

Hotchkiss, A. J., and Gibb, J. W. (1980). Long-term effects of multiple doses of methamphetamine on tryptophan hydroxylase and tyrosine hydroxylase activity in rat brain. *J. Pharmacol. Exp. Ther.* **214**(2), 257–262.

Insel, T. R., Battaglia, G., Johannessen, J. N., Marra, S., and De Souza, E. B. (1989). 3,4-Methylenedioxymethamphetamine ("Ecstasy") selectively destroys brain serotonin terminals in rhesus monkeys. *J. Pharmacol. Exp. Ther.* **249**(3), 713–720.

Insel, T. R., Zohar, J., Benkelfat, C., and Murphy, D. L. (1990). Serotonin in obsessions, compulsions, and the control of aggressive impulses. *Ann. N.Y. Acad. Sci.* **600**, 574–583.

Jacobs, B. L., and Azmitia, E. C. (1992). Structure and function of the brain serotonin system. *Physiol. Rev.* **72**(1), 165–229.

Kleven, M. S., Schuster, C. R., and Seiden, L. S. (1988). Effect of depletion of brain serotonin by repeated fenfluramine on neurochemical and anorectic effects of acute fenfluramine. *J. Pharmacol. Exp. Ther.* **246**, 822–828.

Kogan, F. J., Nichols, W. K., and Gibb, J. W. (1976). Influence of methamphetamine on nigral and striatal tyrosine hydroxylase activity and on striatal dopamine levels. *Eur. J. Pharmacol.* **36**, 363–371.

Kohler, C., Ross, S., Srebo, S., and Ogren, S. (1978). Long-term biochemical and behavioral effects of *p*-chloroamphetamine in the rat. *Ann. N.Y. Acad. Sci.* **305**, 628–663.

Koller, W. C. (1992). When does Parkinson's disease begin? *Neurology (Suppl. 4)* **42**, 27–31.

Koob, G. F., and Bloom, F. E. (1988). Cellular and molecular mechanisms of drug dependence. *Science* **242**, 715–723.

Langston, J. W. (1985). MPTP and Parkinson's disease. *Trends Neurosci.* **8**, 79–83.

Langston, J. W., Langston, E. B., and Irwin, I. (1984). MPTP-induced parkinsonism in human non-human primates—Clinical and experimental aspects. *Acta Neurol. Scand. (Suppl. 100)* **70**, 49–54.

Lewis, S. A., Oswald, I., and Dunleavy, D. L. F. (1971). Chronic fenfluramine administration: Some cerebral effects. *Br. Med. J.* **3**, 67–70.

Lorens, S. A. (1978). Some behavioral effects of serotonin depletion depend on method: A comparison of 5,7-dihydroxytryptamine, *p*-chlorophenylalanine, *p*-chloroamphetamine, and electrolytic raphe lesions. *Ann. N.Y. Acad. Sci.* **305**, 532.

Losonczy, M., Davidson, M., and Davis, K. (1987). The dopamine hypothesis of schizophrenia. In "Psychopharmacology: The Third Generation of Progress" (H. Y. Meltzer, ed.), pp. 715–726. Raven Press, New York.

Malven, P. V. (1993). "Mammalian Neuroendocrinology." CRC Press, Boca Raton, Florida.

McCann, U. D., and Ricaurte, G. A. (1991). Lasting neuropsychiatric sequelae of (±)methylenedioxymethamphetamine ('Ecstasy') in recreational users. *J. Clin. Psychopharmacol.* **11**(6), 302–305.

Mefford, I., Baker, T., Boehme, R., Foutz, A., and Ciaranello, R. (1983). Narcolepsy: Biogenic amine deficits in an animal mode. *Science* **22**, 632–639.

Meltzer, H. Y., and Lowy, M. T. (1987). The serotonin hypothesis of depression. In "Psychopharmacology: The Third Generation of Progress" (H. Y. Meltzer, ed.), pp. 513–526. Raven Press, New York.

Messing, R. B., Phebus, L., Fisher, L. A., and Lytle, L. D. (1976). Effects of *p*-chloroamphetamine on locomotor activity and brain 5-hydroxyindoles. *Neuropharmacology* **15**, 157–163.

Messing, R., Pettibone, D., Kaufman, N., Lytle, L. (1978). Behavioral effects of serotonin neurotoxins. *Ann. N.Y. Acad. Sci.* **305**, 480–496.

Moore, R. Y., and Bloom, F. E. (1978). Central catecholamine neuron systems: Anatomy and physiology of the norepinephrine and epinephrine systems. *Annu. Rev. Neurosci.* **2**, 113–168.

Moore, R. Y., Halaris, A. E., and Jones, B. E. (1978). Serotonin neurons of the midbrain raphe: Ascending projections. *J. Comp. Neurol.* **180**, 417–438.

Myers, C. S., and Wagner, G. C. (1992). Force lever performance of rats with methamphetamine-induced dopamine lesions and pre-treated with vitamin E. *Soc. Neurosci.* **18**(2), 1070.

Nencini, P., Woolverton, W. L., and Seiden, L. S. (1988). Enhancement of morphine-induced analgesia after repeated injection of methylenedioxymethamphetamine. *Brain. Res.* **457**, 136–142.

Ogren, S. O., Ross, S. B., and Baumann, L. (1975). 5-Hydroxytryptamine and learning: Long-term effects of *p*-chloroamphetamine on acquisition. *Med. Biol.* **53**, 165–168.

Ogren, S. O., Kohler, C., Ross, S. B., and Srebro, B. (1976). 5-Hydroxytryptamine depletion and avoidance acquisition in the rat. Antagonism of the long-term effects of *p*-chloroamphetamine with a selective inhibitor of 5-hydroxytryptamine uptake. *Neurosci. Lett.* **3**, 341–347.

Oswald, I., Lewis, S. A., Dunleavy, D. L. F., Brezinova, V., and Briggs, M. (1971). Drugs of dependence though not of abuse: Fenfluramine and imipramine. *Br. Med. J.* **3**, 70–73.

Pinder, R. M., Brogden, R. N., Sawyer, P. R., Speight, T. M., and Avery, G. S. (1975). Fenfluramine: A review of its pharmacological properties and therapeutic efficacy in obesity. *Drugs* **10**, 241–323.

Ricaurte, G. A., Schuster, C. R., and Seiden, L. S. (1980). Long-term effects of repeated methylamphetamine administration on dopamine and serotonin neurons in the rat brain: A regional study. *Brain. Res.* **193**, 153–163.

Ricaurte, G. A., Seiden, L. S., and Schuster, C. R. (1983). Increased dopamine metabolism in the rat neostriatum after toxic doses of *d*-methylamphetamine. *Neuropharmacology* **22**(12A), 1383–1388.

Ricaurte, G. A., DeLanney, L. E., Irwin, I., and Langston, J. W. (1988). Toxic effects of 3,4-methylenedioxymethamphetamine on central serotonergic neurons in the primate: Importance of route and frequency of drug administration. *Brain Res.* **446**, 165–168.

Ricaurte, G., Markowska, A., Wenk, G., Hatzidinutriou, G., Wlos, J., Olton, D. (1993). MDMA, serotonin and memory. *J. Pharmacol. Exp. Ther.* **266**, 1097–1105.

Ricaurte, G. A., Molliver, M. E., Martello, M. B., Katz, J. L., Wilson, M. A., and Martello, A. L. (1991). Dexfenfluramine neurotoxicity in brains of non-human primates. *Lancet* **338**, 1487–1488.

Richards, J. B., Baggott, M. J., Sabol, K. E., and Seiden, L. S. (1993). A high dose methamphetamine regimen results in long-lasting deficits on performance of a reaction time task. *Brain Res.*, in press.

Robinson, T. E., Castaneda, E., and Whishaw, I. Q. (1993). Effects of cortical serotonin depletion induced by 3,4-methylenedioxymethamphetamine (MDMA) on behavior, before and after additional cholinergic blockade. *Neuropsychopharmacology* **8**(1), 77–85.

Roth, R., Wolf, M., and Deutch, A. (1987). Neurochemistry of midbrain dopamine systems. *In* "Psychopharmacology: The Third Generation of Progress" (H. Y. Meltzer, ed.), pp. 81–94. Raven Press, New York.

Rowland, N. E., and Carlton, J. (1986). Neurobiology of an anorectic drug: Fenfluramine. *Progr. Neurobiol.* **27**, 13–62.

Samanin, R., Gumulka, W., and Valzelli, L. (1970). Reduced effect of morphine in midbrain raphe lesioned rats. *Eur. J. Pharmacol.* **10**, 339–343.

Sanders-Bush, E., Bushing, J. A., and Sulser, F. (1972). Long-term effects of *p*-chloroamphetamine on tryptophan hydroxylase activity and on the levels of 5-hydroxytryptamine and 5-hydroxyindole acetic acid in brain. *Eur. J. Pharmacol.* **20**, 385–388.

Schechter, M. D. (1990). Functional consequences of fenfluramine neurotoxicity. *Pharmacol. Biochem. Behav.* **37**, 623–626.

Schmidt, C. J. (1987). Neurotoxicity of the psychedelic amphetamine, methylenedioxymethamphetamine. *J. Pharmacol. Exp. Ther.* **240**(1), 1–7.

Schneider, J., Pope, A., Simpson, K., Tagert, J., Smith, M. and DiStefano, L. (1992). Recovery from experimental parkinsonism with GM, ganglion treatment. *Science* **256**, 843–846.

Schuster, C. R., Lewis, M., and Seiden, L. S. (1986). Fenfluramine: Neurotoxicity. *Psychopharmacol. Bull.* **22**, 148–151.
Seiden, L. S. (1990). MOMA and related compounds. *Ann. N.Y. Acad. Sci.* **600**, 711–715.
Seiden, L. S., and Dykstra, L. A. (1977). "Psychopharmacology: A Biochemical and Behavioral Approach." Van Nostrand, New York.
Seiden, L. S., Fischman, M. W., and Schuster, C. R. (1976). Long-term methamphetamine induced changes in brain catecholamines in tolerant rhesus monkeys. *Drug Alcohol Dep.* **1**(3), 215–219.
Seiden, L. S., Sabol, K. E., and Ricaurte, G. A. (1993a). Amphetamine: Effects on catecholamine systems and behavior. *Annu. Rev. Pharmacol. Toxicol.* **33**, 639–677.
Seiden, L. S., Woolverton, W. L., Lorens, S. A., Williams, J. E., Corwin, R. L., Hata, N., Olimski, M. (1993b). Behavioral consequences of partial monoamine depletion in the CNS after methamphetamine-like drugs: The conflict between pharmacology and toxicology. *NIDA Res. Monogr.* **136**, 34–52.
Sheard, M. H., and Davis, M. (1976). *p*-Chloroamphetamine: Short and long term effects upon shock-elicited aggression. *Eur. J. Pharmacol.* **40**, 295–302.
Siever, L. J., Kahn, R. S., Lawlor, B. A., Trestman, R. L., Lawrence, T. L., and Coccaro, E. F. (1991). Critical issues in defining the role of serotonin in psychiatric disorders. *Pharmacol. Rev.* **43**(4), 509–525.
Steel, J. M., and Briggs, M. (1972). Withdrawal depression in obese patients after fenfluramine treatment. *Br. Med. J.* **3**, 26–27.
Stone, D. M., Stahl, D. C., Hanson, G. R., and Gibb, J. W. (1986). The effects of 3,4-methylenedioxymethamphetamine (MDMA) and 3,4-methylenedioxyamphetamine (MDA) on monoaminergic systems in the rat brain. *Eur. J. Pharmacol.* **128**, 41–48.
Takemori, A. E., Tulunay, F. C., and Yano, I. (1975). Differential effects on morphine analgesia and naloxone antagonism by biogenic amine modifiers. *Life Sci.* **17**, 21–28.
Tulunay, F. C., Yano, I., and Takemori, A. E. (1976). The effect of biogenic amine modifiers on morphine analgesia and its antagonism by naloxone. *Eur. J. Pharmacol.* **35**, 285–292.
Vorhees, C. V., Schaefer, G. J., and Barrett, R. J. (1975). *p*-Chloroamphetamine: Behavioral effects of reduced cerebral serotonin in rats. *Pharmacol. Biochem. Behav.* **3**, 279–284.
Walsh, S. L., and Wagner, G. C. (1992). Motor impairments after methamphetamine-induced neurotoxicity in the rat. *J. Pharmacol. Exp. Ther.* **263**, 617–622.
Winslow, J. T., and Insel, T. R. (1992). Serotonergic modulation of rat pup ultrasonic vocal development: Studies with 3,4-methylenedioxymethamphetamine (MDMA). *J. Pharmacol. Exp. Ther.* **254**, 212–220.
Wise, R. A. (1987). The role of reward pathways in the development of drug dependence. *Pharmacol. Ther.* **35**, 227–263.
Woodward, E., Jr. (1970). Clinical experience with fenfluramine in the United States. *In* "Amphetamine and Related Compounds" (E. Costa and S. Garattini, eds.), pp. 685–691. Raven Press, New York.
Zigmund, M. J., and Stricker, E. M. (1974). Ingestive behavior following damage to central dopamine neurons: Implications for homeostasis and recovery of function. *In* "Neuropsychopharmacology of Monoamines and Regulatory Enzymes" (E. Usdin, ed.), pp. 385–402. Raven Press, New York.

Karen J. Axt
Laura A. Mamounas
Mark E. Molliver

11

Structural Features of Amphetamine Neurotoxicity in the Brain

I. INTRODUCTION

Amphetamine derivatives, like other psychotropic drugs, have the potential to affect neuronal function adversely, especially when administered chronically or in large doses. Such undesired drug effects may be manifested by alterations in behavior, cognition, and mental function, and may result from abnormalities in biochemical, morphological, or physiological parameters. The potential for compensation or recovery from such changes in brain function is contingent on the nature of the morphological or functional changes that occur, as well as on whether a drug produces structural damage to neurons. Other chapters in this volume depict many of the adverse biochemical and behavioral effects associated with the use of amphetamine derivatives; this chapter discusses several structural abnormalities in the brain that have been observed following administration of selected amphetamine derivatives.

A. Neurotoxicity: Level of Analysis

Drug-induced alterations of brain structure have been analyzed by numerous anatomic methods and at different levels of resolution. The behavioral consequences of structural damage in the nervous system depend primarily on the specific systems that are affected. Thus, the functional deficits that are associated with morphologic abnormalities in the central nervous system (CNS) may be understood not in terms of neuronal morphology alone, but in the context of brain organization and the integrity of neuronal circuitry. A systematic assessment of neurotoxicity caused by amphetamine administration begins from the perspective of regional analysis of neuronal injury or changes. The initial, first-order evaluation of neurotoxicity should provide a broad overview of the spatial distribution of anatomic lesions, including identification of brain areas that are damaged and specification of neuronal systems that are affected (sensory, motor, etc.). The next stage of data acquisition is at the cellular (i.e., neuronal or glial) level, to detect cytopathology indicative of toxicological damage. The microscopic examination of brain sections may reveal selective vulnerability of particular types of neurons in terms of size (large or small), morphologic characteristics (e.g., pyramidal or stellate cells), or neurotransmitter. In addition to the identification of specific neuron classes that are vulnerable (or resistant) to amphetamine toxicity, determining whether damage occurs selectively within particular *cellular compartments,* for example, the cell soma, neuronal processes (axons or dendrites), or more restricted sites (e.g., dendritic spines or axonal boutons), is crucial. Nonneuronal cells also respond to neuronal injury, for example, astrocytes, microglia, perivascular cells, or capillary endothelial cells. A glial reaction may serve as an indicator of neuronal damage or may produce altered neuronal function itself. Alternatively, the possibility should be considered that glial cells may be activated or injured directly by neurotoxic agents. A perivascular inflammatory response (vasculitis) has been observed after amphetamine administration to humans or to nonhuman primates (Citron *et al.,* 1970; Margolis and Newton, 1971; Rumbaugh *et al.,* 1971a,b; Bostwick, 1981) and may result from ischemia of the blood vessel walls, causing a breakdown of the blood–brain barrier. Damage to small blood vessels or alterations of the blood–brain barrier may lead to secondary neuronal injury or dysfunction. Finally, subcellular changes in neuronal structure may occur after administration of different cytotoxic agents; structural alterations at the subcellar level may precipitate (or reflect) further damage to the nerve cell, including abnormal cytoskeletal components, mitochondria, vesicular components of the neuron, endoplasmic reticulum, or endosomes. In addition to extensive evaluation of regional manifestations of damage to serotonergic and dopaminergic neurons, anatomic studies of amphetamine neurotoxicity in the past have focused primarily on the cellular level, including analysis of damage to particular neuronal compartments. In comparison, examination of drug-induced changes in nonneuronal cells, particularly microglia, has been minimal.

B. Neuroanatomic Methods: Parameters of Neurotoxicity

A variety of histological methods is available to analyze structural parameters of neurotoxicity: neuronal injury, active degeneration, denervation, and reinnervation or axonal sprouting. Several variants of the classic Nissl stain permit the localization of ribosomes and granular (rough) endoplasmic reticulum (ER), especially where it forms stacks ("Nissl substance"). The loss of Nissl staining after axonal injury (chromatolysis) reflects a redistribution of rRNA and marked reorganization of the cellular machinery for protein synthesis. Thus, these stains provide information about the status of neuronal cell bodies, as well as about cytological changes such as chromatolysis, but they are nonselective with respect to cell type and of no use in evaluating axons or dendrites. In contrast, immunocytochemistry and fluorescence histochemistry (e.g., formaldehyde-induced histofluorescence) are used to identify specific neurochemical markers in tissue sections, and therefore have been extremely useful in revealing the specific neuron types or compartments that are affected, as well as revealing altered morphology of neuronal processes and changes in cytochemical and neurotransmitter levels. Specific neurotransmitter reuptake sites can be labeled with ligands that bind selectively to the monoamine recognition site on the uptake carrier. This labeling technique, which uses sensitive autoradiographic detection, has several advantages; it provides indices of the integrity of axonal membranes in addition to regional localization of axonal damage. The utilization of tracer dyes to assess axonal transport provides information regarding the functional and structural integrity of neuronal projections. Staining methods employing heavy metals such as silver have been used to identify degenerating axons, axon terminals, and neuronal cell bodies. These silver methods have varying success because of technical limitations; the low sensitivity for certain transmitter systems (especially for monoaminergic neurons, which have small-caliber unmyelinated axons), variations in reliability, and dependence on restricted survival times after injury can hamper the interpretation of results. However, silver methods are quite reliable for large diameter axons and have enjoyed a resurgence of interest for studies of neurotoxicity (Beltramino *et al.*, 1993). Finally, the responses of glial cells can be studied using Nissl stains, electron microscopic identification, and antisera to glial proteins. This chapter reviews the results, limitations, and validity of many of these morphological methods as they have been applied to questions of neurotoxicity caused by amphetamine analogs.

An important feature of neuronal responses to injury is the temporal evolution of structural and functional changes. Therefore, to characterize a neurotoxic drug response, an important consideration is assessing structural and biochemical alterations in the brain at a variety of survival times after drug administration. Particular cellular indices of neuronal injury are expressed at different time points, and certain responses may mask others. Since some methods provide valuable information only at appropriate time intervals after drug administration, the time course of neuronal alterations should be characterized. Neuronal responses to injury often include compensatory changes such as axon regeneration or collateral sprouting; these manifestations of recovery form an important component of the neurotoxic response and must be considered in the interpretation of structural changes. Further, regeneration of neuronal processes in the CNS may be either appropriate or anatomically aberrant, the latter resulting in dysfunctional neuronal circuitry. Since neurons may exhibit delayed responses to injury, the long-term life cycle of the neuron also must be considered as an aspect of neurotoxic degeneration. This chapter reviews structural features of amphetamine neurotoxicity, with special attention to the temporal component of morphological changes.

II. BASIC ANATOMY OF ASCENDING SEROTONERGIC AND DOPAMINERGIC PROJECTIONS

Amphetamine analogs most prominently affect serotonergic and dopaminergic projections in the brain. Therefore, before analyzing the anatomic correlates of amphetamine neurotoxicity, a brief overview of the anatomy of serotonergic and dopaminergic systems in forebrain is necessary. The regional distribution of serotonin (5-hydroxytryptamine, 5-HT) and dopamine (DA) axons, variations in axon density, axon morphology, and origins of projections are discussed. This introduction to dopamine and serotonin projections in forebrain summarizes anatomic data from the rat, unless otherwise noted. For detailed descriptions of the anatomy of dopaminergic and serotonergic systems, the reader is referred to several reviews (5-HT: Steinbusch, 1984; Törk, 1990; Jacobs and Azmitia, 1992; DA: Björklund and Lindvall, 1984; Lindvall and Björklund, 1984).

A. Serotonin

1. Distribution of 5-HT Axon Terminals in Forebrain

Employing the histofluorescence method, early studies of indoleamine projections revealed a broad serotonergic innervation in forebrain (Fuxe, 1965; Andén et al., 1966) originating from cells in circumscribed nuclei in brainstem (Dahlström and Fuxe, 1964). Serotonergic innervation of cerebral cortex was characterized as uniform and lacking in regional differentiation (Moore et al., 1978). Subsequently, sensitive immunocytochemical techniques revealed that, although highly divergent, the serotonergic innervation of forebrain exhibits an intricate organization with laminar and regional differences in terminal density (Lidov *et al.*, 1980; Steinbusch, 1981). The organization of 5-HT axons has been described in numerous laboratory animals, including rat (Köhler *et al.*, 1980; Lidov *et al.*, 1980; Steinbusch, 1981), cat (Mulligan and Török, 1988), monkey (Morrison *et al.*, 1982; Hornung *et al.*, 1990; Wilson and Molliver, 1991a), and ferret (Voigt and De Lima, 1991). Highly characteristic laminar patterns of 5-HT innervation are observed in rat olfactory bulb (McClean and Shipley, 1987), hippocampus (Köhler, 1982), and lateral entorhinal area (Köhler *et al.*, 1980); the variations in 5-HT axon density are related to the distribution of neurons in these regions. Neocortical areas with distinctive 5-HT innervation patterns include parietal cortex, in which a dense band of axons innervates layer V (Blue *et al.*, 1988), and primary visual cortex of the squirrel monkey, in which layer IVc receives a dense 5-HT innervation (Morrison *et al.*, 1982). In the marmoset, a relatively dense band of 5-HT axons is found in layer IV of several neocortical areas (Hornung *et al.*, 1990). In other areas of rat forebrain, the density of 5-HT axons varies considerably across subnuclear divisions such as amygdaloid complex (Steinbusch, 1984) and septum (Köhler *et al.*, 1982; Gall and Moore, 1984).

2. Morphology of 5-HT Axons

In multiple species, serotonergic axons in forebrain have been grouped into at least three distinct morphologic classes (Köhler *et al.*, 1981; Kosofsky and Molliver, 1987; Mulligan and Török, 1988; Hornung *et al.*, 1990). One class of 5-HT axon is rather thick in diameter, relatively straight, and exhibits few varicosities; these large-caliber 5-HT axons are thought to be preterminal axons (Mulligan and Török, 1988). A second class of 5-HT axon, designated as fine axons, is the most abundant 5-HT axon type in forebrain and is characterized by a thin caliber (~ 0.2 μm) and small (≤ 1 μm) fusiform or granular varicosities. Fine axons are distributed extensively throughout forebrain (cf. Mamounas *et al.*, 1991), with regional and laminar variations in axon density. A third 5-HT axon type possesses large spherical varicosities ($\sim 1-2.5$ μm), connected by thin axonal segments; these beaded axons have a highly restricted distribution, in contrast to the fine axons just described. In some areas of forebrain, beaded axons densely innervate specific layers, for example, in dentate gyrus, olfactory bulb, and lateral entorhinal area (Köhler *et al.*, 1980; Mamounas *et al.*, 1991), whereas few beaded axons are found in most other regions of forebrain. 5-HT axons have been reported to form specialized associations with individual neurons (pericellular arrays) in rat hippocampus and septum (Köhler, 1982; Köhler *et al.*, 1982; Gall and Moore, 1984). These pericellular arrays, or baskets, are formed exclusively by beaded 5-HT axons, as described in the cat by Mulligan and Török (1988) and later substantiated in rat and marmoset (Freund et al., 1990; Hornung et al., 1990; Axt and Molliver, 1991a). The neuronal targets of these pericellular baskets made by beaded 5-HT axons are primarily GABA-ergic interneurons (Tork et al., 1988; DeFilipe et al., 1991; Halasy et al., 1992; Hornung and Celio, 1992). The size of the varicosities that distinguish beaded axons depends on the region in which they are located, as does the axonal caliber and degree of convolution. For example, the beaded axons in olfactory bulb and lateral entorhinal area are more tortuous and have slightly thicker intervaricose segments and somewhat smaller varicosities, compared with those found in parietal cortex or area CA1 of hippocampus (cf. Mamounas et al., 1991). The distinct distributions of fine and beaded axons suggest that beaded axons innervate restricted, highly specific targets such as small clusters of neurons and their dendritic fields, whereas fine axons are more likely to contact large numbers and many types of neurons, producing a more divergent effect. These observations have led investigators to postulate that these major classes of 5-HT axons form functionally separate serotonergic systems within the brain (Kosofsky and Molliver, 1987; Mulligan and Törk, 1988).

3. Origins of 5-HT Projections

Ascending 5-HT projections in the CNS originate predominantly from two prominent raphe nuclei in midbrain and rostral pons: the dorsal raphe nucleus (DRN or B7) within the periaqueductal gray and contiguous caudally with the B6 group along the floor of the fourth ventricle, and the median raphe nucleus (MRN or B8) in the pontine tegmentum which lies ventral to the medial longitudinal fasciculus. The DRN includes the largest number of 5-HT neurons and contains several subdivisions. A smaller third group (B9) comprises 5-HT neurons located bilaterally in the ventral tegmentum, dorsal to the medial lemniscus. Several studies have shown that DRN and MRN projections have separate but overlapping terminal distributions (O'Hearn and Molliver, 1984; Imai et al., 1986; Kosofsky and Molliver, 1987; Wilson and Molliver 1991b). Evidence also exists for a topographic organization of neuronal projections arising from MRN and DRN. Distinct zones of the DRN have been found to project to differing areas of forebrain. For example, neurons in the lateral DRN project to lateral geniculate nucleus of thalamus (Pasquier and Villar, 1982), but not to neostriatum (Seinbusch et al., 1981). Neostriatum receives 5-HT afferents from the dorsomedial and ventromedial DRN (Steinbusch et al., 1981). The MRN exhibits topographic organization in its projections to limbic areas; for example, medium-sized cells, which lie along the midline were shown to project to entorhinal area, whereas smaller cells which are situated along the periphery of the nucleus project to hippocampus (Köhler, 1982).

The two major morphological classes of serotonergic axons (fine and beaded axons) have been associated with different cells of origin. In the rat, the two major 5-HT axon types have been demonstrated to originate typically from separate raphe nuclei. Anterograde transport of a lectin tracer

(PHA-L) from the DRN or MRN revealed that the majority of 5-HT axons in neocortex arising from the DRN are of the fine type, whereas the neocoritcal axons from the MRN are predominantly of the beaded type (Kosofsky and Molliver, 1987). As discussed in a later section, the selective neurotoxicity of amphetamine derivatives has been utilized to determine the cellular origin of fine and beaded 5-HT axons in several brain regions; briefly, fine axons in fronto-parietal cortex (Mamounas and Molliver, 1988) and main olfactory bulb (Molliver and Mamounas, 1991) originate in the DRN, and beaded axons in these regions originate from cells in the MRN. In summary, the DRN and MRN have distinct, yet overlapping, projections to forebrain, which differ with respect to axon morphology and terminal distribution. Because of the unique distributions of fine and beaded axons and probable differences in their neuronal targets, separate and distinct functions for DRN and MRN projections have been hypothesized.

B. Dopamine

1. Distribution of DA Axon Terminals in Forebrain

Dopaminergic innervation of rat forebrain is not as widespread as serotonergic innervation, but is restricted to a few specific terminal fields (Thierry et al., 1973; Fuxe et al., 1974; Lindvall et al., 1974). Only those regions of forebrain that have been studied in the context of amphetamine neurotoxicity are discussed here. The striatum receives the largest dopaminergic innervation; caudate–putamen (dorsal striatum) is the most densely innervated and best studied dopaminergic terminal field. DA axons are also dense in basal forebrain areas (ventral striatum), including olfactory tubercle, nucleus accumbens, and bed nucleus of stria terminalis. The nomenclature of "ventral striatum" and "dorsal striatum" is used by Björklund and Lindvall (1984) based on the formulation of ventral striatum by Heimer (1978). Several cortical and limbic regions receive moderate dopaminergic innervation; however, many areas of dorsolateral neocortex contain no significant DA input (e.g., primary visual cortex and lateral occipital cortex; Descarries et al., 1987). The DA projections to neocortex were not fully appreciated in early studies because of difficulties in differentiating DA from norepinephrine (NE) axons since catecholamine fluorescence does not easily distinguish these two neurotransmitters. As methods were developed further and were compared with biochemical data, evidence showed that DA projections extended beyond the neostriatum, olfactory tubercle, and hypophysis (Fuxe, 1964; Ungerstedt, 1971). In fact, researchers have demonstrated that, in certain neocortical areas, DA and NE axons innervate complementary layers (Lewis et al., 1979).

The main dopaminergic projection to neocortex is to frontal lobe. Within this region, three DA projections have been defined: the pregenual anteromedial, supragenual, and suprarhinal projections (cf., Lindvall and Björklund, 1984). Dopaminergic axons are found in circumscribed layers of anteromedial and supragenual cortical areas (Lindvall *et al.*, 1974; Berger *et al.*, 1976; Lewis *et al.*, 1979; Morrison *et al.*, 1979; Descarries *et al.*, 1987; Van Eden *et al.*, 1987); in anterior cingulate cortex (supragenual), dopaminergic axons are concentrated in layers II and III, whereas in anteromedial frontal cortex, DA axons are found in layers II through VI, with highest density in layers V and VI. DA axons of the suprarhinal projection preferentially terminate in deeper layers of cortex. Dopaminergic innervation of ventral limbic cortical areas (ventral entorhinal cortex and piriform cortex) is primarily to layers II and III, yet differs from the pattern of innervation in anterior cingulate cortex because the axons are arranged in distinct clusters (Lindvall *et al.*, 1974). In addition to projections to cerebral cortex are prominent DA inputs to other limbic areas such as the septal region and amygdala. In lateral septal nucleus, DA axons form pericellular arrays (Gall and Moore, 1984) similar to those formed by 5-HT axons in the same region (Gall and Moore, 1984; Mulligan and Török, 1988).

2. Origins of DA Projections

Dopaminergic cell bodies reside in a subset of the catecholaminergic nuclei described by Dahlström and Fuxe (1964) and designated by the letter A (adrenergic). The majority of DA axons in forebrain originate from two mesencephalic nuclei: the ventral tegmental area (VTA or A10) and substantia nigra (A9) (which also encompasses the retrorubral nucleus, A8). From these nuclei emerge two long ascending projections, the mesostriatal and mesocorticolimbic systems (cf. review by Björklund and Lindvall, 1984). Other dopaminergic cell groups (A11–A16 and retina) within the CNS have discrete and comparatively short projections. Since these latter cell groups have not been the focus of amphetamine neurotoxicity studies, this summary addresses only the mesostriatal and mesocorticolimbic systems. According to the nomenclature of Björklund and Lindvall (1984), the meso*striatal* system comprises DA projections to the following terminal fields: "dorsal striatum" (caudatoputamen) and "ventral striatum," which includes olfactory tubercle, nucleus accumbens, and bed nucleus of stria terminalis. To simplify the discussion, the term "striatum" will refer to caudatoputamen. Substantia nigra and VTA project heavily to striatum where the axons form an extremely dense terminal plexus. Despite the uniform appearance of DA innervation of striatum, the projections to this region have a complex anatomic organization (Fallon and Moore, 1978; Fallon and Loughlin, 1982). Dopaminergic innervation of dorsolateral striatum originates almost exclusively from substantia nigra (and A8). Different sectors of substantia nigra project to topographically and histochemically distinct regions of the striatum, the so-called patch and matrix divisions. DA neurons in dorsal pars compacta project to the matrix of striatum, whereas DA neurons located in the ventral aspect of the pars compacta and those in the pars reticulata project to sectors of striatum that correspond to the patches (Gerfen *et al.*, 1987a; Gerfen, 1992). Neurons in A8 project to the matrix.

Dopaminergic neurons of VTA (A10) project to ventro-medial striatum, and only to the matrix (Gerfen et al., 1987a). The mesostriatal DA projections to nucleus accumbens, bed nucleus of stria terminalis, and olfactory tubercle originate in the VTA.

The mesocorticolimbic systems comprise projections to frontal cortex, temporal cortex, and various limbic structures including amygdala and septum. The mesocortical DA projections have very circumscribed targets within anteromedial prefrontal cortex, anterior cingulate cortex, entorhinal cortex, and piriform cortex, as previously described. Excepting anterior cingulate cortex, these projections arise predominantly from neurons in VTA and, to a lesser degree, from neurons in medial substantia nigra. DA projections to anterior cingulate cortex arise from substantia nigra. Like the mesostriatal DA projections, mesocortical DA projections tend to be organized topographically (Fallon and Loughlin, 1982).

3. Morphology of DA Axons

Like 5-HT axons, DA axons in forebrain have differing morphologies which, for the most part, are associated with particular cells of origin. DA axons in the striatal matrix originating from neurons in the VTA, retrorubal nucleus (A8), and substantia nigra dorsal pars compacta have a smaller caliber and smaller varicosities (type A axons) than do the DA axons in the striatal patches or striosomes (type B axons) that arise from neurons in the ventral pars compacta and pars reticulata (Gerfen et al., 1987a). Type B DA axons are slightly thicker and have "frequent," somewhat larger varicosities. These projections are further differentiable biochemically: the substantia nigra neurons with smaller caliber type A axons contain the 28 kDa calcium binding protein, calbindin (Gerfen et al., 1987b). DA axons in cerebral cortex have been studied in less detail, although those in anterior cingulate cortex are described as having "a much finer appearance" than those in the anteromedial and suprarhinal systems, which have "clearly visible intervaricose segments and irregularly spaced varicosities" (Van Eden et al., 1987). The latter varicose axons originate in the VTA (A10), whereas the former finer axons arise from substantia nigra (A9). Consistent with this dichotomy, Lindvall et al. (1974) described the same two DA fiber types in cerebral cortex using different terms. In frontal and entorhinal cortex, axons have a "smooth appearance"; in anterior cingulate cortex, axons with "closely spaced varicosities" were observed.

III. TOXIC EFFECTS OF AMPHETAMINE DERIVATIVES: EVIDENCE FOR PERSISTENT AXON LOSS

Numerous studies have reported that several amphetamine derivatives exert acute biochemical effects in 5-HT and DA neurons, including release of monoamine; inhibition of metabolic enzymes tryptophan hydroxylase

(TPH), tyrosine hydroxylase (TH), and monoamine oxidase (MAO); inhibition of monoamine reuptake; and a resultant depletion of the monoamine neurotransmitters. DA and 5-HT neurons are affected differentially by analog of amphetamine, depending on the chemical structure; the parent compound, amphetamine (phenylisopropylamine), and the N-methylated derivative, methamphetamine, exert significant effects on dopaminergic neurons, whereas the halogen and methoxy ring-substituted derivatives have a preferential effect on serotonergic neurons. If administered in high or repeated doses, many, but not all, of the drugs in this class also cause a variety of long-lasting deficits in DA and 5-HT neurons, including decreases in monoamine levels, active reuptake of monoamine, and loss of TPH and TH activities. The incurred biochemical deficits usually last for months and, therefore, very likely reflect a toxic and degenerative effect of these drugs on serotonergic and dopaminergic axon terminals and/or neuronal somata.

The anatomic and histological experiments reviewed in this section complement the biochemical studies by providing morphologic evidence of denervation. In cases of drug-induced neurotoxicity, denervation (the persistent loss of axonal projections) most likely results from chemical axotomy or from damage to or death of neuronal somata. Several anatomic methods have been applied to discern monoamine axon loss. (1) Histofluorescence or immunocytochemical methods may be utilized to reveal changes in axon density, although these techniques rely on the presence of the neurotransmitter or other transmitter-related antigens. The immunocytochemical method also affords exquisite morphologic detail (i.e., classes of axons affected), as well as regional localization. (2) Autoradiographic analysis of the number of monoamine uptake sites within a specified brain region also provides a measure of axon density. (3) Finally, a method that exploits an axonal function has been utilized to test axon integrity: demonstration of axonal transport with retrograde tracers. Damage or death of cell bodies can occur by direct insult or by retrograde degeneration following axotomy. The viability of 5-HT and DA somata after administration of amphetamine derivatives also has been examined by several methods. (1) Histofluorescence or immunocytochemical staining of the cell bodies provides information on cell number and transmitter synthesizing capabilities. (2) Stains for Nissl substance reveal cytopathological changes, such as chromatolysis, as well as cell number. (3) Silver stains specific for neuronal degeneration, given appropriate survival times and careful interpretation, may demonstrate cell death in certain situations.

A. Regional Specificity of Axon Loss

1. Regional Distribution of 5-HT Axon Loss

Serotonergic neurons, although moderately sensitive to high doses of methamphetamine and even higher doses of amphetamine, are particularly vulnerable to ring-substituted amphetamines. Subsequent to biochemical studies, which reported decreases in 5-HT levels in rat brain after administration of *p*-chloroamphetamine (PCA) and fenfluramine (FEN) (Sanders-Bush *et al.*, 1975; Clineschmidt *et al.*, 1976), Lorez *et al.*, (1978) used the monoamine histofluorescence method to demonstrate a decreased indoleamine fluorescence in various brain regions after PCA and *p*-chloromethamphetamine (PCMA), including cingulate cortex, parietal cortex, septum, and hippocampus. Immunocytochemical studies using antisera to 5-HT have revealed marked decreases in 5-HT axon density in most regions of rat forebrain 2 wk after administration of several ring-substituted amphetamines: PCA (Mamounas and Molliver, 1988; Mamounas *et al.*, 1991), methylenedioxyamphetamine (MDA) and its *N*-methylated derivative methylenedioxymethamphetamine (MDMA) (O'Hearn *et al.*, 1988), and FEN (Appel *et al.*, 1989; Molliver and Molliver, 1990). The decreased density of 5-HT axons in neocortex persisted for at least 2 mo after PCA and MDA treatment (Mamounas *et al.*, 1991) and, in some brain regions, as long as 9 mo to 1 yr after PCA treatment (Molliver and Axt, 1990; Mamounas and Molliver, 1991).

Regional differences in the pattern of 5-HT axon loss and axon sparing have been described in the rat after administration of ring-substituted amphetamine derivatives; this pattern is consistent for all the drugs (Mamounas and Molliver, 1988; O'Hearn *et al.*, 1988; Appel *et al.*, 1989; Molliver and Molliver, 1990; Mamounas *et al.*, 1991), although the degree of axon loss depends on the drug administered. High doses of *d*-methamphetamine produce regional differences in axon loss similar to those observed with the ring-substituted amphetamines, although with this drug there is significant variability between individual animals with regard to the extent of denervation (Axt and Molliver, 1991b). Parietal and occipital cortex are quite denervated after treatment with amphetamine analogs, especially in deeper layers (IV–V). Supragranular layers of neocortex exhibit scattered sparing of axons (Mamounas *et al.*, 1991). The extent of axon sparing differs among regions of the hippocampal formation (Fig. 1; Mamounas *et al.*, 1991); area CA1 is the most denervated and CA3 exhibits the most sparing. In dentate gyrus, the molecular layer is more denervated than the borders of the granule cell layer (Fig. 1). Similarly, main olfactory bulb has a striking pattern of axon loss and sparing; 5-HT axons in the glomerular layer are resistant to PCA and MDA, whereas those in the infraglomerular layers are ablated (Mamounas *et al.*, 1991). The dense patch of 5-HT axons found in layer III of lateral entorhinal cortex (Köhler *et al.*, 1980) is also resistant to the neurotoxic effects of these drugs (Mamounas *et al.*, 1991), as are the pericullular arrays in neocortex, hippocampus, and septum (Axt and Molliver, 1991a).

In monkeys, regional variations in 5-HT axon loss also have been observed following MDMA and FEN treatment. Racemic MDMA produces a

CONTROL   MDA (20 mg/kg)

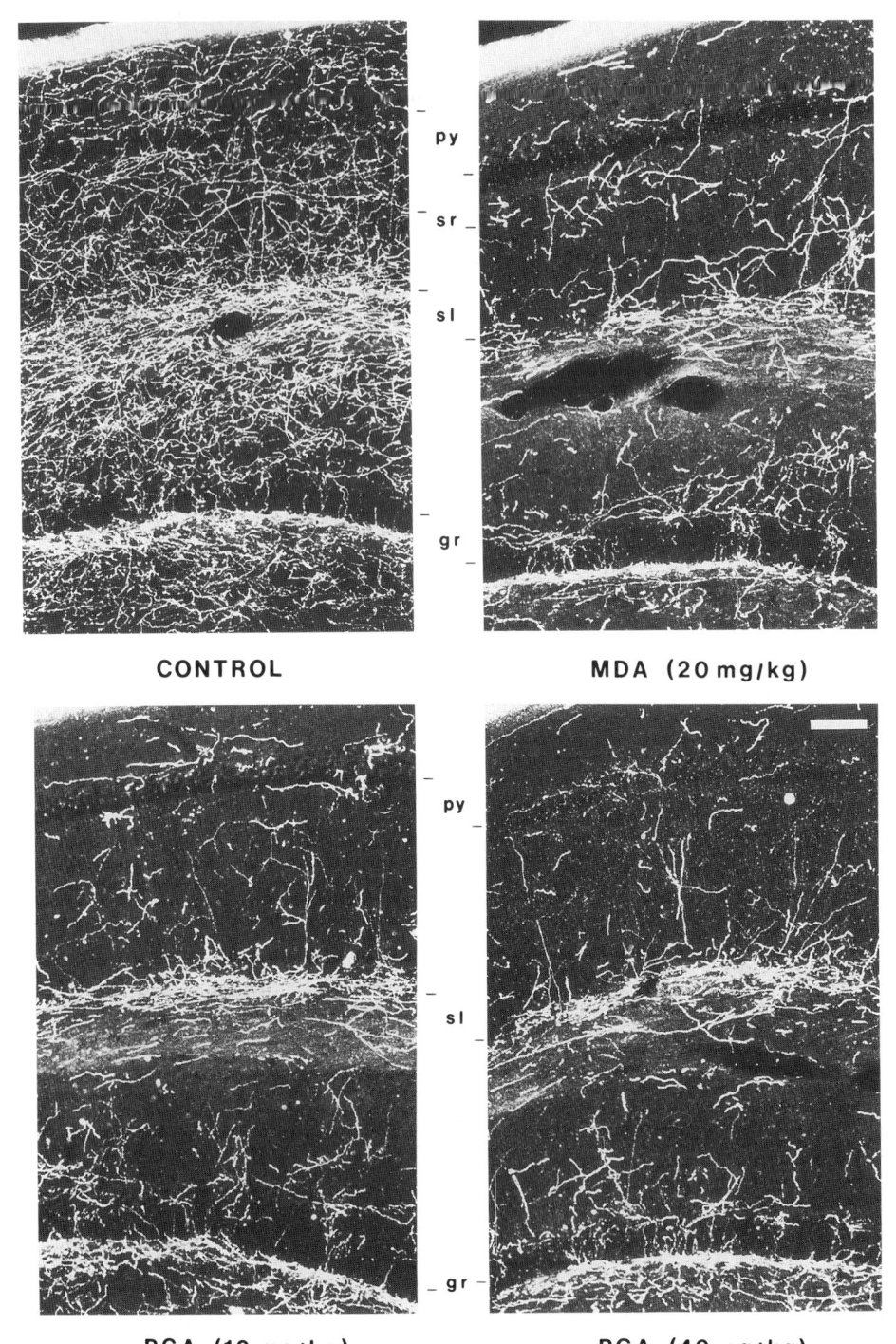

PCA (10 mg/kg)   PCA (40 mg/kg)

robust decrease in 5-HT immunoreactive axons in many cerebrocortical areas of the macaque (Wilson et al., 1989), while sparing axons in (for example) outer layers of somatosensory cortex (area 3b) and layer IV of primary visual cortex (area 17). *d*-Fenfluramine, administered to squirrel monkeys, causes a long-lasting (2 wk) loss of 5-HT-immunoreactive axons in cerebral cortex (Ricaurte et al., 1991b).

In studies of amphetamine neurotoxicity, immunohistochemical methods provide information beyond that available from biochemical assays of neurotransmitter levels. In addition to demonstrating the loss of neurotransmitter, immunohistochemical methods have the advantage of revealing the regional distribution of a neurotoxic effect within subnuclei or cortical layers. Moreover, the microscopic examination of tissue sections incubated in antisera against specific neuronal markers (e.g., 5-HT) provides high resolution anatomic data that can identify specific neurons and neuronal compartments that may be damaged. Limitations of immunocytochemistry include the qualitative nature of data analysis and a steep sensitivity curve for antigen detection (due to signal amplification), so moderate but widespread decreases in 5-HT levels within neurons or axons may not be readily evident (i.e., any signal over the detection threshold rapidly reaches maximal staining intensity). In addition to the level of neurotransmitter, however, the morphological integrity of the axons is an important parameter in assessing neurotoxicity. A particular advantage of immunocytochemistry is the ability to detect cytochemical changes that are highly localized to small areas and, thus, are unlikely to be appreciated by chemical assays of tissue homogenates. Sharply localized regional differences in axon density or in neuronal markers may be obvious in immunocytochemical preparations but not in biochemical studies, as exemplified by the effects of PCA in rat hypothalamus (Fig. 2). The distribution of 5-HT axon loss and sparing in hypothalamus after PCA treatment is similar to that observed after MDA (O'Hearn et al., 1988) and FEN administration (D. C. Molliver and M. E. Molliver, unpublished observations). The ventromedial, dorsomedial, and arcuate hypothalamic nuclei are relatively denervated 2 wk after PCA treat-

---

FIGURE 1 Innervation of CA1 and dentate gyrus of dorsal hippocampus by 5-HT-immunoreactive axons. Darkfield photomicrographs of parasagittal sections reveal a high density of fine axons in CA1 of saline-treated control rat and a prominent band of beaded axons in the hilus subjacent to the granule cell layer of dentate gyrus. Fine axons in CA1 are less frequently observed 2 wk after methylenedioxyamphetamine (MDA, 20 mg/kg × 8) or *p*-chloroamphetamine (PCA, 10–40 mg/kg × 2), but beaded axons along granule cell layer of dentate gyrus are spared, as are beaded and preterminal axons in stratum lacunosum of CA1. Note that MDA and PCA produce similar laminar patterns of denervation and sparing in dorsal hippocampus. Layers of CA1 are indicated as py, stratum pyramidale; sr, stratum radiatum; sl, stratum lacunosum. In dentate gyrus, gr, granule cell layer. Scale bar, 100 μm. Reprinted with permission from Mamounas et al. (1991). Copyright © 1991 by Wiley-Liss, a division of John Wiley & Sons, Inc.

PCA   Control

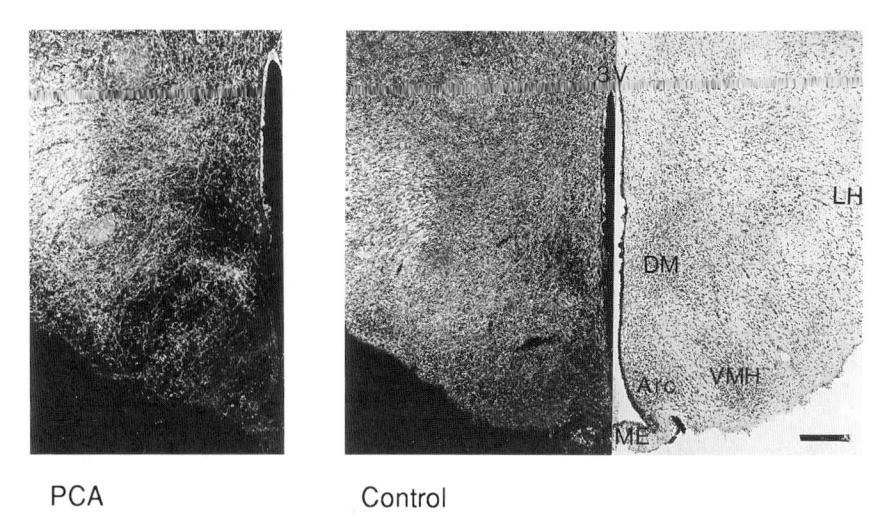

FIGURE 2 Darkfield photomicrographs of 5-HT-immunoreactive axons in rat hypothalamus. Serotonin innervation of control (saline-injected) hypothalamus is moderately dense in most subnuclei. The locations of several hypothalamic nuclei are indicated in the contralateral side of an adjacent Nissl-stained section from the same control rat. A notable decrease in 5-HT axon density is observed in dorsomedial hypothalamic nucleus (DM), ventromedial hypothalamic nucleus (VM), and arcuate hypothalamic nucleus (Arc) of the p-chloroamphetamine (PCA) treated rat (2-wk survival). In contrast, 5-HT immunoreactivity is increased in preterminal 5-HT axons of medial forebrain bundle which courses through the lateral hypothalamic nucleus (LH). 3V, Third ventricle; ME, median eminence. Scale bar, 300 μm.

ment (Fig. 2). 5-HT axon density also is reduced somewhat in periventricular nucleus and the dorsal hypothalamic area. In contrast, 5-HT axons in the lateral hypothalamus (LH) appear relatively resistant to drugs such as PCA and MDA; 5-HT immunoreactivity is more intense in LH (Fig. 2; O'Hearn *et al.*, 1988). One interpretation for the particularly dense staining of 5-HT axons in LH is that the medial forebrain bundle (MFB), a major ascending pathway that contains many preterminal 5-HT axons, courses through this region. After ablation of serotonergic terminal axons, the preterminal axons are likely to accumulate 5-HT proximal to the site of injury, as intra-axonal constituents (including 5-HT, its biosynthetic enzyme TPH, and other anterogradely transported molecules) dam up (cf., Massari *et al.*, 1978a). This striking difference between medial and lateral hypothalamic nuclei in the density of 5-HT-immunoreactive axons after PCA lesion underscores the importance of anatomic considerations in biochemical studies of the neurotoxic effects of drugs. For instance, the selection of appropriate anatomic landmarks for dissection is critical for interpretation of biochemical studies; if the hypothalamic tissue blocks used for chemical assays include LH, 5-HT levels might appear normal or artificially high because of accumulated 5-HT in dilated axons of passage. Since the surviving preterminal axons may accumulate high levels of neurotransmitter in cases of distal axotomy, the combined use of chemical assays to measure levels of transmitter and morphologic criteria to determine cellular localization is warranted for correct interpretation of drug effect.

2. 5-HT Cell Bodies

Although prolonged effects of amphetamine derivatives on 5-HT axon terminals have been observed repeatedly, controversy has remained over whether or not this class of drugs has a neurotoxic effect on 5-HT cell bodies. Serotonergic cell bodies have been reported to be affected adversely by PCA and FEN in one raphe nucleus, B9 (Harvey and McMaster, 1975; Harvey et al., 1975,1977). Using both a silver stain and a stain for Nissl substance, these investigators observed cytological changes in B9 cell bodies which they interpreted as indicative of degeneration; the conclusions were based specifically on the observation of dark, irregularly shaped, silver-stained neurons and neurons with "intense" staining of Nissl substance. This cytotoxicity reportedly was prevented by pretreatment with the 5-HT uptake inhibitor fluoxetine (Harvey et al., 1977). The interpretation of neuronal degeneration in the CNS is not a universally agreed on, straight forward matter with well-established criteria. As a technical note, the reaction of a neuronal cell body to transection of its axon in the peripheral nervous system (PNS) is termed "the retrograde reaction" or "chromatolysis," and is characterized by disaggregation of rough ER and loss of "Nissl bodies," resulting in a decrease in Nissl staining intensity (Brodal, 1982). In the CNS, however, the reaction to axotomy is less consistent than in the PNS; researchers have not established that all neurons with central axons undergo typical chromatolysis after axonal injury (Lieberman, 1971).

The initial reports of cell body damage were supported by Massari et al. (1978b), who found a decrease in the number and intensity of histofluorescent 5-HT neurons in B9 53 days after PCA treatment. TPH activity similarly was found to be reduced in B9 as long as 60 days after PCA administration (Neckers et al., 1976). The results of all these studies were questioned later, however, when other investigators failed to observe a change in any raphe nuclei, including B9, at survival times ranging from 2 to 42 days (Lorez et al., 1976,1978; Sotelo and Zamora, 1978). The lack of a toxic effect of PCMA on B9 neurons was substantiated by qualitative electron microscopic analysis (Lorez et al., 1976). Further, the hyperchromatic (Nissl-stained) neurons described by Harvey and co-workers as degenerating neurons were interpreted by Powers et al. (1979) as possible artifacts, since these neurons were observed equally often in controls, and were not associated with cell loss. Despite this refutation, however, Powers and colleagues (1979) did observe a slight cell loss in B9 of a small number of the PCA-treated rats. In all these studies, the DRN and MRN (B7 and B8) were reported to show no signs of degeneration (i.e., with silver or Nissl stains), and exhibited merely a transitory decrease in TPH activity. Immunocytochemical studies, using antisera to 5-HT, also have indicated that neurons in B7, B8, and B9 display normal immunoreactivity and morphology 2 wk after PCA, MDA, or MDMA treatment (Mamounas and Molliver, 1988; O'Hearn et al., 1988), although no actual cell counts were made in these studies. In one study, cells in the DRN were counted 4 mo after MDMA administration, and no loss of 5-HT immunopositive cells was noted (Scallet et al., 1988), although several DRN cells in the drug-treated group contained inclusion bodies. We have observed no alteration in the number of Nissl-stained DRN neurons in the rat as late as 12 mo after PCA administration (Mamounas et al., 1992). Preliminary studies in squirrel monkeys suggested that the persistent (18 mo) 5-HT axon loss caused by MDMA may be the result of cell loss in DRN, MRN, and B9 (Ricaurte et al., 1991a). However, the cell loss was modest; control for age-related cell loss first must be assessed before these results can be validated.

Although Harvey and co-workers (1975,1977) did show that forebrain 5-HT levels were reduced after PCA or FEN treatment, attributing this loss to a lesion of B9 cells is not consistent with current knowledge of ascending 5-HT projections. Massari et al. (1978b) questioned whether a neurotoxic effect limited to the B9 cell group, which has a very minor and circumscribed projection to forebrain (Massari and Sanders-Bush, 1975), could account for the widespread and severe decreases in forebrain 5-HT produced by PCA. Rather, Massari and co-workers (1978b) speculated that PCA must be specifically toxic to axon terminals originating from somata in B7 and/or B8, which are more highly collateralized.

3. Morphological Selectivity: 5-HT Axons

That regional variations occur in the depletion of 5-HT and loss of uptake sites after treatment with amphetamine derivatives is well established (Sanders-Bush et al., 1975; Köhler et al., 1978; Ricaurte et al., 1980; Commins et al., 1987b; Kleven and Seiden, 1989; Champney and Matthews, 1991). The regional variations in biochemical measures of neurotoxicity are reflected by immunocytochemical studies (O'Hearn et al., 1988; Molliver and Molliver, 1990; Axt and Molliver, 1991b). One explanation for the regional differences in 5-HT depletions and axon loss became apparent when, of the two morphological classes of 5-HT axons innervating forebrain, one class (the fine axons) was found to be particularly vulnerable whereas the beaded axons were resistant (cf. Fig. 3; Mamounas and Molliver, 1988,1991). A detailed regional analysis in PCA- and MDA-treated rats revealed that areas rich in beaded axons—notably the glomerular layer of olfactory bulb, the borders of the granule cell layer in dentate gyrus, area CA3 of hippocampus, layer III of lateral entorhinal cortex, and ventricular 5-HT axon plexus—revealed more axon sparing and most of the spared

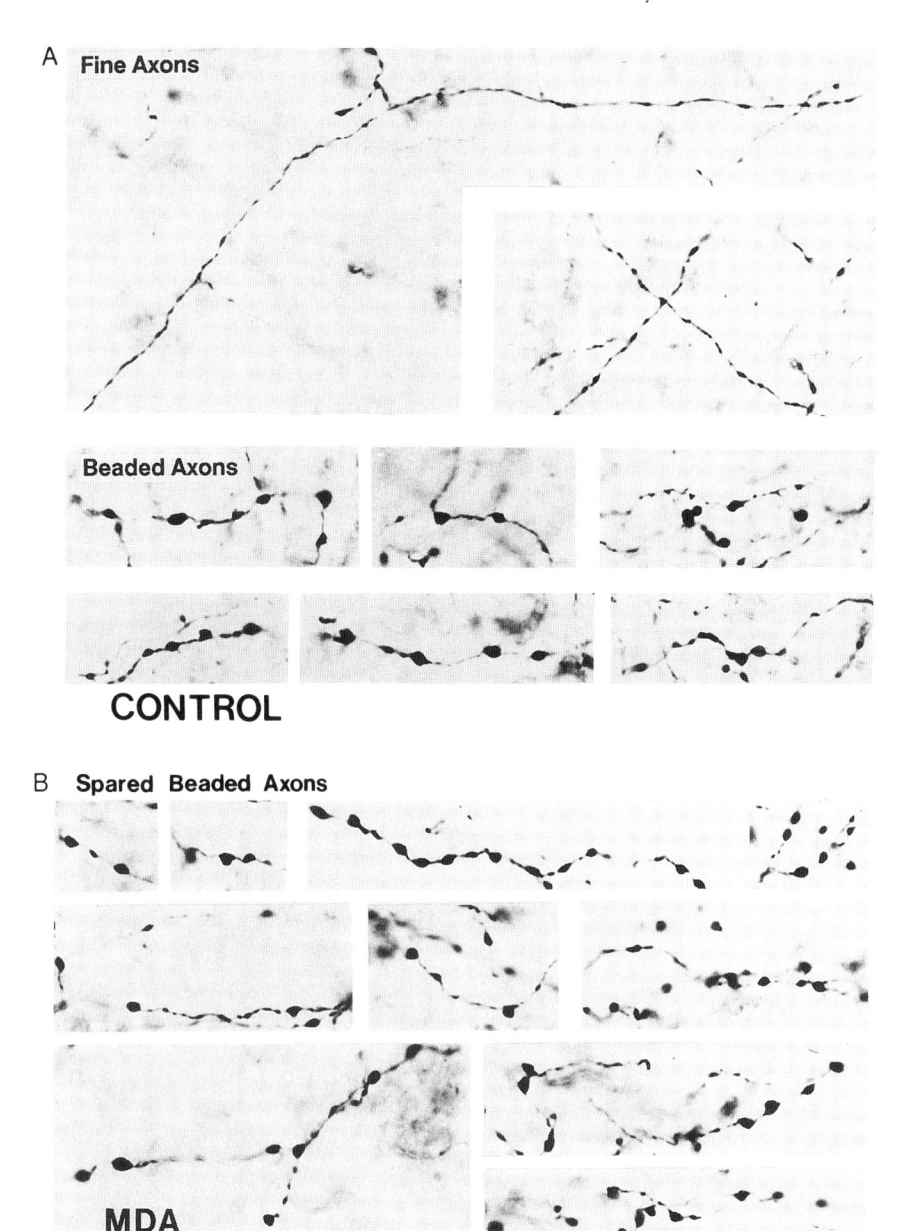

FIGURE 3 Brightfield photomicrographs demonstrating the morphology of 5-HT-immunoreactive axons in dorsal hippocampus of the rat. (A) Fine and beaded serotonergic axons of control (saline-injected) rats. (B) Spared beaded axons of methylenedioxyamphetamine (MDA) treated rats (10 mg/kg × 8). Fine axons have notably smaller varicosities (≤ 1 μm) than do beaded axons (typically 2–3 μm); spared axons after MDA have the same morphology and distribution as beaded axons in control rats. Scale bar, 10 μm. Adapted with permission from Mamounas et al. (1991). Copyright © 1991 Wiley-Liss, a division of John Wiley & Sons, Inc.

axons were beaded axons (Mamounas *et al.*, 1991). Doses of PCA as high as 40 mg/kg did not produce a notable decrease in the density of beaded axons (cf. Fig. 1), although a subtle effect on beaded axons was observed occasionally after a 20 mg/kg dose of PCA (Mamounas *et al.*, 1991). Similar patterns of axonal sparing have been observed after administration of other amphetamine analogs, specifically MDMA (O'Hearn *et al.*, 1988), FEN (Molliver and Molliver, 1990), and methamphetamine (Axt and Molliver, 1991b). As noted earlier, the beaded 5-HT axons that make up the pericellular arrays (or baskets) are spared after PCA administration (Axt and Molliver, 1991a). Further, studies of basal forebrain regions, such as basolateral and lateral amygdaloid nuclei, insular cortex, and olfactory tubercle, where significant axon sparing is observed after PCA treatment, revealed that the spared axons have the same characteristic beaded morphology (Molliver and Axt, 1990). Thus, the regional differences in long-lasting 5-HT depletions measured biochemically may result largely from the regional differential distribution of fine and beaded 5-HT axons and the selective vulnerability of fine axons to the neurotoxic effects of amphetamines.

Despite the relative sparing of beaded axons, conditions can be achieved in which some of the beaded axons are ablated, that is, the selective vulnerability of fine axons is not absolute, but preferential. Vulnerability of beaded 5-HT axons may depend, in part, on the regimen of PCA treatment employed. In certain brain areas such as olfactory bulb and ventricular plexus, beaded 5-HT axons are highly resistant to damage, independent of the treatment regimen. Under certain treatment conditions, however, (dose, warm ambient temperatures, old or large animals), a loss of some 5-HT-immunopositive beaded axons (presumably arising from the MRN) occurs in neocortex and hippocampus. Any such loss of beaded axons is always much less pronounced than the loss of fine 5-HT axons (Mamounas *et al.*, 1991; other unpublished observations from this laboratory).

The distribution of fine and beaded axons in cerebral cortex of macaques is slightly different from that in rats, yet the differential vulnerability of 5-HT axon types to MDMA neurotoxicity is similar (Wilson *et al.*, 1989). Thus, the distribution of spared axons in nonhuman primates differs from that in rats but is consistent with results predicted by the differential vulnerability of 5-HT axons. For example, in dentate gyrus of macaque, beaded axons are abundant in the molecular layer and are spared after MDMA. Also, in the macaque, few beaded axons are found along the granule cell layer of dentate gyrus; therefore, unlike in the rat, a band of spared axons is not observed after MDMA administration (Wilson *et al.*, 1989).

The differential vulnerability of fine and beaded 5-HT axons to amphetamine derivatives further underscores the distinction between DRN and MRN projections, suggesting biochemical differences between these two neuronal systems. Specific factors that cause fine axons to be vulnerable and beaded axons to be resistant to the neurotoxic effects of this class of drugs have yet to be determined, but several possibilities have been raised (Mamounas et al., 1991), including surface-to-volume ratios, neuronal firing rates, metabolic activity, or differing numbers of affinities of 5-HT uptake sites. Finally, the drug resistance of certain 5-HT projections with discrete targets (i.e., beaded axons) indicates that these 5-HT-mediated functions are likely to be spared from the neurotoxic effects of amphetamine derivatives.

4. Regional Distribution of DA Axon Loss

The majority of studies of the neurotoxic effects of amphetamine analogs on the dopaminergic system have been biochemical, not histological. However, data gleaned from both approaches indicate that the dopaminergic system is more sensitive to amphetamine and methamphetamine than to the ring-substituted amphetamines. Regional analyses suggest that neostriatum is the dopaminergic terminal field most vulnerable to the neurotoxic effects of amphetamines. Using antisera directed against TH, Trulson et al. (1985) reported a significant decrease in immunoreactivity in rat striatum 60 days after methamphetamine treatment. In the mouse, (+)-amphetamine produced a preferential decrease in catecholamine histofluorescence in the dorsal aspect of striatum at 8 days of survival (Jonsson and Nwanze, 1982). Reductions in catecholaminergic histofluorescence in ventral striatum, nucleus accumbens, and olfactory tubercle (VTA terminal fields) were not observed in mice after (+)-amphetamine administration (Jonsson and Nwanze, 1982), suggesting a specific vulnerability of the projection from substantia nigra. High doses of methamphetamine administered to rats produces long-lasting depletions of DA in frontal pole, amygdala, and nucleus accumbens (areas that receive minor inputs from substantia nigra), whereas olfactory tubercle and septum (areas that are innervated exclusively by VTA) remain rather resistant to methamphetamine-induced DA depletions (Seiden et al., 1988). Jonsson and Nwanze (1982) also noted no decrease in catecholamine histofluorescence in hypothalamus or median eminence.

In contrast to the comparable effects of amphetamine and methamphetamine on dopaminergic axons, the effect of these drugs on dopaminergic cell bodies is less clear. In the rat, methamphetamine reportedly causes a dramatic loss of TH-immunoreactive neurons in pars compacta of substantia nigra (Trulson et al., 1985) and a decrease in DA levels in substantia nigra (Ricaurte et al., 1980), although Nissl-stained material revealed no change in the number or appearance of neurons in either substantia nigra or VTA (Ricaurte et al., 1982). In contrast, amphetamine does not appear to cause adverse long-lasting effects on any dopaminergic cell bodies in the rat, as measured by TH immunocytochemistry or by the Fink–Heimer silver method (Ryan et al., 1990). Histofluorescence studies in the mouse also suggest that amphetamine causes no loss of cell bodies in either substantia nigra or VTA (Nwanze and Jonsson, 1981; Jonsson and Nwanze, 1982). The results of these studies may reflect a difference in the neurotoxic effects (or potencies) of methamphetamine and amphetamine, but further investigations are warranted to resolve these differences. In summary, amphetamine and methamphetamine appear to have a preferential neurotoxic effect on DA axon terminals originating from substantia nigra, while sparing the mesocorticolimbic projections from VTA.

5. Morphological Selectivity: DA Axons

To date, the neurotoxic effects of amphetamine and methamphetamine have not been addressed with regard to possible differential effects on morphologic subtypes of DA axons. Some indication exists, however, that compartments such as the patch–matrix organization of the striatum are affected differentially by these drugs. One study of the acute effects of methamphetamine (i.e., 90-min survival) has indicated that methamphetamine produces an acute decrease in DA histofluorescence in the matrix compartment of striatum (Fukui et al., 1986). Another study has demonstrated that, 1 day after cessation of amphetamine infusion, the pattern of TH-immunoreactive axons in rat striatum (i.e., those axons resistant to the effects of the drug) is nearly identical to the striatal patches, labeled by leu-enkephalin (Ryan et al., 1988). Thus, the results of these two studies suggest that the DA projection to striatal matrix is affected in these animals, whereas that to the patches is resistant. To review, the DA axons in matrix of dorsolateral striatum arise from neurons in pars compacta of substantia nigra, whereas those in the patches arise from neurons in the ventral pars compacta and pars reticulata (Gerfen et al., 1987a). Although not reflective of persistent changes in DA innervation of striatum, the results of Fukui et al. (1986) and Ryan et al. (1988) suggest a differential pharmacological sensitivity of the type A dopaminergic axons originating from dorsal pars compacta of substantia nigra (cf. Gerfen et al., 1987a) to methamphetamine and amphetamine. Further, the results are in concert with the finding that methamphetamine causes a persistent decrease in the number of TH-immunoreactive neurons in pars compacta (Trulson et al., 1985). The similarity between DA type A axons (fine caliber axons with small varicosities; Gerfen et al., 1987a) and the 5-HT fine axons, which are vulnerable to the toxic effects of amphetamines, is also very intriguing. Additional anatomic studies are warranted to determine whether this subset of DA axons is, indeed, vulnerable to the neurotoxic effects of amphetamine and methamphetamine. Neurons in "dorsal" substantia nigra that contain calbindin D-28k have been reported to be resistant to the neurotoxic effects of the dopaminergic neurotoxin 1-methyl-4-phenyl-1,2,3,6-tetrahydropyridine (MPTP) in mice (Iacopino et al., 1992), and also appear to be mostly spared in humans with Parkinson's disease (Yamada et al., 1990). Type A dopaminergic axons, which are potentially more vulnerable to the neurotoxic effects of amphetamine and methamphetamine, originate from DA cell bodies (in pars compacta) containing calbindin 28 kDa (Gerfen et al., 1987b). It is likely, therefore, that any protection against neuronal injury afforded by the presence of this calcium binding protein does not extend to these drugs. With respect to the serotonergic projections, neither DRN nor MRN neurons contain calbindin D-28k (K. J. Axt, unpublished observations).

B. Loss of Monoamine Uptake Sites

Despite the histofluorescence and immunocytochemical evidence for axon loss after administration of amphetamines, these methods only indicate a loss of neurotransmitter (or biosynthetic enzyme, such as TH) in the axons. Another method of evaluating axon density after drug treatment, independent of the ability of those axons to synthesize and store neurotransmitter, is quantification of the number of monoamine uptake sites. The primary means of removing monoamines from the synapse after its release is reuptake into the axon terminal (cf. Iversen, 1973; Fuller, 1985). 5-HT uptake sites are most abundant at the 5-HT axon terminal (Kuhar and Aghajanian, 1973). Therefore, assessing the number of uptake sites in a given brain region provides a parameter of the density and integrity of axons in that area. In evaluating changes in the number of 5-HT uptake sites as a parameter of neurotoxicity, one critical consideration is the survival time after drug administration; amphetamine derivatives inhibit 5-HT uptake, so at short survival times (e.g., 18 hr) trace amounts of the drug may be present in the brain that could artifactually alter ligand binding (cf. Zaczek *et al.*, 1990).

Autoradiographic studies using two tritiated 5-HT uptake inhibitors, cyanoimipramine or paroxetine, have demonstrated a significant loss of the number of 5-HT uptake sites weeks after methamphetamine (Kovachich *et al.*, 1989; Brunswick *et al.*, 1992), FEN (Appel *et al.*, 1990), or MDMA (Battaglia *et al.*, 1991) treatment. A decrease in the number of striatal DA uptake sites after methamphetamine administration also has been demonstrated autoradiographically using triatiated mazindol (Brunswick *et al.*, 1992; Eisch *et al.*, 1992). The latter study described a preferential decrease in mazindol binding in ventral striatum, a distribution of DA axon loss that coincides with the biochemical data but differs markedly from the histofluorescence studies in amphetamine-treated mice (Jonsson and Nwanze, 1982). These studies verify biochemical demonstrations of persistent decreases in the V_{max}, not the K_m, for active 5-HT and DA uptake (Ricaurte *et al.*, 1980; Wagner *et al.*, 1980; Commins *et al.*, 1987b), as well as decreases in paroxetine binding (B_{max}) in tissue homogenates (Battaglia *et al.*, 1987; Marcusson *et al.*, 1988; Zaczek *et al.*, 1990; Ricaurte *et al.*, 1992), after administration of amphetamine analogs. Barring any reduction in the density of uptake carriers per axon, a decreased number of labeled uptake sites is a good indicator of axonal loss. Autoradiographic and synaptosomal studies must be evaluated carefully in this regard, however, since ligands have differing specific affinities for uptake sites, as well as for other sites. In the case of paroxetine binding, a comparison with other indices of 5-HT function suggests that this ligand binds to sites other than the 5-HT uptake carrier. Regional analysis of [^{3}H]paroxetine binding after treatment with 5-HT neurotoxins reveals a portrait of damage disparate from other parameters of 5-HT innervation (e.g., 5-HT levels or [^{3}H]5-HT uptake). Specifically in striatum no correlation exists between decreases in paroxetine binding and 5-HT uptake after 5,7-dihydroxytryptamine (DHT) lesions (Gobbi et al., 1990); notable differences exist in the effect of 5,7-DHT on these two parameters in hippocampus and hypothalamus. Similarly, decreases in paroxetine binding and 5-HT levels in hippocampus, striatum, and cortex were not well correlated after MDMA or 5,7-DHT treatment (Scheffel and Ricaurte, 1990), although other areas such as olfactory tubercle and hypothalamus exhibited a good correlation between these indices of 5-HT function. In both these studies, the investigators suggested that paroxetine may have an affinity for sites on nonserotonergic neuronal elements. [^{3}H]Imipramine binding also is correlated poorly with 5-HT levels in these same brain regions 1 mo after 5,7-DHT lesion (Brunello et al., 1982). Based on comparatively larger decreases in 5-HT levels, Scheffel and Ricaurte (1990) suggested that [^{3}H]paroxetine binding might underestimate the magnitude of 5-HT axon damage in these few areas. In contrast, Battaglia et al. (1987), using the same dose of MDMA, observed a more profound decrease in [^{3}H]paroxetine binding than in 5-HT levels in hippocampus, striatum, and cortex, and suggested that evaluating 5-HT content would underestimate the extent of damage. The contradiction arising from the results of these studies is puzzling and is unlikely to be due to different survival times (1 wk: Scheffel and Ricaurte, 1990; 2 wk: Battaglia et al., 1987). More importantly, however, the resultant conflict in interpreting the value of these parameters for neurotoxicity assessment should signal caution. Different parameters of neuronal function may exhibit distinct responses to injury or pharmacological perturbation, and quantitative comparisons can be misleading unless the mechanisms and the relative magnitudes of such responses under various conditions are well understood.

Based on their paroxetine binding data, Battaglia et al. (1987) concluded that MDA and MDMA exhibited little regional specificity of neurotoxicity, in contrast with the results of immunocytochemical studies (O'Hearn et al., 1988). Later, these investigators (Battaglia et al., 1991) concluded that MDMA-induced loss of 5-HT uptake sites in certain brain regions did not support the hypothesis of differential vulnerability of axon types or projections. Such a conclusion is likely to be incorrect for several reasons. First, in no study have the distribution of 5-HT uptake binding sites and 5-HT-immunoreactive axons been compared directly in the context of amphetamine neurotoxicity; thus, the separate contributions of fine and beaded 5-HT axons to paroxetine binding is unknown. Second, the sensitivity and resolution of the method (autoradiography) is not high enough to detect subtle intraregional changes (e.g., within a cortical layer), and therefore warrants conservative interpretations. The ability to detect a small number of axons with this method depends, in part, on the density and distribution of the resistant axons; because of the threshold of detection, if the spared axons are distributed diffusely they are less likely to be detected than if they are spaced closely. As another example of the lower sensitivity of radioactive ligand binding, [^{3}H]paroxetine binding to membranes has been reported as "not detectable" (Marcusson et al., 1988) in frontal cortex of rats treated with a dose of PCA at which active [^{3}H]5-HT uptake (Marcusson et al., 1988) and 5-HT-immunoreactive axons (cf., Fig. 8) are still measurable. Further, the comparison of decreases in the number of uptake sites in "sensorimotor" cortex with those in a region such as entorhinal cortex (Battaglia et al., 1991) is potentially misleading since lateral entorhinal cortex contains a dense band of beaded 5-HT axons that are drug resistant (Mamounas et al., 1991), but which extends only about 1 mm through this region (Köhler et al., 1980) and, therefore, may be missed easily or sampled unexpectedly and inconsistently. In addition, coronal sections through entorhinal cortex are tangential to cortical layers and, therefore, are likely to include a preponderance of layer I axons, which are drug sensitive. This contrasts with a coronal section through sensorimotor cortex, which is perpendicular to the six cortical layers, of which layer I obviously is a less prominent portion. Finally, the density of uptake sites on fine and beaded axons may or may not be equivalent.

In summary, decreases in 5-HT and DA uptake sites have been observed after administration of amphetamine analogs using autoradiographic analysis of radioactive ligands. Although these studies provide a quantitative assessment of regional axon loss, a comparison of the magnitude of neuronal damage with other quantitative measures is not necessarily useful, since indices of neurotoxicity vary with respect to each other in their quantitative response to injury. Instead, the measurement of binding of radioligands to uptake sites is useful for comparing the relative neurotoxic effects of a class of drugs, such as amphetamine derivatives, and serves to corroborate other measures of axon damage.

C. Loss of Axonal Transport

The persistent loss of 5-HT axonal markers observed after administration of amphetamine derivatives can be substantiated further by the demonstration of functional loss of axonal transport. If axonal transport, anterograde or retrograde, is compromised after drug treatment, the structure and function of the axon terminals cannot be maintained. Many vital functions of the neuron are dependent on axonal transport, including renewal of axonal membrane and cytoskeletal proteins, delivery to the terminal of neurotransmitter or the machinery for its metabolism, recycling of the membrane, and transfer of substances back to the cell soma for degradation in lysosomes (cf Grafstein and Forman, 1980). Retrograde axonal transport methods have been utilized in several studies to assess the cytoskeletal integrity of serotonergic axons after treatment with amphetamine neurotoxins such as PCA and MDA (Fritschy et al., 1988; Mamounas and Molliver, 1988; Axt and Molliver, 1991c; Molliver and Mamounas, 1991; Haring et al., 1992). Whereas 5-HT immunocytochemical procedures depend on the presence of 5-HT to evaluate neuronal damage, the loss of axonal transport after drug treatment provides an independent measure of axonal degeneration or impaired neuronal function. In these studies, a retrogradely transported marker compound (e.g., a fluorescent dye such as Fluoro-Gold) is injected into a selected area of the brain where 5-HT axon terminals normally are found. In untreated animals, the intact axon terminals at the injection site take up and retrogradely transport the marker to cell bodies of origin in the raphe nuclei. After pretreatment with amphetamine neurotoxins, serotonergic axons that have degenerated or are damaged severely by the drug will not transport the marker; their cell bodies in the raphe nuclei will not be labeled. Thus, a reduction in the number of retrogradely labeled neurons after drug treatment indicates that the axons arising from that cell group either have degenerated or have impaired axonal transport capabilities. Moreover, as mentioned previously, amphetamine neurotoxins selectively damage a subset of the 5-HT axon terminals in forebrain while a morphologically distinct group of axon terminals is spared preferentially; measurements of axonal transport capabilities after neurotoxic lesions allow the differentiation between raphe neurons with axons that are damaged by the neurotoxins, and those neurons that remain unaffected.

After pretreatment with PCA or MDA, retrograde transport studies from neocortex (Mamounas and Molliver, 1988), hippocampus (Axt and Molliver, 1991c; Haring et al., 1992), olfactory bulb (Molliver and Mamounas, 1991), and the trigeminal motor nucleus (Fritschy et al., 1988) reveal a severe loss of neuronal labeling in the DRN. In most areas examined, these drugs appear selectively toxic to dorsal raphe projections, sparing the axonal projections from other raphe cell groups; MRN projections to neocortex (Mamounas and Molliver, 1988) or olfactory bulb (Molliver and Mamounas, 1991) remain intact after administration of amphetamine neurotoxins, as do the 5-HT axonal projections from raphe obscurus and raphe pallidus to the trigeminal motor nucleus (Fritschy et al., 1988). The distribution of raphe neurons (at one representative level) projecting to fronto-parietal cortex in control and PCA-treated rats is depicted in Fig. 4, as is the distribution of all 5-HT-immunopositive neurons at this level of brainstem. The dramatic loss of retrograde transport of dye from fronto-parietal cortex to DRN neurons in PCA-treated rats is presented in Fig. 5.

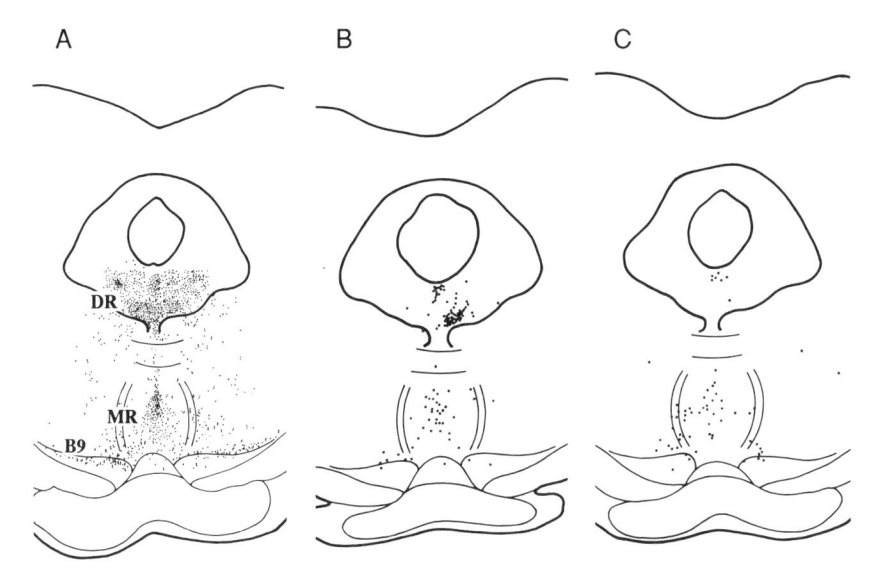

FIGURE 4 Diagrams of transverse sections through rat midbrain, depicting 5-HT neurons and raphe neurons that project to fronto-parietal cortex. Each dot represents one neuron. (A) Distribution of 5-HT-immunoreactive neurons in midbrain of a control rat. Data are combined from two adjacent sections from one animal. Abbreviations: DR, dorsal raphe nucleus; MR, median raphe nucleus; B9, 5-HT cell group of Dahlström and Fuxe (1964). (B) Distribution of cortically projecting neurons in midbrain of control rats. Data are derived from two rats; retrogradely labeled cells are superimposed (combined) from two adjacent sections from each rat. (C) Distribution of cortically projecting neurons in midbrain of p-chloroamphetamine (PCA) treated rats. Data are derived as described for B. After PCA, a dramatic decrease in the number of labeled cells in DR, but not in MR, is observed. Adapted with permission from Mamounas and Molliver (1988).

The highly selective effect of the amphetamine neurotoxins on dorsal raphe axons may not be evident in all brain regions. In hippocampus, 5-HT axons originating from median raphe are also susceptible to damage by PCA, although to a lesser extent than the dorsal raphe axons in this area (Haring *et al.*, 1992; K. J. Axt and M. E. Molliver, unpublished observations). Anterograde transport experiments confirm the loss of axons in hippocampus arising from the MRN. After pretreatment with PCA, PHA-L injections in the median raphe result in fewer labeled beaded axons in hippocampus (Haring *et al.*, 1992). Additional studies are necessary to determine whether hippocampal beaded 5-HT axons are indeed vulnerable to PCA or whether a component of the MRN projection to hippocampus does not have the beaded morphology.

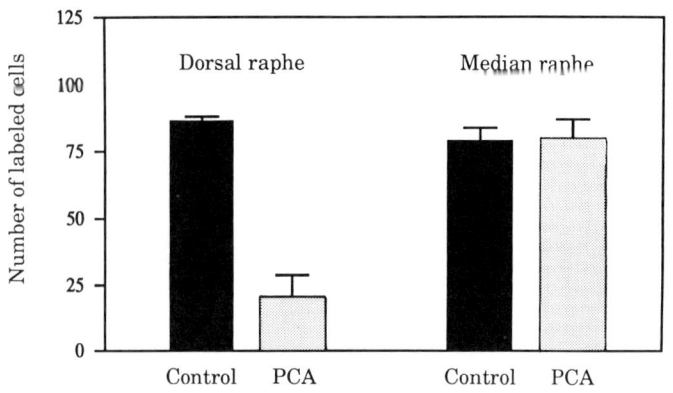

FIGURE 5 Effect of prior (2-wk) *p*-chloroamphetamine (PCA) administration on the number of retrogradely labeled cells in midbrain raphe nuclei of rats receiving fluoro-gold injections into fronto-parietal cortex. Fluoro-gold-labeled cells were counted in every sixth (20 μm) section through dorsal raphe (DRN) and median raphe (MRN) nuclei from each rat. Control, $n = 4$; PCA, $n = 4$. The number of retrogradely labeled cells in DRN of PCA-treated rats is significantly less than in controls ($p < .001$), whereas no difference exists between control and PCA-treated rats in the number of retrogradely labeled cells in MRN. Adapted with permission from Mamounas and Molliver (1988).

IV. MORPHOLOGICAL EVIDENCE OF AXONAL DAMAGE AND DEGENERATION

The persistent decreases in axon density and deficits in axonal function observed after administration of amphetamine analogs are likely to be the result of neurotoxic actions of these drugs. Therefore, determining whether axonal damage indicative of axonal degeneration occurs is important. Several anatomic methods are available to evaluate neuronal degeneration. Since the studies of Cajal (1928), pathological features of axonal damage and degeneration have been recorded, including axonal swelling and pleomorphic dilatations. The cause for such structural perturbations in axons is likely to be disrupted axonal flow and/or disruption of the cytoskeletal proteins that provide shape and physical support for the axons. Morphological studies of axons, either by histofluorescence or immunohistochemistry, hours to days after drug treatment have revealed changes that are indicative of axonal damage or degeneration. Further, degenerating axons and somata can be labeled specifically using silver impregnation methods; degenerating somata display lysosomal inclusions which are revealed using electron microscopy. Nonneuronal markers of neuronal damage, such as reactive gliosis, also have been evaluated and combined with morphological studies of the axonal systems of interest.

A. Evidence of Structural Changes in Axons

Axonal swelling, contortion, and fragmentation all have been described after lesions of peripheral nerves, as well as after intracerebral injections of the monoamine neurotoxins 6-hydroxydopamine (6-OHDA) (Ungerstedt, 1968), 5,6-DHT (Baumgarten et al., 1972), and 5,7-DHT (van Luijtelaar et al., 1989). Several studies have addressed the issue of axonal swelling after administration of amphetamine derivatives. The first studies to reveal structural damage to 5-HT axons caused by amphetamine derivatives utilized the Falck–Hillarp histofluorescence method; swelling and increased histofluorescence of nonterminal 5-HT axons was observed 3 days after administration of PCMA or PCA to rats (Lorez et al., 1976; Fuxe et al., 1978). Lorez and colleagues (1976) concluded, however, that degeneration of 5-HT terminals did not occur, since axon terminals in the suprachiasmatic nucleus and ventricular plexus did not display any signs of degeneration. This errant conclusion arose in part because of an unfortunate and misleading choice of terminal fields; these investigators later described diminished histofluorescence in several other terminal fields at both 3 and 42 days (Lorez et al., 1978). Given current immunocytochemical data, the reason Lorez et al. (1976) observed no signs of axon degeneration in ventricular plexus and suprachiasmatic nucleus is clearer; these areas are innervated almost exclusively by drug-resistant beaded axons (Mamounas et al., 1991; L. A. Mamounas, unpublished observations).

Immunohistochemistry is more sensitive than histofluorescence and consistently reveals a greater number of axons. Therefore, extensive axonal damage and denervation are demonstrated more readily with this method. Like histofluorescence, however, one of the limitations of the immunocytochemical method for depicting denervation or degeneration is that the technique requires the presence of the antigen within the axons. When a drug depletes the antigen, in this case the neurotransmitter or its synthesis enzymes, the absence of staining is not sufficient to conclude that denervation or degeneration has occurred. If the precise time after drug exposure is used for analysis (a time after acute depletion has occurred, and when transmitter levels have recuperated sufficiently that axons are visible), then degenerating axons might be observed. Thus, immunocytochemical methods have revealed highly abnormal morphology in many serotonergic terminal fields 2–3 days after parenteral administration of MDA, MDMA, FEN, PCA, and methamphetamine to rats (O'Hearn et al., 1988; Molliver and Molliver, 1990; Axt et al., 1992a; 1993), 18 hr after oral MDMA (Scallet et al., 1988), and after parenteral FEN administration to squirrel monkeys (Ricaurte et al., 1991b; Wilson et al., 1993). Depicted in Fig. 6 are examples of the pathology observed in rat neocortex after administration of some of these drugs. The drug-induced structural perturbations of 5-HT-

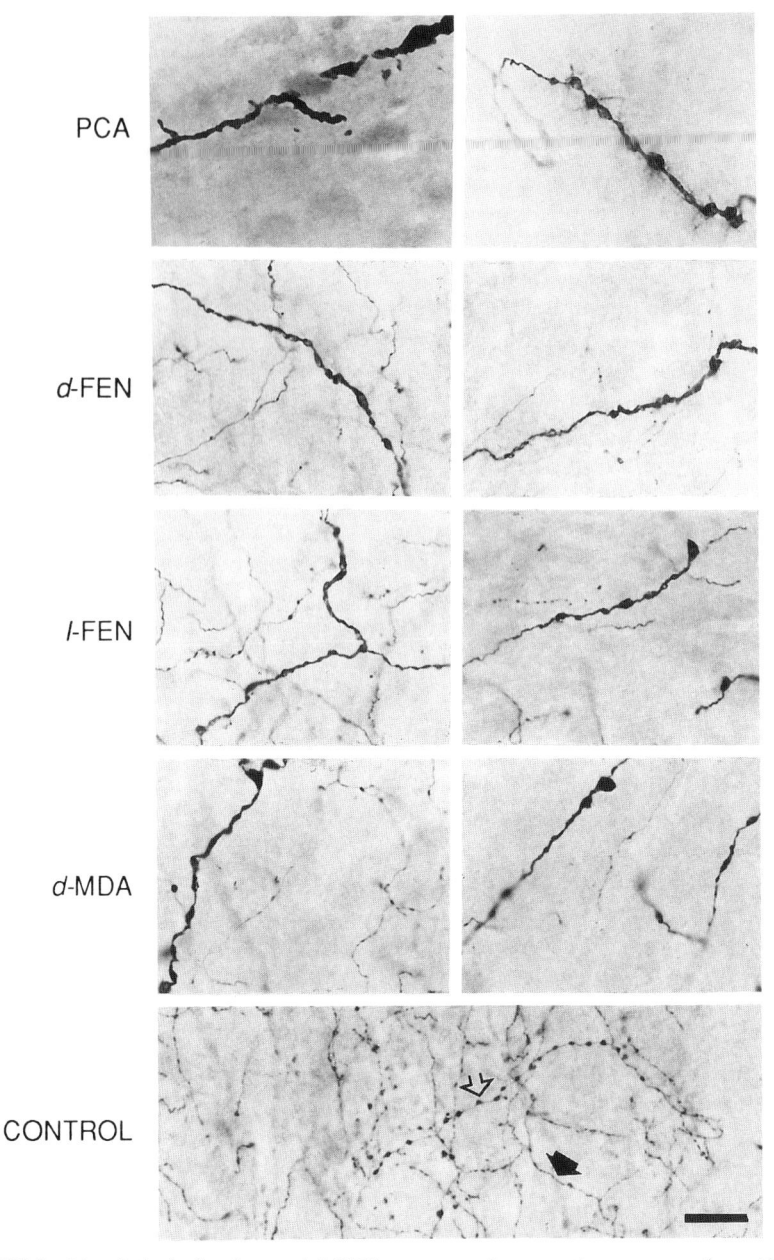

FIGURE 6 Morphologically abnormal 5-HT-immunoreactive axons in neocortex of rats 48 or 72 hr after the last injection of the following 2-day regimens of amphetamine derivatives: *p*-chloroamphetamine (PCA), 10 mg/kg × 2; *d*-fenfluramine (*d*-FEN) 5 mg/kg × 4; *1*-FEN, 5 mg/kg × 4; *d*-methylenedioxyamphetamine (*d*-MDA), 20 mg/kg × 4, or saline control. Note the increased caliber and large varicose areas along the axon terminals of drug-treated brains, compared with the small caliber of normal fine axons (closed arrow) and beaded axons (open arrow) of the control brain. The swollen varicosities of the drug-damaged axons are significantly larger than the normal varicosities on either fine or beaded axons in control brains or beaded axons in drug-treated brains. Scale bar, 20 μm.

immunoreactive axons include increased axon caliber, large pleomorphic varicosities, tangled appearance, bulbous stumps, and occasional sprout-like protrusions. Notably, no alterations in the morphology of TH-immunoreactive axons have been observed in striatum of rats treated with MDMA (Scallet *et al.,* 1988; 18-hr survival), nor with any of the other ring-substituted amphetamines (K. J. Axt, unpublished observations; 48- to 72-hr survival). Further, the damage to serotonergic axons of passage observed by Lorez *et al.* (1976) after PCMA treatment also has been observed after both oral and parenteral FEN (Sotelo, 1991), and after parenteral PCA, methamphetamine, and isomers of MDA (Axt *et al.,* 1992a; 1993). The prevelance of damaged 5-HT axons in both terminal fields and fiber bundles is significantly less at 1–2 wk than at 2–3 days after drug treatment, although profiles of 5-HT axon pathology are still seen more often at these later survival times after drug treatment than at any time in controls. Two similar types of morphological change may reflect temporally different processes in axonal degeneration. Acute swelling of axon terminals is observed within 1–3 days of the insult but rapidly becomes less prevalent as the debris is cleared. Later phases may reveal bulbous preterminal stumps of axons that remain in continuity with the cell body of origin. Additional data are needed to distinguish which bulbous stumps reflect a dying-back process of further degeneration and which may be associated with regenerative sprouting.

One study that demonstrated axonal swelling in frontal cortex of the rat after FEN treatment suggested that the "thickening" of axons was "due to abnormal antigenicity of the 5-HT nerve fiber that was merely the result of the 5-HT uptake-blocking effect of the drug" (Kalia, 1991). This conclusion was based on the finding that the abnormal morphology was found to "recover" by 15 days. The postulated enhanced antigenicity of 5-HT axons following drug treatment may be reasonable with respect to the intensity of staining, but certainly does not explain the change in axonal morphology. The idea that the "thick fibers" described by Kalia (1991) might reflect "an alteration in the quality and quantity of antigenic sites on the surface of the membrane of the 5-HT-containing nerve terminal" is not plausible or consistent with other facts for several reasons: the antigen (5-HT) has not been found to be altered so it would bind more of the antisera, 5-HT would not be expected to sit attached to the external surface of the axonal membrane, and, 2–3 days after drug administration when axon swelling is visible, amphetamine derivatives such as fenfluramine no longer inhibit 5-HT reuptake. As early as 4 hr after treatment with methamphetamine (Axt and Molliver, 1991b), PCA (Molliver *et al.,* 1988; Fig. 8), or MDA (M. E. Molliver, unpublished observations), when the pharmacological actions of the drug are ongoing, 5-HT immunoreactivity is virtually absent except in beaded 5-HT axons, which exhibit normal immunoreactivity and morphology. These findings indicate that the 5-HT released from nerve endings does not persist immediately outside the axons to make them look "swollen."

Serotonin that is not taken back up into the nerve terminal would be deaminated rather rapidly by MAO within the surrounding glial cells. The issue of whether the appearance of axonal swelling might be attributed to accumulation of extracellular 5-HT can be resolved by using an antibody against TPH, the synthetic enzyme for 5-HT (which is located intracellularly). By this approach, we have demonstrated, in rats treated with MDA, that TPH-immunoreactive 5-HT axons display swelling and fragmentation identical to that seen using an antibody against 5-HT (Axt et al., 1993). Thus, axonal swelling detected with a marker that is not the transmitter but a large protein confirms that the extremely large axonal swellings accurately reflect structural pathology indicative of nerve terminal degeneration. Finally, the potent 5-HT uptake inhibitor fluoxetine causes neither a decrease in 5-HT axon density nor any axonal swelling 3 days after administration; in fact, fluoxetine prevents both the decrease in 5-HT-immunoreactive axon density and cytopathology produced by PCA at 3 days (Axt et al., 1993).

Histofluorescence studies in both mouse (Jonsson and Nwanze, 1982) and rat (Ellison et al., 1978) have demonstrated morphologically abnormal DA axons in striatum 1 or 6 days after (+)-amphetamine infusion. Methamphetamine causes a similar type of damage to striatal DA axons in the rat (Lorez, 1981). Consistent with these findings, TH-immunoreactive axons in striatum exhibit a distorted and swollen appearance one day after (+)-amphetamine infusion (Ryan et al., 1988,1990) in two species of rat. In contrast to an apparent difference in the intensity of TH immunoreactivity between patch and matrix compartments of striatum, swollen TH-positive axons were observed in both compartments (Ryan et al., 1988) after (+)-amphetamine treatment. Electron microscopic evaluation of striatum also revealed damage to TH-immunoreactive profiles (Ryan et al., 1990). These investigators found that, in frontal cortex, TH-positive axons displayed no signs of structural damage at the light or electron microscopic level.

B. Silver Impregnation Studies

Degenerating axons can be stained with a variety of silver impregnation methods. Historically, these methods were used in conjunction with surgical axotomy or other lesions of pathways or cell groups to trace axonal projections in the brain. In studies of 5-HT axon projections, however, controversies concerning 5-HT innervation of forebrain arose because the methods were not sufficiently sensitive to stain many 5-HT axon terminals (cf. Conrad et al., 1974). The advantages and limitations of these methods have been reviewed (Beltramino et al., 1993). Because of the limited sensitivity of silver methods for 5-HT axons, silver degeneration studies of amphetamine neurotoxicity have been used primarily for the DA system. There are reports of axon degeneration seen after administration of amphetamine derivatives that primarily affect 5-HT axons, however. The Fink–Heimer silver impregnation method (cf. Fink and Heimer, 1967) revealed an argyrophilic reaction in striatum of rats treated with amphetamine or methamphetamine (Ricaurte et al., 1982,1984; Ryan et al., 1990), as well as MDA, MDMA, and PCA (Ricaurte et al., 1985; Commins et al., 1987a,b; Scallet et al., 1988; Slikker et al., 1988). In these studies, fine grains of reduced silver dot the neuropil of caudatoputamen, suggesting impregnation of axon terminals. Although the presence of argyrophilic neuronal elements does not indicate the neurotransmitter system that is affected, the drugs used in these studies deplete DA and/or 5-HT and cause a decrease in dopaminergic or serotonergic axon density in striatum; thus, the results have been interpreted as showing degeneration of DA or 5-HT nerve terminals. No regional variation in silver staining was reported in rat striatum, in contrast to immunocytochemical studies of methamphetamine and amphetamine that suggested an acute preferential effect on the DA projections to striatal matrix (Fukui et al., 1986; Ryan et al., 1988). In gerbil caudatoputamen, silver staining also has been observed, but only in adult animals and not in juveniles (Teuchert-Noodt and Dawirs, 1991). Further, a mosaic-type pattern of silver deposition was described in adult gerbil striatum, which the authors suggested might reflect the patch–matrix topography of this brain region.

The results of silver staining in neocortex are more difficult to interpret. Although methamphetamine is thought to have a selective effect on DA and 5-HT axons, Commins and Seiden (1986) noted degenerating pyramidal neurons in somatosensory cortex, suggesting an effect on non-monoaminergic systems. Interestingly, this neuronal degeneration was prevented by pretreating the animals with a TH inhibitor, a treatment that prevents the drug-induced depletion of DA and 5-HT. A similar pattern of neuronal degeneration in somatosensory cortex was also found after PCA (Commins et al., 1987a), MDMA (Commins et al., 1987b), and amphetamine (Ryan et al., 1990) treatment. Preliminary studies (Jensen et al., 1991) using the more sensitive de Olmos cupric silver method (cf. Carlsen and de Olmos, 1981) indicate that very high doses of MDMA cause degeneration of neurons as well as axon terminals in neocortex of the rat. However, these investigators found that the pattern of degeneration did not match the pattern of serotonergic innervation of neocortex precisely, and that the degeneration was blocked only partially by the 5-HT uptake inhibitor fluoxetine, a drug that prevents the serotonergic deficits caused by most amphetamine derivatives. These studies of neocortex in the rat suggest that nonmonoaminergic systems might be sensitive to the neurotoxic actions of these drugs, although the mechanisms for such an effect have yet to be clarified.

In the gerbil, a restricted area of prefrontal cortex is affected by methamphetamine. Specifically, using the Gallyas silver technique (Gallyas et al., 1980) to stain nerve terminal degeneration, silver deposition was observed in layers II–III of prefrontal cortex 3–7 days after methamphetamine administration to juvenile gerbils (Wahnschaffe and Esslen, 1985; Teuchert-Noodt and Dawirs, 1991). This region corresponds to part of the mesocortical projection, presumably originating from the VTA. No silver deposition was noted in olfactory tubercle, nucleus accumbens, entorhinal cortex, amygdala, or median eminence of juvenile gerbils (Wahnschaffe and Esslen, 1985). In contrast, degeneration was not observed in prefrontal cortex of adult gerbils (Teuchert-Noodt and Dawirs, 1991). Amphetamine causes apparent axon degeneration in agranular insular cortex and in layers II–III of motor cortex of the rat (Ryan et al., 1990). No abnormal TH (or 5-HT)-immunoreactive axons were observed in these regions, however, suggesting that the silver-stained axon terminals may not be monoaminergic. Further, the terminal degeneration seen in layers II–III of motor cortex differs from that described in medial prefrontal cortex of the juvenile gerbil; the pattern of degeneration does not match either DA or 5-HT innervation of this region. Whether nonmonoaminergic axons also are affected in prefrontal cortex of the juvenile gerbil is unknown. The findings in striatum and neocortex of the gerbil, but not the rat, raise the possibility that subsets of DA axons, with differing cells of origin (Tassin et al., 1978; Gerfen et al., 1987a), may be differentially vulnerable to the neurotoxic effects of methamphetamine.

C. Reactions of Glial Cells

Several studies have indicated that treatments that produce neuronal trauma or cell death also result in an astrocytic response typified by increased production of an intermediate gliofilament, glial fibrillary acidic protein (GFAP), by type I astrocytes (Björklund et al., 1986; Strömberg et al., 1986; Miyake et al., 1988; Ogawa et al., 1989; Kindy et al., 1992). GFAP is a structural protein that is produced normally in type I astrocytes and, therefore, is histologically a specific marker for visualization of these cells. During gliosis, reactive type I astrocytes produce increased amounts of GFAP, a process that is accompanied by hypertrophy, an increase in cell size and number of processes. Increases in the number of GFAP-containing astrocytes (hyperplasia) after injury have been reported, but whether these increases reflect actual cell division or the recruitment of nonexpressing astrocytes to express GFAP is a matter of debate. Although the function and significance of astrocytic GFAP production is not understood, the increased expression of this protein and the associated morphological changes in astrocytes provide a nonneuronal indicator of neuronal injury that can be measured both biochemically and histologically. Therefore, demonstration of an astrocytic response provides a nonneuronal marker and complementary tool for assessing the axonal injury and degenerative process produced by amphetamine analogs.

Preliminary studies with MDMA have suggested that biochemical levels of GFAP rise in several areas of neocortex (O'Callaghan *et al.*, 1991). Only in frontal pole was the GFAP response blocked by pretreatment with the 5-HT uptake inhibitor fluoxetine, however, suggesting that a component of the GFAP response in other cortical areas was the result of nonserotonergic effects (O'Callaghan *et al.*, 1991). Immunocytochemical staining of GFAP reveals a slight alteration in astrocytes in frontal pole of PCA- (and MDA-) treated rats (Axt *et al.*, 1993), an area that sustains the greatest 5-HT axon loss and normally has one of the highest densities of 5-HT axons (cf. Fig. 8). In other brain regions in which PCA causes 5-HT axon loss, however, no obvious GFAP response is noted (Axt *et al.*, 1993). The astrocytic response in frontal pole of PCA-treated rats is not hyperplastic, but mildly hypertrophic; a comparison with other astrocytic markers indicated no increase in the number of astrocytes in PCA-treated rats, but a larger number of astrocytes displayed GFAP immunoreactivity. The "reactive" astrocytes are slightly thicker and more immunoreactive, suggesting they contain more GFAP, and they often exhibit more immunoreactive processes than controls, but do not display the robust hypertrophy of reactive astrocytes seen in a stab wound (Miyake *et al.*, 1992; Axt *et al.*, 1993) or in intraventricular 6-OHDA (Strömberg *et al.*, 1986; Ogawa *et al.*, 1989) models. However, in comparing the astrocytosis after intraventricular administration of neurotoxic substances (such as 6-OHDA) to the astrocytosis found after parenteral administration of chemical toxins, the physical trauma incurred from the injection also must be considered (cf. Strömberg *et al.*, 1986). Although reactive gliosis has not been evaluated in striatum of amphetamine- or methamphetamine-treated rats, electron microscopic material from striatum of amphetamine-treated rats has been reported to contain dark profiles resembling glial processes (Ryan *et al.*, 1990). Another electron microscopic study (Dastur *et al.*, 1985) indicated that fenfluramine administration to neonatal rats resulted in intralysosomal inclusions in both neurons and oligodendrocytes. Whether these cytoplasmic effects relate to the neurotoxic effects observed in adults, or to some transient interaction with phospholipids of very young animals, is unknown.

Currently no satisfactory explanation is available for the relatively mild nature of astrocytosis observed after administration of amphetamines that appear to cause profound serotonergic degeneration and denervation. One suggested explanation (Axt *et al.*, 1993), however, is that the small fraction of the neuropil affected by these drugs, namely the fine 5-HT axons, is not sufficient to evoke a glial response that is measurable with current methods. Peripheral administration of MPTP to mice produces an increase in GFAP immunoreactivity in striatum (O'Callaghan and Jensen, 1992) that is accompanied by a cupric silver degeneration reaction, suggesting that DA terminal degeneration can induce a glial response. Of course the density of DA axons in striatum (and, thus, the consequent volume of axonal loss after MPTP treatment) is much greater than in any serotonergic axon plexus in forebrain. In addition, the increase in reactive GFAP-immunopositive astrocytes in striatum caused by DA terminal loss after 6-OHDA injection is not nearly as robust as that caused by cell loss after kainic acid lesions (Ogawa et al., 1989), indicating that degenerating nerve terminals are not as effective at eliciting astrocytosis as are degenerating cell bodies.

V. REINNERVATION: SEROTONERGIC PROJECTIONS

Amphetamine derivatives produce a significant loss of fine 5-HT axons throughout forebrain, while having little apparent effect on neuronal somata in the raphe nuclei. The relative sparing of cell bodies after administration of amphetamine derivatives raises the possibility that 5-HT axons might exhibit a regenerative response at long survival times. Several studies have demonstrated that monoaminergic neurons respond to the chemical injury of their axons by sprouting and reinnervating target areas (Björklund et al., 1973; Azmitia et al., 1978; Zhou and Azmitia, 1986; Date et al., 1990; Fritschy and Grzanna, 1992). After MDMA or methamphetamine administration to nonhuman primates, however, 5-HT innervation in forebrain has shown few signs of recovery (Insel et al., 1989; Woolverton et al., 1989; Ricaurte et al., 1991a,1992); depletion of 5-HT persists in rhesus and squirrel monkeys, reportedly as long as 1.5–4 yr after drug treatment. In contrast, biochemical studies in the rat report a recovery of 5-HT levels and numbers of 5-HT reuptake sites 2–4 mo after MDMA or FEN administration (Battaglia et al., 1988; Zaczek et al., 1990). Whether these increases in biochemical indices of axonal function in the rat reflect regeneration of 5-HT axons, as suggested (Battaglia et al., 1988; Zaczek et al., 1990), or represent a functional up-regulation within existing axons is not readily apparent, however, especially since several biochemical parameters do not recover in parallel (Zaczek et al., 1990).

Immunocytochemical studies in the rat corroborate the conclusion that 5-HT axons reinnervate forebrain target areas after denervation by amphetamine analogs. The denervation of rat forebrain after drugs such as PCA and MDA is followed by a slow progressive reinnervation (Molliver et al., 1989; Molliver and Axt, 1990; Mamounas and Molliver, 1991). The time course and pattern of regeneration of 5-HT axons is as follows. Between 1 and 2 mo after administration of MDA (20 mg/kg $\times$ 8) or PCA (10 mg/kg $\times$ 2), robust sprouting of 5-HT axons occurs, primarily in frontal pole (Molliver et al., 1989; Mamounas and Molliver, 1991; Fig. 8). From 2 to 6 mo, 5-HT-immunoreactive axons continue to increase in density in frontal cortex; progressively greater numbers of axons are observed in parietal cortex. Parietal cortex is more reinnervated than occipital cortex during this period. Thus, a rostro-caudal gradient of neocortical reinnervation is observed (Molliver et al., 1989; Mamounas and Molliver, 1991); rostral neocortex is reinnervated sooner and much more completely than caudal neocortex. Moreover, ventral areas of forebrain are reinnervated earlier and more completely than dorsal areas; for example, in rats with parietal cortex that is still relatively denervated 9 mo after PCA treatment, piriform cortex and amygdaloid nuclei are well reinnervated (Molliver and Axt, 1990). The rostro-caudal and ventro-dorsal reinnervation gradients may be related to the proximity of target areas such as frontal pole and piriform cortex to surviving 5-HT preterminal axons and cell bodies (i.e., the distance over which regenerating axons might travel).

The rostro-caudal and ventro-dorsal gradients of 5-HT reinnervation of neocortex after administration of amphetamine derivatives is, in several ways, similar to the pattern of growth of 5-HT axons into developing cerebral cortex in rats. During perinatal development, 5-HT axons innervate rostral and ventral areas of cerebral cortex earlier than more caudal and dorsal areas (Lidov and Molliver, 1982). Further, in MDA-treated rats, reinnervating axons initially exhibit a bilaminar pattern in fronto-parietal and parietal cortex (long tangential axons form parallel bands in layers I and VI); these axons later branch into the middle layers of neocortex (Molliver *et al.*, 1989). This pattern of regeneration reiterates the normal development of 5-HT axons, which form two sheets of axons above and below the cortical plate before arborizing into the middle of the cortex (Lidov and Molliver, 1982). The gradient of recovery strongly suggests that the progressive reappearance of 5-HT-immunoreactive axons months after amphetamine-induced denervation results from regrowth of axons, not simply from recovery of 5-HT levels within intact axons.

Although during the first several months after drug treatment the sprouting of 5-HT axons is robust, the reinnervation of rat neocortex remains incomplete. This phenomenon is particularly evident in caudal areas of dorsal neocortex, in which 5-HT axon density does not reach control levels, even as long as 8 mo after administration of PCA (Mamounas and Molliver, 1991; Axt *et al.*, 1992b) or MDA (Molliver *et al.*, 1989). The degree of reinnervation in caudal areas of neocortex is related to the degree of initial denervation at 2 wk. Since toxicity is temperature dependent, rats that are treated with 20 mg/kg PCA in a warm ambient temperature are denervated severely at 2 wk (i.e., preterminal axons are nearly absent in frontal pole) and exhibit minimal sprouting of 5-HT axons at 4 mo, even in frontal pole (Axt *et al.*, 1992b). In contrast, drug regimens that cause less severe denervation in rostral areas of neocortex, for example, MDA (Molliver *et al.*, 1989) or lower doses of PCA (Axt *et al.*, 1992b), result in a more robust and complete reinnervation of neocortex at longer survival times. When observed 4 mo after a low dose of PCA (2.5 mg/kg), 5-HT axon density in frontal pole is indistinguishable from age-matched controls, whereas higher doses result in progressively less reinnervation. By 8 mo survival, when parietal cortex is partially reinnervated after 10 mg/kg PCA, axon density in parietal cortex of 2.5 mg/kg PCA-treated rats is indistinguishable from that of controls (Axt et al., 1992b). Although, by 8 mo after 2.5 mg/kg PCA, the density of 5-HT axons in these more rostral areas of neocortex approaches or equals that observed in control rats, the most caudal portion of dorsal neocortex (occipital pole) remains relatively denervated (Axt et al., 1992b). The results of these studies suggest that the decreased capacity to regenerate after larger doses of PCA may depend on both the magnitude of the insult to 5-HT neurons and the distance over which the axons must travel to their targets (Axt et al., 1992b).

The morphology of the reinnervating axons (2–6 mo after drug treatment) is similar to that found in control animals (Molliver et al., 1989; Mamounas and Molliver, 1991; Fig. 9). Specifically, the 5-HT axons that reinnervate neocortex resemble the fine axons ablated by these drugs; beaded axons do not sprout and expand their terminal aborizations into denervated brain regions. The reinnervation by axons with the *fine axon* morphology raises the possibility that functional innervation may be restored. However, whether the 5-HT axons that reinnervate these cortical areas successfully reach their targets locally, establish the original neuronal circuitry, or form synaptic contacts is not known. Also, whether the reinnervating axons originate from cells with ablated processes or, alternatively, from homotypic collateral sprouting of preterminal and spared terminal axons from other neuronal somata has yet to be determined.

When rats were examined at very long survival times to determine whether areas such as occipital pole eventually were reinnervated, the normal pattern of 5-HT innervation was found not to be re-established after administration of PCA; occipital pole was not reinnervated by 12 mo survival (Mamounas and Molliver, 1991). Moroever, the regeneration process is aborted and regresses between 6 and 12 mo after PCA treatment (Mamounas and Molliver, 1991); in frontal and parietal cortex, where robust sprouting is observed initially (between 2 and 6 mo after PCA treatment), the density of 5-HT axons gradually decreases between 6 and 12 mo after drug treatment. This delayed loss of 5-HT axons in cerebral cortex is accompanied by increasing numbers of structurally abnormal axons. The morphology of the abnormal axons at these long survival times has been described as thick and tortuous, and includes convoluted tangles or twisted knots (Mamounas and Molliver, 1991; Mamounas et al., 1992). In rats treated with PCA at 2 mo of age, tangled 5-HT axons are not observed during the first 6 mo after drug administration (i.e., in rats aged 4–8 mo) (Mamounas and Molliver, 1991; Mamounas et al., 1992). From 6 to 12 mo after PCA treatment, however, the tangled axons increase in frequency, so by 1 yr after drug treatment (i.e., in 14-month-old rats), these abnormal axons form a large proportion of the 5-HT plexus in neocortex. Thus, despite the slow yet robust regenerative process, the normal 5-HT innervation pattern is not re-established, and aberrant axon morphology develops.

One possible interpretation of the secondary (or delayed) axon loss observed 1 yr after PCA treatment might be that older rats are unable to continue the regenerative processes, that is, age per se may prohibit complete reinnervation. To test this possibility, 10-month-old rats were administered a comparable dose regimen of PCA and were allowed to survive for 4 mo, to 14 mo of age. These older rats exhibited a regenerative response comparable to that observed in young adult rats (i.e., those treated at age 2 mo and allowed to survive 4 mo); the reinnervation density and the morphology of the sprouting axons in the older rats was similar to that observed in the young adult rats (Mamounas *et al.*, 1992). These results indicate that (1) older rats (10–14 mo of age) retain the capacity for 5-HT axon regeneration and (2) the regression in axon density and development of abnormal axons observed 1 yr after PCA treatment are more dependent on the protracted regenerative process than on the age of the animal.

The degenerative changes, specifically the highly convoluted and thickened 5-HT axons, observed 1 yr after PCA treatment (14-month-old rats) resemble aberrant 5-HT axons normally found in aged (2-year-old) untreated rats (van Luijtelaar *et al.*, 1989; Mamounas and Molliver, 1991; Mamounas *et al.*, 1992). In untreated young adult rats (2–4 mo of age), convoluted 5-HT axons are extremely rare (van Luijtelaar *et al.*, 1989; Mamounas and Molliver, 1991; Mamounas *et al.*, 1992), but are first apparent in rats that are about 12 mo of age. The prevalence and tortuousness of these abnormal 5-HT axons increases through 2 yr of age (Mamounas *et al.*, 1992), indicating that the appearance of this type of 5-HT axon morphology is a pathological feature of the normal aging process. Further, neocortex of 2-year-old (untreated) rats exhibits a less dense 5-HT plexus than that of 1-year-old (or younger) rats (Mamounas *et al.*, 1992; van Luijtelaar *et al.*, 1992), suggesting that, as rats age, 5-HT axons begin to degenerate and innervation density progressively decreases. In contrast, this normal progression of cytopathology appears to be accelerated in PCA-treated rats; 1 yr after PCA treatment, the number of convoluted 5-HT axons in neocortex is 3-fold higher than that found in age-matched controls (i.e., 14-month-old rats) (Mamounas *et al.*, 1992). Therefore, the combined observations of premature development of age-related morphological changes and the late secondary axon loss strongly suggest that, in PCA-treated rats, serotonergic neurons and their projections undergo an accelerated process of aging.

In summary, the extremely slow time course of the recovery, the morphology of the regenerating axons, and the spatial distribution of the sprouting axons (copying the pattern of perinatal development) support the conclusion that amphetamine derivatives are neurotoxic and cause fine 5-HT axons to degenerate. The neurotoxic action of this class of drugs is manifested further in the observations of a second delayed axon loss that follows an aborted regenerative response. This secondary degenerative phase observed at the axon terminals may reflect a delayed, adverse, retrograde response within the neuron somata, appearing only after long survival times.

VI. SUMMARY

Unequivocal evidence for denervation and axonal degeneration is not expected to be found using one method alone. The absence of neurotransmitter, measured biochemically or histologically, does not by itself indicate axon loss, although functional deficits are likely. This is exemplified by administration of the TPH inhibitor *p*-chlorophenylalanine, which irreversibly binds to the enzyme (Jéquier *et al.*, 1967) thereby reducing 5-HT levels for up to 1 wk, but causes no structural damage to the axons (Axt *et al.*, 1993). Further, measurement of biosynthetic enzyme activity that appears normal may not indicate normal recovery from a drug insult, since transmitter turnover still may be abnormal. Such a phenomenon has been observed in striatum of mice treated with MPTP, in which TH levels and immunoreactivity were observed to recover much more quickly than DA levels (Mori *et al.*, 1988). This disparity apparently was caused in part by an abnormally high DA catabolism rate, which the authors speculated might be the result of a long-lasting disruption of DA storage within the axons. Thus, by combining a variety of biochemical and histological methods for assessing axonal integrity and function, a constellation of evidence for nerve terminal damage can be gathered. Only then can the breadth of neurotoxic drug effects be realized and the conclusion be reached that axonal degeneration has occurred.

In conjunction with biochemical data, the anatomic data reviewed here indicate that amphetamine derivatives are neurotoxic to serotonergic and dopaminergic nerve terminals, while having relatively little effect on their cell bodies. Anatomic studies of the effects of amphetamine analogs on the serotonergic system demonstrate that (1) these drugs produce a prolonged decrease in axon density, (2) the decreased density of immunoreactive axons is substantiated by a decreased number of reuptake sites located on those axons, (3) retrograde transport from axonal terminals to cell bodies is impaired, (4) the effects are restricted to a select population of neuronal projections, and (5) this persistent denervation is the result of axonal degeneration, evidenced by structural alterations indicative of disrupted cytoskeleton within the axons, argyrophilia, and mild glial reactivity. Further, serotonin neurons respond to this chemical insult by sprouting new, morphologically homotypic axons. This regenerative response proceeds very slowly (over months) and eventually is aborted, resulting in a secondary phase of axon loss and degenerative morphologic changes.

A. Phases of Neurotoxicity

The neurotoxic effects of amphetamine derivatives on serotonergic axons can be summarized by dividing the morphologic correlates into five

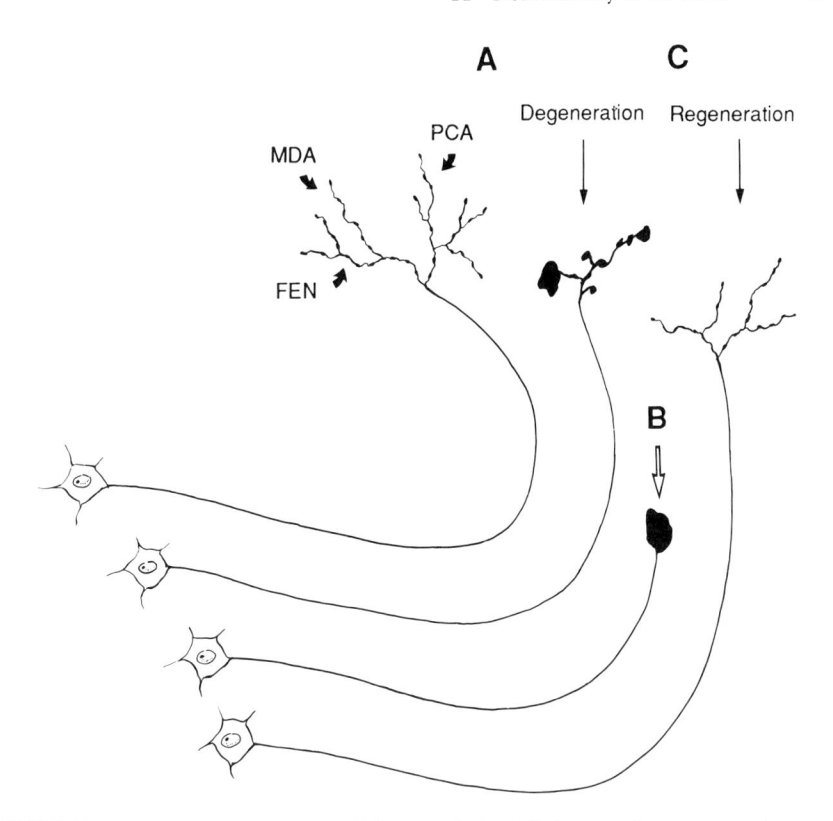

FIGURE 7 Schematic representation of the morphological changes that occur to the axon terminals of a 5-HT neuron projecting from dorsal raphe nucleus (DRN) to neocortex after administration of an amphetamine analog. (A) The process of amphetamine-induced degeneration of the axon terminal at 2–3 days. (B) The resultant denervation (open arrow) at which stage (2 wk) the preterminal axon ends with cytoskeletal and other proteins accumulating to form a bulbous stump, and no axon terminal is observed in the cortex. (C) The later (~ 4 mo) *regeneration* of the axon terminal, presumably into the same cortical area. FEN, fenfluramine; MDA, methylenedioxyamphetamine; PCA, *p*-chloroamphetamine. Artist: *Mi Young Toh*.

temporal phases, as we have suggested previously (Molliver *et al.*, 1990): (1) an acute depletion phase (minutes to hours), in which 5-HT is released from the axons and 5-HT synthesis is inhibited; (2) a degeneration phase (days), in which 5-HT axons actively degenerate or recuperate from the drug insult; (3) a denervation phase (weeks to months), in which 5-HT innervation density is decreased in many brain regions, (4) a regeneration phase (months), a long period of time during which 5-HT axons are found slowly to reinnervate previously denervated brain regions; and (5) accelerated aging (> mo after drug treatment). Phases 2–4 of the reaction of 5-HT axons to neurotoxic amphetamines are depicted schematically in Fig. 7. A DRN neuron and its fine axon innervating an area of neocortex are shown with

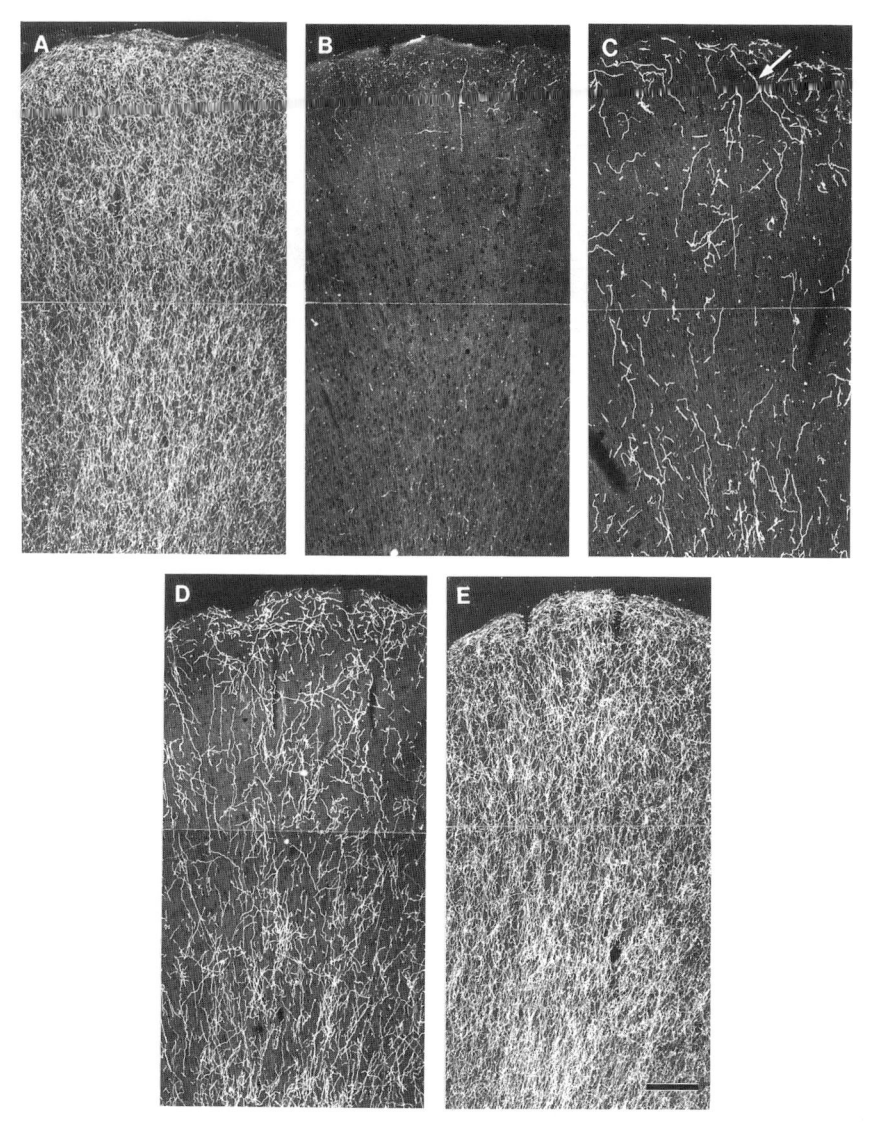

FIGURE 8 Phases of *p*-chloroamphetamine (PCA) neurotoxicity in rat frontal cortex (FCTX). Darkfield photomicrographs of 5-HT-immunoreactive axons from control (A) and PCA-treated rats at 4-hr (B), 3-day (C), 2-wk (D), and 4-mo (E) survivals. 5-HT is depleted from nearly all axons in FCTX at 4 hr, but by 3 days several preterminal axons regain 5-HT levels, although most of these axons morphologically appear damaged (see Fig. 7B). By 2 wk, a number axons have regained their ability to synthesize and store 5-HT, but most axons have been ablated. At 4 mo, 5-HT axons have regenerated into FCTX and approach the density of control rats. Scale bar, 200 μm.

various amphetamine derivatives acting at the axon terminal. Next to that axon is a representation of the events at this axon 2–3 days after drug treatment, during the degeneration phase. The axon has increased caliber and Molliver, 1991b). Several studies indicate that the morphological class of 5-HT axons (fine axons) that is susceptible to the neurotoxic effects of other proteins have accumulated. No terminal axon is observed in neocortex at this stage. In the fourth phase (regeneration), by 2–4 mo after drug treatment, the neuron has sprouted a new axon terminal, presumably into its original target area, with morphology comparable to that prior to drug insult although perhaps with fewer axon collaterals.

The first four phases of the neurotoxic action of amphetamine analogs on 5-HT axons are depicted photographically in Fig. 8; serotonergic innervation of frontal cortex is shown in a control rat and at various time points after PCA administration. At 4 hr, very few 5-HT-immunoreactive axons are visible except some thicker preterminal axons. No beaded axons innervate this area of frontal cortex. However, in regions of forebrain innervated by beaded 5-HT axons, the beaded axons would be visible at this time (cf. Axt and Molliver, 1991b). Several studies indicate that the morphological class of 5-HT axons (fine axons) that is susceptible to the neurotoxic effects of amphetamine derivatives selectively is depleted of transmitter acutely; 4 hr after administration of PCA, MDA, FEN, and methamphetamine, beaded 5-HT-immunoreactive axons are visible, retaining their normal morphology, and fine axons are absent (Molliver *et al.*, 1988; Molliver and Molliver, 1990; Axt and Molliver, 1991b; M. E. Molliver, unpublished observations). These results indicate a direct relationship between the pharmacological action of these drugs and neurotoxic actions. Low doses of methamphetamine, which produce no long-lasting serotonergic deficits, also cause a significant decrease in the density of fine 5-HT-immunoreactive axons at 90 min and 4 hr (Fukui *et al.*, 1989; Axt and Molliver, 1991b).

By 3 days after drug treatment, some fine axons have recovered levels of 5-HT and therefore are immunoreactive, although in PCA-treated rats most of these axons exhibit abnormal morphology indicative of axon degeneration (Figs. 6,9). Profiles of swollen and distorted axons are visible most frequently from 1 to 4 days, after which the incidence of pathology subsides (O'Hearn *et al.*, 1988; Molliver and Molliver, 1990; Axt *et al.*, 1993). Between 10 days and 2 wk, the density of fine axons has stabilized and remains relatively unchanged through 1 mo. At this stage the brain regions are considered denervated; very few fine axons display degenerating profiles, but appear morphologically similar to fine axons in control brains (Molliver and Axt, 1990; Mamounas and Molliver, 1991; Fig. 7). In frontal pole, the number of 5-HT-immunoreactive axons is greater at 2 wk than at 3 days (Fig. 6). However, that this evidence represents an early stage of regeneration is not clear. Although that possibility cannot be ruled out, some recovery of 5-HT levels is likely to be ongoing at 3 days in preterminal axons (which are

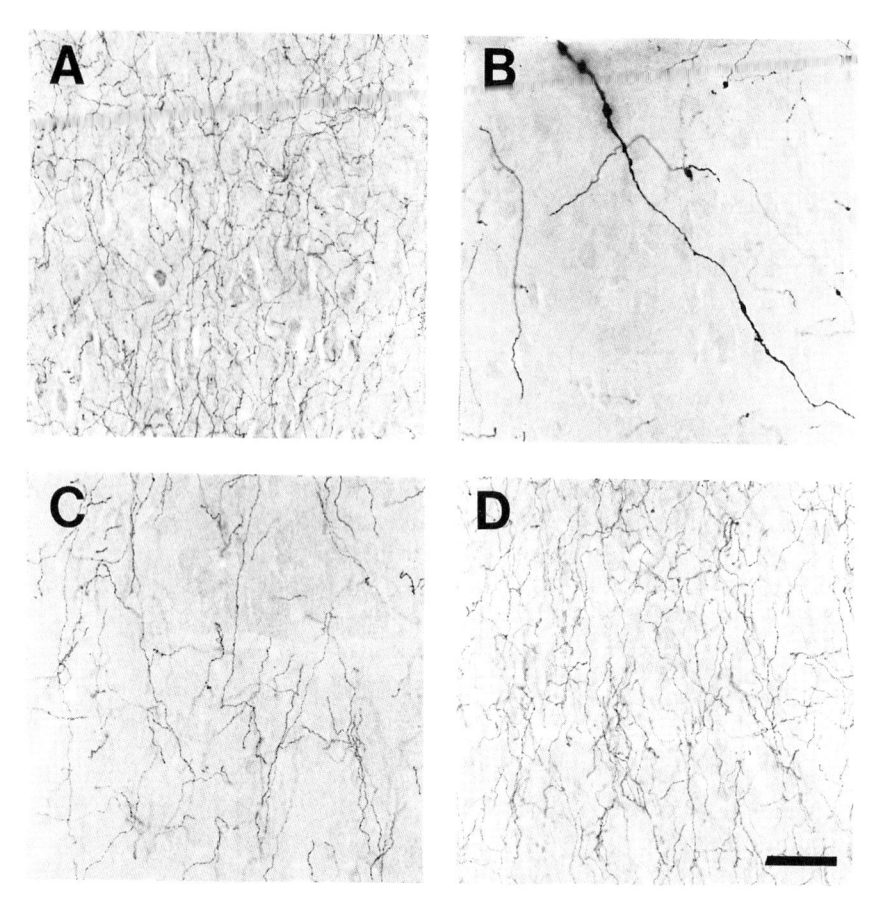

FIGURE 9 5-HT axon morphology in frontal cortex at various survival times after *p*-chloroamphetamine (PCA). (A) 5-HT-immunoreactive axons in control frontal cortex appear coarser than in, for example, parietal cortex, and axon density is rather high. (B) Of the 5-HT axons that are immunoreactive at 3 days, many exhibit an abnormal morphology, suggestive of axon degeneration. The most prominent axon in B is indicated by the arrow in Fig. 8c. (C) 5-HT-immunoreactive axons display normal morphology 2 wk after PCA administration, but axon density is reduced greatly compared with the control. (D) At 4-mo survival, 5-HT axon density has increased and the morphology of the reinnervating axons resembles that of controls. Scale bar, 50 μm.

abundant in this brain region); these preterminal axons simply are not yet visible at this early time point, and regain detectable 5-HT levels by 2 wk. Finally, by 2 mo survival, reinnervation of several areas of forebrain has begun and continues through 4 and 6 mo survival (Molliver *et al.*, 1989; Molliver and Mamounas, 1991).

B. Future Directions

The pattern of 5-HT axon loss 2 wk after drug administration reflects the differing distribution of fine and beaded 5-HT axons; the more restricted and specific neuronal connections of beaded 5-HT axons are retained after amphetamine administration, whereas the more global innervation by the fine axons is lost. Considering that fine and beaded axon projections are likely to have different functions in the brain, future studies should be designed to elucidate whether any behavioral or physiological deficits can be ascribed to these distinct projections and, in turn, how the morphological selectivity of amphetamine analogs would predict behavioral deficits resulting from drug administration. The pharmacological differences between fine and beaded (and, more generally, DRN and MRN) axons in neocortex also present a direction for future investigation into the physiological and biochemical differences that might exist between the morphological subtypes. The differential vulnerability of fine and beaded 5-HT axons may be analyzed through electron microscopic examination of the axons, the identification of proteins or subcellular components in cell soma or axons that might protect the beaded type neurons, or the development of ligands specific to possible variations in 5-HT carrier proteins on the axonal membranes.

If future studies of the DA nigrostriatal projections were to reveal a differential vulnerability of type A and type B axons to the neurotoxic effects of amphetamine or methamphetamine, a pattern of monoaminergic axon vulnerability and resistance to chemical toxins based on axon morphology or nuclear origin would be complete. Specifically, as mentioned, the amphetamine-sensitive 5-HT axons originate in the DRN (Mamounas and Molliver, 1988) and have a fine caliber with small fusiform varicosities (Mamounas et al., 1991); this axonal morphology is similar to both the DA type A axons (of substantia nigra pars compacta), which may be affected acutely by amphetamines (Fukui et al., 1986; Ryan et al., 1988), and the NE axons originating in locus coeruleus, which are selectively vulnerable to the neurotoxic effects of N-(2-chloroethyl)-N-ethyl-2-bromobenzylamine (DSP4) (Fritschy and Grzanna, 1989). In addition, the monoaminergic axons resistant to these toxins—5-HT beaded axons originating from median raphe, the DA type B axons originating in ventral pars compacta and pars reticulata of substantia nigra, and non-locus coeruleus NE axons—all have characteristically larger varicosities and, in the case of 5-HT and NE axons, very fine intervaricose segments. The pharmacological, physiological, physical, or molecular basis for the differential effect of these neurotoxins is unknown, but may be related to specific factors such as comparable morphology, metabolism, or differences in numbers or affinities of the monoamine reuptake carriers. The curious relationship between monoamine axon projections from circumscribed nuclei and their vulnerability to peripherally administered neurotoxins and capacity for regeneration (Molliver et al., 1989; Molliver and Axt, 1990; Fritschy and Grzanna, 1992) is a fruitful area of research.

Future directions for anatomic studies of the neurotoxicity of amphetamine analogs also may include evaluation of neuronal changes at the subcellular or molecular level. For example, alterations in expression of structural proteins or reorganization or phosphorylation of these proteins may be visualized using *in situ* hybridization, immunocytochemistry, or electron microscopy, providing morphological evidence of axonal damage and capacity to regenerate. Monoaminergic neurons appear to exhibit properties of damage and recuperation that differ from those of other CNS or PNS systems, and thereby provide a unique system in which to study these processes. In addition, examination of neuronal–glial interactions as they pertain to the neurotoxic effects of amphetamine analogs may provide insight into processes of neuronal degeneration and regeneration.

ACKNOWLEDGMENTS

We thank Patrice Carr for expert technical assistance. This work was supported by U.S. Public Health Service research grants NS15199 and DA04431, and by NIDA contract 271-90-7408.

REFERENCES

Andén, N.-E., Dahlström, A., Fuxe, K., Larsson, K., Olson, L., and Ungerstedt, U. (1966). Ascending monoamine neurons to telencephalon and diencephalon. *Acta Physiol. Scand.* **67,** 313–326.

Appel, N. M., Contrera, J. F., and De Souza, E. B. (1989). Fenfluramine selectively and differentially decreases the density of serotonergic nerve terminals in rat brain: Evidence from immunocytochemical studies. *J. Pharmacol. Exp. Ther.* **249,** 928–943.

Appel, N. M., Mitchell, W. M., Contrera, J. F., and De Souza, E. B. (1990). Effects of high-dose fenfluramine treatment on monoamine uptake sites in rat brain: Assessment using quantitative autoradiography. *Synapse* **6,** 33–44.

Axt, K. J., and Molliver, M. E. (1991a). Beaded serotonergic axons are not typically associated with calbindin- or parvalbumin-positive neurons in rat brain. *Soc. Neurosci. Abstr.* **17,** 1180.

Axt, K. J., and Molliver, M. E. (1991b). Immunocytochemical evidence for methamphetamine-induced serotonergic axon loss in the rat brain. *Synapse* **9,** 302–313.

Axt, K. J., and Molliver, M. E. (1991c). Dorsal and median raphe projections to rap hippocampus are differentially affected by *p*-chloroamphetamine. *Third IBRO World Congr. Neurosci. Abstr.,* 287. Aug. 4–9. Montréal, Canada.

Axt, K. J., Mullen, C. A., and Molliver, M. E. (1992a). Cytopathologic features indicative of 5-hydroxytryptamine axon degeneration are observed in rat brain after administration of *d*- and *l*-methylenedioxyamphetamine. *Ann. N.Y. Acad. Sci.* **648,** 244–247.

Axt, K. J., Mamounas, L. A., and Molliver, M. E. (1992b). Regeneration of 5-HT projections to neocortex: Magnitude of *p*-chloroamphetamine-induced denervation influences reinnervation. *Soc. Neurosci. Abstr.* **18,** 629.

Axt, K. J., Mullen, C. A., and Molliver, M. E. (1993). Immunocytochemical evidence for structural damage to serotonergic axons in rat brain following administration of methamphetamine, *p*-chloramphetamine, or isomers of methylenedioxyamphetamine. *Neuroscience,* submitted for publication.

Azmitia, E. C., Buchan, A. M., and Williams, J. H. (1978). Structural and functional restoration by collateral sprouting of hippocampal 5-HT axons. *Nature (London)* **274,** 374–376.

Battaglia, G., Yeh, S. Y., O'Hearn, E., Molliver, M. E., Kuhar, M. J., and De Souza, E. B. (1987). 3,4-Methylenedioxymethamphetamine and 3,4-methylenedioxyamphetamine destroy serotonin terminals in rat brain: Quantification of neurodegeneration by measurement of [^{3}H]paroxetine-labeled serotonin uptake sites. *J. Pharmacol. Exp. Ther.* **242,** 911–916.

Battaglia, G., Yeh, S. Y., and De Souza, E. B. (1988). MDMA-induced neurotoxicity: Parameters of degeneration and recovery of brain serotonin neurons. *Pharmacol. Biochem. Behav.* **29,** 269–274.

Battaglia, G., Sharkey, J., Kuhar, M. J., and De Souza, E. B. (1991). Neuroanatomic specificity and time course of alterations in rat brain serotonergic pathways induced by MDMA (3,4-methylenedioxymethamphetamine): Assessment using quantitative autoradiography. *Synapse* **8,** 249–260.

Baumgarten, H. G., Lachenmayer, L., and Schlossberger, H. G. (1972). Evidence for a degeneration of indoleamine containing nerve terminals in rat brain, induced by 5,6-dihydroxytryptamine. *Z. Zellforsch,* **125,** 553–569.

Beltramino, C. A., de Olmos, J., Gallyas, S., Heimer, L., and Zaborsky, L. (1993). Silver staining as a tool for neurotoxic assessment. *NIDA Res. Monogr.* **136,** 101–132.

Berger, B., Thierry, A. M., Tassin, J. P., and Moyne, M. A. (1976). Dopaminergic innervation of the rat prefrontal cortex: A fluorescence histochemical study. *Brain Res.* **106,** 133–145.

Björklund, A., and Lindvall, O. (1984). Dopamine-containing systems in the CNS. In "Handbook of Chemical Neuroanatomy" (A. Björklund and T. Hökfelt, eds.), Vol. 2, Classical Transmitters in the CNS, Part I, pp. 55–122. Elsevier, Amsterdam.

Björklund, A., Nobin, A., and Stenevi, U. (1973). Regeneration of central serotonin neurons after axonal degeneration induced by 5,6-dihydroxytryptamine. *Brain Res.* **50,** 214–220.

Björklund, H., Olson, L., Dahl, D., and Schwarcz, R. (1986). Short- and long-term consequences of intracranial injections of the excitotoxin, quinolinic acid, as evidenced by GFA immunocytochemistry of astrocytes. *Brain Res.* **371,** 267–277.

Blue, M. E., Yagaloff, K. A., Mamounas, L. A., Hartig, P. R., and Molliver, M. E. (1988). Correspondence between 5-HT$_2$ receptors and serotonergic axons in rat neocortex. *Brain Res.* **453,** 315–328.

Bostwick, D. G. (1981). Amphetamine induced cerebral vasculitis. *Hum. Pathol.* **12,** 1031–1033.

Brodal, A. (1982). Anterograde and retrograde degeneration of nerve cells in the central nervous system. In "Histology and Histopathology of the Nervous System" (W. Haymaker and R. D. Adams, eds.), pp. 276–362. Charles C. Thomas, Springfield, Illinois.

Brunello, N., Chuang, D. M., and Costa, E. (1982). Different synaptic location of mianserin and imipramine binding sites. *Science* **215,** 1112–1115.

Brunswick, D. J., Benmansour, S., Tejani-Butt, S. M., and Hauptmann, M. (1992). Effects of high-dose methamphetamine on monoamine uptake sites in rat brain measured by quantitative autoradiography. *Synapse* **11,** 287–293.

Cajal, S. R.-y (1928). "Regeneration and Degeneration of the Central Nervous System" (transl. R. M. May), Vol. 2. Oxford University Press, London.

Carlsen, J., and de Olmos, J. (1981). A modified cupric silver technique for the impregnation of degenerating neurons and their processes. *Brain Res.* **208,** 426–431.

Champney, T. H., and Matthews, R. T. (1991). Pineal serotonin is resistant to depletion by serotonergic neurotoxins in rats. *J. Pineal Res.* **11,** 163–167.
Citron, B. P., Halpern, M., McCarron, M., Lundyere, G. D., McCormick, R., Pincus, I. J., Tatter, D., and Haverback, B. S. (1970). Necrotizing angietus associated with drug abuse. *N. Engl. J. Med.* **283,** 1003–1011.
Clineschmidt, B. V., Totaro, J. A., McGuffin, J. C., and Pflueger, A. B. (1976). Fenfluramine: Long-term reduction in brain serotonin (5-hydroxytryptamine). *Eur. J. Pharmacol.* **35,** 211–214.
Commins, D. L., and Seiden, L. S. (1986). α-Methyltyrosine blocks methylamphetamine-induced degeneration in the rat somatosensory cortex. *Brain Res.* **365,** 15–20.
Commins, D. L., Axt, K. J., Vosmer, G., and Seiden, L. S. (1987a). Endogenously produced 5,6-dihydroxytryptamine may mediate the neurotoxic effects of *para*-chloroamphetamine. *Brain Res.* **419,** 253–261.
Commins, D. L., Vosmer, G., Virus, R. M., Woolverton, W. L., Schuster, C. R., and Seiden, L. S. (1987b). Biochemical and histological evidence that methylenedioxymethamphetamine (MDMA) is toxic to neurons in the rat brain. *J. Pharmacol. Exp. Ther.* **241,** 338–345.
Conrad, L. C. A., Leonard, C. M., and Pfaff, D. W. (1974). Connections of the median and dorsal raphe nuclei in the rat: An autoradiographic and degeneration study. *J. Comp. Neurol.* **156,** 179–206.
Dahlström, A., and Fuxe, K. (1964). Evidence for the existence of monoamine neurons in the central nervous system. I. Demonstration of monoamines in the cell bodies of brain stem neurons. *Acta Physiol. Scand. (Suppl. 232)* **62,** 1–55.
Dastur, D. K., Thakkar, B. K., and Desai, P. R. (1985). Experimental neurotoxicity of the anorectic fenfluramine. I. A fine structural model for cerebral lysosomal storage and neuroglial reaction. *Acta Neuropathol. (Berlin)* **67,** 142–154.
Date, I., Felten, D. L., and Felten, S. Y. (1990). Long-term effect of MPTP in the mouse brain in relation to aging: Neurochemical and immunocytochemical analysis. *Brain Res.* **519,** 266–276.
DeFilipe, J., Hendry, S. H. C., Hashikawa, T., and Jones, E. G. (1991). Synaptic relationships of serotonin-immunoreactive terminal baskets on GABA neurons in the cat auditory cortex. *Cerebral Cortex* **1,** 1047–3211.
Descarries, L., Lemay, B., Doucet, G., and Berger, B. (1987). Regional and laminar density of the dopamine innervation in adult rat cerebral cortex. *Neuroscience* **21,** 807–824.
Eisch, A. J., Gaffney, M., Wiehmuller, F. B., O'Dell, S. J., and Marshall, J. F. (1992). Striatal subregions are differentially vulnerable to the neurotoxic effects of methamphetamine. *Brain Res.* **598,** 321–326.
Ellison, G., Eison, M. S., Huberman, H. S., and Daniel, F. (1978). Long-term changes in dopaminergic innervation of caudate nucleus after continuous amphetamine administration. *Science* **201,** 276–278.
Fallon, J. H., and Loughlin, S. E. (1982). Monoamine innervation of the forebrain: Collateralization. *Brain Res. Bull.* **9,** 295–307.
Fallon, J. H., and Moore, R. (1978). Catecholamine innervation of the basal forebrain: IV Topography of the dopamine projection to the basal forebrain and neostriatum. *J. Comp. Neurol.* **180,** 545–572.
Fink, R. P., and Heimer, L. (1967). Two methods for selective silver impregnation of degenerating axons and their synaptic endings in the central nervous system. *Brain Res.* **4,** 369–374.
Freund, T. F., Gulyás, A. I., Acsády, L., Görcs, T., and Tóth, K. (1990). Serotonergic control of the hippocampus via local inhibitory interneurons. *Proc. Natl. Acad. Sci. U.S.A.* **87,** 8501–8505.
Fritschy, J.-M., and Grzanna, R. (1989). Immunohistochemical analysis of the neurotoxic effects of DSP-4 identifies two populations of noradrenergic axon terminals. *Neuroscience* **30,** 181–197.

Fritschy, J.-M., and Grzanna, R. (1992). Restoration of ascending noradrenergic projections by residual locus coeruleus neurons: Compensatory response to neurotoxin-induced cell death in the adult rat brain. *J. Comp. Neurol.* **321,** 421–441.

Fritschy, J.-M., Lyons, W. E., Molliver, M. E., and Grzanna, R. (1988). Neurotoxic effects of p-chloroamphetamine on the serotonergic innervation of the trigeminal motor nucleus: A retrograde transport study. *Brain Res.* **473,** 261–270.

Fukui, K., Kariyama, H., Kashiba, A., Kato, N., and Kimura, H. (1986). Further confirmation of heterogeneity of the rat striatum: Different mosaic patterns of dopamine fibers after administration of methamphetamine or reserpine. *Brain Res.* **382,** 81–86.

Fukui, K., Nakajima, T., Kariyama, H., Kashiba, A., Kato, N., Tohyama, I., and Kimura, H. (1989). Selective reduction of serotonin immunoreactivity in some forebrain regions of rats induced by acute methamphetamine treatment: Quantitative morphometric analysis by serotonin immunocytochemistry. *Brain Res.* **482,** 198–203.

Fuller, R. W. (1985). Drugs altering serotonin synthesis and metabolism. In "Neuropharmacology of Serotonin" (A. R. Green, ed.), pp. 1–20. Oxford University Press, New York.

Fuxe, K. (1964). Cellular localization of monoamines in the median eminence and infundibular stem of some mammals. *Z. Zellforsch. Mikrosk. Anat.* **61,** 710–724.

Fuxe, K. (1965). Evidence for the existence of monoamine neurons in the central nervous system. IV. Distribution of monoamine nerve terminals in the central nervous system. *Acta Physiol. Scan. (Suppl. 247)* **64,** 39–85.

Fuxe, K., Hökfelt, T., Johansson, O., Jonsson, G., Lidbrink, P., and Ljungdahl, A. (1974). The origin of the dopamine nerve terminals in limbic and frontal cortex. Evidence for mesocortical-dopamine neurons. *Brain Res.* **82,** 349–355.

Fuxe, K., Ögren, S.-O., Agnati, L. F., Jonsson, G., and Gustafsson, J.-Å. (1978). 5,7-Dihydroxytryptamine as a tool to study the functional role of central 5-hydroxytryptamine neurons. *Ann. N.Y. Acad. Sci.* **305,** 346–369.

Gall, C., and Moore, R. Y. (1984). Distribution of enkephalin, substance P, tyrosine hydroxylase, and 5-hydroxytryptamine immunoreactivity in the septal region of the rat. *J. Comp. Neurol.* **225,** 212–227.

Gallyas, F., Wolff, J. R., Böttcher, H., and Zaborsky, L. (1980). A reliable and sensitive method to localize terminal degeneration and lysosomes in the central nervous system. *Stain Technol.* **55,** 299–306.

Gerfen, C. R. (1992). The neostriatal mosaic: Multiple levels of compartmental organization in the basal ganglia. *Ann. Rev. Neurosci.* **15,** 285–320.

Gerfen, C. R., Herkenham, M., and Thibault, J. (1987a). The neostriatal mosaic. II. Patch and matrix-directed mesostriatal dopaminergic and non-dopaminergic systems. *J. Neurosci.* **7,** 3915–3934.

Gerfen, C. R., Baimbridge, K. G., and Thibault, J. (1987b). The neostriatal mosaic. III. Biochemical and developmental dissociation of patch-matrix mesostriatal systems. *J. neurosci.* **7,** 3935–3944.

Gobbi, M., Cervo, L., Taddei, C., and Mennini, T. (1990). Autoradiographic localization of [^{3}H]paroxetine specific binding in the rat brain. *Neurochem. Int.* **16,** 247–251.

Grafstein, B., and Forman, D. S. (1980). Intracellular transport in neurons. *Physiol. Rev.* **60,** 1167–1283.

Halasy, K., Miettinen, R., Szabat, E., and Freund, T. F. (1992). GABAergic interneurons are the major postsynaptic targets of median raphe afferents in the rat dentate gyrus. *Eur. J. Neurosci.* **4,** 144–153.

Haring, J. H., Myerson, L., and Hoffman, T. L. (1992). Effects of parachloroamphetamine upon the serotonergic innervation of the rat hippocampus. *Brain Res.* **577,** 253–260.

Harvey, J. A., and McMaster, S. E. (1975). Fenfluramine: evidence for a neurotoxic action on midbrain and a long-term depletion of serotonin. *Psychopharmacol. Commun.* **1,** 217–228.

Harvey, J. A., McMaster, S. E., and Yunger, L. M. (1975). p-Chloroamphetamine: Selective neurotoxic action in brain. *Science* **187,** 841–843.

Harvey, J. A., McMaster, S. E., and Fuller, R. W. (1977). Comparison between the neurotoxic and serotonergic-depleting effects of various halogenated derivatives of amphetamine in the rat. *J. Pharmacol. Exp. Ther.* **202**, 581–589.

Heimer, L. (1970). The olfactory cortex and ventral striatum. *In* "Limbic Mechanisms—The Continuing Evolution of the Limbic System Concept" (K. E. Livingston and O. Hornykiewicz, eds.), pp. 95–187. Plenum Press, New York.

Hornung, J.-P., and Celio, M. R. (1992). The selective innervation by serotoninergic axons of calbindin-containing interneurons in the neocortex and hippocampus of the marmoset. *J. Comp. Neurol.* **320**, 457–467.

Hornung, J.-P., Fritschy, J.-M., and Törk, I. (1990). Distribution of two morphologically distinct subsets of serotonergic axons in the cerebral cortex of the marmoset. *J. Comp. Neurol.* **297**, 165–181.

Iacopino, A., Christakos, S., German, D., Sonsalla, P. K., and Altar, C. A. (1992). Calbindin-D28k-containing neurons in animals models of neurodegeneration: Possible protection from excitotoxicity. *Mol. Brain Res.* **13**, 251–261.

Imai, H., Steindler, D. A., and Kitai, S. T. (1986). The organization of divergent axonal projections from the midbrain raphe nuclei in the rat. *J. Comp. Neurol.* **243**, 363–380.

Insel, T. R., Battaglia, G., Johanssen, J., Marra, S., and De Souza, E. B. (1989). 3,4-Methylenedioxymethamphetamine ("Ecstasy") selectively destroys brain serotonin nerve terminals in rhesus monkeys. *J. Pharmacol. Exp. Ther.* **249**, 713–720.

Iversen, L. L. (1973). Neuronal and extraneuronal catecholamine uptake mechanisms. *In* "Frontiers in Catecholamine Research" (E. Usdin and S. H. Snyder, eds.), pp. 403–408. Pergamon, Oxford.

Jacobs, B. L., and Azmitia, E. C. (1992). Structure and function of the brain serotonin system. *Physiol. Rev.* **72**, 165–229.

Jensen, K. F., Miller, D. B., Olin, J. K., Haykal-Coates, N., and O'Callaghan, J. P. (1991). Characterization of methylenedioxymethamphetamine (MDMA) neurotoxicity using the De Olmos cupric silver stain and GFAP immunohistochemistry: Changes in specific neocortical regions implicate serotonergic and non-serotonergic targets. *Soc. Neurosci. Abstr.* **17**, 1429.

Jéquier, E., Lovenberg, W., and Sjoerdsma, A. (1967). Tryptophan hydroxylase inhibition: The mechanism by which *p*-chlorophenylalanine depletes rat brain serotonin. *Mol. Pharmacol.* **3**, 274–278.

Jonsson, G., and Nwanze, E. (1982). Selective (+)-amphetamine neurotoxicity on striatal dopamine nerve terminals in the mouse. *Br. J. Pharmacol.* **77**, 335–345.

Kalia, M. (1991). Reversible, short-lasting, and dose-dependent effect of (+)-fenfluramine on neocortical serotonergic axons. *Brain Res.* **548**, 111–125.

Kindy, M. S., Bhat, A. N., and Bhat, N. R. (1992). Transient ischemia stimulates glial fibrillary acid protein and vimentin gene expression in the gerbil neocortex, striatum, and hippocampus. *Mol. Brain Res.* **13**, 199–206.

Kleven, M. S., and Seiden, L. S. (1989). D-, L- and DL-Fenfluramine cause long-lasting depletions of serotonin in rat brain. *Brain Res.* **505**, 351–353.

Köhler, C. (1982). On the serotonergic innervation of the hippocampal region: An analysis employing immunohistochemistry and retrograde fluorescent tracing in the rat brain. *In* "Cytochemical Methods in Neuroanatomy" (S. L. Palay and V. Chan-Palay, eds.), pp. 387–405. Liss, New York.

Köhler, C., Ross, S. B., Srebro, B., and Ögren, S.-O. (1978). Long-term biochemical and behavioral effects of *p*-chloroamphetamine in the rat. *Ann. N.Y. Acad. Sci.* **305**, 645–663.

Köhler, C., Chan-Palay, V., Haglund, L., and Steinbusch, H. W. M. (1980). Immunohistochemical localization of serotonin nerve terminals in the lateral entorhinal cortex of the rat: Demonstration of two separate patterns of innervation from the midbrain raphe. *Anat. Embryol.* **160**, 121–129.

Köhler, C., Chan-Palay, V., and Steinbusch, W. (1981). The distribution and orientation of serotonin fibers in the entorhinal and other retrohippocampal areas. An immunohistochemical study with anti-serotonin antibodies in the rat's brain. *Anat. Embryol.* **161,** 237–264.

Köhler, C., Chan-Palay, V., and Steinbusch, H. (1982). The distribution and origin of serotonin-containing fibers in the septal area: A combined immunohistochemical and fluorescent retrograde tracing study in the rat. *J. Comp. Neurol.* **209,** 91–111.

Kosofsky, B. E., and Molliver, M. E. (1987). The serotonergic innervation of cerebral cortex: Different classes of axon terminals arise from dorsal and median raphe nuclei. *Synapse* **1,** 153–168.

Kovachich, G. B., Aronson, C. E., and Brunswick, D. J. (1989). Effects of high-dose methamphetamine administration on serotonin uptake sites in rat brain measured using [^{3}H]cyanoimipramine autoradiography. *Brain Res.* **505,** 123–129.

Kuhar, M. J., and Aghajanian, G. K. (1973). Selective accumulation of ^{3}H-serotonin by nerve terminals of raphe neurons: An autoradiographic study. *Nature (London)* **241,** 187–189.

Lewis, M. S., Molliver, M. E., Morrison, J. H., and Lidov, H. G. W. (1979). Complementarity of dopaminergic and noradrenergic innervation in anterior cingulate cortex of the rat. *Brain Res.* **164,** 328–333.

Lidov, H. G. W., and Molliver, M. E. (1982). An immunohistochemical study of serotonin neuron development in the rat: Ascending pathways and terminal fields. *Brain Res. Bull,* **8,** 389–430.

Lidov, H. G. W., Grzanna, R., and Molliver, M. E. (1980). The serotonergic innervation of the cerebral cortex in the rat—An immunocytochemical analysis. *Neuroscience* **5,** 207–227.

Lieberman, A. R. (1971). The axon reaction: A review of the principal features of perikaryal responses to axon injury. *Int. Rev. Neurobiol.* **14,** 49–124.

Lindvall, O., and Björklund, A. (1984). General organization of cortical monoamine systems. In "Monoamine Innervation of Cerebral Cortex" (L. Descarries, T. R. Reader, and H. H. Jasper, eds.), pp. 9–40. Liss, New York.

Lindvall, O., Björklund, A., Moore, R. Y., and Stenevi, U. (1974). Mesencephalic dopamine neurons projecting to neocortex. *Brain Res.* **81,** 325–331.

Lorez, H. (1981). Fluorescence histochemistry indicates damage of striatal dopamine nerve terminals in rats after multiple doses of methamphetamine. *Life Sci.* **28,** 911–916.

Lorez, H., Saner, A., Richards, J. G., and Da Prada, M. (1976). Accumulation of 5HT in nonterminal axons after *p*-chloro-N-methyl-amphetamine without degeneration of identified 5HT nerve terminals. *Eur. J. Pharmacol.* **38,** 79–88.

Lorez, H. P., Saner, A., and Richards, J. G. (1978). Evidence against a neurotoxic action of halogenated amphetamines on serotonergic B9 cells. A morphometric fluorescence histochemical study. *Brain Res.* **146,** 188–194.

Mamounas, L. A., and Molliver, M. E. (1988). Evidence for dual serotonergic projections to neocortex: Axons from the dorsal and median raphe nuclei are differentially vulnerable to the neurotoxin *p*-chloroamphetamine (PCA). *Exp. Neurol.* **102,** 23–36.

Mamounas, L. A., and Molliver, M. E. (1991). Aberrant reinnervation of rat cerebral cortex by serotonergic axons after denervation by *p*-chloroamphetamine (PCA). *Soc. Neurosci. Abstr.* **17,** 1181.

Mamounas, L. A., Mullen, C. A., O'Hearn, E., and Molliver, M. E. (1991). Dual serotonergic projections to forebrain in the rat: Morphologically distinct 5-HT axon terminals exhibit differential vulnerability to neurotoxic amphetamine derivatives. *J. Comp. Neurol.* **314,** 558–586.

Mamounas, L. A., Axt, K. J., and Molliver, M. E. (1992). Abnormal morphology of regenerated 5-HT axons in rat cerebral cortex one year after ablation by *p*-chloroamphetamine (PCA): Accelerated aging of serotonergic projections. *Soc. Neurosci. Abstr.* **18,** 629.

Marcusson, J. O., Berström, M., Eriksson, K., and Ross, S. B. (1988). Characterization of [^{3}H]paroxetine binding in rat brain. *J. Neurochem.* **50,** 1783–1790.

Margolis, M. T., and Newton, T. H. (1971). Methamphetamine ("Speed") arteritis. *Neuroradiology* **2**, 179–182.

Massari, V. J., and Sanders-Bush, E. (1975). Synaptosomal uptake and levels of serotonin in rat brain areas after p-chloroamphetamine or B-9 lesions. *Eur. J. Pharmacol.* **33**, 419–422.

Massari, V. J., Tizabi, Y., and Sanders-Bush, E. (1978a). Evaluation of the neurotoxic effects of p-chloroamphetamine: A histological and biochemical study. *Neuropharmacology* **17**, 541–548.

Massari, V. J., Tizabi, Y., Gottsfeld, Z., and Jacobowitz, D. M. (1987b). A fluorescence histochemical and biochemical evaluation of the effect of p-chloroamphetamine on individual serotonergic nuclei in the rat brain. *Neuroscience* **3**, 339–344.

McClean, J. H., and Shipley, M. T. (1987). Serotonergic afferents to the rat olfactory bulb. I. Origins and laminar specificity of serotonergic inputs in adult rat. *J. Neuroscience* **7**, 3016–3028.

Miyake, T., Hattori, T., Fukuda, M., Kitamura, T., and Fujita, S. (1988). Quantitative studies on proliferative changes of reactive astrocytes in mouse cerebral cortex. *Brain Res.* **451**, 133–138.

Molliver, M. E., and Axt, K. J. (1990). Specific regional patterns of serotonergic innervation in ventral areas of forebrain in the rat: Selective denervation and reinnervation after administration of p-chloroamphetamine (pCA). *Soc. Neurosci. Abstr.* **16**, 1032.

Molliver, M. E., and Mamounas, L. A. (1991). Dual serotonergic projections to rat olfactory bulb: p-Chloroamphetamine (PCA) selectively damages dorsal raphe axons while sparing median raphe axons. *Soc. Neurosci. Abstr.* **17**, 1180.

Molliver, D. C., and Molliver, M. E. (1990). Anatomic evidence for a neurotoxic effect of ($\pm$)-fenfluramine upon serotonergic projections in the rat. *Brain Res.* **511**, 165–168.

Molliver, M. E., Stratton, K., Carr, P., Grzanna, R., and Barbaran, J. (1988). Contrasting *in vitro* and *in vivo* effects of p-chloroamphetamine (PCA) on 5-HT axons: Immunocytochemical studies in hippocampal slices. *Soc. Neurosci. Abstr.* **14**, 210.

Molliver, M. E., Mamounas, L. A., and Carr, P. (1989). Reinnervation of cerebral cortex by 5-HT axons after denervation by psychotropic amphetamine derivatives. *Soc. Neurosci. Abstr.* **15**, 417.

Molliver, M. E., Berger, U. V., Mamounas, L. A., Molliver, D. C., O'Hearn, E., and Wilson, M. A. (1990). Neurotoxicity of MDMA and related compounds: Anatomic studies. *Ann. N.Y. Acad. Sci.* **600**, 640–664.

Moore, R. Y., Halaris, A. E., and Jones, B. E. (1978). Serotonin neurons of the midbrain raphe: Ascending projections. *J. Comp. Neurol.* **180**, 417–438.

Mori, S., Fujitake, J., Kuno, S., and Sano, Y. (1988). Immunohistochemical evaluation of the neurotoxic effects of 1-methyl-4-phenyl-1,2,3,6-tetrahydropyridine (MPTP) on dopaminergic nigrostriatal neurons of young adult mice using dopamine and tyrosine hydroxylase antibodies. *Neurosci. Lett.* **90**, 57–62.

Morrison, J. H., Molliver, M. E., Grzanna, R., and Coyle, J. T. (1979). Noradrenergic innervation patterns in three regions of medial cortex: An immunofluorescence characterization. *Brain Res. Bull.* **4**, 849–857.

Morrison, J. H., Foote, S. L., Molliver, M. E., Bloom, F. E., and Lidov, H. G. W. (1982). Noradrenergic and serotonergic fibers innervate complementary layers in monkey primary visual cortex: An immunocytochemical study. *Proc. Natl. Acad. Sci. U.S.A.* **79**, 2401–2405.

Mulligan, K. A., and Törk, I. (1988). Serotonergic innervation of cat cerebral cortex. *J. Comp. Neurol.* **270**, 86–110.

Neckers, L. M., Bertilsson, L., Koslow, S. H., and Meek, J. L. (1976). Reduction of tryptophan hydroxylase activity and 5-hydroxytryptamine concentration in certain rat brain nuclei after p-chloroamphetamine. *J. Pharmacol. Exp. Ther.* **196**, 333–338.

Nwanze, E., and Jonsson, G. (1981). Amphetamine neurotoxicity on dopamine nerve terminals in the caudate nucleus of mice. *Neurosci. Lett.* **26**, 163–168.

O'Callaghan, J. P., and Jensen, K. F. (1992). Enhanced expression of glial fibrillary acidic protein and the cupric silver degeneration reaction can be used as sensitive and early indicators of neurotoxicity. *Neurotoxicology* **13**, 113–122.

O'Callaghan, J. P., Jensen, K. F., and Miller, D. B. (1991). Characterization of methylenedioxymethamphetamine (MDMA) neurotoxicity using assays of serotonin and glial fibrillary acidic protein (GFAP): Changes in specific neocortical regions implicate serotonergic and non-serotonergic targets. *Soc. Neurosci. Abstr.* **17**, 1429.

Ogawa, M., Araki, M., Nagatsu, I., and Yoshida, M. (1989). Astroglial cell alteration caused by neurotoxins: Immunocytochemical observations with antibodies to glial fibrillary acidic protein, laminin, and tryosine hydroxylase. *Exp. Neurol.* **106**, 187–196.

O'Hearn, E., and Molliver, M. E. (1984). Organization of raphe-cortical projections in rat: A quantitative retrograde study. *Brain Res. Bull,* **13**, 709–726.

O'Hearn, E., Battaglia, G., De Souza, E. B., Kuhar, M. J., and Molliver, M. E. (1988). Methylenedioxyamphetamine (MDA) and methylenedioxymethamphetamine (MDMA) cause selective ablation of serotonergic axon terminals in forebrain: Immunocytochemical evidence for neurotoxicity. *J. Neurosci.* **8**, 2788–2803.

Pasquier, D. A., and Villar, M. J. (1982). Specific serotonergic projections to the lateral geniculate body from the lateral cell groups of the dorsal raphe nucleus. *Brain Res.* **249**, 142–146.

Powers, J. M., Mann, G. T., Jones, R., Ward, J. W., Elsea, J. R., and Smith, H. M. (1979). A reassessment of the significance of dark neurons in serotonergic cell groups. *Neuropharmacology* **18**, 383–389.

Ricaurte, G. A., Schuster, C. R., and Seiden, L. S. (1980). Long-term effects of repeated methylamphetamine administration on dopamine and serotonin neurons in the rat brain: A regional study. *Brain Res.* **193**, 153–163.

Ricaurte, G. A., Guillery, R. W., Seiden, L. S., Schuster, C. R., and Moore, R. Y. (1982). Dopamine nerve terminal degeneration produced by high doses of methylamphetamine in rat brain. *Brain Res.* **235**, 93–103.

Ricaurte, G. A., Seiden, L. S., and Schuster, C. R. (1984). Further evidence that amphetamines produce long-lasting dopamine neurochemical deficits by destroying dopamine nerve fibers. *Brain Res.* **303**, 359–364.

Ricaurte, G. A., Bryan, G., Strauss, L., Seiden, L., and Schuster, C. (1985). Hallucinogenic amphetamine selectively destroys brain serotonin nerve terminals. *Science* **229**, 986–988.

Ricaurte, G. A., Katz, J. L., and Hatzidimitriou, G. (1991a). Cell body loss underlies persistent serotonergic deficits induced by (±)3,4-methylenedioxymethamphetamine (MDMA) in primates. *Soc. Neurosci. Abstr.* **17**, 1182.

Ricaurte, G. A., Molliver, M. E., Martello, M. B., Katz, J. L., Wilson, M. A., and Martello, A. L. (1991b). Dexfenfluramine neurotoxicity in brains of non-human primates. *Lancet* **338**, 1487–1488.

Ricaurte, G. A., Martello, A. L., Katz, J. L., and Martello, M. B. (1992). Lasting effects of (±)-3,4-methylenedioxymethamphetamine (MDMA) on central serotonergic neurons in nonhuman primates: Neurochemical observations. *J. Pharmacol. Exp. Ther.* **261**, 616–622.

Rumbaugh, C. L., Bergeron, R. T., Fang, H. C., and McCormick, R. (1971a). Cerebral angiographic changes in the drug abuse patient. *Radiology* **101**, 335–344.

Rumbaugh, C. L., Bergeron, R. T., Scanlon, R. L., Teal, J. S., Segall, H. D., and Fang, H. C. (1971b). Cerebrovascular changes secondary to amphetamine abuse in the experimental animal. *Radiology* **101**, 345–351.

Ryan, L. J., Martone, M. E., Linder, J. C., and Groves, P. M. (1988). Continuous amphetamine administration induces tyrosine hydroxylase immunoreactive patches in the adult rat neostriatum. *Brain Res. Bull.* **21**, 133–137.

Ryan, L. J., Linder, J. C., Martone, M. E., and Groves, P. M. (1990). Histological and ultrastructural evidence that d-amphetamine causes degeneration in neostriatum and frontal cortex of rats. *Brain Res.* **518**, 67–77.

Sanders-Bush, E., Bushing, J. A., and Sulser, F. (1975). Long-term effects of p-chloro amphetamine and related drugs on central serotonergic mechanisms. *J. Pharmacol. Exp. Ther.* **192**, 33–41.

Scallet, A. C., Lipe, G. W., Ali, S. F., Holson, R. R., Frith, C. H., and Slikker, Jr., W. (1988). Neuropathological evaluation by combined immunohistochemistry and degeneration-specific methods: Application to methylenedioxymethamphetamine. *Neurotoxicology* **9**, 529–538.

Scheffel, U., and Ricaurte, G. A. (1990). Paroxetine as an in vivo indicator of 3,4-methylenedioxymethamphetamine neurotoxicity: A presynaptic serotonergic positron emission tomography ligand? *Brain Res.* **527**, 89–95.

Seiden, L. S., Commins, D. L., Vosmer, G., Axt, K., and Marek, G. (1988). Neurotoxicity in dopamine and 5-hydroxytryptamine terminal fields: A regional analysis in nigrostriatal and mesolimbic projections. *Ann. N.Y. Acad. Sci.* **537**, 161–172.

Slikker, W., Jr., Ali, S. F., Scallet, A. C., Frith, C. H., Newport, G. D., and Bailey, J. R. (1988). Neurochemical and neurohistological alterations in the rat and monkey produced by orally administered methylenedioxymethamphetamine (MDMA). *Toxicol. Appl. Pharmacol.* **94**, 448–457.

Sotelo, C. (1991). Immunocytochemical study of short- and long-term effects of dl-fenfluramine on the serotonergic innervation of the rat hippocampal formation. *Brain Res.* **541**, 309–326.

Sotelo, C., and Zamora, A. (1978). Lack of morphological changes in the neurons of the B-9 group in rat treated with fenfluramine. *Curr. Med. Res. Opin.* (Suppl. 1) **6**, 55–62.

Steinbusch, H. W. M. (1981). Distribution of serotonin-immunoreactivity in the central nervous system of the rat—Cell bodies and terminals. *Neuroscience* **6**, 557–618.

Steinbusch, H. W. M. (1984). Serotonin-immunoreactive neurons and their projections in the CNS. *In* "Handbook of Chemical Neuroanatomy" (A. Björklund, T. Hökfelt, and M. J. Kuhar, eds.), Vol. 3, Classical Transmitters and Transmitter Receptors in the CNS, Part II, pp. 68–125. Elsevier, Amsterdam.

Steinbusch, H. W. M., Nieuwenhuys, R., Verhofstad, A. A. J., and Van Der Kooy, D. (1981). The nucleus raphe dorsalis of the rat and its projection upon the caudatoputamen. A combined cytoarchitectonic, immunohistochemical and retrograde transport study. *J. Physiol. (Paris)* **77**, 157–174.

Strömberg, I., Björklund, H., Dahl, D., Jonsson, G., Sundström, E., and Olson, L. (1986). Astrocyte responses to dopaminergic denervations by 6-hydroxydopamine and 1-methyl-4-phenyl-1,2,3,6-tetrahydropyridine as evidenced by glial fibrillary acidic protein immunocytochemistry. *Brain Res. Bull.* **17**, 225–236.

Tassin, J. P., Bockaert, J., Blanc, G., Stinus, L., Thierry, A. M., Lavielle, S., Prémont, J., and Glowinski, J. (1978). Topographic distribution of dopaminergic innervation and dopaminergic receptors of the anterior cerebral cortex of the rat. *Brain Res.* **154**, 241–251.

Teuchert-Noodt, G., and Dawirs, R. R. (1991). Age-related toxicity in prefrontal cortex and caudate-putamen complex of gerbils (*Meriones unguiculatus*) after a single dose of methamphetamine. *Neuropharmacology*, **30**, 733–743.

Thierry, A. M., Stinus, L., Blanc, G., and Glowinski, J. (1973). Some evidence for the existence of dopaminergic neurons in the rat cerebral cortex. *Brain Res.* **50**, 230–234.

Törk, I. (1990). Anatomy of the serotonergic system. *Ann. N.Y. Acad. Sci.* **600**, 9–35.

Törk, I., Hornung, J.-P., and Somogyi, P. (1988). Serotonergic innervation of GABA-ergic neurons in the cerebral cortex. *Neurosci. Lett.* (Suppl.) **30**, S131.

Trulson, M. E., Cannon, M. S., Faegg, T. S., and Raese, J. D. (1985). Effects of chronic methamphetamine on the nigral–striatal dopamine system in rat brain: Tyrosine hydroxylase immunocytochemistry and quantitative light microscopic studies. *Brain Res. Bull.* **15**, 569–577.

Ungerstedt, U. (1968). 6-Hydroxydopamine induced degeneration of central monoamine neurons. *Eur. J. Pharmacol.* **5,** 107–110.
Ungerstedt, U. (1971). Stereotaxic mapping of the monoamine pathways in the rat brain. *Acta Physiol. Scand.* **367,** 1–48.
Van Eden, C. G., Hoorneman, E. M. D., Buijs, R. M., Matthijssen, M. A. H., Geffard, M., and Uylings, H. B. M. (1987). Immunocytochemical localization of dopamine in the prefrontal cortex of the rat at the light and electron microscopic level. *Neuroscience* **22,** 849–862.
van Luijtelaar, M. G. P. A., Steinbusch, H. W. M., and Tonnaer, J. A. D. M. (1989). Similarities between aberrant serotonergic fibers in the aged and 5,7-DHT denervated young adult rat brain. *Exp. Brain Res.* **78,** 81–89.
van Luijtelaar, M. G. P. A., Tonnaer, J. A. D. M., and Steinbusch, H. W. M. (1992). Aging of the serotonergic system in the rat forebrain: An immunocytochemical and neurochemical study. *Neurobiol. Aging* **13,** 201–215.
Voigt, T., and De Lima, A. D. (1991). Serotonergic innervation of the ferret cerebral cortex. I. Adult pattern. *J. Comp. Neurol.* **314,** 403–414.
Wagner, G. C., Ricaurte, G. R., Seiden, L. S., Schuster, C. R., Miller, R. J., and Westley, J. (1980). Long-lasting depletions of striatal dopamine and loss of dopamine uptake sites following repeated administration of methamphetamine. *Brain Res.* **181,** 151–160.
Wahnschaffe, U., and Esslen, J. (1985). Structural evidence for the neurotoxicity of methylamphetamine in the frontal cortex of gerbils (*Meriones unguilculatus*): A light and electron microscopical study. *Brain Res.* **337,** 299–310.
Wilson, M. A., and Molliver, M. E. (1991a). The organization of serotonergic projections to cerebral cortex in primates: Regional distribution of axon terminals. *Neuroscience* **44,** 537–553.
Wilson, M. A., and Molliver, M. E. (1991b). The organization of serotonergic projections to cerebral cortex in primates: Retrograde transport studies. *Neuroscience* **44,** 555–570.
Wilson, M. A., Ricaurte, G. A., and Molliver, M. E. (1989). Distinct morphologic classes of serotonergic axons in primates exhibit differential vulnerability to the psychotomimetic drug 3,4-methylenedioxymethamphetamine. *Neuroscience* **28,** 121–137.
Wilson, M. A., Mamounas, L. A., Fasman, K. H., Axt, K. J., and Molliver, M. E. (1993). Reactions of 5-HT neurons to drugs of abuse: Neurotoxicity and plasticity. *NIDA Res. Monogr.* **136,** 155–187.
Woolverton, W. L., Ricaurte, G. A., Forno, L. S., and Seiden, L. S. (1989). Long-term effects of chronic methamphetamine administration in rhesus monkeys. *Brain Res.* **486,** 73–78.
Yamada, T., McGeer, P. L., Baimbridge, K. G., and McGeer, E. G. (1990). Relative sparing in Parkinson's disease of substantia nigra dopamine neurons containing calbindin-D28k. *Brain Res.* **526,** 303–307.
Zaczek, R., Battaglia, G., Culp, S., Appel, N. M., Contrera, J. F., and De Souza, E. B. (1990). Effects of repeated fenfluramine administration on indices of monoamine function in rat brain: Pharmacokinetic, dose response, regional specificity and time course data. *J. Pharmacol. Exp. Ther.* **253,** 104–112.
Zhou, F. C., and Azmitia, E. C. (1986). Induced homotypic collateral sprouting of serotonergic fibers in hippocampus. II. An immunocytochemistry study. *Brain Res.* **373,** 337–348.

IV
USE AND ABUSE

Una D. McCann
George A. Ricaurte

12
Use and Abuse of Ring-Substituted Amphetamines

I. INTRODUCTION

The ring-substituted amphetamine class of drugs is composed of a large number of compounds with a wide range of pharmacological properties. Depending on the location and composition of the ring substitution, drugs in this class include those with predominantly stimulant (e.g., *p*-chloroamphetamine, PCA), hallucinogenic (e.g., 2,5-dimethoxy-4-methylamphetamine, DOM), or appetite suppressant (e.g., *N*-ethyl-α-methyl-*m*-trifluoromethylphenethylamine, fenfluramine) effects. Other ring-substituted amphetamines that have been synthesized and used by humans include TMA (3,4,5-trimethoxyamphetamine) and DOB (2,5-dimethoxy-4-bromoamphetamine), two analogs with predominantly hallucinogenic properties; PMA (*p*-methoxyamphetamine) and PMMA (*p*-methoxymethamphetamine), analogs with hallucinogenic and stimulant properties; and MDMA (3,4-methylenedioxymethamphetamine), MDE (methylenedioxyethamphetamine), MDA (Methylenedioxyamphetamine), and MBDB [N-methyl-1-(1,3-benzodioxol-5-yl)-2-butanamine], drugs that have stimulant and hallucinogen-like qualities in addition to unique pharmacological and psychological properties (Shulgin, 1990).

Of the various ring-substituted amphetamines, MDMA and its congeners have received the most recent attention, largely because of reports of the increased popularity of MDMA in a variety of recreational settings (Randall, 1992). The psychoactive properties of MDMA first were described by Shulgin and Nichols (1978), who stated that the drug evoked "an easily controlled altered state of consciousness, with sensual overtones," and proposed that MDMA might be useful as a psychotherapeutic adjunct. In the 1980s, this possibility was explored in a variety of uncontrolled settings (Greer, 1985; Greer and Strassman, 1985; Downing, 1986; Greer and Tolbert, 1986). Knowledge of the unique psychoactive profile of MDMA spread quickly to nonprofessional circles, where it gained popularity as a recreational drug (Cohen, 1985; Dowling et al., 1987; Peroutka, 1987). In 1985, out of concern for its abuse potential and possible neurotoxic effects, the Drug Enforcement Administration (DEA) classified MDMA as a Schedule I controlled substance and, on an emergency basis, restricted its use (Lawn, 1985; Barnes, 1988). Despite these restrictions, reports of the popularity of MDMA have increased (Anonymous, 1992; Henry et al., 1992; Abbot and Concar, 1992). Most recently, MDMA has been described as the drug of choice for use in large social settings in the United States and England ("Raves;" Randall, 1992). Because of increasing recreational use of MDMA, this drug is the primary focus of this chapter. However, the use and abuse of other ring-substituted amphetamines also is discussed briefly.

II. METHYLENEDIOXY METHAMPHETAMINE

A. Behavioral Effects

Four studies have been published investigating the subjective effects of MDMA (Downing, 1986; Greer and Tolbert, 1986; Peroutka et al., 1988; Liester et al., 1992). Downing reported the effects of a single exposure to MDMA in 21 healthy volunteers with previous MDMA experience. Subjects were allowed to request their own dose of MDMA, which ranged between 1.75 and 4.18 mg/kg of body weight (average dose 2.5 mg/kg or 175 mg for an individual weighing 70 kg). Acute effects (time of drug ingestion until 3 hr postdrug) included euphoria, increased physical and emotional energy, heightened sensual awareness, and decreased appetite. The majority of subjects experienced trismus and exhibited increased deep tendon reflexes and gait instability; a significant proportion (30–40%) demonstrated impaired judgment and difficulty in performing mathematical calculations. No significant or lasting untoward physical symptoms were noted in this group.

The study by Greer and Tolbert (1986) was also prospective, and involved the evaluation of subjective effects of MDMA in 29 patients in a clinical setting. Subjects received an initial dose of MDMA of 75–150 mg, with an opportunity to receive a second dose of 50 or 75 mg approximately 2 hr later (one subject requested and received higher doses). Reported effects in this group were similar to those reported in normal volunteers. Most patients reported increased closeness with others, enhanced communication, positive changes in attitudes and feelings, and cognitive benefits. Cognitive benefits mentioned included improved ("expanded") perspective and improved self-examination ("intrapsychic communication"). Five subjects using low dose MDMA (50 mg) reported enhanced creative writing skills following MDMA administration.

In a retrospective study of college students who had used MDMA recreationally, Peroutka and colleagues (1988) asked subjects to fill out questionnaires regarding the acute and subacute effects of MDMA. Acute effects that were experienced by the majority of subjects included, in decreasing order of frequency, a sense of "closeness" with others, trismus, tachycardia, bruxism, dry mouth, and increased alertness. Subacute effects were reported less frequently; drowsiness and muscle aches, or fatigue were the most frequent, reported by 36 and 32% of individuals, respectively. Lingering effects reported by less than 25% of individuals included a sense of "closeness" with others, depression, tight jaw muscles, and difficulty concentrating.

In another retrospective study, Liester and colleagues (1992) evaluated 20 psychiatrists who had taken MDMA previously, using a semistructured interview. The majority of subjects reported many of the positive MDMA effects just listed, in addition to altered time perception, decreased defensiveness and fear, reduced sense of alienation from others, changes in visual perception, and decreased aggression. Transient adverse effects reported in the majority of subjects included decreased desire to perform mental or physical tasks, decreased appetite, and trismus. The authors of the study expressed the view that MDMA had psychotherapeutic potential.

B. Biochemical Effects

As might be expected, given its structural similarities to amphetamine, MDMA is a potent monoamine-releasing agent and monoamine uptake inhibitor, both *in vitro* (Nichols *et al.*, 1982; Johnson *et al.*, Schmidt *et al.*, 1987; Steele *et al.*, 1987) and *in vivo* (Yamamoto and Spanos, 1988). Like other phenylethylamines, MDMA-induced monoamine release occurs via a carrier-mediated Ca^{2+}-independent process (Schmidt *et al.*, 1987). Unlike that of many amphetamines, however, MDMA-induced monoamine release is relatively selective for serotonin, with less potent dopamine-releasing properties (Schmidt, 1987). The sympathomimetic properties of MDMA are, most likely, largely the results of its actions on central nervous system

(CNS) catecholaminergic systems. Although the mechanisms underlying the unique psychoactive properties of MDMA (e.g., social closeness and enhanced communication) are not known, its behavioral effects have been postulated to be indirect and mediated by serotonin release (Nichols et al., 1982).

However, two lines of evidence suggest that serotonin release alone is not responsible for the unique pharmacological profile of MDMA. First, MDMA users who concurrently take fluoxetine, a drug that blocks serotonin release and uptake (Schmidt et al., 1987; Hekmatpanah and Peroutka, 1990), report that the singular reinforcing effects of MDMA are unaltered (McCann and Ricaurte, 1993). Second, individuals who take fenfluramine or PCA, two other potent serotonin-releasing agents, do not report psychoactive effects similar to those associated with MDMA (Rowland and Carlton, 1986; Van Praag and Korf, 1973). Other possible mechanisms responsible for the psychoactive properties of MDMA include influences of other monoamines (dopamine, norepinephrine), acting alone or in concert with serotonin; direct interactions with $5-HT_2$ or other brain receptors (Battaglia and De Souza, 1989); or actions at a unique brain site (Nichols et al., 1986).

C. Use Patterns

Perhaps at the center of controversy over MDMA is the notion that a drug with abuse potential has potential use as a psychotherapeutic adjunct (particularly since MDMA also is known to have neurotoxic potential; see subsequent discussion). Indeed, as noted earlier, prior to government restriction on its use, MDMA was used by professionals in a variety of psychotherapeutic settings, with particular utility in couples therapy (Greer and Strassman, 1985; Greer and Tolbert, 1986; Downing, 1986). Frequency and dose of MDMA used in the course of psychotherapy differed depending on the type of therapy and the particular therapist involved. A typical regimen used during the first therapeutic session would involve the ingestion of 75–125 mg MDMA, with a booster dose of 50 mg offered 1.5–3 hr later (Eisner, 1989). Subsequent MDMA "sessions" would be conducted based on the wishes of the therapist and the patient, at least 1 wk after the initial MDMA experience. Unfortunately, largely due to the Schedule I status on MDMA, controlled studies comparing the efficacy of MDMA with that of other accepted treatments have not been done; reports of MDMA efficacy are, therefore, anecdotal and retrospective.

No formal epidemiological surveys have been done regarding typical recreational MDMA use patterns. With the exception of isolated reports (McCann and Ricaurte, 1991), MDMA has not been associated with escalating use patterns. In a survey of college students who use MDMA for nonprofessional purposes, individuals reported that they preferred using

MDMA on weekends (rather than more frequently) and on an episodic, intermittent basis (Peroutka, 1990).

A more recent pattern that has arisen in the United States and Europe is the use of MDMA in large social gatherings called "Raves" (Randall, 1992). Raves, which are typically conducted in large warehouses or dance halls, reportedly are attended regularly by thousands of partygoers dressed in festive costumes. The rave phenomenon itself consists of all night dancing to electronically generated music with computer-generated videos, while partygoers drink amino acid-laced beverages "for energy" and use MDMA as their drug of choice (Abbot and Concar, 1992; Randall, 1992). Because of the illicit nature of MDMA use in a recreational setting, verifying doses used is difficult. However, the most frequently reported preferred recreational doses are similar to those used in psychotherapy, ranging from 100 to 180 mg.

D. Adverse Consequences

Since no controlled studies using MDMA in humans have been conducted, information regarding adverse consequences associated with MDMA derives from uncontrolled retrospective sources, usually in the form of case reports. Untoward consequences of MDMA use have fallen into three general categories: neuropsychiatric difficulties, general medical complications, and possible neurotoxicity. Each of these effects is considered in turn.

1. Neuropsychiatric Difficulties

Adverse neuropsychiatric manifestations associated with MDMA include those experienced while under the influence of MDMA, as well as those that occur after the acute drug experience. Short-term adverse neuropsychiatric effects that have been reported with MDMA include confusion, anxiety, insomnia (Greer, 1985), panic attacks (Whitaker-Azmitia and Aronson, 1989), and acute psychosis (Creighton *et al.*, 1991; McGuire and Fahy, 1991). Enduring ill effects associated with MDMA include flashbacks (Creighton, *et al.*, 1991), chronic psychosis (McGuire and Fahy, 1991; Schifano, 1991), major depressive disorder (Benazzi and Mazzoli, 1991; McCann and Ricaurte, 1991), memory disturbance (McCann and Ricaurte, 1991), and panic disorder (McCann and Ricaurte, 1992; Pallanti and Mazzi, 1992). These enduring adverse effects of MDMA persist well beyond the period required to metabolize MDMA fully (i.e., months); several cases required psychiatric intervention with medications prior to clinical improvement. In some instances, neuropsychiatric difficulties appeared more likely to occur in predisposed individuals who took high doses of MDMA (e.g., see McCann and Ricaurte, 1991). In others (McCann and Ricaurte, 1992), a single "typical" dose of MDMA was sufficient to induce persistent ill effects (panic attacks).

2. Medical Complications

Adverse medical effects reported with MDMA use include cardiac arrhythmias and asystole (Dowling *et al.*, 1987; Henry *et al.*, 1992), cardiovascular collapse (Suarez and Reimersma, 1988), rhabdomyalysis, disseminated intravascular coagulation, hyperthermia (Brown and Osterloh, 1987; Chadwick *et al.*, 1991; Campkin and Davies, 1992; Henry *et al.*, 1992; Screaton *et al.*, 1992) acute renal failure (Fahal *et al.*, 1992), and hepatotoxicity (Henry *et al.*, 1992). The possibility that other drugs of abuse or impurities in the MDMA preparation played a role in these medical complications must be considered, since all these reports were retrospective and involved MDMA obtained from illicit sources.

3. Neurotoxicity

In addition to its association with neuropsychiatric and medical complications, MDMA is well-established as a serotonergic neurotoxin in a number of animal species, including rats (Stone *et al.*, 1986; Battaglia *et al.*, 1987; Commins *et al.*, 1987; Schmidt, 1987; O'Hearn *et al.*, 1988; Slikker *et al.*, 1988; 1989), guinea pigs (Commins *et al.*, 1987), mice (Stone *et al.*, 1986), and monkeys (Ricaurte *et al.*, 1988a; Slikker *et al.*, 1988, 1989; Insel *et al.*, 1989; Wilson *et al.*, 1989). Further, evidence in nonhuman primates suggests that MDMA-induced neurotoxicity may be permanent (Ricaurte *et al.*, 1992). Although, for obvious reasons, neurochemical and neuropathological studies have not been performed in human MDMA users, the findings in nonhuman primates raise concern that MDMA also may be neurotoxic in humans. This concern is intensified when one considers that the dose of MDMA that damages serotonin neurons in monkeys (Ricaurte *et al.*, 1988b) is close to that typically taken by recreational users, and that the dose–response curve for MDMA neurotoxicity in primates is steep (Ricaurte and McCann, 1992).

By necessity, currently available methods for detecting serotonin function in living humans are indirect and include peripheral measures [e.g., platelet serotonin receptor measurements; concentration of 5-hydroxy indole acetic acid (5-HIAA), the major metabolite of serotonin in urine or blood samples], neuroendocrine challenges (e.g., the prolactin response to L-tryptophan, a putative serotonin-mediated hormonal response), and measurements of 5-HIAA concentrations in lumbar cerebrospinal fluid (CSF) samples. Neuroimaging techniques hold promise for the detection of CNS serotonergic neurotoxicity, but unfortunately, at present no presynaptic serotonin-specific radioligands suitable for such studies are available. An unexplored method for detecting serotonergic neurotoxicity is pharmacological challenge, during which an individual with suspected neurotoxicity receives a serotonin-specific drug (e.g., *p*-chlorophenylalanine, PCPA) and is evaluated for abnormal behavioral responses. Since none of the current available methodology for detecting human serotonergic neurotoxicity is conclusive, it will be essential to develop converging lines of evidence to establish the neurotoxic potential of MDMA in humans.

Results of the three published studies that have evaluated MDMA users for evidence of serotonin neurotoxicity have been conflicting. Two studies evaluated the lumbar CSF 5-HIAA levels of MDMA users. One of these reported reductions in CSF 5-HIAA (Ricaurte *et al.*, 1990) whereas the other did not (Peroutka *et al.*, 1987). Since neither of these studies was conducted in a controlled setting, and since many factors are known to influence CSF 5-HIAA levels (Post *et al.*, 1980; Curzon *et al.*, 1972; Stanley *et al.*, 1985; Mendels *et al.*, 1972; Bowers and Gerbode, 1968; Asberg *et al.*, 1976; Post and Goodwin, 1974), the lack of uniformity in these data is not surprising. A third study probed for serotonergic functional abnormalities using the prolactin response to L-tryptophan (Price *et al.*, 1988). Although an altered prolactin response was suggested in MDMA users, the findings did not reach statistical significance. Definitive studies to determine whether humans are susceptible to the neurotoxic effects of MDMA have yet to be reported.

III. OTHER RING-SUBSTITUTED AMPHETAMINES CURRENTLY USED CLINICALLY OR RECREATIONALLY

A. 2,5-Dimethoxy-4-methylamphetamine

DOM is also known on the street as STP (serenity, tranquility, and peace). This drug first appeared on the recreational drug scene in the late 1960s. At low doses (less than 3 mg), DOM is said to have mescaline-like properties. At higher doses, frank hallucinosis and unpleasant side effects including nausea, diaphoresis, and tremor occur, and may persist for 8–24 hr (Snyder *et al.*, 1967). At a present, illicit use of DOM is sporadic; no medical use of DOM is recognized.

B. 2,5-Dimethoxy-4-bromoamphetamine

DOB, like DOM, has both "hallucinogenic" and sympathomimetic properties, although this drug is better known for its hallucinogenic effects. Because of its high potency, DOB often is sold impregnated in blotter paper, as was LSD (Shulgin, 1981). Because equal distribution of DOB is difficult to insure in such preparations, certain segments of the blotter sometimes contain disproportionately high doses of DOB. Possibly because of this particular method of DOB distribution (and, thus, occasional very high doses of DOB), numerous reports of bad experiences with the drug have been made

(Delliou, 1980). Like DOM, DOB has never achieved great popularity in the United States, possibly because of its association with untoward side effects. Neither DOM nor DOB has been reported to cause neurotoxic changes in the brain, most likely because their primary actions are postsynaptic whereas MDMA acts presynaptically.

C. Mescaline

Although not technically an amphetamine (because it is not a isopropylamine), mescaline (3,4,5-Trimethoxyphenethylamine) is included in this chapter because it is structurally related to the drugs being considered and often is regarded as the prototypical hallucinogen. Mescaline is a naturally occurring ring-substituted phenethylamine that is derived from the cactus plant *Lophora williamsii* or *Anhalonium lewinii*. The dried tops of the cactus plant, often referred to as peyote buttons, contain mescaline and are known for their hallucinogenic properties. Peyote buttons have been used by Mexican and American Indian shamans for centuries and, in the 1800s, were adopted by American Indian tribes for use in religious ceremonies. Today, the ingestion of peyote buttons continues to be an integral part of religious ceremonies of the Native American Church, a religious sect of the American Plains Indians with a following of over 200,000 people. Mescaline is rarely abused (e.g., is not used for other than ceremonial purposes by the Plains Indians). No report of death due to mescaline overdose has ever been made (Karch, 1993), nor have reports been made on the neurotoxic potential of mescaline.

D. 3,4-Methylenedioxyamphetamine

MDA is a ring-substituted amphetamine with both sympathomimetic and hallucinogenic properties. Although synthesized in 1910 (Mannich and Jacobsohn, 1910), MDA did not become popular as a recreational drug until the 1960s, when it was also known as "the love drug." Like MDMA, MDA has been advocated as a potential psychotherapeutic adjunct (Naranjo *et al.*, 1967; Yensen *et al.*, 1976). Depending on the isomer taken, the ratio of sympathomimetic to hallucinogenic effects differs; the D isomer possesses more hallucinogenic properties than the L isomer. On the street, MDA usually is sold in its racemic form. When taken orally, the effects of MDA, to a large degree, are similar to those of MDMA (although many experienced MDMA users maintain that the effects of the two can be distinguished). MDA has been implicated in a number of sudden deaths, possibly due to cardiac arrhythmia (the D isomer is extremely arrythmogenic in rats; Karch, 1993). High doses of MDA also can cause convulsions and hyperthermia. Although largely superseded by MDMA, MDA is still available from illicit sources. The last MDA associated death was reported in 1990 (Nichols *et al.*, 1990). Like MDMA, MDA causes neurotoxic injury to brain serotonin neurons (Ricaurte *et al.*, 1985).

E. 3,4-Methylenedioxy-*N*-ethylamphetamine

MDE, also known as "Eve," became popular as a recreational drug shortly after MDMA was placed on Schedule I as a restricted drug (Dowling *et al.*, 1987). The psychoactive properties of MDE are said to be similar to those of MDMA, although the drug is possibly less potent at typically used doses. One report has been made of an MDE-associated death in an individual with an enlarged heart. MDE, like MDMA, has been found to damage serotonin neurons in the brains of experimental animals, although the drug is approximately four times less potent in eliciting this effect (Ricaurte *et al.*, 1987).

F. *N*-Methyl-1-(1,3-benzodixol-5-yl)-2-butanamine

The psychoactive effects of MBDB have been described as different from those of typical hallucinogens and generally similar to those of MDMA, with less euphoria (Nichols *et al.*, 1986). Preclinical studies using discriminative stimulus techniques also indicate that MBDB, like MDA, produces effects similar to those of MDMA and different from those of other stimulants [(+)-amphetamine, (+)-methamphetamine, and cocaine)] and other hallucinogens (LSD, DOM, and mescaline) (Nichols and Oberlender, 1990). Little is known regarding the epidemiology of MBDB use, but its use appears to be sporadic and only by cognoscentes of MDMA who are seeking similar experiences. MBDB has been reported to be neurotoxic to serotonin neurons in rats, although to a lesser degree than MDMA (Nichols and Oberlender, 1989).

G. *p*-Methoxymethamphetamine

Although little has been reported on the psychoactive effects of PMMA, in recent years PMMA has been reported to be sold illicitly as a substitute for MDMA (Steele *et al.*, 1992). In rats, PMMA produces neurotoxic effects on serotonin neurons, but to a lesser degree than MDMA (Steele *et al.*, 1992).

H. *p*-Methyoxyamphetamine

PMA, like many of the ring-substituted amphetamines, possesses both stimulant and hallucinogenic properties. Because of a number of deaths associated with PMA in the 1970s (Cimbura, 1974), PMA was classified as a restricted drug and its use dissipated rapidly. Studies in rats have demonstrated that use of this drug leads to long-term depletions of brain serotonin, but that it is not as potent a neurotoxin as MDMA (Steele *et al.,* 1992).

I. *p*-Chloroamphetamine

In the early 1970s, PCA was considered for use in the treatment of major depression because of its known actions on serotonin neurons. In a controlled trial in 15 patients with depression, PCA (75 mg daily for 4 wk) was found to improve mood, cognitive abilities, and sleep, and to increase interest in work and the environment (Van Praag and Korf, 1973). Shortly thereafter (Harvey *et al.,* 1975), PCA was determined to be a selective serotonin neurotoxin, resulting in no further efforts to use PCA clinically.

J. Fenfluramine

Racemic fenfluramine [N-ethyl-α-methyl-m-(trifluoromethyl)phenethylamine] and its D or *dextro* isomer, dexfenfluramine, are ring-substituted amphetamines that currently are approved for clinical use. Fenfluramine is used as an anorectic agent for the treatment of obesity (see Rowland and Carlton, 1986). Fenfluramine is widely accepted to have very little liability for abuse, although isolated reports have been made of experienced drug users who have taken fenfluramine purely for its psychoactive effects (Levin, 1973). The anorectic properties of fenfluramine are thought to be secondary to its serotonin-releasing properties and, to a lesser degree, to its serotonin reuptake inhibiting properties (Fuller *et al.,* 1988). Like MDMA, fenfluramine causes rapid depletions in brain concentrations of serotonin, (5-HIAA), and tryptophan hydroxylase (Steranka and Sanders-Bush, 1979; Garattini, 1980). In rats given a single dose of fenfluramine, significant depletions of serotonin are still present 30 days later (Harvey and McMaster, 1975). The report that nonhuman primates treated with dexfenfluramine (at doses approximating those given to obese patients) also suffer lasting depletions of serotonin has given rise to the concern that fenfluramine is neurotoxic in humans (Ricaurte *et al.,* 1991).

IV. SUMMARY AND IMPLICATIONS

The ring-substituted amphetamine class of compounds includes a number of used and abused drugs with a wide range of properties ranging from predominantly stimulant to predominantly hallucinogenic. Several ring-substituted analogs have been found to be neurotoxic to brain serotonin neurons, brain dopamine neurons, or both. Although the extent of use of the most popular recreational drug in this class, MDMA, is difficult to determine because of its restricted status, reports have been made that the use of

MDMA is increasing. Because MDMA is known to be neurotoxic to serotonin neurons in the brains of experimental animals, including nonhuman primates, these reports have raised the concern that MDMA may pose a public health threat. The recreational use of other ring-substituted amphetamines is sporadic and difficult to estimate, given the restricted status of many of these compounds. In contrast, the extent of use of fenfluramine, a ring-substituted amphetamine used clinically as an anorectic agent, is known. The racemic form of fenfluramine has been prescribed for the treatment of obesity in 90 countries (including the United States) to approximately 50 million patients (Derome-Tremblay and Nathans, 1989). More recently, dexfenfluramine, the isomer with more potent anorectic effects, has been prescribed to 5 million people (Blundell, 1992). Since fenfluramine and dexfenfluramine are known serotonin neurotoxins in nonhuman primates (Schuster *et al.*, 1986; Ricaurte *et al.*, 1991), prospective controlled studies in individuals being treated with dexfenfluramine are indicated to determine whether humans also sustain injury to brain serotonin systems. These prospective studies should probably be preceded by retrospective studies of individuals already exposed to fenfluramine.

The possibility that MDMA and several other drugs in this class may be of therapeutic value in psychotherapy also deserves further exploration. However, until the risk–benefit ratio has been determined, the experimental use of ring-substituted amphetamines in a psychotherapeutic setting would be ill advised. One potential mechanism for expediting studies using ring-substituted amphetamines in clinical settings is through the synthesis of nontoxic analogs or the combination of the use of a "therapeutic" agent (e.g., MDMA) with a drug that prevents neurotoxicity (e.g., fluoxetine). Case reports suggest that, when combined with fluoxetine, MDMA retains many of its reinforcing psychoactive properties, although these reports must be confirmed in a controlled setting (McCann and Ricaurte, 1993).

In addition to their potential for use in clinical settings, the ring-substituted amphetamines have potential value as research tools. For example, those compounds with neurotoxic properties may be useful in the study of neuronal degeneration and regeneration. Further, studies in humans who have been exposed to high doses of neurotoxic ring-substituted amphetamines could shed light on the role of serotonin in normal brain function, as well as in neuropsychiatric diseases in which serotonin has been implicated (e.g., major depression, anxiety disorders, Alzheimer's disease).

REFERENCES

Abbot, A., and Concar, D. (1992). A trip into the unknown. *New Scientist* **29**, 30–34.
Anonymous (1992). Drug culture. *Lancet* **339**, 117.
Asberg, M., Thoren, P., and Traskman, L. (1976). "Serotonin depression"—A biochemical subgroup within the affective disorders? *Science* **191**, 478–480.

Barnes, D. M. (1988). New data intensify the agony over ecstasy. *Science* **239**, 864–866.
Battaglia, G., and De Souza, E. B. (1989). Pharmacologic profile of amphetamine derivatives at various brain recognition sites. *NIDA Res. Monogr.* **94**, 240–258.
Battaglia, G., Yeh, S. Y., O'Hearn, E., Molliver, M. E., Kuhar, M. J., and DeSouza, E. B. (1987). 3,4-Methylenedioxymethamphetamine and 3,4-methylenedioxyamphetamine destroy serotonin terminals in rat brain: quantification of neurodegeneration by measurement of [^{3}H]paroxetine-labeled serotonin uptake sites. *J. Pharmacol. Exp. Ther.* **243**(3), 911–916.
Benazzi, F., and Mazzoli, M. (1991). Psychiatric illness associated with "ecstasy". *Lancet* **338**, 1520.
Blundell, J. (1992). Letter to the editor. *Lancet* **339**, 360.
Bowers, M. B., and Gerbode, F. A. (1968). Relationship of monoamine metabolites in human cerebrospinal fluid to age. *Nature (London)* **19**, 1256–1257.
Brown, C., and Osterloh, J. (1987). Multiple complications from recreational ingestion of MDMA ("Ecstasy"). *J. Am. Med. Assoc.* **258**, 780–781.
Campkin, N. T. A., and Davies, U. M. (1992). Another death from ecstasy. *J. R. Soc. Med.* **85**, 61.
Chadwick, I. S., Linsey, A., Freemont, A. J., Doran, B., and Curry, P. D. (1991). Ecstasy, 3,4-methylenedioxymethamphetamine (MDMA): A fatality with coagulopathy and hyperthermia. *J. R. Soc. Med.* **84**, 371.
Cimbura, G. (1974). PMA deaths in Ontario. *Can. Med. Assoc. J.* **110**, 1263–1267.
Cohen, S. (1985). They call it Ecstasy. *Drug Abuse Alcoholism Newsl.* **14**(6), 1–3.
Commins, D. L., Vosmer, G., Virus, R., Woolverton, W., Schuster, C., and Seiden, L. (1987). Biochemical and histological evidence that methylenedioxymethylamphetamine (MDMA) is toxic to neurons in the rat brain. *J. Pharmacol. Exp. Ther.* **241**, 338–345.
Creighton, F. J., Black, D. L., and Hyde, C. E. (1991). Ecstasy psychosis and flashbacks. *Br. J. Psychiatry* **159**, 713–15.
Curzon, G., Joseph, M. H., and Knott, P. J. (1972). Effects of immobilization and food deprivation on rat brain tryptophan metabolism. *J. Neurochem.* **19**, 1967–1974.
Delliou, D. (1980). Bromo-DMA: New hallucinogenic drug. *Med. J. Aust.* **83**.
Derome-Tremblay, M., and Nathans, C. (1989). Fenfluramine studies. *Science* **243**, 991.
Dowling, G. P., McDonough, E. T., and Bost, R. O. (1987). "Eve" and "Ecstasy": A report of five deaths associated with the use of MDEA and MDMA. *J. Am. Med. Assoc.* **257**(12), 1615–1617.
Downing, J. (1986). The psychological and physiological effects of MDMA on normal volunteers. *J. Psychoact. Drugs* **18**, 335–340.
Eisner, B. (1989). "Ecstasy: The MDMA Story." Ronin Publishing, Berkeley, California.
Fahal, I. M., Sallomi, D. F., Yaqoob, M., and Bell, G. M. (1992). Acute renal failure after ecstasy. *Br. Med. J.* **305**, 9.
Fuller, R., Snoddy, H., and Robertson, D. (1988). Mechanisms of effect of d-fenfluramine on brain serotonin in rats: Uptake inhibition versus release. *Pharmacol. Biochem. Behav.* **30**, 715–721.
Garattini, S. (1980). Recent studies on anorectic agents. *Trends Pharmaceut. Sci.* **1**, 354–356.
Greer, G. (1985). Using MDMA in psychotherapy. *Advances* **2**(2), 55–57.
Greer, G., and Strassman, R. J. (1985). Information on Ecstasy. *Am. J. Psychiatry* **142**(11), 1391.
Greer, G., and Tolbert, R. (1986). Subjective reports of the effects of MDMA in a clinical setting. *J. Psychoact. Drugs* **18**(4), 319–327.
Harvey, J., and McMaster, S. (1975). Fenfluramine: Evidence for a neurotoxic action on midbrain and a long-term depletion of serotonin. *Psychopharmacol. Commun.* **1**, 217–228.
Harvey, J. A., McMaster, S. E., and Yunger, L. H. (1975). p-Chloroamphetamine: Selective neurotoxic action in the brain. *Science* **187**, 841–843.
Hekmatpanah, C. R., and Peroutka, S. J. (1990). 5-Hydroxytryptamine uptake blockers attenu-

ate the 5-hydroxytryptamine-releasing effect of 3,4-methylenedioxymethamphetamine and related agents. *Eur. J. Pharmacol.* **177**, 95.

Henry, J. A., Jeffreys, K. J., and Dawling, S. (1992). Toxicity and deaths from 3,4-methylenedioxymethamphetamine ("ecstasy"). *Lancet* **340**, 384–387.

Insel, T. R., Battaglia, G., Johannessen, J. N., Marra, S., and DeSouza, E. B. (1989). (+)3,4-Methylenedioxymethamphetamine (MDMA; "Ecstasy") selectively destroys brain serotonin terminals in rhesus monkey. *J. Pharmacol. Exp. Ther.* **249**, 713–720.

Johnson, M. P., Hoffman, A. H., and Nichols, D. E. (1986). Effects of the enantiomers of MDA, MDMA and related analogues on [^{3}H]-serotonin and [^{3}H]-dopamine release from superfused rat brain slices. *European J. Pharmacology* **132**, 269–276.

Karch, S. B. (1993). Synthetic stimulants. *In* "The Pathology of Drug Abuse," pp. 165–218. CRC Press, Boca Raton, Florida.

Lawn, J. C. (1985). Schedules of controlled substances: Temporary placement of 3,4-methylenedioxymethamphetamine (MDMA) into Schedule I. *Fed. Reg.* **50**(105), 23118–23120.

Leister, M. B., Grob, C. S., Bravo, G. L., and Walsh, R. N. (1992). Phenomenology and sequelae of 3,4-methylenedioxymethamphetamine use. *J. Nerv. Ment. Dis.* **180**(6), 345–352.

Levin, A. (1973). Abuse of fenfluramine. *Br. Med. J.* **2**, 49.

Mannich, C., and Jacobsohn, W. (1910). Hydroxyphenylalkylamines and dihydroxyphenylalkylamines. *Berichte* **43**, 189.

McCann, U. D., and Ricaurte, G. A. (1991). Lasting neuropsychiatric sequelae of (±) methylenedioxymethamphetamine ("ecstasy") in recreational users. *J. Clin. Psychopharmacol.* **11**(5), 302–305.

McCann, U. D., and Ricaurte, G. A. (1992). MDMA ("Ecstasy") and panic disorder: Induction by a single dose. *Biol. Psychiatry* **32**(10), 950–953.

McCann, U. D., and Ricaurte, G. A. (1993). Reinforcing Subjective Effects of (±) 3,4-methylenedioxymethamphetamine (MDMA; "Ecstasy") may be separable from its neurotoxic actions: clinical evidence. *J. Clin. Psychopharmacol.* **13**(3), 214–217.

McGuire, P., and Fahy, T. (1991). Chronic paranoid psychosis after misuse of MDMA ("ecstasy"). *Br. Med. J.* **302**, 697.

Mendels, J., Frazer, A., Fitzgerald, R. G., Ramsey, T. A., and Stokes, J. (1972). Biogenic amine metabolites in cerebrospinal fluid of depressed and manic patients. *Science* **175**, 1380–1382.

Naranjo, C., Shulgin, A., and Sargeant, T. (1967). Evaluation of 3,4-methyldioxyamphetamine (MDA) as an adjunct to psychotherapy. *Med. Pharmacol. Exp.* **17**, 359–364.

Nichols, D. E., and Oberlender, R. (1989). Structure–activity relationships of MDMA-like substances. *NIDA Res. Monogr.* **94**, 1–29.

Nichols, D. E., and Oberlander, R. A. (1990). Structure–activity relationships of MDMA and related compounds: A new class of psychoactive drugs? *Ann. N.Y. Acad. Sci.* **600**, 613–625.

Nichols, D. E., Hoffman, A. J., Oberlander, R. A., Jacob, P., III, and Shulgin, A. T. (1982). Effects of certain hallucinogenic amphetamine analogues on the release of [^{3}H]serotonin from rat brain synaptosomes. *J. Med. Chem.* **25**, 530–535.

Nichols, D. E., Hoffman, A. J., Oberlander, R. A., Jacob, P., and Shulgin, A. T. (1986). Derivatives of 1-(1,2-benzodioxol-5-yl)-2-butanamine: Representatives of a novel therapeutic class. *J. Med. Chem.* **29**, 2009–2015.

Nichols, G., Davis, C., Corrigan, C., and Ransdell, J. (1990). Death associated with abuse of a "designer drug." *Kentucky Med. Assoc. J.* **88**, 600–603.

O'Hearn, E., Battaglia, G., DeSouza, E. B., Kuhar, M. J., and Molliver, M. E. (1988). Methylenedioxyamphetamine (MDA) and methylenedioxy-methamphetamine (MDMA) cause ablation of serotonergic axon terminals in forebrain: Immunocytochemical evidence. *J. Neurosci.* **8**, 2788–2803.

Pallanti, S., and Mazzi, D. (1992). MDMA (Ecstasy) precipitation of panic disorder. *Biol. Psychiatry* 32(1), 91–94.
Peroutka, S. J. (1987). Incidence of recreational use of (±)3,4 methylenedioxymethamphetamine (Ecstasy) on an undergraduate campus. *N. Engl. J. Med.* 317(24), 1542–1543.
Peroutka, S. J. (1990). In "Ecstasy: The Clinical, Pharmacological and Neurotoxicological Effects of the Drug MDMA". S. J. Peroutka, ed., Preface, p. xii. Norwell, Kluwer Academic Publishers, MA.
Peroutka, S. J., Pascoe, N., and Faull, K. F. (1987). Monoamine metabolites in the cerebrospinal fluid of recreational users of 3,4-methylenedioxymethamphetamine (MDMA; "Ecstasy"). *Res. Commun. Subs. Abuse* 8, 125–138.
Peroutka, S. J., Newman, H., and Harris, H. (1988). Subjective effects of 3,4-methylenedioxymethamphetamine in recreational users. *Neuropsychopharmacology* 1(4), 273–277.
Post, R. M., and Goodwin, F. K. (1974). Effects of amytriptyline and imipramine on amine metabolites in the cerebrospinal fluid of depressed patients. *Arch. Gen. Psychiatry* 30, 234–239.
Post, R. M., Ballenger, J. C., and Goodwin, F. K. (1980). Cerebrospinal fluid studies of neurotransmitter function in manic and depressive illness. In "Neurobiology of Cerebrospinal Fluid" (J. H. Wood, ed.), pp. 685–717. Plenum Press, New York.
Price, L. H., Ricaurte, G. A., Krystal, J., and Heninger, G. (1988). Neuroendocrine and mood response to intravenous L-tryptophan in (±) 3,4-methylenedioxymethamphetamine (MDMA) users. *Arch. Gen. Psychiatry* 46, 20–22.
Randall, T. (1992). Ecstasy-fueled "rave" parties become dances of death for English youths. *J. Am. Med. Assoc.* 268(12), 1505–1506.
Ricaurte, G. A., and McCann, U. D. (1992). Neurotoxic amphetamine analogues: Effects in monkeys and implications for humans. *Ann. N.Y. Acad. Sci.* 648, 371–382.
Ricaurte, G. A., Bryan, G., Strauss, L., Seiden, L., and Schuster, C. R. (1985). Hallucinogenic amphetamine selectively destroys brain serotonin nerve terminals. *Science* 229, 986.
Ricaurte, G. A., Finnegan, K. F., Nichols, D. E., DeLanney, L. E., Irwin, I., and Langston, J. W. (1987). Methylenedioxyethylamphetamine (MDE), a novel analogue of MDMA, produces long-lasting depletion of serotonin in the rat brain. *Eur. J. Pharmacol.* 137, 265–268.
Ricaurte, G. A., Forno, L. S., Wilson, M. A., DeLanney, L. E., Irwin, I., Molliver, M. E., and Langston, J. W. (1988a). (±)3,4-Methylenedioxymethamphetamine (MDMA) selectively damages central serotonergic neurons in non-human primates. *J. Am. Med. Assoc.* 260(1), 51–55.
Ricaurte, G. A., DeLanney, L. E., Irwin, I., and Langston, J. W. (1988b). Toxic effects of MDMA on central serotonergic neurons in the primate: Importance of route and frequency of drug administration. *Brain Res.* 446, 165–168.
Ricaurte, G. A., Finnegan, K. T., Irwin, I., and Langston, J. W. (1990). Aminergic metabolites in cerebrospinal fluid of humans previously exposed to MDMA: Preliminary observations. *Ann. N.Y. Acad. Sci.* 600, 699–710.
Ricaurte, G. A., Molliver, M. E., Martello, M. B., Katz, J. L., Wilson, M. A., and Martello, A. L. (1991). Dexfenfluramine neurotoxicity in brains of non-human primates. *Lancet* 338, 1487–1488.
Ricaurte, G. A., Katz, J. L., and Martello, M. B. (1992). Lasting effects of (±)3,4-methylenedioxymethamphetamine on central serotonergic neurons in non-human primates. *J. Pharmacol. Exp. Ther.* 261(2), 616–622.
Rowland, N., and Carlton, J. (1986). Neurobiology of an anorectic drug: Fenfluramine. *Progr. Neurobiol.* 27, 13–62.
Schifano, F. (1991). Chronic atypical psychosis associated with MDMA ("ecstasy") abuse. *Lancet* 338(8778), 1335.
Schmidt, C. J. (1987). Neurotoxicity of the psychedelic amphetamine, methylenedioxymethamphetamine. *J. Pharmacol. Exp. Ther.* 240(1), 1–7.
Schmidt, C. J., Levin, J. A., and Lovenberg, W. (1987). In vitro and in vivo neurochemical effects

of methylenedioxymethamphetamine on striatal monoamine systems in the rat brain. *Biochem. Pharmacol.* **36**(5), 747–755.
Schuster, C. R., Lewis, M. and Seiden, L. M. (1986). Fenfluramine neurotoxicity. *Psychopharmacol. Bull.* **22**, 148–151.
Screaton, G. R., Cairns, H. S., Sarner, M., Singer, M., Thrasher, A., and Cohen, S. L. (1992). Hyperpyrexia and rhabdomyolysis after MDMA ("ecstasy") abuse. *Lancet* **339**, 677–678.
Shulgin, A. (1981). Profiles of psychedelic drugs: DOB. *J. Psychedelic Drugs* **13**, 99.
Shulgin, A. T., and Nichols, D. E. (1978). Characterization of three new psychotomimetics. In "The Pharmacology of Hallucinogens" (R. C. Stillman and R. E. Willette, eds.), pp. 74–83. Pergamon Press, New York.
Slikker, W., Ali, S. F., Scallet, C., Frith, C. H., Newport, G. D., and Bailey, J. R. (1988). Neurochemical and neurohistological alterations in the rat and monkey produced by orally administered methylenedioxymethamphetamine (MDMA). *Toxicol. Appl. Pharmacol.* **94**, 448–457.
Shulgin, A. T. History of MDMA (1990). In "Ecstasy: The Clinical Pharmacological, and Neurotoxicological Effects of the Drug MDMA" S. J. Peroutka ed., Kluwer Publishers, Norwell, Massachusetts.
Slikker, W., Jr., Holson, R. R., Ali, S. F., Kolta, M. G., Paule, M. G., Scallet, A. C., McMillan, D. E., Bailey, J. R., Hong, J. S., and Scalzo, F. M. (1989). Behavioral and neurochemical effects of orally administered MDMA in the rodent and nonhuman primate. *Neurotoxicology* **10**(3), 529–549.
Snyder, S., Failace, L., and Hollister, L. (1967). 2,5-Dimethoxy-4-methyl-amphetamine (STP): A new hallucinogenic drug. *Science* **158**, 669–670.
Stanley, M., Traskman-Bendz, L., and Dorovin-Zis, K. (1985). Correlation between aminergic metabolites simultaneously obtained from human CSF and brain. *Life Sci.* **37**, 1279–1286.
Steele, T. D., Nichols, D. E., and Yim, G. K. W. (1987). Stereochemical effects of 3,4-methylenedioxymethamphetamine (MDMA) and related amphetamine derivatives on inhibition of uptake of [^{3}H]monoamines into synaptosomes from different regions of rat brain. *Biochem. Pharmacol.* **36**(14), 2297–2303.
Steele, T. D., Katz, J. L., and Ricaurte, G. A. (1992). Evaluation of the neurotoxicity of N-methyl-1-(4-methoxyphenyl)-2-aminopropane (*para*-methoxymethamphetamine, PMMA). *Brain Res.* **589**, 349–352.
Steranka, L., and Sanders-Bush, E. (1979). Long-term effects of fenfluramine on central serotonergic mechanisms. *Neuropharmacology* **18**, 895–903.
Stone, D. M., Stahl, D. S., Hanson, G. L., and Gibb, J. W. (1986). The effects of 3,4-methylenedioxymethamphetamine (MDMA) and 3,4-methylenedioxyamphetamine on monoaminergic systems in the rat brain. *Eur. J. Pharmacol.* **128**, 41–48.
Suarez, R. V., and Riemersma, R. (1988). "Ecstasy" and sudden cardiac death. *Am. J. Forens. Med. Pathol.* **9**(4), 339–341.
Van Praag, H. M., and Korf, J. (1973). 4-Chloroamphetamine: Chance and trend in the development of new antidepressants. *J. Clin. Pharmacol.* **13**, 3–14.
Whitaker-Azmitia, P. M., and Aronson, T. (1989). Ecstasy: (MDMA)-induced panic. *Am. J. Psychiatry* **146**(1), 119.
Wilson, M. A., Ricaurte, G. A., and Molliver, M. E. (1989). Distinct morphologic classes of serotonergic axons in primates exhibit differential vulnerability to the psychotropic drug (+)3,4-methylenedioxymethamphetamine. *Neuroscience* **28**, 121–137.
Yamamoto, B. K., and Spanos, L. J. (1988). The acute effects of methylenedioxymethamphetamine on dopamine release in the awake-behaving rat. *Eur. J. Pharmacol.* **148**, 195–203.
Yensen, R., DiLeo, F. B., Rhead, J. C., Richards, W. A., Soskin, R. A., Turek, B., and Kurland, A. A. (1976). MDA-assisted psychotherapy with neurotic outpatients: A pilot study. *J. Nerv. Ment. Dis.* **163**, 233–245.

Burt Angrist

13
Amphetamine Psychosis: Clinical Variations of the Syndrome

I. INTRODUCTION

First reported in 1938 (Young and Scoville, 1938) and originally considered a rare condition, amphetamine psychosis, as seen most commonly by clinicians, is a paranoid or paranoid–hallucinatory psychosis in a setting of clear consciousness in which formal aspects of thought are relatively intact but in which delusions and hallucinations frequently evoke intense fear. The condition usually is encountered in the context of ongoing abuse of high doses of amphetamine and usually clears spontaneously in several days to a week. Some case histories illustrate these features:

> A 19-year-old who had used amphetamine intravenously for 2 yr was away from New York for a month. On returning he learned that some friends recently had been arrested. He injected amphetamine in somewhat larger amounts than he was used to ($\frac{1}{4}$ "spoon" per injection; $\frac{1}{4}$ spoon usually being equal to 125–250 mg drug). Thereafter, while walking on the street, he began to feel that people were following him. He then "realized" that he was suspected of having called the police and caused his

friends' arrests. On the street, every impression reinforced this idea. For example, he heard someone greeting another person say, "Hey man," but since "man" is also slang for police, he assumed that people were really talking about him, and saying that he was an undercover narcotics agent. People on the street seemed to be "closing in" on him so he ran to a girlfriend's hotel. He went into her room asking her to "hide" him and shortly thereafter saw a "whole crowd of people with movie cameras on the roof next door." He felt the hotel was about to be "stormed" and ran down the hall knocking on every room, asking to be let in to be "saved." This behavior led to the police being called and his being taken to a psychiatric hospital. Symptoms resolved completely after 2–3 days.

An 18-year-old who had taken amphetamine orally for a year hitchhiked to New York and took 100 mg orally before going to a nightclub. There he argued with a man and left. On the street he became preoccupied with the idea that the person with whom he had argued might have called friends to "get" him. Thus, anyone on the street might have been sent after him. He went into a bar where people seemed to be looking at him in a sinister, amused way. He left the bar and on the street was particularly frightened by people with "flat" faces (since the person with whom he had argued originally had a "flat" face). He went to his hotel and barricaded the door with a bed. Soon he heard a radio being played in the hall, and assumed it was being played loudly to drown out his screams as he was being murdered. He crawled out on the window ledge (on the third story) and, while out on the ledge, heard voices saying, "Let's get him now." At this point, he decided that he had to run and ran out of the hotel with an open penknife in his hand. He met a policeman while still holding the knife and asked to be taken to a psychiatric hospital. Entering the hospital, he felt he was being pursued by a large crowd of people. His symptoms cleared completely in 3 days.

Despite these case histories, exceptions to the generalizations about amphetamine psychosis noted initially also occur. For example, (1) researchers can now say that the condition is not rare and certainly is not an idiosyncratic reaction to the drug; (2) the condition has been reported after acute ingestions in patients who are not chronic abusers; (3) the psychosis has been reported, although infrequently, after low doses of amphetamine; and (4) persistent psychosis of very long duration has been reported.

With respect to symptomatology, exceptions to the usual pattern also occur: (1) confusional reactions occasionally are seen; (2) nonparanoid presentations, particularly of disorganized or bizarre behavior or of pathological emotional lability have occurred; and (3) formal disorganization of thought, although usually not prominent, has been noted by some investigators.

These inconsistencies and exceptional presentations simply may represent natural variations in the syndrome. The purpose of this chapter is to examine this variability and some of its potential substrates.

II. DATA SOURCES

Amphetamine psychosis is the only drug-induced psychosis to be studied prospectively under experimental conditions. Thus, two data sets are available, each of which has strengths and weaknesses.

1. Naturally Occurring Cases

Premorbid psychiatric status usually is not known and may influence the quality of acute symptomatology and/or duration of symptoms. Reported dosage is open to question (and may be impossible to ascertain if black market preparations are used). Use of other drugs may cloud the clinical picture. Unacknowledged persistent drug use may affect duration of symptoms. Despite these weaknesses, the fact that amphetamine abuse was once quite common means that chances of observing unusual response patterns are maximized.

2. Experimentally Induced Cases

Premorbid psychiatric status is known and patients with concurrent Axis I disorders are excluded. Drug dose is known precisely and access to other drugs presumably is controlled. However, for ethical reasons, these studies have, with a few exceptions, only been done in experienced abusers. Thus, conclusions with respect to the role of tolerance or sensitization must be limited since nearly all subjects can be considered "tolerant" or "sensitized" to some degree.

Most of all, less than 50 subjects have participated in these studies. Thus, unusual or idiosyncratic responses are unlikely to have been documented in this otherwise more definitive database.

III. INDIVIDUAL VARIATIONS IN RESPONSE TO CENTRAL NERVOUS SYSTEM STIMULANTS

Heavy cocaine abusers often take several grams or more daily. At the peak of the United States amphetamine epidemic, abusers frequently reported injections of several hundred milligrams; single injections of up to 1 gm occasionally were claimed (Kramer *et al.*, 1967). In contrast to such large doses, extreme sensitivity to these agents also has been noted, and has caused concern or even alarm in the medical community. Interestingly, the early history of both cocaine and amphetamine indicated substantial enthusiasm for the therapeutic potential of each drug, punctuated by concerns because of instances of unpredictable toxicity.

Koller's use of cocaine in ophthalmology as the first local anesthetic in October 1884 was a sensational development in medicine, which led to rapid extension of the use of the drug in ear–nose–throat (ENT) surgery and to the other surgical applications. By November 29, 1884, *Le Progres Medicale* stated, "All medical journals resound at the moment with news of this triumph of healing. It is scarcely two months since Dr. Koller of Vienna published for the first time the happy attribute (of cocaine) as a local anesthetic for the eye—and already publications on the subject are so numerous and results so uniform that there exists a whole bibliography" (quoted in Becker, 1963).

In the context of this enthusiastic and widespread medical use, reports of toxicity emerged. By the next year, Freud (1885) wrote, "I must stress, even more emphatically than before, the diversity of individual reactions to cocaine." In the first volume of the 1887 *Lancet,* the following cases were reported: (1) cocaine poisoning after accidental ingestion causing "nausea, throbbing, and feeling of bursting in his head, failure of eyesight, loss of the use of his legs, incoherence of speech, and confusion of ideas" (Kilham, 1887); (2) seizures after instillation of cocaine into the bladder (Anonymous, 1887); and (3) transient blindness and incoherence after injection into breast tissue (Roberts, 1887). An article on "Cocaine Dosage and Cocaine Addiction" in the same 1887 volume of the *Lancet* opened as follows: "The recent sad story of the Russian surgeon's suicide from sorrow or remorse due to his belief that a patient had died from an overdose of cocaine points a moral, the import of which demands more than a passing notice" (Matteson, 1887).

Similarly, when amphetamine was introduced as a treatment for narcolepsy (Prinzmetal and Bloomberg, 1935), a period of enthusiastic exploration of the therapeutic potential of this drug ensued. The first review article on "Benzedrine Sulfate Therapy, the Present Status" (Reifenstien and Davidoff, 1939), published less than 4 yr later, contained 115 references to the use of amphetamine in 22 separate conditions. The same article also documented (1) blood pressure increase from 110/60 to 200/100 after 10 mg intravenously, (2) unconsciousness and seizures after 30 mg orally, and (3) coma of 36 hr duration, convulsions, and circulatory collapse with eventual recovery after injection of 140 mg. However, Reifenstien and Davidoff also indicated, "We have given a patient 200 mg orally without untoward effects. Patients with orthostatic hypotension have received 150 mg of the drug daily for 6 months." These investigators concluded that "although many untoward, paradoxical, and unpredictable effects of benzedrine sulfate occur, the exceedingly alarming reactions just mentioned must be considered to represent idiosyncrasy to the drug—because of the large number of patients in whom no such serious effect has appeared."

Sensitivity to psychosis-inducing effects of amphetamine is also clearly quite variable. Kalant's (1973) review specifically noted a severe psychosis after a single dose of only 10 mg racemic amphetamine intravenously, whereas another patient became "mildly excited" but not psychotic after 630 mg by mouth. Clearly very marked individual variations exist in sensitivity of the effects of central nervous system (CNS) stimulants.

IV. POSSIBLE SUBSTRATES FOR VARIABLE SENSITIVITY TO CENTRAL NERVOUS SYSTEM STIMULANTS

A. Genetic Variables

In mice, strain-dependent differences in sensitivity to amphetamine lethality were noted in the early 1960s (Weaver and Kerley, 1962). Studies of substrates for genetic differences in responsiveness to amphetamine have shown higher dopamine turnover [as measured by homovanillic acid (HVA) increase after probenecid administration or by dopamine depletion after α-methyltyrosine pretreatment] in an amphetamine-sensitive than in an unresponsive strain (Jori and Garattini, 1973). Subsequent studies have shown that two different amphetamine-sensitive and -insensitive mouse strains differ in *number* of dopamine neurons in the substantia nigra and ventral tegmental areas (Reis *et al.*, 1979). In both these studies (Jori and Garattini, 1973; Reis *et al.*, 1979), brain amphetamine levels were measured and shown not to differ in responsive and unresponsive animals, that is, genetic differences in response were pharmacodynamic, not secondary to pharmacokinetic variables.

In humans, genetic control of pharmacodynamic responsivity to amphetamine was demonstrated in a placebo-controlled study of the effects of 0.3 mg/kg *d*-amphetamine administered intravenously to 13 identical twin pairs. Behavioral and neuroendocrine responses differed widely between the pairs but were strikingly concordant within each pair. Only one measure of many—elation—correlated with plasma amphetamine levels (Nurnberger *et al.*, 1981).

B. Tolerance

In animals, tolerance develops to autonomic and anorexigenic effects of amphetamine (Kosman and Unna, 1968; Lewander, 1972). The mechanism of this tolerance is not a metabolic or pharmacokinetic one since tolerant and naive animals have similar amphetamine half-lives and distribution (Lewander, 1972; Kuhn and Schanberg, 1977).

In humans, tolerance is thought to occur to anorectic effects since weight loss stops after several weeks of treatment. Tolerance is presumed to occur to cardiovascular effects because of the very large doses that abusers frequently survive (although cerebrovascular accidents with and without death have been reported; Harrington *et al.*, 1983). In Connell's (1958) series of 42 patients, as many cases in which blood pressure was normal were seen as those in which it was elevated. Connell concluded, "there are no diagnostic physical signs of amphetamine intoxication." Tolerance development may, indeed, be necessary for some abusers to survive the doses they take. In a study in which amphetamine psychosis was induced experimentally in abusers, 4 of 16 subjects were unable to complete the protocol without nausea and vomiting. Of these 4, 2 were found to have misrepresented their drug history and not to have been regular users, whereas a third individual had been abstinent for 2.5 yr (Bell, 1972). Tolerance in humans is not the result of increased amphetamine metabolism, since chronic abusers show metabolic patterns similar to those of drug-naive individuals (Anggard *et al.*, 1973).

Possible tolerance to euphorogenic effects of amphetamine is suggested by anecdotal comments by some abusers to the effect that the drug no longer seems as pleasant. Such tolerance theoretically could be a basis for the dose escalation that almost inevitably occurs in those recreational users who develop more severe abuse patterns and dependence. My clinical impression, however, is that dose escalation results from the desire to experience more intense effects rather than an attempt to recapture euphoric experiences that no longer occur.

C. Sensitization

1. Preclinical data

Repeated intermittent administration of a fixed dose of amphetamine (or other CNS stimulant) is now well recognized to result in alterations of the behavioral response in which most (but not all) indices of sensitivity appear augmented. This "sensitization" is an extraordinarily robust phenomenon and, as noted by several investigators, has been observed in every mammalian species studied to date (Segal and Janowsky, 1978; Segal *et al.*, 1981; Segal and Schuckit, 1983; Post and Contel, 1983; Robinson and Becker, 1986).

Reviews of stimulant-induced sensitization include those by Segal (1975), Segal and Janowsky (1978), Segal *et al.* (1981), Segal and Schuckit (1983), Post and Contel (1983), Robinson and Becker (1986), Segal and Kuczenski (1987), and Kuczenski and Segal (1988). The biological basis of this sensitization has been studied intensively and a number of possible mechanisms proposed, but currently controversy exists over each proposed mechanism (see Segal and Janowsky, 1978; Segal *et al.*, 1981; Segal and Schuckit, 1983; Post and Contel, 1983; Robinson and Becker, 1986; Kuczenski and Segal, 1988; Segal and Kuczenski, 1992).

2. Clinical Data

Sensitization to the psychosis-inducing effects in humans could be profoundly important to the clinical spectrum of psychosis induced by amphetamine. Several parameters logically could be expected to be influenced by such sensitization, including a lower minimal dose required for inducing psychosis as well as a longer duration of psychosis. Spontaneous psychoses have been reported in amphetamine abusers in the context of stressful life events (Utena, 1966; Sato *et al.*, 1983), a phenomenon that could be considered an index of sensitization to the extent that psychosis occurs in the absence of amphetamine ingestion. In this context, the similarities between many forms of stress and the effects of amphetamine and, indeed, their interchangeability should be noted (Antelman *et al.*, 1980; Antelman and Chiodo, 1983).

Finally, since sensitization might occur to some types of symptomatology and not others, the possibility of different symptoms or symptom patterns in long-term abusers compared with relatively naive individuals must be evaluated. For example, might such symptoms as olfactory hallucinations or sudden delusional perceptions be more likely to occur in long-term stimulant abusers? Since sensitization is such an important issue, the evidence supporting its possible occurrence in humans is reviewed here.

Ellinwood (1967, 1972) stressed that amphetamine psychosis is an evolving process in which progressively abnormal behaviors develop. For example, (1) an initial "intense feeling of curiosity" ultimately led to behaviors such as that described here: "I read magazines looking at the periods with a jeweler's glass for codes—they were to help me solve the mystery." (2) In some patients, "philosophic concerns, usually unsophisticated dealing with beginnings, meanings, and essences" led to "Eureka" experiences ("I suddenly discovered how the world began."). (3) A "ubiquitous feeling of being watched" became the precursor to paranoid delusions in over half the patients (Ellinwood, 1967). Elsewhere, Ellinwood (1972) stressed the same theme: "Visual hallucinations start with fleeting glimpses of just recognizable images in the peripheral vision; later they become fully formed and stable. Auditory hallucinations begin with the perception of simple noises."

Such descriptions are certainly consistent with the concept of sensitization over time. However, considering the long half-life of amphetamine in humans, evolving psychotic behavior in a single "run" also could be linked to increasing cumulative dose and rising amphetamine plasma levels (Segal *et al.*, 1981; Segal and Schuckit, 1983). Over a longer time period encompassing numerous "runs," dose escalation makes conclusions about sensitization risky, unless dose is specifically known to be constant. Other investigators have shared these concerns (Segal and Janowsky, 1978).

Kramer (1972) described a progressively increased vulnerability to psychosis that develops over months of abuse:

> The paranoia does not usually start during the first few months of high-dose intravenous use. When it does finally begin, it is mild, easily controlled, and is largely dissipated upon waking after crashing; and it usually doesn't start again until after two or three days on a new run. As time goes on, it may start earlier in a run and may persist to some extent even after crashing. In some instances the first injection after a period of sleep will bring about a return of the paranoia. Once an individual has experienced amphetamine paranoia, it will rather readily return even after a prolonged period of abstinence.

Such a pattern is suggestive of a sensitization process. However, individual case history data were not given. As noted earlier, such data are crucial to avoid confounding the escalation of dosage that occurs so commonly with true sensitization.

Bell (1973) reported a study in which amphetamine psychosis was reproduced experimentally in experienced abusers. In this study, 16 professed amphetamine-dependent subjects with variable periods of abstinence received intravenous amphetamine in doses individualized to raise blood pressure by 50%. The doses ranged from 55 to 260 mg (640 mg in one case) and were given within an hour. In 4 subjects, infusions had to be stopped because of nausea and vomiting. As noted earlier, 2 of those 4 had misrepresented their drug history and were not regular users; a third had been abstinent for 2.5 yr, and the fourth subject "stated that he always experienced nausea when he took amphetamine, but he had persisted despite this." Of the 12 regular users who completed the protocol, all developed psychosis within 1–90 hr. These psychoses lasted from 1–2 days in 9 subjects and for 6 days in 2 others. Psychosis was intermittently present for 26 days in the 12th patient, at which point he "revealed that he had managed to take the drug secretly despite supervision."

Since only experienced users developed psychosis and since these psychoses emerged so promptly, this paper is cited often as strong evidence for behavioral sensitization. However, in the 12 subjects who became psychotic, the dose administered was *less* than the reported usual dose taken (suggesting sensitization) in 5, but *greater* than the patients' reported maximum daily dose in 7. Thus, in 7 of 12 patients, a threshold may have been met or exceeded and sensitization need not be inferred.

Sato *et al.* (1983) described 16 patients who developed psychosis during methamphetamine (METH) abuse and then were abstinent from 1 to 60 mo (longer periods usually the result of being in jail) and developed similar psychoses soon after relapse to METH use (1–6 injections). However, the same concerns can be raised with these data as with Bell's (1973) study. Of these individuals, 10 took their usual doses (which had caused psychosis in the past) and 2 took larger doses. However, 4 patients *did* become psychotic after 20–50% of their usual prior dose. Sato *et al.* (1983) indicated that in these 4 patients the lower doses ranged from 5 to 20 mg. If true, these are very low doses indeed for induction of psychosis, but black market METH was used, so dose estimation is a problem. Questioning whether this study indicates sensitization or simply exceeding a threshold seems reasonable.

Although some reservations about the clinical data just presented that suggest amphetamine sensitization have been noted, recent data on sensitization to psychosis-inducing effects of cocaine appear to be more quantitative and clear. Satel *et al.* (1991b) reported a questionnaire study of cocaine-induced paranoid experiences. These investigators studied 50 patients in rehabilitation for primary cocaine dependence. Of these, 34 (68%) acknowledged paranoid experiences that were not trivial. These individuals checked windows and doors (often repeatedly) or hid. Over one-third of those who experienced paranoia armed themselves. Three-fourths of the patients who became paranoid (26) said these experiences clearly worsened with continued use and 20 described a more rapid onset. Brady *et al.* (1991) did a similar questionnaire study of 55 patients, also in a rehabilitation program for primary cocaine dependence. Of these patients, 29 (53%) had experienced cocaine psychosis; 21 of these 29 reported psychosis with less drug and with increasing frequency over time. In this group, 14 stated that they were now unable to use cocaine at all without becoming paranoid and 15 noted that psychosis now occurred with a more rapid onset. No subject described decreased psychosis over time. These two studies (Brady *et al.*, 1991; Satel *et al.*, 1991b) are consistent and mutually reinforcing. In both, dose escalation was considered explicitly and rejected as a cause for increasing psychosis vulnerability. Sensitization to the psychosis-inducing effects of cocaine does appear to occur. This finding reinforces the possibility that sensitization to amphetamine may occur as well. Reservations expressed with respect to the clinical data suggesting sensitization to amphetamine are not meant to imply that such sensitization does not occur. Indeed, I believe it probably does, but the data are criticizable primarily because escalation of dosage was not discussed explicitly.

Alternatively, the data on cocaine sensitization also could be interpreted as indicating that sensitization develops much more robustly to cocaine than to amphetamine. However, this possibility is difficulty to reconcile with the fact that cross-sensitization between cocaine and amphetamine has been shown to occur in preclinical studies (Segal and Kuczenski, 1987).

D. Premorbid Psychiatric Status

1. Schizophrenia

Ellinwood (1967,1972) found that 25% of a series of amphetamine abusers had pre-existing schizophrenia. Indeed, with the exception of antisocial personality disorder (45%), schizophrenia was the most common comorbid psychiatric diagnosis. Potential important implications of this finding are as follows:

1. Psychosis after low doses of amphetamine might represent activation of pre-existing schizophrenia, as occurs in CNS stimulant "challenge" studies in schizophrenia (40% of schizophrenics who receive CNS stimulants show increased psychotic symptoms after doses that do not cause psychosis in normals). A comprehensive review of this subject has been published (Lieberman *et al.*, 1987).
2. Some symptomatology considered to be amphetamine induced may "really" be the result of an interaction between schizophrenia and amphetamine effects, that is, pre-existing schizophrenia may color the quality of symptomatology that emerges in such patients.
3. Pre-existing schizophrenia is also a possible basis for very prolonged psychosis occasionally reported after amphetamine abuse (Tatetsu, 1964; Utena, 1966).
4. Pre-existing schizophrenia also could be a basis for spontaneous psychosis occasionally reported to occur after stress in amphetamine abusers (Utena, 1966; Sato *et al.*, 1983).

2. Bipolar Disorder

A switch into mania has been described after low dose amphetamine administration (Bunney, 1978). L-DOPA also has been shown to induce mania with regularity in bipolar patients (Murphy et al., 1971). Thus, speculating that bipolar patients also might show several symptoms:

1. enhanced vulnerability to usually subpsychogenic doses
2. reactions colored by a propensity to manic symptomatology
3. psychosis disproportionate in duration to the effects of amphetamine
4. spontaneous or stress-induced psychosis without drug ingestion

is not unreasonable.

V. CLINICAL VARIATIONS AND UNRESOLVED ISSUES IN AMPHETAMINE PSYCHOSIS

With these substrates for variability in the response to CNS stimulants and potential pitfalls in interpretation of phenomonology in mind, several ad hoc questions can be posed regarding variations in the clinical spectrum of amphetamine psychosis and possible unresolved issues.

A. Why Was Amphetamine Psychosis Originally Considered a Rare Condition and Later Seen Frequently?

Kalant (1973) was able to collect only 71 reports of amphetamine psychosis in the 20 years after it was reported first (Young and Scoville, 1938), yet in the next 5 yr, well over 100 cases were noted. By 1966–1968, approximately 1 amphetamine-related admission per week was seen at Belleview Psychiatric Hospital in New York (Angrist and Gershon, 1969). Two factors probably account for this increase in incidence. (1) Amphetamine abuse was, in fact, becoming increasingly prevalent between 1937 and 1970. The history of amphetamine abuse is discussed further in Chapter 15 and by Griffith (1966), Grinspoon and Hedblom (1975), Sadusk (1966), and Angrist and Sudilovsky (1978). (2) Many patients with amphetamine psychosis almost certainly were misdiagnosed. Connel (1958) particularly emphasized this possibility and cited as an example a letter from a prominent British consulting psychiatrist about a patient who "persistently denied symptoms which he apparently had until immediately before admission here. We are therefore compelled to discharge him. I am quite sure, however, that he remains a chronic schizophrenic. You may also be interested to know that he is still consuming the contents of amphetamine inhalers." Connell's first conclusion in his monograph (1958) was, "Psychosis associated with amphetamine usage is much more common in this country than would be expected from reports in the literature."

B. Is Amphetamine Psychosis an Idiosyncratic Reaction to the Drug or a Manifestation of Latent Psychosis?

This question can be answered with a definite "no." Connell, as early as 1958, noted that "apparently normal and well-adjusted individuals may develop this reaction." The strongest data addressing this issue, however, are from prospective studies in which amphetamine psychosis was induced in abusers who were screened for pre-existing psychotic disorders. Griffith et al. (1972a) observed psychosis in 6 of 9 subjects. Angrist and Gershon (1970,1972) and Angrist et al. (1972,1974) observed psychosis with loss of insight in 5 of 9 subjects. An additional 3 had transient psychotic symptoms but knew these symptoms were drug induced. The remaining subject denied psychotic symptoms but appeared to have transient visual hallucinations. Bell (1973) documented psychosis in 12 of 16 subjects. Of these, 3 had pre-existing schizophrenia. If these subjects are excluded, an amphetamine psychosis occurred in 11 of 13 patients.

Thus, in total, 25 of 31 amphetamine users developed psychotic symptoms and 22 of 31 became frankly psychotic. As Griffith et al. (1972a) commented, "A paranoid reaction must be considered a probable complication of high-dose amphetamine abuse."

C. Is a Minimum Dose Required for Amphetamine Psychosis?

As noted earlier, the most likely source of data to identify patients with unusual sensitivity to amphetamine psychosis consists of anecdotal reports of naturally occurring cases from the time when amphetamine was prescribed more widely and abused more frequently. Individuals who unexpectedly developed psychoses after low doses of amphetamine might be expected to avoid the drug thereafter. Such individuals might not be represented in prospective studies in experienced abusers. However, in fact, the findings from anecdotal reports and from prospective studies are rather consistent.

Connell (1958) specifically commented that "reliable evidence about the dose was difficult to obtain." In his literature review of cases reported prior to his study, 50 mg was the lowest reported dose. In his series, the minimal total dose was 20–60 mg. He concluded that "the writer would be disinclined to incriminate amphetamine as an agent in the production of a psychotic reaction in doses less than 50 mg." Kalant (1973) collected reports of 10 cases at <50 mg/day chronically and 2 cases after single doses of 30 mg orally and 10 mg intravenously. The latter case was not trivial. "She began to scream that she was being watched and that the doctors wanted to kill her under orders from the church. . . . Later she developed visual and auditory hallucinations. . . . These acute psychotic symptoms subsided after 2 days" (Ruiz Ogara, 1954; cited by Kalant, 1973). Gold and Bowers (1978) gave a detailed description of a patient with a psychosis with auditory hallucinations and beliefs that the "FBI had agents watching me" and that family and friends were impostors. This psychosis occurred after only 25 mg dextroamphetamine and persisted for over a month. In studies of experimentally induced psychosis, one of Bell's patients developed paranoid psychosis and auditory hallucinations lasting for 1 day after a dose of 55 mg.

Thus, the data rather consistently indicate that amphetamine psychosis occasionally occurs at doses of about 50 mg and occurs very rarely with less. Gold and Bowers (1978) apparently reached the same conclusion, as indicated by the title of their case report: "Neurobiological vulnerability to low-dose amphetamine psychosis."

D. Is a Minimal Duration of Use Required for Amphetamine Psychosis?

Because stimulant psychoses usually are seen in long-term heavy abusers, researchers often assume that some chronicity of use is a sine qua non for psychosis development. As emphasized in the preceding section, however, stimulant abusers often escalate their dose over time. Thus, sorting out the contribution of duration of use and dosage vis a vis psychosis vulnerability frequently becomes difficult.

The data that exist suggest that dosage is the more critical factor. Historically, much of these data come from the time when amphetamine inhalers were still marketed. These inhalers contained very large doses of amphetamine. The benzedrine inhaler, for example, contained 350 mg racemic amphetamine base, equal to 561 mg sulfate (Connell, 1958). The contents were well known to be ingestible. By 1947, Monroe and Drell estimated that 25% of a prison population in a military stockade abused these inhalers, and psychoses were documented (Monroe and Drell, 1947).

Connell (1958) and Kalant (1973) documented instances of amphetamine psychosis in relatively naive users. Connell presented 8 patients who became psychotic after single doses. None were considered addicted. Kalant's literature review collected 54 cases of acute amphetamine toxicity in which psychotic symptoms occurred in 30. None of these patients had used amphetamine for more than a month and, in most cases, a single large dose was taken. Thus, amphetamine psychosis clearly can occur after single doses and after brief exposure to the drug.

Some animal data indicate long-standing increases in sensitivity to amphetamine after a single dose (Segal, 1975; Robinson and Becker, 1986). Thus, documenting whether psychoses have occurred in completely naive users after a first exposure would be a particular interest. Such cases have been reported. Some of these cases were atypical because of the very low doses that induced the psychosis, suggesting unusual vulnerability (Gold and Bowers, 1978; Ruiz Ogara, 1954, cited by Kalant 1973). (Both these reactions are described in more detail in Section V,C). However, in Connell's series, among the 8 patients who were not addicted and became psychotic after single doses were 2 who appeared to have minimal prior exposure. One had taken a strip of inhaler given by a friend 2 yr prior; no further use is noted in the detailed case history in the appendix. Another was given 75 mg amphetamine by mistake in a hospital. However, that these patients had never had any other prior exposure is not stated explicitly.

In our studies of experimentally induced amphetamine psychosis in which heavy stimulant use was a requirement for inclusion, one subject later admitted that he had misrepresented his past history of claimed intravenous methylphenidate abuse. Thus, by including him, we inadvertently administered high doses of amphetamine to a subject with almost no prior exposure to CNS stimulants and no history of abuse of these agents. During these experiments, he developed a rather florid and typical psychosis that is described in detail in the next section.

E. Is the Symptomatology of Amphetamine Psychosis Similar or Different in Relatively Naive and Chronic Abusers?

Connell (1958) divided the 42 patients he saw into three groups: (1) 8 patients who were not addicts and "took single doses of the drug only," (2) "four patients who had also been taking alcohol at the time of the development of the psychosis," and (3) "thirty patients who had been taking regular doses of the drug for more than a month." In comparing their symptomatology, he concluded that "there is . . . no useful difference in the complaints of patients in the three different groups." Indeed, the presenting complaints of Connell's first group seem rather characteristic and typical of any series of patients with amphetamine psychosis. Of the 8, 6 were "afraid of a gang or persons who were going to do a personal injury."

Similarly, Kalant (1973) considered the symptoms of patients who had become psychotic after using amphetamine for less than a month (and in most cases after a single large dose) to be "indistinguishable from those seen in chronic abusers of amphetamines." Thus, most data suggest that symptomatology is similar in relatively naive patients and chronic abusers when an amphetamine psychosis occurs. Our single, nearly naive subject also developed a rather typical syndrome:

> He was a 24-year-old who, for the past 3 years since discharge from the army, had lived alternately at his parents home, in communes, and with friends in Greenwich Village. During this time, he experimented with marijuana, peyote, (6 times), opiates (opium and intravenous heroin a few times without addiction), and occasional barbiturates. He later denied stimulant use during this period but initially told the investigator that he had taken intravenous methylphenidate to gain access to the studies. He was the friend of a CNS stimulant abuser who had been a subject.
>
> He had, in fact, been treated with methylphenidate as a child when he was 10–11 years old. He stated the treatment was not helpful in getting him to "sit and learn" and was discontinued after a month. Thus, his actual past stimulant exposure was

"one blue pill" of methylphenidate (presumably 10 mg) daily for about a month, 13–14 years previously. During the in-patient observation period, prior to the experiment, he had borrowed $4.50 from a nurse's aid.

During the experiment (Angrist et al., 1974), he received a cumulative dose of 465 mg over just under 24 hr without striking effects except feeling "self conscious," perceiving "bad vibes," and at times becoming somewhat sullen. Abruptly he heard a gang enter the ward (presumably sent by the nurse's aid) and heard them threatening to kill him. He sprang at the investigator menacingly, then stopped and began to speak of "set ups" and "traps." Explanations that his experiences were amphetamine induced were rejected with sardonic mock agreement. "Oh sure! Ha! Is that the way it's going to be?" At other times he became panicky and tearfully begged the investigator to tell him what was really going on. "Please! Just tell me. I always respected you." Finally he concluded that he wouldn't be told for his own protection.

He insisted on standing in a place on the ward where he could watch the hallway. He saw gang members at the water fountain. He jumped at every movement and interpreted the investigator's gestures as "signals" to the gang. When another attempt was made to tell him his experiences were caused by the amphetamine, he said, "OK, I lose!" and spoke of being "jumped," tied to a bed, "knocked out," and being taken to a state hospital.

Four hours later he told another physician that his sweater was a radio tuned in to him. "I won't talk to you as long as you're wearing that radio. Get it off. Oh, God, please get it off!" Patients were gang members in disguise. At times he confronted the "gang." "If you want to beat me up—OK—get it over with." He stood up (holding his genitals) and shouted, "OK, come and get me. Let's get this over with."

He then asked to be left alone and wept continuously at a table. Thereafter symptomatology waned, but the same themes persisted. Papers on a bulletin board "turned into" a gangster in a white raincoat. He saw the doors in the hall open and close and pleaded with the investigator to acknowledge his danger and finally concluded that others were too slow to see the evidence. At times he still became agitated and threatening, saying "If I get hurt, you get hurt."

He felt he was accused of stealing some money that was (actually) taken on the ward and heard someone say, "The thief is back, hide your money," when people actually passed by talking. He confronted a nursing staff member. "What's that about money that was stolen?" but was taken aback when the nurse said, "I didn't hear about it."

After about 25 hr, psychotic symptoms waned. He spoke of feeling tired and an hour later still heard his name called from time to time.

Thus, although the subject was not completely naive to stimulants, having received treatment with methylphenidate for 1 mo 13–14 years previously, this experiment does document a rather typical psychotic syndrome in the context of minimal past exposure.

F. Does Amphetamine Abuse Lead to Spontaneous or Stress-Induced Psychosis?

Utena (1966) (commenting on "Peculiarities in the clinical features of methamphetamine psychosis") noted:

> Another peculiarity in the residual state is a tendency for earlier symptoms to recur, which are usually induced by some kind of stress, either physical or psychological.

Such tendency to relapse was found in one-fourth of the residual cases who were admitted to Matzuzawa Mental Hospital. Compared with other types of intoxication psychoses, this is really an unusual phenomenon, since the causative agent no longer exists and an enhanced vulnerability alone remains.

Sato *et al.* (1983) reported such a psychosis in one patient, the only case of its type ever documented in detail:

> A forty-two year old man who had been addicted to intravenous methamphetamine (MAP) since age thirty-seven was taking injections daily. "In March 1977, the day after a MAP injection, he felt that someone was hiding in the ceiling and people were walking beneath the floor. He felt panic, smelled foul odors in food, and was convinced he was under surveillance of 'people in the background' using a strange machine." . . . "Under strict observation by relatives and a policeman, the MAP was discontinued and the above symptoms disappeared. One month later, he reinjected one-fifth of the previous single amount (15 mg) of MAP. A second psychotic episode similar to the first recurred immediately. Again the symptoms were relieved by abstinence from MAP, under intensive observation by the relatives. Thereafter he kept in fair mental condition. On the night of April 19, 1979, he suddenly shouted, "I can't breathe, no, no - foul," simultaneously becoming very violent. His family members denied that he had taken any stimulant prior to this third psychotic episode. He was admitted to the Takaoka hospital the next day where he received treatment with neuroleptics for 28 days. As he remained very suspicious and jealous when he was discharged, follow-up in the outpatient clinic continued for 9 months. Thereafter, the family did not notice any mental abnormality until the fourth episode occurred in January 1981. At that time, he was found standing at midnight, bowing toward the ceiling stereotypically. On readmission to the hospital there were paranoid delusions and auditory hallucinations. Both he and his wife denied any MAP reuse."

This report by Sato *et al.* (1983) is considered by some investigators to constitute conclusive proof of the development of spontaneous psychoses in amphetamine abusers. The clinical symptomatology is well documented, but two potential pitfalls exist. First, the patient may have developed a psychotic disorder coincidentally that was unrelated to amphetamine ingestion. A hospitalization of 28 days, neuroleptic treatment, and need for outpatient follow-up for 9 months thereafter (presumably with continued neuroleptic treatment) is consistent with such a possibility. This criticism almost could be considered "unfair," since it could be applied to any single case. Nonetheless, since 25% of amphetamine abusers have been found to have pre-existing schizophrenia (Ellinwood 1967,1972), the possibility of premorbid vulnerability to psychosis cannot be dismissed lightly.

The second concern is that, although Sato *et al.* (1983) specifically addressed the issue of continued methamphetamine use, they did not specifically exclude the possibility with toxicology data. Similarly, Utena (1966) did not present toxicology data for the 25% of patients who showed stress-induced psychotic relapses. This issue is of particular concern because continued surreptitious amphetamine abuse is very common in this population. Connell (1958) specifically noted this "one patient . . . sewed the contents of amphetamine inhalers in the hem of her skirt and one patient obtained an inhaler whilst actually in an observation ward." No less than 10 of his 42 patients "were known to have taken the drug while in the hospital, at one time or another." Bell (1973) also encountered a patient in his study who was psychotic intermittently for 26 days, "then revealed that he had managed to take the drug secretly despite supervision." Two studies of psychiatric in-patients done at a time when amphetamine abuse was common found 15% of urine samples to contain amphetamine (Rockwell and Ostwald, 1968; Robinson and Wolkind, 1970). In the absence of urine toxicology data to supplement the observations of Utena (1966) and the report by Sato et al. (1983), spontaneous psychosis due to past amphetamine exposure must be considered unproven.

G. Can Amphetamine Cause Persistent Psychosis?

In the post-World War II amphetamine epidemic in Japan, psychoses that persisted long after amphetamine withdrawal were seen and interpreted as a persistent drug effect. Tatetsu (1964) noted that 14.4% "stayed in hospitals for more than five years because of their symptoms after the withdrawal of injections." In a subsequent publication (Tatetsu, 1972) Tatetsu explicitly rejected pre-existing schizophrenia as a substrate:

> It has been suggested that amphetamine addicts with this schizophrenia-like picture may actually be schizophrenics. Unlike true schizophrenics, however, the intoxicated patients keep smooth contact with others around them and maintain fairly good rapport—. They are also less unnatural in their posture and less awkward in their movements than schizophrenics.

Most modern investigators might question quality of rapport, unnatural posture, and awkward movement as reliable discriminators in diagnosing schizophrenia. However, these criteria do suggest that Tatetsu used a rather narrow concept of schizophrenia since such features usually are seen in rather severe and chronic patients.

Tatetsu (1972) also compared the incidence of schizophrenia in parents and siblings of methamphetamine psychotics, schizophrenics, and the general population. Patients with methamphetamine psychosis had a higher incidence of schizophrenia in these first degree relatives than the general population, but a rate lower than that found in the relatives schizophrenics.

Utena (1966) indicated that "in most . . . cases, the apparent psychotic symptoms subsided in a week or a month", but that "in a small proportion, the disorder took a deleterious course and eventually developed into a protracted schizophrenia-like state. Of all the cases in Matzuzawa Mental Hospital, about 5% belonged to this group." Utena also rejected pre-existing schizophrenia as an explanation: "It seems too simple an explanation to say that these protracted psychotic cases should be diagnosed as having been schizophrenic from the beginning."

In contrast, western investigators generally have observed shorter durations of psychosis and have considered persistent psychotic symptoms to indicate a pre-existing psychotic disorder. Connell (1958) was unambiguous:

> Patients with amphetamine psychosis recover within a week unless there is demonstrable cause for continuance of symptoms, e.g., continued excretion of the drug or hysterical prolongation of the symptoms.—Patients whose symptoms continue after the urine has been shown to contain no amphetamine should be excluded from the diagnosis and considered as probable schizophrenics.

Bell (1965) concurred that uncomplicated amphetamine psychosis cleared rapidly whereas amphetamine intoxication superimposed on a schizophrenic substrate did not:

> The patients in this series can be divided into two distinct groups: The nonschizophrenics had psychotic episodes that cleared within 10 days after the withdrawal of amphetamines. They did not exhibit thought disorder, and it was not uncommon for them to experience vivid visual hallucinations. The other group of subjects suffered psychosis that lasted for months and all the characteristics of their illness, including the presence of thought disorder and the relative absence of visual hallucinations were typical of schizophrenia.

Gold and Bowers (1978), as mentioned earlier, reported a single case, remarkable not only for the low dosage that precipitated a psychosis but for its duration, which was longer than a month. In this case, urine toxicology was done and found to be negative at a time when the patient was still symptomatic. Follow-up data on this patient might both test Connell's (1958) contention that "patients whose symptoms continue after the urine has been shown to contain no amphetamine should be . . . considered as probable schizophrenics" and help clarify the role of premorbid state in prolonging symptom duration.

Satel *et al.* (1991a), studying cocaine addicts, noted that "in our sample of 100 cocaine dependent males, none reported paranoia extending beyond the crash phase. Conversely in a series of 66 cocaine-using schizophrenic and bipolar patients admitted for disorganized behavior, all remained psychotic for at least 42 days."

Our own experience has been consistent with that of other Western investigators, that is, prolonged psychosis after amphetamine ingestion has not been observed. In experiments in which amphetamine psychosis was induced prospectively in nonpsychotic subjects, none had symptoms that persisted for more than 6 days (Bell, 1973) and clearing was usually more prompt (Angrist and Gershon, 1970; Griffith *et al.*, 1972; Angrist *et al.*, 1972). As noted earlier, however, the number of subjects that have participated in such studies was small and rare response patterns might have been missed.

However, note also that prolonged exacerbations of psychopathology have not occurred in modern "challenge" studies in which amphetamine and methylphenidate have been administered to rigorously diagnosed schizophrenic patients. This result may, in fact, be caused by the fact that lower doses of stimulant agents are administered in such studies than usually are taken by abusers. [For reviews of this area of schizophrenia research, see Segal and Janowsky (1978); Angrist and van Kammen (1984), and Lieberman et al. (1987).]

The problem is a difficult one and, without knowledge of a patient's premorbid state, is fraught with circulatory. Given a chronic psychosis after a drug exposure, one can conclude equally that the crucial element is the drug or the premorbid state, a situation analogous to deciding whether the chicken or the egg came first. Currently, our assessment is that the potential for amphetamine to cause long-standing psychosis is unproven, given that the premorbid state was not known and that continued drug use was not excluded in patients in whom this effect was observed.

H. Does Pre-existing Schizophrenia Affect the Quality of Symptoms That Occur after Amphetamine Abuse?

Ellinwood (1972) specifically noted that the quality of the delusions in amphetamine-intoxicated schizophrenics and nonschizophrenics often differed. "Patients who had more stable personalities tended to have delusions of persecution that were more reality oriented (for example: persecution by federal narcotics agents). The more schizophrenic individuals were likely to be persecuted by Martians, evil spirits and devils." Our experience generally has been quite similar. The following two clinical examples are of amphetamine-intoxicated individuals who, after clearing, showed residual schizophrenic symptomatology.

> A 24-year-old man with a 3-year history of intravenous amphetamine abuse injected an unknown amount of the drug and felt he was being followed by Jewish policemen. He heard people talking about him, saw purple shapes in passing cars, felt he could control traffic lights, and sang in the streets that he loved people and "lived on their hate". He was brought to the hospital by the police after bizarre behavior in a subway. Hallucinations cleared after 1–2 days and delusions cleared over 3–4 days. Residual blunted affect and illogical thinking persisted without diminution until discharge after 12 days.
>
> A 44-year-old man with a 6- to 7-year history of intravenous amphetamine abuse was brought to the hospital by police after being found wandering nude. He explained to the ward physician that someone had put something in his clothes, perhaps lice, that made him itch and that he "had to tear them off and throw them to the four winds North, South, East, and West."
>
> He had injected $\frac{1}{8}$ of a "spoon" of amphetamine that morning and also acknowledged a number of symptoms that occurred each time he took amphetamine. These were (1) voices saying that he was "a demigod and immortal," "saying not to go out with women, to have sex with men," (these were induced telepathically by homosexual amphetamine abusers that he knew, whose sexual advances he had refused), saying that they could "read his mind," and the voices of his dead mother and

brother telling him to go to the hospital; (2) ideas of influence; that is, of someone controlling his body "like a puppet, a my-size puppet;" "at the same time the voices would reinforce the influencer(s)," for example, saying "Touch your penis;" and (3) tactile hallucinations, such as fingers on his legs and throat. These symptoms diminished and cleared after 12 days.

However, although schizophrenics often had bizarre delusions, not all patients with bizarre delusions were considered schizophrenic after acute symptoms cleared. For example:

> An 18-year-old with a 3-year history of amphetamine abuse spent the entire weekend "just shooting and shooting." He began to feel that everything had begun to take on a special significance. "Everything was there for me to do something about it. It was all a big game." He felt that there were two forces in the world corresponding to good and evil and that these somehow were represented by his left and right sides. He felt that the right side was clearly "his" side because "things came more naturally to me on the right side" (disregarding the fact that he was right-handed). Television appeared to apply to him specifically. While on a walk in Central Park he saw a fire and fire trucks. This, he felt was a "signal to either take off my clothes and jump over the wall" (which, he explained, "would help everyone and make them OK") or to "stand by myself." In this state, he asked a policeman to take him to a psychiatric hospital. When seen the day after his admission, he felt that he was still "definitely involved" in the news on television and showed some metonymic use of words, that is, when asked about his involvement in rock music, he said, "I sold my guitar down the river." Thereafter symptoms declined rapidly over 4 days and he showed no signs of residual schizophrenia.

Note also that Segal and Janowsky (1978) and Janowsky and Risch (1979) studied the reports of nonschizophrenic subjects who developed amphetamine psychosis and found that both World Health Organization symptoms that correlated with a diagnosis of schizophrenia and Schneider's first rank symptoms were noted frequently.

I. Variations and Range of Acute Clinical Symptoms of Amphetamine Psychosis

1. Preponderance of the Paranoid/Paranoid–Hallucinatory Syndrome and Less Typical Variations

Based on a review of 30 published cases and the 42 patients he observed, Connell (1958) characterized the syndrome of amphetamine psychosis as "primarily a paranoid psychosis with ideas of reference, delusions of persecution, and auditory and visual hallucinations in a setting of clear consciousness." Kalant (1973) corroborated Connell's observations in a literature review of 94 patients for whom clinical descriptions were provided. Both Kalant and Connell found paranoid delusions to occur in just over 80% of patients, and hallucinations of various modalities in 60–70%. Tactile hallucinations were noted in 12% by both Connell and Kalant, and olfactory hallucinations in less than 10% in both series. Disorientation was seen in 7% by both investigators.

Ellinwood (1972) also commented on both the characteristic syndrome and its clinical variations:

> In spite of great individual variability, amphetamine psychosis usually produces a fairly distinct syndrome characterized by delusions of persecution, ideas of reference, visual and auditory hallucinations, changes in body image, and hyperactivity and excitation. Disorientation and clouding of memory are not part of the picture and, in fact, one of the remarkable features of this psychosis is the hypermnesia for the psychotic episode.
>
> Because the incidence of symptoms corresponds closely in all the reported series, it appears statistically that amphetamine psychosis (in the later stages) is a fairly well-delineated syndrome. Because of the marked individual variability, however, statistics are misleading. Chronic amphetamine reactions are much like syphilis in that they can mimic any number of psychiatric disorders including hypomania, depression, emotional lability, and obsessive–compulsive reactions; the most constant and characteristic form, however, remains the paranoid schizophreniform syndrome.

2. Nonparanoid Presentations

Because the focus of this chapter is clinical variations of the amphetamine psychosis syndrome, some atypical presentations are noted here, in addition to the more typical patterns of the patients described in the introduction, (Angrist and Gershon, 1969):

1. *Confusional states* One patient was picked up by the police trying to force his way into a stranger's apartment. He told them it was his mother's apartment in another state. A second "played drums" on garbage cans, tore down a section of wire fence, and, when apprehended, asked the police for amphetamine.
2. *Emotional lability syndromes* A woman cut her wrist deeply after being "rejected" by a man she had just met that evening. A second woman cut her wrists in response to the delusion that the effects of amphetamine on her circulation had eroded the bone structure of her face and that she was disfigured.
3. *Bizarre sexual behavior* After ingesting mephentermine, a patient lay on the sidewalk and asked passing women to step over him. In a subsequent admission, he masturbated in a church (Angrist *et al.*, 1970).
4. *Destructive outbursts* One patient broke a window in a restaurant and another broke several musical instruments. Neither could explain why subsequently.
5. *Unmotivated assaults* A patient (who had been a mugger off drugs) walked abreast of people he did not know on the street and then spun and hit them. He did this 3 or 4 times, then came to the hospital voluntarily. A second patient saw his roommate sleeping and felt that

he looked "dead." He then took a detached doorknob and beat his head with it until restrained by a second roommate.

3. Clouding of Consciousness

Examples of confusional states were just presented. Disorientation, although not characteristic of the amphetamine psychosis syndrome, also was encountered by both Connell and Kalant in 7% of patients. The most explicit discussion of this issue was by Connell:

> With regard to the absence of disorientation in the present study, the writer has no doubt that amphetamine, in sufficient doses, can produce disorientation. In fact, several case histories suggested that disorientation might have been present at some time during the psychotic episode but was short-lived and no longer present when the patient eventually came under psychiatric care.

Connell then noted the prior comments of Mayer-Gross (1951):

> Differences in symptoms found with various drugs have been attributed to differences in the personality of the intoxicated. This may be true for an early stage or for cases of very mild intoxication. The delirious picture, on the other hand, is probably common to all intoxicants when their effect is most severe. Between these two ends of the scale is a stage in which probably each drug shows certain special features." (Mayer-Gross, 1951)

4. Affect in Amphetamine Psychosis

In a review of Connell's monograph, Slater (1959) commented that one of the qualities that might distinguish amphetamine psychosis from schizophrenia was "the brisk emotional reaction usually in the direction of anxiety." In contrast, Griffith *et al.* (1972a), in experimental inductions of the psychosis in which 5–10 mg *dextro*-amphetamine was given hourly, observed that

> all volunteers appeared depressed when the cumulative dose of dextroamphetamine exceeded 50 mg. They became quite hypochondriacal, spent most of their time in bed and lost interest in their surroundings and their usual activities . . . subjects who were previously quite verbal and relatively trusting became quite taciturn, reserved and negativistic.

Different effects on emotionality were observed by Bell (1973), who administered methamphetamine intravenously in doses individualized to raise blood pressure 50%. Under these circumstances, patients became psychotic while still experiencing acute elation. Bell suggested that the difference in emotional response in the two studies might be the result of differences in the dosing regimen used by Griffith *et al.* (1972a) and the one he employed. "In fact, the relatively slow oral administration of the drug in their study probably confused the issue."

Our findings may provide some support for Bell's speculations (Angrist and Gershon 1972; Angrist *et al.*, 1974). The dosing regimen we used was somewhat more acute than that used by Griffith *et al.*, (1972a) (up to 50 mg racemic amphetamine/hr) but not nearly as acute as that used by Bell. Under these circumstances, some subjects showed intense and labile affective responses whereas the affect of others became blunt and constricted. One subject, for example (described earlier), became panicky and tearful in response to acute paranoid delusions whereas two others became noticeably more emotionally monotonic and less animated. In addition to dose regimen, of course, individual differences in response may play an important role. However, in general, our dose regimen was intermediate between those used by Griffith *et al.* (1972a) and by Bell (1973), and effects on emotionality described by each of these two investigators were seen.

5. Thought Disorder in Amphetamine Psychosis

Neither Griffith *et al.* (1972a) nor Bell (1973) observed thought disorder during their prospective studies. Indeed, Bell (1965) (see Section V,G) indicated previously that thought disorder did not occur in amphetamine psychosis.

In contrast, during our prospective studies, we observed thought disorder of three types:

1. Diffuseness, inability to focus on a point and loss of goal direction—For example, a patient who had received a cumulative dose of 430 mg *l*-amphetamine said, when asked how he felt, "agitated and annoyed." Asked why, he responded, "It's a ridiculous thing like the marijuana laws, or birth control. That's totally ridiculous! It's like a thunderstorm in the forest. It affects young trees. There is a balance of nature. You mess with the balance of nature, you lose buffalo, you lose birds. Things become extinct. For man, you lose philosophies."
2. Pressured "philosophic" rambling that appeared disorganized and bizarre—This subject felt he had received special enlightenment and had become a prophet. An example of his written productions is shown in Fig. 1.
3. Idiosyncratic speech—A young man who, after acute paranoid delusions had cleared, showed no evidence of residual schizophrenia said (of his brother's drug use), "My brother has been playing with the fires of hell" and responded to the proverb "People who live in glass houses should not throw stones" with "If you throw stones you risk your life. Living in a glass house would shatter your whole being." Another example of this idiosyncratic speech is the third case described in Section V,H, who said, "I sold my guitar down the river."

Yet another example of thought disorder with pervasive predicative identification and illogicality was reported in detail by Siomopoulis (1976):

> A few months ago he came to believe that "twisted" hair in his head may cause cancer of the brain and schizophrenia, and he had to "untwist" it to prevent occur-

> Even though being used as a profit of the Energy Force that is all Good. ~~This has been~~ ~~I much~~ Reccuing the gift of Knowledge of Truth of creation of mankind. This does not mean that peace comes with it. When it is Ready you shall Recieve untill That Time "I must endure great Soffering would be mine. until my conscious is given all good and peace, at His will

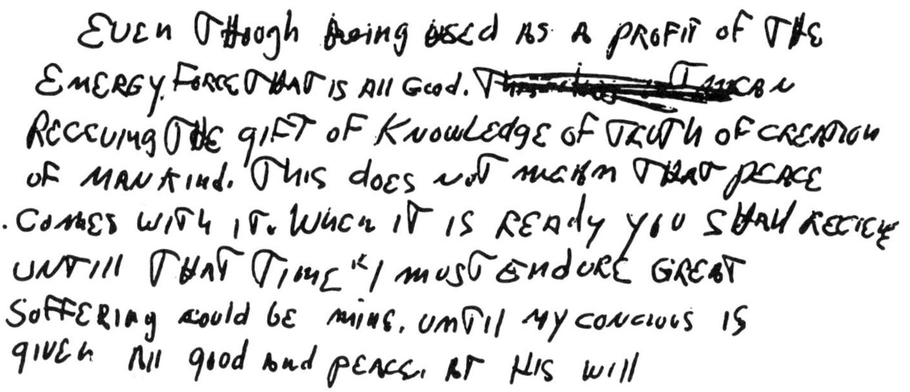

FIGURE 1 Written productions of a nonpsychotic subject who received 590 mg racemic amphetamine over 46 hr. After 475 mg, he began speaking of "revelations," but could not explain what had been revealed, that is, he "received new understanding not given to everyone in this cycle." He alluded to being "a prophet" and began to write excitedly. He then began to speak the things he was writing and ultimately stopped writing and began to "preach" to the ward at large. The content was essentially as shown here. This behavior persisted over 11.75 hr.

rence of these illnesses.—His preoccupation with "hair" was pervasive. He pointed out that hair and sperm under the microscope look like worms; therefore, they are worms; that the testes are bags of worms, and that the intestines and the brain and the whole human body are just big worms.—He called attention to the fact that he was born on July 17, that Caesar's first name was Julius, and that since the Romans celebrated the founding of Rome on the 17th of each month, it was more than likely that Julius Caesar was born on July 17. Since the patient was also Roman (Italian), he reasoned that he was Julius Caesar reincarnated.

A milder example of predicative identification is in the second case history given in Section I in which the patient was particularly frightened of people with "flat" faces because the person with whom he had argued had a "flat" face.

J. What Does Amphetamine Psychosis "Model"?

Connell (1958) first proposed that the syndrome of amphetamine psychosis "may be indistinguishable from acute or chronic paranoid schizophrenia." As noted earlier, Slater (1959) thought that affect differentiated the two conditions. He also noted several other potential discriminators: "past history of psychopathic traits, the rapidity of the onset, the dream-like quality of the experiences, the tendency towards visual hallucination and the brisk emotional reaction usually in the direction of anxiety." He felt that "only the most hyperacute of paranoid schizophrenic states will mimic the syndrome in all its peculiarities."

Bell (1965), after comparing the symptoms of schizophrenic and nonschizophrenic patients with amphetamine intoxication, considered the duration of symptoms, the type of hallucination (auditory or visual), and the presence of thought disorder to discriminate the two patient groups (see Section V,G).

In the prospective studies of Griffith et al. (1972a), thought disorder was not observed. Elsewhere, Griffith et al. (1972b) commented that

> Disorders of thought and bizarre associations were not a prominent feature of the amphetamine psychosis in Bell's series or in our controlled study; yet fragmented and bizarre associations are so typical of schizophrenia that psychiatrists use the term "thought disorder" as a synonym. For this reason it may be more accurate—at least until the dimensions of amphetamine psychosis are better defined—to compare the classical amphetamine psychosis to a *paranoid state* rather than to schizophrenia.

In the 20 years since Griffith made this comment, however, diagnostic criteria have changed. In DSM III-R, thought disorder is no longer part of the definition of paranoid schizophrenia.

Moreover, another interpretation has been proposed that also has considerable merit. Post (1975) pointed out that the response to increasing doses of CNS stimulants progresses in a continuum from activation and euphoria through dysphoria to psychosis. A similar longitudinal evolution occurs in an episode of psychotic mania (Carlson and Goodwin, 1973). Fibiger (1991), noting this similarity, proposed that amphetamine psychosis also might be considered a model of psychotic mania.

Psychotic mania and acute schizophrenia often cannot be distinguished cross-sectionally at a single point in time; the distinction is made longitudinally and based on the interphase functioning and deficits. Since, in our experience, amphetamine psychosis clears without residual clinical deficit, this longitudinal feature also supports Fibiger's concept.

VI. SUMMARY

As noted by Ellinwood (1972),

> In spite of great individual variability, amphetamine psychosis usually produces a fairly distinct syndrome—the incidence of symptoms corresponds closely in all reported series. Because of the marked individual variability, however, statistics are misleading. Chronic amphetamine reactions are much like syphilis in that they can mimic any number of psychiatric disorders.

The purpose of this chapter was to examine this clinical variability.

The early history of both cocaine and amphetamine was noted to indicate substantial enthusiasm for the therapeutic potential of each drug, punctuated by concern or even alarm over instances of idiosyncratic reactions and unpredictable toxicity. Possible substrates for variable sensitivity to CNS stimulants were discussed, including genetic variables, premorbid psychiatric status, tolerance, and, particularly, sensitization. Although past human data suggesting clinically meaningful sensitization to amphetamine effects were considered to be criticizable, more recent studies documenting sensitization to the psychotomimetic effects of cocaine were noted (Brady *et al.*, 1991; Satel *et al.*, 1991). These findings reinforce the possibility that sensitization to amphetamine can occur. Such sensitization could be an important determinant of such factors as minimal dose required for psychosis, duration of psychosis, the possible occurrence of psychosis in the absence of drug ingestion, and possible alterations in symptom patterns with increasing drug exposure.

In this context, clinical data of reported cases of amphetamine psychosis were reviewed and a number of ad hoc questions posed. Amphetamine psychosis was concluded not to be considered the result of idiosyncrasy or of a "latent" psychotic condition and, as Griffith *et al.* (1972a) noted, "a paranoid reaction must be considered a probable complication of high dose amphetamine abuse." A minimal dose of 50 mg or less was found occasionally to be associated with psychoses in both case reports and prospective studies. Data for a minimal duration of use were found to be equivocal. Symptomatology did not appear to differ importantly in experienced and relatively naive abusers. Data suggesting the development of psychoses in the absence of drug ingestion and of persistent psychoses were thought to be criticizable since these reports did not include toxicology studies to rule out unreported drug use. Premorbid psychiatric status was considered to influence the quality of symptoms that emerged during amphetamine intoxication. Finally, the possibility that amphetamine psychosis might mimic psychotic mania as faithfully as schizophrenia was noted.

REFERENCES

Anggard, E., Jonsson, L. E., Hogmark, A. L., and Gunne, L. M. (1973). Amphetamine metabolism in amphetamine psychosis. *Clin. Pharmacol. Ther.* **14**, 870–880.

Angrist, B., and Gershon, S. (1969). Amphetamine abuse in New York City—1966 to 1968. *Sem. Psychiatry* **1**, 195–207.

Angrist, B., and Gershon, S. (1970). The phenomonology of experimentally induced amphetamine psychosis—Preliminary observations. *Biol. Psychiatry* **2**, 95–107.

Angrist, B., and Gershon, S. (1972). Some recent studies on amphetamine psychosis—Unresolved issues. *In* "Current Concepts on Amphetamine Abuse" (E. H. Ellinwood and S. Cohen, eds.), pp. 193–204. DHEW Publication No. (HSM) 72-9085. U. S. Government Printing Office, Washington, D.C.

Angrist, B., and Sudilovsky, A. (1978). Central nervous system stimulants: Historical aspects and clinical effects. *In* "Handbook of Psychopharmacology" (L. L. Iverson, S. D. Iverson, and S. H. Snyder eds.), Vol. 11, pp. 99–165. Plenum Press, New York.

Angrist, B., and van Kammen, D. (1984). CNS stimulants as tools in the study of schizophrenia. *Trends Neurosci.* **7**, 388–390.

Angrist, B., Schweitzer, J., Gershon, S., and Friedhoff A. J. (1970). Mephentermine psychosis: Misuse of the Wyamine inhaler. *Am. J. Psychiatry* **126**, 1315–1317.

Angrist, B., Shopsin, B., and Gershon, S. (1972). Metabolites of monoamines in urine and cerebrospinal fluid after large dose amphetamine administration. *Psychopharmacologia (Berlin)* **26**, 1–9.

Angrist, B., Sathananthan, G., Wilk, S., and Gershon, S. (1974). Amphetamine psychosis: Behavioral and biochemical aspects. *J. Psychiat. Res.* **11**, 13–23.

Anonymous (1887). Alleged toxic effects of cocaine in the bladder. *Lancet* **1**, 1332.

Antelman, S. M., and Chiodo, L. A. (1983). Amphetamine as a stressor. *In* "Stimulants: Neurochemical, Behavioral and Clinical Perspectives" (I. Creese, ed.), pp. 269–299. Raven Press, New York.

Antelman, S. M., Eichler, A. J., Black, C. A., and Kocan, D. (1980). Interchangeability of stress and amphetamine in sensitization. *Science* **207**, 329–331.

Becker, H. K. (1963). "Coca Koller": Carl Koller's discovery of local anaesthesia. *In:* S. Freud, "Cocaine Papers" (1974) (R. Byck, ed.), pp. 261–319. Meridian, New York.

Bell, D. S. (1965). A comparison of amphetamine psychosis and schizophrenia. *Br. J. Psychiatry* **3**, 701–706.

Bell, D. S. (1973). The experimental reproduction of amphetamine psychosis. *Arch. Gen. Psychiatry* **29**, 35–40.

Brady, K. T., Lydiard, R. B., Malcolm, R. M., and Ballenger, J. C. (1991). Cocaine-induced psychosis. *J. Clin. Psychiatry* **52**, 509–512.

Bunney, W. E., Jr. (1978). Psychopharmacology of the switch process in affective illness. *In* "Psychopharmacology: A Generation of Progress" (M. A. Lipton, A. DiMascio, and K. F. Killam, eds.), pp. 1249–1259. Raven Press, New York.

Carlson, G. A., and Goodwin, F. K. (1973). The stages of mania. *Arch. Gen. Psychiatry* **28**, 221–228.

Connell, P. H. (1958). "Amphetamine Psychosis," Maudsley Monographs No. 5. Oxford University Press, London.

Ellinwood, E. H., Jr. (1967). Amphetamine psychosis: Description of the individuals and the process. *J. Nerv. Ment. Dis.* **144**, 273–283.

Ellinwood, E. H., Jr. (1972). Amphetamine psychosis: Individuals, settings and sequences. *In* "Current Concepts on Amphetamine Abuse" (E. H. Ellinwood and S. Cohen, eds.), pp. 143–157. DHEW Publication No. (HSM) 72-9085. U.S. Government Printing Office, Washington, D.C.

Fibiger, H. C. (1991). The dopamine hypothesis of schizophrenia and mood disorders: Contradictions and speculations. *In* "The Mesolimbic Dopamine System: From Motivation to Action" (P. Willner and J. Scheel-Kruger, eds.), pp. 615–637. John Wiley and Sons, Chichester, England.

Freud, S. (1885) Addenda to 'Über Coca.' *In* S. Freud, "Cocaine Papers" (1974) (R. Byck, ed.), pp. 105–118. Meridian, New York.

Gold, M. S., and Bowers, M. B. (1978). Neurobiological vulnerability to low dose amphetamine psychosis. *Am. J. Psychiatry* **135**, 1546–1548.

Griffith, J. D. (1966). A study of illicit amphetamine drug traffic in Oklahoma City. *Am. J. Psychiatry* **123**, 560–569.

Griffith, J. D., Cavanaugh, J., Held, J., and Oates, J. A. (1972a). Dextroamphetamine evaluation of psychomimetic properties in man. *Arch. Gen. Psychiatry* **26**, 97–100.

Griffith, J. D., Fann, W. E., and Oates, J. (1972b). The amphetamine psychosis: Experimental manifestations. *In* "Current Concepts on Amphetamine Abuse" (E. H. Ellinwood and S. Cohen, eds.), pp. 185–191. DHEW Publications No. (HSM) 72-9085. U.S. Government Printing Office, Washington, D.C.

Grinspoon, L., and Hedlbom, P. (1975). "The Speed Culture." Harvard University Press. Cambridge, Massachusetts.

Harrington, H., Heller, H. A., Dawson, D., Caplan, L., and Rumbaugh C. (1983). Intracerebral hemorrhage and oral amphetamine. *Arch. Neurol.* **40**, 503–507.

Janowsky, D. S., and Risch, C. (1979). Amphetamine psychosis and psychotic symptoms. *Psychopharmacology* **65**, 73–77.

Jori, A., and Garattini, S. (1973). Catecholamine metabolism and amphetamine effects on sensitive and insensitive mice. In "Frontiers in Catecholamine Research" (E. Usdin and S. H. Snyder, eds.), pp. 939–941. Pergamon Press, New York.

Kalant, O. J. (1973). "The Amphetamines: Toxicity and Addiction," 2d Ed. Brookside Monographs of the Addiction Research Foundation No. 5. University of Toronto Press, Toronto.

Kilham, C. S. (1887). Case of cocaine poisoning. *Lancet* **1**, 17.

Kosman, M. E., and Unna, K. R. (1968). Effects of chronic administration of amphetamines and other stimulants on behavior. *Clin. Pharmacol. Ther.* **9**, 240–54.

Kramer, J. C. (1972). Introduction to amphetamine abuse. In "Current Concepts on Amphetamine Abuse" (E. H. Ellinwood and S. Cohen, eds.), pp. 177–184. DHEW Publication No. (HSM) 72-9085. U.S. Government Printing Office. Washington, D.C.

Kramer, J. C., Fischman, V. S., and Littlefield, D. C. (1967). Amphetamine abuse: Pattern and effects of high doses taken intravenously. *J. Am. Med. Assoc.* **201**, 305–309.

Kuczenski, R., and Segal, D. S. (1988). Psychomotor stimulant-induced sensitization: Behavioral and neurochemical correlates. In "Sensitization of the Nervous System" (P. Kalvinas, ed.), pp. 175–205. Telford Press, Caldwell, New Jersey.

Kuhn, C. M., and Schanberg, S. (1977). Distribution and metabolism of amphetamine in tolerant animals. In "Cocaine and Other Stimulants" (E. H. Ellinwood and M. Kilby, eds.), pp. 161–177. Plenum Press, New York.

Lewander, T. (1972). Experimental and clinical studies on amphetamine dependence. In "Behavioral and Pharmacological Aspects of Dependence and Reports on Marijuana Research" (H. M. Van Praag, ed.), pp. 69–84. DeErven F. Boun, Haarlem, Netherlands.

Lieberman, J. A., Kane, J. M., and Alvir, J. (1987). Provocative tests with psychostimulant drugs in schizophrenia. *Psychopharmacology* **91**, 415–433.

Matteson, J. B. (1887). Cocaine dosage and cocaine addiction. *Lancet* **1**, 1024–1026.

Mayer-Gross, W. (1951). Experimental psychoses and other mental abnormalities produced by drugs. *Br. Med. J.* **2**, 317–321.

Monroe, R. R., and Drell, H. J. (1947). Oral use of stimulants obtained from inhalers. *J. Am. Med. Assoc.* **135**, 909–915.

Murphy, D. L., Brodie, H. K., Goodwin, F. K., and Bunney, W. E., Jr. (1971). Regular induction of hypomania by L-DOPA in "bipolar" manic-depressive patients. *Nature (London)* **229**, 135–136.

Nurnberger, J., Gershon, E., Jimerson, D., Buschsbaum, M., Gold, P., Brown, G., and Ebert, M. C. (1981). Pharmacogenetics of *d*-amphetamine in man. In "Genetic Research Strategies for Psychobiology and Psychiatry" (E. S. Gershon, S. Matthysse, X. O. Breakfield, and R. D. Ciranello, eds.), pp. 257–268. Boxwood Press, Pacific Grove, California.

Post, R. M. (1975). Cocaine psychosis a continuum model. *Am. J. Psychiatry* **132**, 225–231.

Post, R. M., and Contel, N. P. (1983). Human and animal studies of cocaine: Implications for development of behavioral pathology. In "Stimulants: Neurochemical, Behavioral and Clinical Perspectives" (I. Creese, ed.), pp. 169–203. Raven Press, New York.

Prinzmetal, M., and Bloomberg, W. (1935). Use of benzedrine for the treatment of narcolepsy. *J. Am. Med. Assoc.* **105**, 2051–2054.

Reifenstein, E. C., Jr., and Davidoff, E. (1939). Benzedrine sulfate therapy: The present status. *N.Y. State J. Med.* **39**, 42–57.

Reis, D. J., Baker, H., Fink, J. S., and Joh, T. H. (1979). A genetic control of central dopamine neurons in relation to brain organization, drug responses and behavior. In "Catecholamines: Basic and Clinical Frontiers" (E. Usdin, I. J. Kopin, and J. Barchas, eds.), pp. 23–33. Pergamon Press, New York.

Roberts, A. (1887). Dangers of cocaine. *Lancet* **1**, 780.

Robinson, A. E., and Wolkind, S. N. (1970). Amphetamine abuse amongst psychiatric inpatients: The use of chromatography. *Br. J. Psychiatry* **116,** 643–644.
Robinson, T. E., and Becker, J. B. (1986). Enduring changes in brain and behavior produced by chronic amphetamine administration: A review and evaluation of animal models of amphetamine psychosis. *Brain Res. Rev.* **11,** 157–198.
Rockwell, D. A., and Ostwald, P. (1968). Amphetamine use and abuse in psychiatric patients. *Arch. Gen. Psychiatry* **18,** 612–616.
Ruiz Ogara, C. (1954). Psicosis desencadenada por choque amphetaminico. *Rev. Espan. Oto-Neuro-Oftol.* **13,** 318–323.
Sadusk, J. R. (1966). Non-narcotic addiction size and extent of the problem. *J. Am. Med. Assoc.* **196,** 707–709.
Satel, S. L., Seibyl, J. P., and Charney, D. S. (1991a). Prolonged cocaine psychosis implies underlying major psychopathology. *J. Clin. Psychiatry* **52,** 349–350.
Satel, S. L., Southwick, S. M., and Gawin, F. H. (1991b). Clinical features of cocaine-induced paranoia. *Am. J. Psychiatry* **148,** 485–498.
Sato, M., Chen, C. C., Akiyama, K., and Otsuki, S. (1983). Acute exacerbation of paranoid psychotic state after long term abstinence in patients with previous methamphetamine psychosis. *Biol. Psychiatry* **18,** 429–440.
Segal, D. S. (1975). Behavioral and neurochemical correlates of repeated *d*-amphetamine administration. *In* "Neurobiological Mechanisms of Adaptation and Behavior" (A. J. Mandell, ed.), pp. 247–262. Raven Press, New York.
Segal, D. S., and Janowsky, D. S. (1978). Psychostimulant-induced behavioral effects: possible models of schizophrenia. *In* "Psychopharmacology: A Generation of Progress" (M. A. Lipton, A. DiMascio, and K. F. Killam, eds.), pp. 1113–1123. Raven Press, New York.
Segal, D. S., and Kuczenski, R. (1987). Behavioral and neurochemical characteristics of stimulant-induced augmentation. *Psychopharmacol. Bull.* **23,** 417–424.
Segal, D. S., and Kuczenski, R. (1992). In vivo microdialysis reveals a diminished amphetamine-induced DA response corresponding to behavioral sensitization produced by repeated amphetamine pretreatment. *Brain Res.* **571,** 330–337.
Segal, D. S., and Schuckit, M. A. (1983). Animal models of stimulant-induced psychosis. *In* "Stimulants: Neurochemical, Behavioral and Clinical Perspectives" (I. Creese, ed.), pp. 131–167. Raven Press, New York.
Segal, D. S., Geyer, M. A., and Schuckit, M. A. (1981). Stimulant-induced psychosis: An evaluation of animal models. *In* "Essays in Neurochemistry and Neuropharmacology" (M. B. H. Youdim, W. Lovenberg, D. F. Sharman, and J. R. Lagnado, eds.), Vol. 5, pp. 95–129. John Wiley and Sons, Chichester, England.
Siomopoulis, V. (1976). Thought disorder in amphetamine psychosis: A case report. *Psychosomatics* **17,** 42–44.
Slater, E. (1959). Review of *Amphetamine Psychosis* by P. H. Connell. *Br. Med. J.* **1,** 488.
Tatetsu, S. (1964). Methamphetamine psychosis. *Folia Psych. Neurol. Jap. (Suppl.)* **7,** 377–380.
Tatetsu, S. (1972). Methamphetamine psychosis. *In* "Current Concepts on Amphetamine Abuse" (E. H. Ellinwood and S. Cohen, eds.), pp. 159–161. DHEW Publication No. (HSM) 72-9085. U. S. Government Printing Office, Washington, D.C.
Utena, H. (1966). Behavioral aberrations in methamphetamine-intoxicated animals and chemical correlates in the brain. *Prog. Brain Res.* **21B,** 192–207.
Weaver, L. C., and Kerley, T. L. (1962). Strain difference in response of mice to *d*-amphetamine. *J. Pharmacol. Exp. Ther.* **135,** 240–244.
Young, D., and Scoville, W. B. (1938). Paranoid psychosis in narcolepsy and the possible danger of benzedrine treatment. *Med. Clin. North Am.* **22,** 637–646.

Kyohei Konuma

14
Use and Abuse of Amphetamines in Japan

I. INTRODUCTION

Japan has experienced two major epidemics of amphetamine abuse. The first epidemic occurred between 1947 and 1957, reaching a peak in 1954. The second epidemic began approximately 1970 and continues today, having reached a peak in 1984. In a sense, these outbreaks have been valuable experiences for the Japanese psychiatric community. Because they have been able to accumulate extensive clinical findings on methamphetamine (METH)-related mental disorders, Japanese researchers have explored unique ground in this field compared with the studies conducted in Europe and the United States.

This chapter describes the historical background of the use and abuse of amphetamines since the drugs first were introduced to Japan. The chapter also addresses the work of Tatetsu *et al.* (1956), which played a major role in establishing the concept of METH psychosis in Japan. In Japan, previous reports focused on the fact that METH psychosis and schizophrenia share cross-sectional clinical features and clinical courses of symptom recurrence. As a result, researchers paid little attention to METH dependence itself. However, METH abuse, dependence, and resulting psychosis are an important and difficult theme in psychiatry. I have treated patients with METH-related mental disorders at the National Psychiatric Institute of Shimohusa since the early phase of the second outbreak. Primarily based on my clinical experience, in this chapter I introduce clinical studies on METH-related mental disorders reported in Japan.

II. HISTORICAL BACKGROUND

A. Advent of Amphetamines

Amphetamines first were introduced to Japan as drugs in 1941. Amphetamine was sold under the brand names Zedrin and Agotin; METH was sold under the brand names Philopon, Hospitan, and Neoagotin. These drugs first were used to increase productivity and to treat narcolepsy and depression. However, these drugs were barely known to the public at that time.

Beginning in 1942, Philopon was administered experimentally to members of the Army and Navy, night workers at shipyards and airplane production plants, and nurses who worked the night shift. As World War II intensified, the drug was administered, partly by coercion, to members of the Air Force, airfield construction workers, and workers in wartime industries to increase productivity (Tatetsu et al., 1956).

The amount of amphetamines produced annually during the war remains unknown because of the absence of official data. Studies suggest that substantial amounts of amphetamines were produced during the war for use by the military. At the end of the war, pharmaceutical companies reportedly held large amounts of amphetamines in stock (Hayashi, 1966). After the war, amphetamines were sold in pharmacies as nonprescription drugs described as "antifatigue stimulants" and backed by a large-scale advertising campaign. The drugs were welcomed by Japanese society during this unstable and chaotic postwar period. This marketing of amphetamines resulted in a major outbreak of amphetamine abuse (Hayashi, 1966).

B. Amphetamine Use in the First Epidemic

Shortly after the war, in 1947 or 1948, amphetamine use spread rapidly among people from all walks of life. By approximately 1950, users included students preparing for entrance examinations, white-collar workers, entertainers, and writers who were engaged in irregular overwork. These individuals used amphetamines to enhance their study, night work, and productivity. Amphetamines also were used by delinquent teenagers and adults to stay awake all night while gambling or pursuing other pleasures (Kasamatsu and Kurino, 1971). As the number of users increased, the typical dosage of individual users also escalated. Unsatisfied with the slow effect of tablets, users opted for a more rapid route of administration, that is, injection. These individuals shifted gradually from subcutaneous to intravenous injection to produce a more potent and rapid effect. As a result, the ravages of amphetamine abuse spread rapidly. This problem soon came to the attention of the public. With the growing demand for amphetamines, illicit drugs entered the market in the early 1950s. All drugs sold on the black market were used for injection.

In February 1950, amphetamines were designated as prescription drugs under the Pharmaceutical Affairs Law. To prevent the harmful effects of amphetamine abuse on health and hygiene, the Stimulant Control Law was established in June 1951. This law enforced necessary controls over the manufacture, import, transfer, acceptance, possession, and use of amphetamines. The restrictions imposed on amphetamines by this law were comparable to those on narcotics and opium. Since that time, legally produced amphetamines have been used only for medical and research purposes.

According to Tatetsu *et al.* (1956), almost all the amphetamines seized by the police in 1954 were METH. The content of METH ranged from 0.2 to 11.6 mg/ml solution, and usually was 3.1–3.3 mg. Drugs sold on the black market always contained adulterants, such as caffeine and sodium benzoate, in addition to METH. As the crackdown on illicit drugs intensified, the content of METH dropped. Drugs seized in September 1955 sometimes contained no amphetamines, but only caffeine, sodium benzoate, and sodium chloride. Some drugs also contained antipyrine and ephedrine hydrochloride.

In general, METH users gradually increased their dosage from the initial dose of 1–2 ml. Drug injection was repeated at irregular intervals of several days. Daily dosage varied: over half the patients admitted to Tokyo Metropolitan Matsuzawa Hospital used maximum daily doses of 10–60 ml (Tatetsu *et al.*, 1956). Motives for METH use were its "recommendation by others" in 51.0% of all cases, "curiosity" in 19.4%, "increased productivity" in 16.1%, "imitating others" in 3.2%, and "studying for examinations" in 3.2% (Tatetsu *et al.*, 1956).

C. Development of Methamphetamine Psychosis

Although isolated instances of METH psychosis have been reported since 1946, the incidence has increased dramatically since 1951 or 1952. Tatetsu *et al.* (1956) examined a total of 492 METH dependents, including 224 patients admitted to Tokyo Metropolitan Matsuzawa Hospital and other psychiatric hospitals in Tokyo and nearby prefectures; 148 inmates in juvenile training schools, prisons, and other correctional institutions; and 120 abusers among private railway employees and street dwellers in the Ueno district. On the basis of the results of this study, these researchers published a book entitled *Methamphetamine Psychosis*. This book is a masterpiece of clinical research that describes amphetamine abuse during the first epidemic in Japan. In the following sections, I present from their book the types of abnormal states observed in patients with METH-related mental disorders and their differences from schizophrenia.

The amount of time between initial METH use and onset of marked hallucinations, delusions, and abnormal behavior easily detectable by people around the abuser was 3 mo in 25.0%, 1 yr in 75.0%, and 3 yr in 95.0% of the cases (Tatetsu et al., 1956).

1. Types of Abnormal States Seen in Patients with Methamphetamine-Related Mental Disorders

Tatetsu et al. (1956) and Tatetsu (1972) classified the basic mental states of patients with METH-related mental disorders into two major types, psychopathy-like and psychotic states, as shown in Table I. Clinical courses were divided into two stages: a recovery stage in which psychotic symptoms showed dramatic improvement within 1 mo of abstinence (hospitalization) and a subsequent stationary stage during which the improvement of symptoms slowed, suggesting stabilization of the patient's condition. Incidence rates were shown according to these classifications. Patients 1 mo to 16 yr after abstinence were included in the stationary stage.

Among psychotic states, a cycle of apathetic–exhausted state is seen during METH abuse and for some time after abstinence, sometimes accom-

TABLE I Incidence of Mental Disorders Seen in Hospitalized Patients Who Abused Methamphetamine[a]

Clinical features	Recovery stage		Stationary stage	
	Number of cases	(%)	Number of cases	(%)
Psychopathy-like state	8	(7.6)	39	(40.2)
Psychotic state				
Schizophrenia-like state	20	(19.0)	16	(16.5)
Manic-depressive state	24	(22.9)	28	(28.9)
State combining schizophrenic and manic–depressive features	20	(19.0)	14	(14.4)
Apathetic–exhausted state	33	(31.4)	0	(0.0)
Total	105	(100.0)	97	(100.0)

[a]Data from Tatetsu (1972) and Tatetsu et al. (1956).

panied by another psychotic state. However, if this cycle is not accompanied by another psychotic state, only a psychopathy-like state will remain for a long time after the cycle has been completed. Emphasizing the nonspecific phenomenon of a hallucinatory paranoid state, many researchers discussed similarities between METH psychosis and schizophrenia. However, Tatetsu *et al.* (1956) pointed out that the basic mental states observed in METH dependents during abuse were a manic (depressive) state and an apathetic–exhausted state. Focusing on these two specific phenomena produced by the pharmacological effects of METH, these researchers pointed out similarities between METH psychosis and manic–depressive psychosis. Tatetsu *et al.* (1956) used the term "schizophrenia-like state" to describe the clinical features of a METH psychosis that included not only a hallucinatory paranoid state, but also severe apathy and emotional flatness, hypobulia, and solitude.

The psychopathy-like state is classified into the following four personalities: (1) violent/intimidating, (2) relaxed, (3) frivolous, and (4) childish. These personalities are probably generated by the direct effect of METH abuse or by modification and exaggeration of the original personality traits by abuse.

Tatetsu *et al.* (1956) found that a disturbance of consciousness was absent among patients showing METH-related mental disorders. On the whole, very few researchers reported a disturbance of consciousness during the first epidemic. Exceptions were a small number of researchers who had studied early abusers in the first epidemic. For instance, Takeyama (1974) classified the clinical picture of METH psychosis into the following three types:

1. hallucinatory paranoid type (the presence of hallucinations and delusions without the disturbance of consciousness)
2. delirium type (the presence of clouded consciousness, hallucinations, anxiety, and excitement)
3. hebetude type (lack of initiative and mild stuporous state)

On the other hand, Kasamatsu and Kurino (1971) classified METH psychosis into the following three types:

1. predominance of disturbance of consciousness and delusion
2. predominance of only delusion and hallucination
3. disturbance of consciousness or hallucination and delusion, accompanied by other schizophrenia-like abnormal experience

2. Genetic Study

The greatest contribution of the studies of Tatetsu *et al.* (1956) and Tatetsu (1972) is based on long-term clinical observations of patients with METH psychosis. These investigators found that some parenteral abusers exhibited severe mental disorders and periodic exacerbation of psychosis, producing a stuporous state and ego impairment for 1 mo to 16 yr after

TABLE II Incidence of Schizophrenia in the Families of Methamphetamine Psychotics with Schizophrenia-Like Symptoms Compared with That in the Families of Schizophrenics and in the General Population[a]

		Incidence of schizophrenia	
Proband	Number of subjects	Siblings (%)	Parents (%)
Methamphetamine psychotics	91	4.66	3.44
Schizophrenics	123	13.08	10.66
General population	2085	0.63	0.19

[a]Data from Tatetsu (1972) and Tatetsu et al. (1956).

abstinence. These studies raised the possibility that METH psychosis that produces a schizophrenia-like state is actually schizophrenia. However, Tatetsu et al. (1956) reported that, unlike schizophrenics, patients with METH psychosis maintained smooth relationships and a good rapport with people around them despite a severe deterioration in volition. These researchers studied the incidence of schizophrenia among the siblings and parents of 91 METH psychotics who exhibited severe mental disorders and frequent exacerbation of psychosis in the stationary stage, 123 schizophrenics, and 2085 in-patients in the departments of surgery and internal medicine who served as the general population (Table II). The incidence of schizophrenia in the families of METH psychotics was significantly lower than in the families of schizophrenics, but higher than in the families of the general population. This study demonstrated that METH psychosis that produces a schizophrenia-like state is not genetically identical to schizophrenia and that only a limited correlation exists between the two disorders. The study explained that the higher incidence of schizophrenia in families of METH psychotics than in the families of the general population may have resulted from a small number of schizophrenics included in the group of METH psychotics.

The accumulated clinical studies during the first epidemic, particularly the work of Tatetsu et al. (1956), have contributed to the general acceptance of the following findings on METH psychosis in Japan. Although METH abuse may produce a clinical picture similar to that of schizophrenia, dramatic improvements in symptoms occur within 1 mo of abstinence; nonetheless, nearly 23% of METH abusers continue to exhibit psychotic symptoms, with frequent relapses and aggravation, during a subsequent stationary stage.

D. Use of Amphetamines after the First Epidemic

In Japan, the Stimulant Control Law enforces strict control over production of amphetamines by manufacturers, transfer from manufacturers to

TABLE III Annual Production and Transfer of Legal Methamphetamine Hydrochloride (Philopon) since 1968 and Annual Consumption since 1977 According to Medical and Research Use

Fiscal year	Production (g)	Transfer (g)	Consumption (g)	
			Medical institutions	Designated researchers
1968	298	119		
1969	28	201		
1970	171	156		
1971	191	218		
1972	163	120		
1973	19	146		
1974	181	162		
1975	199	175		
1976	19	158		
1977	391	188	20	99
1978	306	228	16	123
1979	15	283	21	139
1980	756	203	16	146
1981	0	181	12	133
1982	90	330	11	173
1983	791	175	12	176
1984	0	210	12	162
1985	0	133	11	478
1986	15	137	11	586
1987	15	126	9	123
1988	15	136	8	106
1989	200	165	9	98
1990	315	121	8	260

designated medical institutions and researchers, administration to patients, and use for scientific research. Three preparations of METH—bulk powder, 1-mg tablets, and 3-mg/ml parenteral solutions—are currently under production. Sales of amphetamine were halted in 1968.

Table III shows annual production and transfer of legal METH hydrochloride (Philopon) since 1968 and annual consumption since 1977 according to medical and research use. As of 1990, 289 institutions and 579 researchers across the country are designated for amphetamine use. A total of 8 g METH was used for medical purposes in 1990. Medical use of amphetamines is limited to patients with a relatively specific type of narcolepsy in specified institutions.

In Japan, Philopon is used for (1) improvement of narcolepsy, various types of coma, somnolence, clouded consciousness, insulin shock, depression, depressive state, and hebetude in schizophrenia; (2) enhancement of recovery from collapse during and after surgery and awakening from anesthesia; and (3) improvement of acute intoxication from anesthetics and hypnotics. The drug is not used as an appetite suppressant.

E. Spread of Methamphetamine Use during the Second Epidemic

The second epidemic of amphetamine abuse began in approximately 1970. METH was smuggled into Japan and sold on the black market by organized criminal syndicates to finance their activities. METH was circulated mainly as crystals or powder. In an early phase of the epidemic, the drug contained adulterants such as caffeine, sodium benzoate, ephedrine, artificial seasoning (Ajinomoto), sodium chloride, and bleaching powder. Since then, highly pure METH (purity of approximate 99%) has been circulated because the effects of low purity METH do not last long among abusers.

With the outbreak of the second epidemic, METH abuse spread to people who had contact with members of organized criminal syndicates (primary or core abusers), including those engaged in the construction, restaurant, transportation, bar and cabaret, gambling, finance, and real estate businesses. These groups constituted the secondary abusers who began taking METH for its pharmacological effects, that is, excitation of the central nervous system, the ability to stay awake at night, enhancement of concentration and judgment, and relief of fatigue. Subsequently, METH abuse spread rapidly to ordinary people such as factory workers, housewives, students, and public servants, who had contact with the secondary abusers. This group consisted of anomic people who condoned violations of the law provided the abuse was not discovered. These individuals began METH use primarily to pursue the euphoria induced by the drug. However, recent implementation of strong preventive measures against METH abuse, such as an intensification of the crackdown on violations, severe punishment of violators, and the launching of preventive educational campaigns, has led to a gradual decline in METH abuse after a peak in 1984.

F. Patterns of Methamphetamine Use during the Second Epidemic

Konuma (1984) classified the patterns of METH use (Table IV) based on the study of METH dependents treated during the second epidemic. As their motives such as "curiosity" and "recommendation by others" suggest, most abusers began taking METH with a knowledge of the drug they were taking. However, in some unfortunate cases the subjects were deceived and forced to inject METH.

1. Occasional Use

Occasional use of METH usually is seen among early abusers who have the drug injected by their associates once or twice in the course of a night-

TABLE IV Patterns of Methamphetamine Use[a]

Occasional use
Dependent use
  Regular use
  Cyclic use (cycle of run and crash)
  Daily high-dose use
  Compulsive use
  Single high-dose use
  Introspective abstinence

[a]Data from Konuma (1984).

long mah-jongg or gambling session. The frequency of METH use in this group is usually once or twice a month. Since individuals can inject the drug themselves, they often shift to short-term continuous use. They purchase a 0.2- or 0.3-g packet of METH from a dealer once or twice a month and inject the drug at a dose of 20–30 mg, 2–3 times a day, until the stock is exhausted. Abusers in this group may be hospitalized because of psychotic symptoms. After they have recovered and been discharged from a hospital, some patients may be enticed by their associates to resume short-term continuous use.

2. Dependent Use

a. Regular Use The lives of many regular uses are not yet disrupted by their use of METH, and they can carry out their normal work. For instance, a barmaid may inject a single dose of METH every evening before going to work, or a truck driver may have one or two injections to stay awake during long-distance transport on a 2-day shift. Thus, users regularly inject METH at low doses with a limited frequency. At this stage, insomnia and anorexia are not noticeable.

b. Cyclic Use Cyclic users repeat the pattern of frequent injections of METH for a relatively short period of time (a run) followed by a lasting drowsy period (a crash). This pattern is seen frequently among members of organized criminal syndicates and in well-to-do construction contractors and real estate agents. Cyclic users commonly obtain highly pure METH in a "gram packet" (actual content, approximate 0.8 g). Usually, a dose of 30–60 mg is injected 2–6 times a day for several days. During this period, abusers scarcely sleep or eat and drink only large quantities of water. After the run, individuals usually take a steam bath to induce sweat or take alcohol and hypnotics to induce sleep. Because of accumulated fatigue, these people sleep continuously for 1–2 days. After waking up, they consume twice as much food as normal. As soon as they recover from fatigue, these individuals resume their run and crash habit. Usually, a 0.3- to 1.0-g packet is consumed on each "run." At this stage, the entire lives of the abusers revolve around METH; their family and work are abandoned completely.

c. Daily High-Dose Use Some cyclic users shift to daily high-dose use by gradually increasing their dose. Despite daily injections of high doses of 0.3–0.5 g in 4–5 divided doses, certain abusers can maintain a 5- to 6-hr sleep schedule and consume food as usual. Because such daily high-dose users do not accumulate excess fatigue, they "crash" less frequently. When abusers have to perform an important task, they may abstain temporarily from the habit in advance of their duty or assignment. A "crash" frequency occurs as soon as they have abstained. A plausible explanation for this phenomenon is that accumulated fatigue that remains latent during daily high-dose use surfaces after the effects of METH disappear.

d. Compulsive Use Cyclic users shift to compulsive use when they accidentally obtain a substantial amount of METH (5–10 g) or when they have their "last shooting" before serving a prison sentence. Such use also occurs when abusers try to relieve anxiety and tension after fighting or failing to fulfill their responsibilities. These individuals use METH compulsively at high doses of 0.3–1.0 g daily for 3–10 days until their supply is exhausted. The frequency of injection is also high, at 8–10 times a day. Because compulsive use frequently results in acute toxic psychosis, a "crash" usually occurs after abusers are admitted to a hospital.

e. Single High-Dose Use A typical single high-dose user is a male who administers a 0.3- to 1.0-g packet in a single injection and tries to hide the needle mark from his wife by pressing a lighted cigarette against his skin. This pattern of abuse is seen occasionally among METH dependents who resume the habit after abstinence.

f. Introspective Abstinence METH dependents of the types already described may abstain from the drug as they reflect on their own conduct after seeing one of their shooting companions in a psychotic state. Some abusers may discontinue the habit spontaneously during month-long police campaigns, or they may abstain from the habit for more than 1 month while they are hospitalized because of jaundice, emaciation, and psychotic symptoms or while they are incarcerated for crimes they have committed.

Thus, the reason for abstinence is not necessarily the abuser's introspection. However, if users abstain from the habit for more than 1 month, they naturally become more reflective of their conduct. For this reason, such cases are classified as "introspective abstinence."

With the exception of single high-dose use, which is not a typical abuse pattern, METH dependents actually shift from one abuse pattern to another.

As pointed out by Seevers (1974), a major determining factor for the pattern of drug abuse is the availability of the drug to the user. In METH abuse, to a certain extent the occupation of the user determines the availability of the drug. Therefore, certain abuse patterns are common to individuals of specific occupations. As METH dependence progresses to a chronic stage, abusers become drug dealers and use the profit to finance their habit. Such cases are seen frequently in METH dependents. Cyclic and daily high-dose users are common among these abusers.

Regarding the relationship between patterns of abuse and the tolerance of the user to METH, periodic and cyclic use is expected to "increase tolerance" whereas introspective abstinence is expected to "decrease tolerance." Compulsive and single high-dose use probably results in concentrated administration of high doses exceeding the tolerance of the user. These facts are considered among the most important factors in recognizing the onset of psychotic symptoms. The daily high-dose use pattern results from repeated cyclic use, which raises the tolerance of the user so that he or she no longer exhibits anorexia or insomnia during abuse.

III. TYPES OF METHAMPHETAMINE-INDUCED MENTAL DISORDERS

The Expert Committee on Stimulant Addicts (chairman, Nobukatsu Kato) (1985) was commissioned by the Director General of the Pharmaceutical Affairs Bureau, Ministry of Health and Welfare, from 1983 to 1985. The committee used the term "METH-induced mental disorders" to denote mental disorders induced by METH abuse. As shown in Table V, the committee classified METH-induced mental disorders as acute intoxication, dependence syndrome, and psychosis, and established diagnostic criteria according to this classification.

A. Methamphetamine Acute Intoxication

When administered orally, subcutaneously, or intravenously, METH is absorbed rapidly and readily penetrates the blood–brain barrier. This drug excites the central nervous system, giving the user a sense of invigoration and enhanced confidence and generating talkative and cheerful behavior. METH heightens interest, alertness, and vigor and diminishes fatigue. The drug also enhances simple calculation ability but decreased efficiency in complex calculation. Moreover, the sympathomimetic effects of the drug produce the following complications: palpitation due to cardiac hyperfunction, dazzling sensations due to mydriasis, a sensation of piloerection in the head, coldness of the upper and lower extremities due to peripheral vasoconstriction, increased viscosity of saliva, a dry mouth, elevated blood pressure, and increased tendon reflex. These acute effects last for several hours to

TABLE V Classification of Methamphetamine-Induced Mental Disorders[a]

Disorder	Description
METH acute intoxication	Acute intoxication that includes psychoneurological and sympathicotonic symptoms emerging within 1 hr of METH use and rebound phenomena lasting for a few days after disappearance of the drug effects; sometimes "acute syndrome" characterized by acute confusional or hallucinatory state with marked psychomotor excitation will occur
METH dependence syndrome	Dependence syndrome that includes changes in drug-taking behavior and psychoneurological symptoms due to drug effects; no evidence of manifest hallucinatory paranoid state
METH psychosis	"Psychosis" that is composed mainly of hallucinatory paranoid state seen in persons showing "dependence syndrome"
Early recovery type	Symptoms disappear within 1 mo of METH abstinence
Prolonged type	Symptoms last over 1 mo; in some cases symptoms fluctuate from a lull to exacerbation for more than 6 mo after METH abstinence

[a]Data from the Expert Committee, Ministry of Health and Welfare (1985).

half a day, depending on the administered doses. As a reaction to these hyperactivities, a rebound phenomenon develops accompanied by weakness, fatigue, and depression. These symptoms usually disappear after prolonged sleep.

Compulsive METH users may develop an "acute syndrome" with varying degrees of consciousness disturbance and psychomotor excitement. Organic solvent abusers and persons with constitutional hypersensitivity may develop the acute syndrome as a result of METH use once or several times.

B. Methamphetamine Dependence Syndrome

Occasional use is common at an early stage of METH abuse. Since users can inject the drug themselves, they develop dependence relatively rapidly. The following abuse patterns are seen in METH dependents: (1) regular use, (2) cyclic use, (3) daily high-dose use, (4) compulsive use, (5) single high-dose use, and (6) introspective abstinence. The most common pattern is cyclic use. As shown in Fig. 1, in this pattern users repeat a cycle from the first stage to the third at intervals of 1 wk to 10 days. The presence of this three-stage structure frequently provides a diagnostic basis for METH dependence syndrome.

The "run" (first stage) and "crash" (second stage) periods already were described in detail in the discussion of abuse patterns. The subsequent third stage is called the "craving" period. At this stage, abusers have recovered

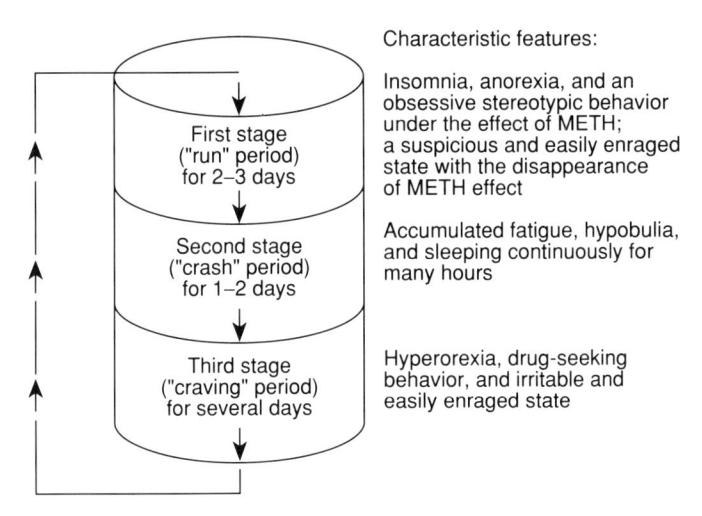

FIGURE 1 Three-stage behavior pattern under cyclic use of methamphetamine (METH).

Diagram contents:

First stage ("run" period) for 2–3 days — Characteristic features: Insomnia, anorexia, and an obsessive stereotypic behavior under the effect of METH; a suspicious and easily enraged state with the disappearance of METH effect

Second stage ("crash" period) for 1–2 days — Accumulated fatigue, hypobulia, and sleeping continuously for many hours

Third stage ("craving" period) for several days — Hyperorexia, drug-seeking behavior, and irritable and easily enraged state

from the "crash" and consume about twice as much food as usual. To obtain METH, they pawn their possessions and resort to lying to get money. If unsuccessful, they become irritated and enraged, intimidating their family, destroying objects, and committing violent acts. This violent behavior during the craving period frequently leads to hospitalization of the abuser.

Usually METH-induced anorexia and insomnia become less noticeable with chronic abuse due to the development of tolerance. In contrast, METH-induced stereotypy and psychotic state are more likely to be manifested with chronic abuse because prolonged use increases sensitivity to these drug effects (the reverse tolerance phenomenon). Once this reverse tolerance develops, it is retained for a long period after abstinence.

At an early stage of METH dependence, stereotypic behavior is not noticeable. However, after 1–2 mo of dependence, a typical abuser exhibits the following obsessive stereotypic behavior: (1) becoming absorbed in gambling, pachinko (pinball game), driving, or sex; (2) becoming engrossed in dismantling and reassembling electric appliances such as tape recorders; (3) becoming obsessed with small pieces of trash and cleaning the house meticulously; (4) taking notes on newspaper articles or copying lyrics from record sleeves; and (5) peeling hangnails until they bleed or endlessly picking at acne. The abuser remains in a good mood for 2–3 hr after METH injection. After the manifestation of psychotic symptoms, this obsessive stereotypic behavior turns to a preoccupation with searching for wiretapping devices and hidden cameras or investigating suspicious matters.

With the disappearance of the effects of the drug, hypersensitivity to the sounds of footsteps and engines may develop. The individual expresses anxiety, insisting that someone is hiding around the house and following him or her, or he or she becomes suspicious and easily enraged, interpreting harmless conversations as personal slander.

In a state of METH dependence, "METH shooting" becomes the center of the life of the abuser. Meals and sleep are taken irregularly, and the incidence of spending nights outside the home increases. Work and family eventually are abandoned. Ogura (1956) already pointed out in the first epidemic that personality changes observed in the course of METH dependence can be divided into two categories: emotional disturbances such as an unstable temperament or sudden enragement, and disturbances of volition such as apathy, laziness, and autistic behavior. Personality changes in these two directions also have been observed in the current epidemic. Diagnosing METH dependence syndrome is relatively easy once a history of behavioral change is obtained from family members living with the patient.

C. Methamphetamine Psychosis

METH psychosis is a METH-induced psychotic state with a predominance of hallucination and delusion. At an early stage, in addition to disturbances of emotion and volition induced by METH dependence, psychotic experiences relapse intermittently with a temporary resolution following sleep. Subsequently, the psychotic experiences no longer disappear, despite temporary abstinence and sleep, although symptom alleviation may occur. The patient may have hallucinatory experiences such as being watched and reported on. Auditory hallucinations also develop. For instance, the patient hears voices discussing a conspiracy or voices that ridicule, instruct, accuse, threaten, or order the patient. Since these hallucinatory experiences are accompanied by ideas of having his or her thoughts read and being watched, wiretapped, or pursued, the patient becomes restive. For instance, he or she becomes hypersensitive to sounds, suddenly rushing out of the house and looking around. Suspecting that someone is hiding in an attic, he or she may remove the ceiling boards. Suspecting a wiretap, he or she searches for electrical devices in the closets and attic. In a fit of delusional jealousy, he may severely beat his wife. He might maintain constant vigilance, holding a weapon and tightly locking up the house.

When this hallucinatory paranoid state becomes chronic, the patient may hear a conversational monologue in response to auditory hallucinations. A state of absent-mindedness and seclusion may be evident. Some patients develop systematic delusion, inquiring and complaining about their concerns to institutions they consider responsible.

Diagnosing METH psychosis is easy if a history of METH abuse and psychotic behavior can be obtained from the patient and his or her associates, including his or her family. Symptoms may fluctuate at the outset of the disease because psychotic experiences are related to the time that has elapsed after injection and the effects of METH. At an early stage, the manifestation of psychosis is usually transient and fluctuating. At this stage, psychotic experiences rarely control the entire personality of the patient. Usually the patient can perceive reality and understand that his or her psychotic experiences are caused by METH use. As psychotic experiences such as hallucinations and delusions become chronic, they can control the entire personality. The contents of hallucinations and delusions are often circumstantial; the patient overreacts to each psychotic experience and puts it into action. The criteria published by Fukushima (1977) are helpful in evaluating the ability of patients with METH-induced mental disorders to take criminal responsibility. In many cases, injection marks are evident. A positive urine test for METH offers important diagnostic evidence. A prolonged hallucinatory paranoid state is sometimes difficult to differentiate from schizophrenia. One major difference from schizophrenia is that, in METH psychosis, relatively good interpersonal relationships and rapport usually are maintained even in patients exhibiting signs of marked hypobulia. According to Wakamatsu (1985), the nature of delusion in METH psychosis could be understood under the circumstances. He pointed out that thought disturbances such as loosening of associations, passivity phenomena, and ideas of thought control and thought hearing are present in METH psychosis, but thought withdrawal and thought insertion are not.

Table VI classifies clinical types of METH-induced mental disorders observed in patients admitted to the National Psychiatric Institute of Shimohusa during the second epidemic of METH abuse. The period is divided into two phases, the first 6 yr (1976–1981), a period during which METH abuse spread rapidly throughout the country, and the subsequent 5 yr

TABLE VI Comparison of Clinical Types of Methamphetamine-Induced Mental Disorders Observed in Patients Admitted to the National Psychiatric Institute of Shimohusa between the Widespread Period (1976–1981) and the Subsequent Stationary Period (1982–1986)

Clinical disorders	Widespread period	Stationary period
Acute syndrome	17.2	22.0
Acute confusional state	9.9	14.0
Acute hallucinatory state	7.3	8.0
METH dependence syndrome	7.8	12.7
METH psychosis	72.9	62.6
Early recovery type	59.9	33.3
Prolonged type	13.0	29.3
Neurosis-like state	2.1	2.7
Total number of cases	192	150

(1982–1986), a stationary period during which METH abuse stabilized or decreased.

Compared with the widespread period, the percentage of patients with early recovery type of METH psychosis dropped during the stationary period whereas the prolonged type increased.

D. Sequelae of Methamphetamine-Induced Mental Disorders

With a prolongation of the second METH abuse epidemic, the percentage of patients suffering from severe sequelae of METH-induced mental disorders is increasing.

Kato (1987) classified sequelae of METH-induced mental disorders into two categories: a residual syndrome in which chronic symptoms continued after abstinence and a relapse phenomenon in which symptoms that had disappeared, such as hallucinations and delusions, promptly returned on reuse of METH or psychological stress.

1. Residual Syndrome

Kato (1987) classified the residual syndrome observed after abstinence into four types. In clinical practice, I have seen patients with frequent relapses of a marked depressive state over a long period after abstinence. I classified these patients as a residual depressive type and added them to Kato's classification, as summarized in Table VII.

Major types of this syndrome are residual neurosis and positive symptom types. Patients of the residual neurosis type are free of positive symptoms such as hallucinations and delusions, but complain about fatigue and nocturnal insomnia, and have psychosomatic complaints and anxiety over a long period of time. The residual positive symptoms type is common among patients who have not received treatment at an early stage of METH psychosis and among those who have had frequent relapses of a psychotic state on reuse of METH. Since their positive symptoms have become chronic, these patients are very refractory to treatment with antipsychotics.

TABLE VII Classification of Residual Syndromes of Methamphetamine-Induced Mental Disorders and Shifts in the Manifestation of Symptoms

Anxiety neurosis-like state or hypochondriac state (residual neurosis type)
Chronic hallucinatory paranoid state (positive symptoms type)
⇓   ⇑
Sensitive state or depressive state (residual depressive type)
⇓   ⇑
Emotional disturbance and amotivational state (negative symptoms type)
Personality changes including antisocial type, shiftless and frivolous type, etc. (personality disorder type)

The residual syndrome is not stationary but undergoes periodic remissions and recurrences. Usually, the syndrome improves gradually with treatment. Also common in clinical practice is observing a shift in the manifestation of symptoms during the course of treatment. For instance, as a hallucinatory paranoid state improves, emotional disturbances and amotivational states emerge as predominant symptoms. Also, the aggravation of depressive and hypersensitive states by insomnia and stress occasionally may lead to auditory hallucinations.

Residual personality changes include antisocial type, shiftless and frivolous type, explosive type, and asocial type. To a certain extent, these personality traits may have existed already before METH abuse, were exaggerated and modified by METH abuse, and then remained after abstinence.

2. Relapse Phenomenon

The hallucinatory paranoid state of METH psychosis can be controlled relatively easily if treated early. However, in some cases, the hallucinatory paranoid state that once seemed to have disappeared recurs readily in the later course of recovery.

Sato (1978; Sato *et al.*, 1983) reported cases who showed symptom recurrence within 1 wk (in some cases almost instantaneously) after the reuse of relatively small amounts of METH. Subsequent clinical studies have demonstrated that a relapse of hallucinatory paranoid state can be triggered easily by alcohol ingestion and psychological stress without the reuse of METH. Sato *et al.* (1983) interpreted this relapse phenomenon biologically in the following manner. METH abuse increases the sensitivity of the brain to METH (reverse tolerance phenomenon). Since this reverse tolerance remains in the brain over a long period after abstinence, the brain overreacts to a small amount of METH, producing the relapse phenomenon.

On the other hand, Morita (1980) conducted a close clinical observation of prisoners serving the term for stimulant offenders, all of whom exhibited sequelae of METH psychosis. Focusing his attention on the fact that hypochondriac and hypersensitive states were observed during symptom remission and a hallucinatory paranoid state was observed during symptom aggravation in the same patients, Morita postulated the pathological states underlying the sequelae of METH-induced mental disorders, as shown in Fig. 2. If METH abuse produced a severe hallucinatory paranoid state that lasted a given length of time, hypersensitive personality and persistent somatic complaints associated with the impairment of the autonomic nervous endocrine system (palpitation, headache, dizziness, tinnitus, and night sweats) developed as sequelae. After a relatively brief respite, this state promptly shifted to a hallucinatory paranoid state triggered by minor psychological stressors.

If we agree with the hypothesis of Morita (1980) that the sequelae of METH-induced mental disorders originate from the same root (prolonged

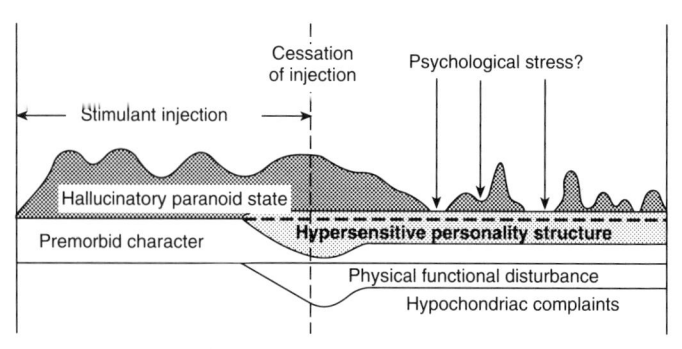

FIGURE 2 Pathological states underlying the sequelae of methamphetamine-induced mental disorders (Morita, 1980).

brain dysfunction), the difference between residual syndrome and relapse phenomenon (the two major types of sequelae) is merely the result of the presence or absence of manifested symptoms. Therefore, drawing a line between these two types is impossible; a transitional state definitely exists between them. The most important clinical issue concerning the sequelae of METH-induced mental disorders is the control of the hallucinatory paranoid state. The analysis of factors causing its relapse and aggravation is important in this regard. Table VIII summarizes these factors, based on my clinical experience. For instance, both physical fatigue and psychological stress occur when the patient is the chief mourner at a relative's funeral, when he or she moves to a new region, and when he or she is released from prison and meets with old acquaintances.

Based on the unique concept of METH psychosis described earlier, relatively long-term use of antipsychotic drugs such as haloperidol is recommended for treatment of METH-induced mental disorders in Japan. Antipsychotic drugs are used for (1) treatment of psychotic symptoms such as hallucinations and delusions, (2) prevention of a relapse of psychotic symptoms with the reuse of METH in the early recovery type (as demonstrated by

TABLE VIII Factors Causing Relapses and Aggravation in Methamphetamine Psychosis

---

Reuse of METH
Use of other drugs of dependence (e.g., alcohol, volatile solvents)
Sudden reduction of doses or cessation of antipsychotic drugs for treatment
Insomnia, nocturnal labor, irregular life-style
Extreme physical fatigue
Psychological stress

Sato, 1978, and Sato *et al.*, 1983), and (3) prevention of spontaneous relapse of symptoms triggered by psychological stress and other factors in the prolonged type. In Europe and the United States, where chronic stimulant psychosis is not recognized, short-term use of antipsychotic drugs is considered sufficient, as advocated by Schuckit (1989).

We can avoid adverse effects of antipsychotics, such as severe stiffness and dystonia pointed out by Schuckit (1989), by maintaining drugs at therapeutic levels until remission of psychotic symptoms, and then gradually reducing the doses to the prevention level and adding anti-Parkinson drugs.

IV. DIFFERENCES BETWEEN JAPAN AND THE WEST IN THE CONCEPTION OF METHAMPHETAMINE PSYCHOSIS

Since the interpretation of the clinical course of METH psychosis after abstinence in Japan differs from that in Europe and the United States, a unique concept of METH psychosis has developed in Japan, as pointed out by Sato (1988). In the first epidemic of amphetamine abuse in Japan, researchers such as Tatetsu *et al.* (1956) reported that symptoms of METH psychosis could continue over a long period of time after abstinence. The conventional idea was that chronic METH abuse produced biological changes from which schizophrenia-like symptoms with a predominant hallucinatory paranoid state developed that lasted for a long period of time after abstinence (Sato, 1988). Tatetsu *et al.* (1956) asserted that the disturbance of consciousness was absent in METH psychosis. Although this view was predominant in the first epidemic, researchers such as Takeyama (1974) and Kasamatsu and Kurino (1971) observed the disturbance of consciousness among their patients. In contrast, all researchers in the second epidemic observed a disturbance of consciousness and reported it in detail. For instance, 9.9% of my patients developed acute confusion; after recovery, they had no memory of what had happened during the confusion (Konuma, 1984).

On the other hand, general research on amphetamine psychosis in Europe and the United States is based on the classic monograph of Connell (1958). As reviewed by Griffith (1977), since the clinical features of amphetamine- and METH-induced psychosis mimic those of the acute paranoid type of schizophrenia in a clear consciousness, distinguishing them is difficult. A positive urine test for amphetamine or METH and recovery within 1 wk of abstinence are accepted as criteria for a differential diagnosis of amphetamine or METH psychosis. Connell claimed that a prolonged psychotic state was the result of reuse of latent amphetamine or hysterical prolongation. However, chronic amphetamine psychosis generally has been understood to occur when true schizophrenics use amphetamine (Bell, 1965; Angrist and Gershon, 1972; Ellinwood, 1972). As pointed out by

Nakatani (1991), explaining prolonged symptoms by mechanisms of hysteria and latent schizophrenia seems to simplistic, even if such cases cannot be excluded entirely.

Bron (1987) classified autonomically developing psychosis ("eigengesetzlich ablaufende Psychose") as one type of drug-induced psychosis. According to Bron, the onset of this type apparently is induced by drug ingestion, but its subsequent recurrence does not require reuse of the drug. Bron's view supports the view of the unique clinical course for METH psychosis that has been developed in Japan.

V. CONCLUSION

Japan has experienced two major outbreaks of amphetamine abuse (mainly METH), the scale of which cannot be compared with that in any other country in the world. In this report, the historical background of the use and abuse of amphetamines in Japan was described. Also, abuse patterns were classified and explained, primarily based on clinical experience during the second epidemic (Konuma, 1984). In addition, researchers in the first epidemic, most notably Tatetsu et al. (1956), already observed the continuance of symptoms of METH psychosis long after abstinence. Since this concept is unique compared with that prevailing in Europe and the United States, this perspective of chronic METH psychosis was explained in detail, based mainly on the works of Tatetsu et al. (1956; Tatetsu, 1972). Finally, the types of METH-induced mental disorders (mental disorders induced by METH abuse) defined by the Expert Committee on Stimulant Addicts were discussed and the syndrome of sequelae of METH-induced mental disorders was described in detail.

REFERENCES

Angrist, B., and Gershon, S. (1972). Some recent studies on amphetamine psychosis—Unresolved issues. In "Current Concepts on Amphetamine Abuse" (E. H. Ellinwood and S. Cohen, eds.), pp. 193–204. DHEW Publication No. (HSM) 72-9085. U.S. Government Printing Office, Washington, D.C.

Bell, D. S. (1965). Comparison of amphetamine psychosis and schizophrenia. Br. J. Psychiatry 111, 701–707.

Bron, B. (1987). Drogenpsychosen. In "Psychiatrie der Gegenwart" (K. P. Kisker, H. Lauter, J.-E. Meyer, C. Müller, and E. Strömgren, eds.), Vol. 3, pp. 345–358. Springer-Verlag, Berlin.

Connell, P. H. (1958). "Amphetamine Psychosis." Oxford University Press, London.

Ellinwood, E. H., Jr. (1972). Amphetamine psychosis: Individuals, settings, and sequences. In "Current Concepts on Amphetamine Abuse" (E. H. Ellinwood and S. Cohen, eds.), pp. 143–158. DHEW Publication No. (HSM) 72-9085. U.S. Government Printing Office, Washington, D.C.

Expert Committee, Ministry of Health and Welfare. (1985). "Study Report of Comprehensive Countermeasures for Methamphetamine-Related Mental Disorders for Fiscal 1984. Narcotics Division, Pharmaceutical Affairs Bureau, Ministry of Health and Welfare of Japan, Tokyo. (in Japanese)

Fukushima, A. (1977). Methamphetamine abuse: Psychopathology and responsibility. *In* "Studies of Criminal Psychology I," pp. 9–27. Kongo Shuppan, Tokyo. (in Japanese)

Griffith, J. D. (1977). Amphetamine dependence: Clinical features. *In* "Drug Addiction, II. Amphetamine, Psychotogen, and Marihuana Dependence" (W. R. Martin, ed.), pp. 277–304. Springer-Verlag, Berlin.

Hayashi, S. (1966). Stimulants addiction. *In* "Encyclopedia of Japanese Psychiatry" (H. Akimoto *et al.*, eds.), Vol. 4(II), pp. 527–546. Kanehara Shuppan, Tokyo. (in Japanese)

Kasamatsu, A., and Kurino, R. (1971). Wake-amine (Philopon) addiction. *In* "Clinical Epidemiology of Drug Abuse" (A. Kasamatsu, T. Hemmi, and K. Takizawa, eds.), pp. 48–63. Ishiyaku Shuppan, Tokyo. (in Japanese)

Kato, N. (1987). Residual syndrome and relapse in patients with methamphetamine related mental disorders. *In* "Information on Psychoactive Substance. Series No. 2, Amphetamines" (Study Team on Drug Abuse Reporting System, ed.), pp. 99–106. Chiba, Japan. (in Japanese)

Konuma, K. (1984). Multiphasic clinical types of methamphetamine psychosis and its dependence. *Psych. Neurol. Jap.* **86**, 315–339. (in Japanese)

Morita, S. (1980). Residual syndrome. *In* "Stimulants Addiction" (I. Yamashita and S. Morita, eds.), pp. 63–88. Kongou Shuppan, Tokyo. (in Japanese)

Nakatani, Y. (1991). Diagnosis and clinical symptomatology of methamphetamine psychosis. *In* "Cocaine and Methamphetamine: Behavioral Toxicology, Clinical Psychiatry, and Epidemiology" (S. Fukui, K. Wada, and M. Iyo, eds.), pp. 333–345. Drug Abuse Prevention Center, Tokyo.

Ogura, H. (1956). Zur Psychopathologie der Chronischen Weckaminvergiftung. *Psych. Neurol. Jap.* **58**, 339–354. (in Japanese)

Sato, M. (1978). Acute reappearance of paranoid state by small amount of methamphetamine injection in 7 cases of chronic methamphetamine psychoses. *Clin. Psychiatry* **20**, 643–648. (in Japanese)

Sato, M. (1988). Amphetamine psychosis and its relationship with schizophrenia. *Clin. Psychiatry* **30**(4), 433–442.(in Japanese)

Sato, M., Chen, C. C., Akiyma, S., *et al.* (1983). Acute exacerbation of paranoid state after long-term abstinence in patients with previous methamphetamine psychosis. *Biol. Psychiatry* **18**, 429–440.

Schuckit, M. A. (1989). "Drug and Alcohol Abuse: A Clinical Guide to Diagnosis and Treatment," 3d Ed., pp. 96–117. Plenum Medical, New York.

Seevers, M. H. (1974). Psychopharmacological etiology of drug dependence. *Clin. Pharm.* **5**, 115–121.

Takeyama, T. (1974). Responsibility of methamphetamine addicts. *In* "Horrible Stimulants" (T. Sugawara, ed.), pp. 277–323. Tokiwayama Bunko, Tokyo. (in Japanese)

Tatetsu, S. (1972). Methamphetamine psychosis. *In* "Current Concept on Amphetamine Abuse" (E. H. Ellinwood and S. Cohen, eds.), pp. 159–161. DHEW Publication No. (HSM) 72-9085. U.S. Government Printing Office, Washington, D.C.

Tatetsu, S., Goto, A., and Fujiwara, T. (1956). "The Methamphetamine Psychosis." Igakushoin, Tokyo. (in Japanese)

Wakamatsu, N. (1985). Psychopathologische Untersuchungen der Weckamin-Psychosen. *Psych. Neurol. Jap.* **87**, 373–396. (in Japanese)

V
EPIDEMIOLOGY

Marissa A. Miller
Arthur L. Hughes

15
Epidemiology of Amphetamine Use in the United States

I. INTRODUCTION

Over the past six decades, epidemics of amphetamine abuse have occurred in technologically advanced countries such as the United States, Japan, and Sweden. As for other health-related problems, describing and quantifying the event becomes necessary to assess its magnitude and to surmount it successfully. Epidemiology is the major tool used in the field of public health to obtain the appropriate data to characterize the population at risk of the disease, to describe the natural history of the disease, to identify risk factors associated with the disease, and to evaluate treatment, prevention, or interventions.

A. Epidemiological Triangle: Agent, Host, Environment

Epidemiology is the study of the distribution and determinants of diseases, injuries, and conditions in defined human populations (Sartwell and

Last, 1980). Epidemiological methods were applied first to infectious disease outbreaks but also can be used to describe chronic and addictive disease epidemics. The central factors influencing the occurrence of disease, whether infectious, noninfectious, or addictive, generally are organized into three areas: agent, host, and environment. This triad is referred to as the epidemiological triangle (Mausner and Kramer, 1985). Each factor must be analyzed and understood to comprehend and predict the patterns of disease.

The epidemiological triangle implies that the presence of a disease agent is not sufficient to cause disease. This fact is manifested in infectious disease outbreaks, when not all exposed persons become clinically ill. Host factors such as previous experience or immunity to the agent influence the expression of disease in the individual, and in the population as a whole as herd immunity. Other host factors such as smoking, alcohol consumption, and attainment of health care affect disease expression. Environmental factors— including biological, social, and physical elements—also influence outbreaks and epidemics. Biological environmental factors include reservoirs of infections, vectors that transmit disease, and sources of ameliorative agents. The social environment may be defined in terms of the overall economic and political organization of a society relevant to exposure to the agent and general health. The physical environment also influences the occurrence of disease by exposing the host to stressors such as crowded living space, poverty, and pollution.

B. Factors Influencing Amphetamine Epidemics

These concepts can be applied to substance abuse epidemics (Kozel *et al.*, 1991). In the context of substance abuse, the agent is considered the availability or presence of the drug, the host is the user, and the environment is the society or community in which the user lives. Similar to infectious diseases, the mere presence of the agent does not necessarily lead to an outbreak or epidemic; the host must be susceptible to use of the drug and must be present in an environment amenable to, or tolerant of, substance abuse.

The factors specifically involved in promulgating an amphetamine epidemic have been identified by Ellinwood (1974) as (1) an initial oversupply of amphetamines in both legal and illegal markets; (2) the initiation or inoculation of large segments of the population; (3) widespread dissemination of knowledge concerning the amphetamine experience; (4) development of a core of chronic abusers; (5) increasing use of the parenteral route of administration; and (6) widespread manufacturing capabilities. These conditions, analogous to those factors influencing infectious and chronic diseases, have impacted the spread of amphetamine abuse and are outlined in this chapter for epidemics in the United States.

II. HISTORY OF EPIDEMIC USE

A. Drug Development and Early Medical Use

Amphetamines are synthetic stimulants that first were produced in 1887 (Caldwell, 1980). The related substance, methamphetamine, was developed approximately 30 years later by modifying the amphetamine structural formula but retaining similar central nervous system (CNS) activity and pharmaceutical uses (Kramer, 1970; Jaffe, 1985). The hypertensive and bronchodilator properties of amphetamines, the ability of these drugs to reverse barbiturate anesthesia, and their utility in treating lung congestion were recognized during the 1930s. Shortly after these discoveries were made, amphetamines were marketed as nasal inhalers, serving to relieve nasal and bronchial congestion associated with colds and hay fever (Lucas, 1985; Cho, 1990).

Identification of the CNS stimulant actions of amphetamine and accounts of abuse followed a few years after the release of the bronchodilator products (Grinspoon and Hedblom, 1975). When the American Medical Association (AMA) recognized amphetamine for treatment of Parkinson's disease and narcolepsy, a mild warning was added that cautioned that "continuous doses higher than recommended" might cause "restlessness and sleeplessness," but physicians were assured that "no serious reactions had been observed" (AMA Council on Drugs, 1963; Fischman, 1990). Not until several decades later were the addictive properties and psychiatric complications of amphetamine fully recognized by the medical community (Lemere, 1966).

Interest in amphetamines increased in both the medical and lay communities. Between 1932 and 1946, the pharmaceutical industry developed a list of over three dozen generally accepted clinical uses for amphetamines, including the treatment of schizophrenia, morphine and codeine addiction, tobacco smoking, heart block, head injuries, infantile cerebral palsy, radiation sickness, low blood pressure, seasickness, and persistent hiccups, among others (Grinspoon and Hedblom, 1975; Lucas, 1985). Many of these claims were fallacious and exaggerated. Amphetamines were promoted as being effective without the risk of addiction. During this period, amphetamine and its derivatives, such as methamphetamine, became available in both oral and intravenous preparations. These factors—public interest, readily available information, and multiple pharmaceutical forms of the drug—set the stage for the epidemic that followed.

B. Inoculation of a Large Population

During the 1940s and 1950s, amphetamines were prescribed liberally as oral medications. The increase in the popularity of amphetamines was influenced by widespread availability, low cost, and long duration of effect (Fischman, 1990). The pattern of oral amphetamine abuse involved patients gradually increasing the dose taken as a consequence of tolerance. Persons who exhibited this pattern included housewives, as a result of being prescribed amphetamines for weight loss; businessmen and professionals, using amphetamines for their antifatigue effects; students, using amphetamines for late night study sessions; and truck drivers, taking amphetamines to stay awake for long-distance hauling (Ellinwood, 1974). The general population had been inoculated with amphetamine-type drugs before the addictive qualities of these drugs were recognized fully and their adverse consequences characterized.

Between the 1930s and the 1970s, the public could obtain amphetamine-like drugs in a variety of over-the-counter nasal inhaler preparations. Abuse involved breaking open the inhalers and ingesting the drug directly, or soaking the fillers in alcohol or coffee. Inhaler use is suspected to have introduced hundreds of thousands of Americans to amphetamine abuse; however, this type of abuse was most prevalent in prison populations and among deviant groups (Monroe and Drell, 1947). The ability to cause euphoria, dysphoria, and psychic stimulation resulted in removal of amphetamine-like drugs from over-the-counter inhaler preparations in 1971. However, amphetamine products remained available in pill, capsule, and injectable form.

C. Amphetamine Use and Abuse during and after World War II

Further expansion of the population at risk occurred during World War II, when methamphetamine and amphetamines were used widely by the United States, British, German, and Japanese military as insomniacs and stimulants to increase alertness during battle and night watches. These drugs also were employed by war-related industries to extend shift work. An estimated 200 million tablets and pills were supplied to American troops during World War II (Grinspoon and Hedblom, 1975).

After World War II, the stage was being set for an amphetamine epidemic in the United States. The general population had previous exposure and experience with these drugs in over-the-counter preparations, and many soldiers returning from World War II had used and formed dependencies on amphetamines. In addition, during the 1940s and 1950s, enormous quantities of these drugs were prescribed without concern for their addictive qualities. College students, athletes, truck drivers, and housewives began using amphetamine for nonmedical purposes (Lucas, 1985; Fischman, 1990). Amphetamine use expanded in the United States during the 1950s as production of the drug increased significantly. These drugs were being marketed to treat obesity, narcolepsy, hyperkinesis, and depression. Illicitly, people were taking them primarily to increase energy, decrease the need for sleep, and elevate mood. Pharmaceutical production reached 3.5 billion tablets (about 20 standard dosage units per United States citizen) in 1958 and 10 billion tablets by 1970 (Grinspoon and Hedblom, 1975).

D. Escalation of Amphetamine Abuse during the 1960s and Development of a Drug Subculture

The 1960s were a period of societal change. Young adults were in search of increased awareness and spiritual development, and hallucinogenic and other drugs were vehicles in this search. Young people flocked to the Haight–Ashbury district of San Francisco, which served as a focal point of both the drug subculture and the changing values of its society. In Haight–Ashbury, "speed," the street name for amphetamine and methamphetamine, began to replace hallucinogenic drugs such as LSD in popularity. The transition to amphetamine use in the hippie culture was influenced by complex interactions among personal, social, and pharmacological variables (Pittel and Hofer, 1970).

Within the 1960s drug subculture, speed use escalated and a shift from oral preparations to intravenous abuse occurred. Tolerance to the effects of methamphetamine develop after continued use, particularly when administered intravenously, leading to an escalation of the dose needed to obtain the same euphoria and stimulation. Serial injections of escalating doses, over several days to several weeks, came to be known as a "speed run" (Smith, 1970). Exhaustion, then depression, accompanied the end of a run, followed by a cycle of readministration of drug that was necessary to mitigate unpleasant side effects and to regain the previous euphoria and high.

Serial intravenous speed users became known as "speed freaks." A public campaign was initiated to inform users of the hazards and harm associated with speed use. In part as a result of the "Speed Kills" campaign, the prevalence of amphetamine and methamphetamine use dropped sharply after 1972. From 1972 to 1977, the characteristics of the user population changed from heavy users to predominantly light to moderate users; a growing proportion of these users were women (Newmeyer, 1978).

E. Oversupply of Amphetamines

1. Diversion of Pharmaceutical Stores

One consequence of excessive production combined with widespread popularity of amphetamines was diversion of pharmaceutical grade drugs to illegal traffic and use. The black market in amphetamines involved diversions from pharmaceutical companies, wholesalers, druggists, and physicians (Grinspoon and Hedblom, 1975). Unscrupulous physicians who already were prescribing intravenous methamphetamine to treat heroin became involved in illegal prescriptions (Lake and Quirk, 1984). In 1963, injectable ampules of methamphetamine voluntarily were removed from sale to retail pharmacies in California by manufacturers.

Over half (and potentially 90%) of the total commercial product is estimated to have been diverted into the black market. In 1966, the Food and Drug Administration (FDA) estimated that more than 25 tons of amphetamine were distributed illegally (Fischman, 1990). By the mid-1960s, the need for intervention and legislative controls over amphetamine production and distribution was clear. The Drug Abuse Control Amendments of 1965 were passed by Congress, requiring increased record-keeping throughout the system of manufacture, distribution, prescription, and sale. However, diversion of pharmaceutical amphetamine to illicit use continued. In 1971, the Justice Department began imposing quotas on legal amphetamine production (Grinspoon and Hedblom, 1975).

2. Clandestine Manufacture of Speed

With the escalation of use during the 1960s came an increase in violence and a diffusion of clandestine manufacturing and distribution of speed outward from Haight–Ashbury to other areas along the West Coast (Smith, 1970). Outlaw motorcycle gangs were reported to be involved heavily in methamphetamine manufacture and distribution [National Narcotics Intelligence Consumers Committee (NNICC), 1992]. Methamphetamine is synthesized relatively easily and can be produced from commercially available materials (Irvine and Chin, 1991).

As a result of increasing controls on the prescribing and marketing of pharmaceutical amphetamines, including passage of the Controlled Substances Act of 1970, which placed amphetamine and some related stimulant drugs on Schedule II and imposed marketing quotas, the clandestine manufacture of methamphetamine became more widespread (Morgan and Kagan, 1978; Puder *et al.*, 1988). Availability of illicitly synthesized methamphetamine varied greatly during the 1970s and 1980s. Between 1972 and 1983, analyses of street samples of drugs purported to be methamphetamine revealed that, until 1974, specimens submitted as methamphetamine were, on average, less than 30% methamphetamine. From 1975 to 1983, the methamphetamine content in samples submitted as methamphetamine increased from 60% to over 95%. For the street samples submitted as stimulants—including those submitted as amphetamine, methamphetamine, or speed—methamphetamine made up a relatively small percentage between 1972 and 1979, but increased to approximately 60% in 1983 (Puder *et al.*, 1988). These data demonstrate the low but increasing prevalence of methamphetamine in the street speed market between the early 1970s and 1980s.

3. "Look-Alike" Speed

Prior to the increase in quality of street speed, the products sold as methamphetamine or speed were most frequently a combination of phenylpropanolamine hydrochloride, ephedrine, and caffeine and were referred to as "look-alike" speed (Lake and Quirk, 1984). The term "look-alike" speed referred to the similarity of both the appearance of these drugs and the CNS effects (Heischober and Miller, 1991). Other constituents also found in products purported to be speed included pseudoephedrine and cocaine.

4. Current Manufacture and Distribution of Illicit Methamphetamine

Since the mid-1980s, virtually all substances marketed illicitly as amphetamine or by street terms—crystal, crank, go, go-fast, zip, or cristy—have contained methamphetamine. By analyzing contaminants found in street methamphetamine samples, investigators have determined that clandestine manufacture of methamphetamine rather than diversion of pharmaceutical products now supplies the illicit marketplace (Puder *et al.*, 1988). The Drug Enforcement Administration (DEA) reports that methamphetamine is the most prevalent clandestinely manufactured controlled substance in the United States. In 1991, 84% of all clandestine laboratories seized were producing methamphetamine. The number of clandestine methamphetamine laboratories seized rose dramatically during the 1980s, from 88 in 1981 to 652 in 1989 (U.S. Department of Justice, 1991a). In 1991 the great majority of laboratories seized were located in California, Texas, and Oregon; in 1992, laboratories also were seized in Arizona, Florida, New Jersey, Pennsylvania, and Washington State [National Institute on Drug Abuse (NIDA), 1992b]. A decrease in the number of methamphetamine laboratories seized occurred in the early 1990s, largely because of the enactment and enforcement of the Chemical Diversion and Trafficking Act of 1988, which placed the distribution of 12 precursor and 8 essential chemicals used in the production of illicit drugs under federal control.

Concomitant with the increase in methamphetamine laboratory seizures during the 1980s was a localized resurgence of methamphetamine abuse. This associated increase was based on the assumption that the introduction of clandestine manufacture of methamphetamine into a community facilitates the development of a market and demand for the drug and, ultimately, the spread of use of the substance (Fischman, 1990; Hall and Broderick, 1991).

III. CURRENT PATTERNS AND TRENDS OF ABUSE

A. Introduction and Overview

Unlike the anecdotal reports that characterized amphetamine use and abuse in earlier decades, presently several large surveys provide current and reliable information on the prevalence and consequences of drug use in the United States. The National Household Survey on Drug Abuse (NHSDA), the Drug Abuse Warning Network (DAWN), and the Monitoring the Future (MTF) Study all serve as major sources of epidemiological information on the patterns and trends of substance abuse and the characteristics of abusers. With the exception of the medical examiner portion of the DAWN system, these studies generate data from national probability samples. The NHSDA produces estimates of drug use among all persons age 12 and over living in households; additionally, data have been collected for noninstitutionalized group quarters since 1991. The MTF Study produces estimates of drug use for high school students and a follow-up of these students; beginning in 1991, data were collected for 8th and 10th graders also. The DAWN surveillance system consist of a medical examiner and a hospital emergency room component and is used for ongoing monitoring of the levels and trends of morbidity and mortality related to drug use among patients and decedents age 6 and over.

Following the surge in amphetamine use experienced during the 1970s, the next smaller increase occurred during the mid- and late 1980s; this increase was geographically based. From the availability of ongoing population-based data has emerged the ability to monitor trends of amphetamine abuse, identify population subgroups most likely to have used amphetamines, and describe the types of health and medical consequences associated with amphetamine and methamphetamine use.

B. National Household Survey on Drug Abuse

1. Description of Survey

Since 1971, the NHSDA has been conducted periodically to provide important information on the level and trends of licit and illicit drug use in the United States. Traditionally, NHSDA data have been collected from a random sample of the civilian noninstitutional population (households), age 12 and over, in the United States. In 1991, the NHSDA sample consisted of 32,594 respondents (an 81% response rate) representing over 202 million persons living in households and noninstitutionalized group quarters such as homeless shelters and college dormitories (NIDA, 1992c). In addition to providing information on current and past use of drugs such as alcohol, sedatives, marijuana, and cocaine, respondents furnished information on their nonmedical use of stimulants. These stimulants are defined largely as amphetamines such as dexedrine and biphetamine (which are used to prevent narcolepsy), tepanil and didrex (which are used as appetite suppressants), and methamphetamine-containing compounds (methedrine, "speed," "ice," and "crank") [Substance Abuse and Mental Health Services Administration (SAMHSA), 1993]. Only nonmedical use of prescription-type stimulants and illicit stimulants is reported; drugs purchased over the counter are not included.

2. Limitations of the NHSDA

The target population of the survey was expanded in 1991 to include noninstitutional group quarters; however, the survey was not designed to cover homeless persons living on the street or persons in institutional quarters such as jails, mental institutions, and nursing homes. Despite this omission, these groups as a whole represent less than 2% of the total population; therefore, a moderate to large difference in drug use prevalence for these groups would not change the overall national rate significantly.

NHSDA survey data are based on self-report, creating a potential for underreporting because of the sensitive nature of some of the questions on illicit drug use. In the case of stimulants, the illicit forms of methamphetamine ("speed," "ice," or "crank") may be underreported not only because they are illegal, but also because of their inclusion at the end of a relatively long list of prescription stimulants through which the respondents are asked to read in the answer sheet portion of the questionnaire.

3. Characteristics of a Methamphetamine User

Data from the 1991 NHSDA for persons using either methedrine (a prescription form of methamphetamine) for nonmedical reasons or the illicit form (i.e., "speed," "ice," or "crank") were combined; these persons were called methamphetamine users for this analysis. In 1991, an estimated 5,235,000 persons had used methamphetamine at least once in their lifetime, accounting for 37% of the reported lifetime stimulant users. Population subgroups most likely to have members involved in use of methamphetamine at least once in their lifetime include those between 26 and 34 years of age (5.1%), males (3.4%), whites (2.9%), persons located in the West census region (11.5%), the unemployed (6.1%), and those with some college work completed (3.4%) (Table I). Lifetime use among persons living in families making less than $7000 a year (2.4%) was not much different than use among persons belonging to families making over $50,000 a year (2.5%). Persons in families making a combined income between $30,000 and $50,000 had the highest prevalence rate (3.0%), but this rate was not statistically different from rates in other income brackets. The estimated median age of first use of all stimulants was 18 years.

Among the four census regions, a statistically significant difference was seen between lifetime methamphetamine use in the West (5.2%) and in each of the other regions. Rates in the Northeast and North Central regions of the country were about the same (Fig. 1).

C. Drug Abuse Warning Network

1. Description of Surveillance System

DAWN provides national surveillance capabilities for medical emergencies and deaths caused by drug abuse. DAWN data on hospital emergency

TABLE I Estimated Prevalence of Lifetime Use of Methamphetamine[a,b]

Demographic attribute	Methamphetamine use (%)	95% CI[c]	Sample size
Total	2.6	2.3, 2.9	32594
Age			
12–17	1.1	0.8, 1.6	8005
18–25	4.5	3.8, 5.4	7937
26–34	5.1	4.3, 6.0	8126
35+	1.5	1.2, 1.9	8526
Sex			
Male	3.4	2.8, 4.0	14422
Female	1.9	1.6, 2.2	18172
Race/ethnicity[d]			
White non-Hispanic	2.9	2.5, 3.3	15648
Black non-Hispanic	0.8	0.6, 1.2	8050
Hispanic	2.2	1.6, 3.0	7916
Labor force status[e]			
Employed full time	3.5	3.0, 4.1	12849
Employed part time	2.7	2.0, 3.6	3030
Unemployed	6.1	4.4, 8.4	2157
Not in labor force	0.8	0.6, 1.1	6553
Adult education[e]			
Less than high school	2.4	1.9, 3.0	6422
High school graduate	2.8	2.3, 3.5	8442
Some college	3.4	2.6, 4.5	5502
College graduate	2.3	1.6, 3.3	4223

[a]Methamphetamine includes methadrine, speed, ice, and crack.
[b]Data from the 1991 National Household Survey on Drug Abuse.
[c]CI, Confidence interval.
[d]Data from persons reporting race/ethnicity as "other" are not included. Hispanics include persons of any race.
[e]Persons age 12–17 are excluded.

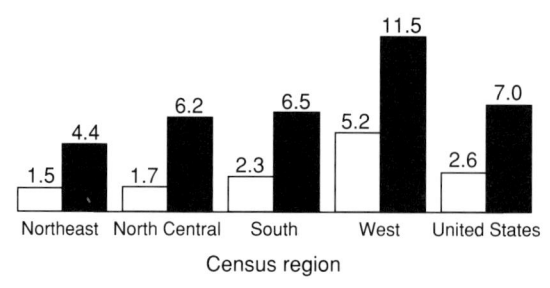

FIGURE 1 Estimated prevalence of lifetime use of methamphetamine (□) or any stimulants (■) for each census region and the United States in 1991. Data are given as the percentage of the population in the region or the United States. Note that "methamphetamine" includes methedrine, speed, ice, and crack. Stimulants include methamphetamines and other amphetamines. Data from the 1991 National Household Survey on Drug Abuse.

room episodes and medical examiner decedents resulting from the abuse of licit and illicit drugs are collected through a voluntary reporting system. DAWN identifies consequences of drug use and emerging patterns of use. For the purpose of reporting to the DAWN system, drug abuse is the nonmedical use of a drug or substance for psychic effect, dependence, or suicide attempt or gesture (NIDA, 1992a). Quarterly and annual emergency room estimates are produced from a probability sample of about 500 responding hospitals (implemented with the 1988 data, with a 78% response rate in 1991) and auxiliary data obtained from the American Hospital Association (AHA). The target population consists of about 5100 nonfederal hospitals operating in the contiguous United States that are general surgical and medical, short-stay (patient stays less than 30 days) facilities with a 24-hr emergency room. An estimated 400,000 drug abuse episodes (all drug-related emergency room visits) occurred nationwide in DAWN-eligible hospitals in 1991.

2. Limitations of DAWN

Hospital emergency room data are based on medical records and typically reflect self-reporting by the patient that frequently is not validated by toxicology tests. Since the same individual may be a patient at the emergency room on more than one occasion, DAWN estimates represent the number of visits (or episodes) and not the number of persons involved. Since the scope of the survey is restricted to hospitals, urgent care facilities and physicians offices that may accept and treat patients for drug abuse problems are not included.

3. Characteristics of Methamphetamine/Speed Episodes

In 1991, 4980 episodes involving use of methamphetamine/speed were reported (ranking 19th among all drug-related episodes). (NIDA, 1992a). Other specified amphetamines ranked 44th (2331 episodes). Three primary metropolitan statistical areas in California (San Francisco, San Diego, and Los Angeles–Long Beach), in addition to the Seattle metropolitan statistical area, account for 40% of all the methamphetamine/speed episodes in 1991. Methamphetamine/speed was reported mostly in combination with alcohol (1586 episodes; 32%), followed by cocaine (925 episodes; 19%), heroin/morphine (303 episodes; 6%), and marijuana/hashish (280 episodes; 6%). More recent emergency room data (from 1988 to the first half of 1992) indicate a downward trend in all amphetamine episodes, including those of methamphetamine/speed (Fig. 2). In fact, a statistically significant decrease in methamphetamine/speed episodes occurred from 1988 (8992 cases) to 1991 (4980 cases).

The average age of persons involved in methamphetamine/speed emergency room episodes over this 4-yr period is 28; approximately 40% of these cases are reported by persons between the ages of 26 and 34. Other subgroups showing up in emergency rooms across the nation in relatively large numbers are males (63% of all methamphetamine/speed episodes) and

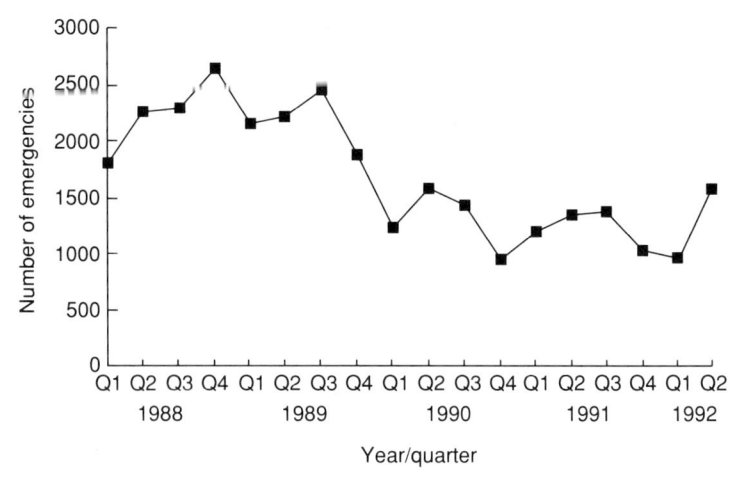

FIGURE 2 Estimated number of methamphetamine/speed emergencies in the contiguous United States by quarter (Q11988–Q21992). Note that quarterly estimates for 1992 are preliminary. Data from the Drug Abuse Warning Network (October 1992 data files).

whites (75%). Many patients use the substance for enjoyment (33%) or because they are dependent (40%; 73% combined). These patients usually visit the emergency room because they have experienced an unexpected reaction (34% of the methamphetamine/speed episodes), an overdose (22%), or some type of chronic effect (21%). Nearly 32% of these patients use a needle to administer the drug, whereas the same percentage take the drug orally or intranasally; however, the route of administration was unknown for about 25% of these types of episodes, so some of these percentages may be underestimated.

D. Monitoring the Future Study

1. Description of Study

The MTF Study provides information on the changing life-styles, values, and preferences of American youth on an annual basis. Included in the MTF Study is the estimation of the prevalence of both licit and illicit drug use. Each year since 1975, between 15,000 and 19,000 senior year students from a probability sample of 123 to 136 public and private high schools have participated in this survey. Over the years, the school response rate has ranged from 66 to 80%, while the student response rate has averaged around 80%. Schools that refused to participate were replaced with ones that were similar in size, geographic location, and demographic characteristics. In addition to sampling seniors, the 1991 survey launched the beginning of the inclusion of data from a comparable sample of 8th and 10th graders. All grade school data are collected in the contiguous United States during the spring of each year. Questions from the MTF Study solicit responses on use of amphetamines referred to as uppers, ups, speed, bennies, dexies, pep pills, and diet pills (Bachman et al., 1991).

2. Limitations of MTF

Since the survey is conducted in secondary school classrooms, students who drop out are not included. Using data from the 1991 NHSDA, Gfroerer (1992) showed illicit drug use prevalence rates among 16- to 18-year-old dropouts to be much higher than those among 12th graders from either the MTF or NHSDA sample. Beginning with the 1982 survey, the MTF questionnaire was reworded to put more emphasis on excluding over-the-counter stimulants such as those used to lose weight or stay awake (Bachman et al., 1991).

3. Findings

Lifetime, past year, and past month use of stimulants (or amphetamines) among high school seniors has declined significantly from 1982 to 1992 (Fig. 3). Follow-up data from college students 1–4 yrs beyond high school show similar downward trends from 1986 to 1992 (NIDA, 1993a). During this same time period, the percentage of seniors indicating that great risk is involved in taking amphetamines regularly (perceived harm) has increased gradually from 65% in 1982 to 72% in 1992 (NIDA, 1993b). This increase in awareness of hazards associated with stimulant use is consistent with decreasing use over time.

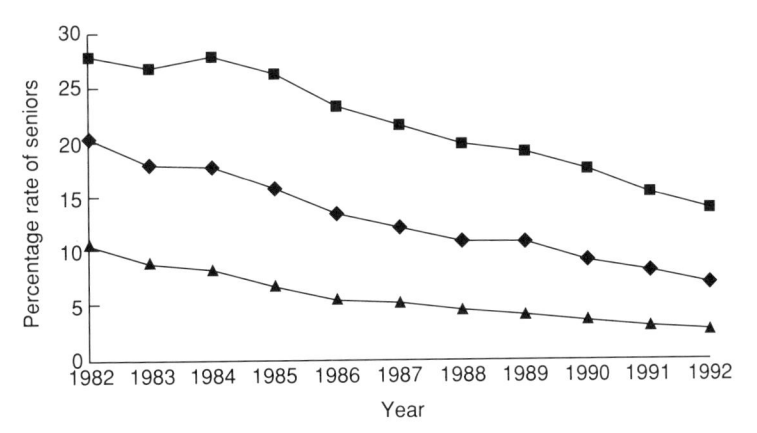

FIGURE 3 Estimated prevalence of lifetime (■), past year (♦), and past month (▲) use of stimulants among high school seniors (1982–1992). Data from the Monitoring the Future Study.

TABLE II Estimated Prevalence of Stimulant Use among 8th, 10th, and 12th Graders in 1991 and 1992[a,b]

	8th Grade (%)		10th Grade (%)		12th Grade (%)	
Drug use	1991	1992	1991	1992	1991	1992
Lifetime	10.5	10.8	13.2	13.1	15.4	13.9[c]
Past year	6.2	6.5	8.2	8.2	8.2	7.1[c]
Past month	2.6	3.3[c]	3.3	3.6	3.2	2.8

[a]Data from the Monitoring the Future Study.
[b]Approximate sample sizes are, for 8th graders: 17,500 in 1991, 18,600 in 1992; for 10th graders: 14,800 in 1991, 14,800 in 1992; and for 12th graders: 15,000 in 1991, 15,800 in 1992.
[c]Difference between 1991 and 1992 is statistically significant at the .05 level.

For 1991, the percentage of 12th and 10th graders reporting current stimulant use (use during the past month) is about the same, at 3.2 and 3.3%, respectively, and is slightly higher than that of 8th graders. However, notable increases in prevalence from 1991 to 1992 can be seen in current use among 8th graders (from 2.6% in 1991 to 3.3% in 1992) (Table II).

E. Ethnographic and Field Study Data

1. Pilot Methamphetamine Field Study

A small-scale field and ethnographic study of drug-treatment clients who had abused methamphetamine, conducted in 1988 in San Diego, Portland, and Dallas, demonstrated the increase in abuse of methamphetamine during the middle to late 1980s, particularly on the West Coast (NIDA, 1989). The demographic profile of communities tracked in this study showed an abusing population that was predominantly white, low to middle income, high-school educated, and young adult, generally ranging in age from 20 to 35 years. Most of the clients in this study reported intravenous methamphetamine administration. Increases in drug use indicators also were reported for methamphetamine during the mid-1980s through the Community Epidemiology Work Group, a network of state and local drug abuse experts representing 20 cities and metropolitan areas across the United States (NIDA, 1986).

2. Hawaiian "Ice" Outbreak

While increases in methamphetamine use were being noted on the mainland of the United States, Hawaii was experiencing increased use of a new dosage form of methamphetamine (Miller and Tomas, 1989; Miller, 1991).

A sharp increase in drug abuse indicators was recorded between 1986 and 1989 for a drug that, on the street, was referred to as "ice," crystal, shabu (Japanese), or batu (Filipino for rock). This substance had the appearance of a large, usually clear, crystal resembling broken fragments of glass or rock candy (Drug Enforcement Administration, 1990). "Ice," the name adopted by the popular press, is of high purity (90–100%) and exclusively the *d* isomer (a more active pharmacological form) of methamphetamine hydrochloride salt (Jaffe, 1985). In Hawaii, the drug is almost exclusively smoked in a glass pipe. The hydrochloride salt is sufficiently volatile to vaporize in a pipe so it can be inhaled (Chiang and Hawks, 1989; Cook *et al.*, 1991). This route of administration allows rapid absorption into the bloodstream with onset of effects similar to those experienced with intravenous administration (Cho, 1990; Perez-Reyes *et al.*, 1991).

The characteristics of this drug—high potency, high purity, and ability to administer by smoking—significantly contributed to the escalation of its use. In addition, the pattern of abuse that developed contributed to the development of rapid tolerance and addiction. Hawaiian "ice" typically is smoked in runs or periods of continuous use averaging 3–8 days in length, with 1–2 days between runs during which time the user would "crash" into deep prolonged sleep. Users reporting this use pattern became addicted rapidly and experienced numerous adverse medical, social, and physiological consequences (Miller, 1991).

This crystalline form of methamphetamine is believed to have been present on Oahu for 15 or more years, but was used in small exclusive groups and gangs. "Ice" became known by other groups in the early 1980s, but was scarce and expensive. "Ice" was introduced on a larger scale in the mid- and late 1980s. The principal trafficking groups are Filipino, Vietnamese, and Chinese (U.S. Department of Justice, 1991b). The epidemic started in West Honolulu in an area of working class and public housing projects and a large Filipino community. Use rapidly spread to a broad spectrum of the Hawaiian population, including ethnic minorities, both genders, and people of all ages and all socioeconomic classes. Field study data collected by NIDA described this outbreak and suggested that this epidemic peaked in 1989 (Miller, 1991). Drug abuse indicators and drug seizures decreased after 1989, but have begun to increase in 1992, suggesting a resurgence in "ice" use (NIDA, 1992b).

Prior to 1990, all the "ice" entering Hawaii originated from Asian sources; the "ice" form of methamphetamine was known to be produced in Hong Kong, Korea, Japan, Taiwan, Thailand, and the Philippines. Attempts to smuggle "ice" from Taiwan and Korea into Hawaii can be documented back to the mid-1980s (DEA, 1989). The importation and distribution of "ice" in Hawaii has been linked to Asian and Hawaiian criminal organizations and gangs.

By 1989, limited distribution of "ice" had occurred on the West Coast of the United States. In 1990, increased amounts of "ice" had found their way to California and, subsequently, to other limited locations (NNICC, 1992). This increase in availability of "ice" was believed to be the result of clandestine laboratories operating in California. During 1990, seven clandestine "ice" laboratories were seized nationwide, six of them in California. Domestically manufactured "ice" began to be supplied to distributors in Hawaii. This domestic "ice" was compensating for a disruption of major trafficking organizations that had been smuggling "ice" from South Korea.

IV. CONCLUSION

The occurrence of amphetamine epidemics over the past six or seven decades has been influenced by factors involving host, agent, and environment. The United States, similar to other countries that have experienced amphetamine epidemics, has a cultural emphasis on personal productivity and has been involved in changing societal values. Significant population migration also has been a factor in the occurrence of these epidemics. Our general cultural acceptance of the use of these drugs is shown by the wide distribution of use, including use by females.

In the United States, the patterns of abuse are geographically specific and are location influenced. At present, the highest prevalence of amphetamine and methamphetamine abuse is in cities along the West Coast and in Hawaii, but abuse also occurs elsewhere in the country. The patterns of abuse and the characteristics of the abuser population vary in California, Hawaii, and the remainder of the nation. In general, population subgroups that are most likely to abuse methamphetamine are whites, persons in their mid-twenties to early thirties, and males. The abusing population has broadened in the late 1980s and early 1990s in Hawaii and California, with growing representation in all socioeconomic classes and among women. Diversification of use among ethnic minorities and an increase in special populations, including adolescents and gay males, using methamphetamine (Newmeyer, 1988) have been seen. In addition, an increase in smoking, which is most prevalent in Hawaii, and in inhalation (NIDA, 1989, 1992a) of amphetamine also appears to be occurring.

Over the more than 60 years of availability, amphetamine and methamphetamine have had periodic resurgences in popularity and use. With the emergence of the new smokable potent form of methamphetamine, and with significant production and distribution systems in place, these drugs continue to pose a significant public health threat.

REFERENCES

AMA Council on Drugs (1963). New drugs and developments in therapeutics. *J. Am. Med. Assoc.* **183**, 362–363.

Bachman, J. G., Johnston, L. D., and O'Malley, P. M. (1991). "Monitoring the Future: Questionnaire Responses From the Nation's High School Seniors 1988." Survey Research Center, Institute for Social Research, The University of Michigan, Ann Arbor.

Caldwell, J. (1980). "Amphetamines and Related Stimulants: Chemical, Biological, Clinical, and Sociological Aspects." CRC Press, Boca Raton, Florida.

Chiang, N., and Hawks, R. (1989). "Pyrolysis Studies—Cocaine, Phencyclidine, Heroin, and Methamphetamine." Technical Review Brief. National Institute on Drug Abuse, Rockville, Maryland.

Cho, A. K. (1990). Ice: A new dosage form of an old drug. *Science* 249, 631–634.

Cook, C. E., Jeffcoat, A. R., Perez-Reyes, M., Sadler, B. M., Hill, J. M., White, W. R., and McDonald, S. (1991). Plasma levels of methamphetamine after smoking of methamphetamine hydrochloride. *In* "Problems of Drug Dependence, 1990," Proceedings of the 52nd Annual Scientific Meeting of The Committee on Problems of Drug Dependence, pp. 578–579. National Institute on Drug Abuse Research Monograph Series No. 105. DHHS Publication No. (ADM)91-1753. U.S. Government Printing Office, Washington, D.C.

Drug Enforcement Administration (1990). A special report on "ice." *In* "Epidemiologic Trends in Drug Abuse," Proceedings of the Community Epidemiology Work Group, December, 1989, pp. 69–83. DHHS Publication No. 721-757:20058. U.S. Government Printing Office, Washington, D.C.

Ellinwood, E. H. (1974). Epidemiology of stimulant abuse. *In* "Drug Use" (E. Josephson and E. Carroll, eds.), pp. 303–329. Hemisphere Publishing, Washington, D.C.

Fischman, M. W. (1990). History and current use of methamphetamine in the United States. *In* "Cocaine and Methamphetamine: Behavioral Toxicology, Clinical Psychiatry and Epidemiology," Proceedings from Japan–U.S. Scientific Symposium '90 on Drug Dependence and Abuse, Tokyo, Japan, pp. 239–250.

Gfroerer, J. (1992). "An Overview of the National Household Survey on Drug Abuse and Related Methodological Research," American Statistical Association, 1992 Proceedings of the Section on Survey Research Methods. American Statistical Association, Alexandria, Virginia, pp. 464–469.

Grinspoon, L., and Hedblom, P. (1975). "The Speed Culture. Amphetamine Use and Abuse in America." Harvard University Press, Cambridge, Massachusetts.

Hall, J. N., and Broderick, P. M. (1991). Community networks for response to abuse outbreaks of methamphetamine and its analogs. *In* "Methamphetamine Abuse: Epidemiologic Issues and Implications" (M. A. Miller and N. J. Kozel, eds.), pp. 72–83. National Institute on Drug Abuse Research Monograph Series No. 115. DHHS Publication No. (ADM)-91-1836. U.S. Government Printing Office, Washington, D.C.

Heischober, B., and Miller, M. A. (1991). Methamphetamine abuse in California. *In* "Methamphetamine Abuse: Epidemiologic Issues and Implications" (M. A. Miller and N. J. Kozel, eds.), pp. 60–71. National Institute on Drug Abuse Research Monograph No. 115. DHHS Publication No. (ADM)91-1836. U.S. Government Printing Office, Washington, D.C.

Irvine, G. D., and Chin, L. (1991). The environmental impact and adverse health effects of the clandestine manufacture of methamphetamine. *In* "Methamphetamine Abuse: Epidemiologic Issues and Implications" (M. A. Miller and N. J. Kozel, eds.), pp. 33–46. National Institute on Drug Abuse Research Monograph No. 115. DHHS Publication No. (ADM)91-1836. U.S. Government Printing Office, Washington, D.C.

Jaffe, J. H. (1985). Drug addiction and drug abuse. CNS sympathomimetics: Amphetamine, cocaine and related drugs. *In* "The Pharmacological Basis of Therapeutics" (A. G. Gilman, L. S. Goodman, T. W. Rall, and F. Murad, eds.), pp. 550–554. McMillan, New York.

Kozel, N. J., Sanborn, J. S., and Kennedy, N. J. (1991). A conceptualization of addictive disease epidemiology as compared to infectious and chronic disease epidemiology. Presented at the International Symposium on Drug Dependence: From the Molecular to the Social Level, Mexico City, Mexico.

Kramer, J. C. (1970). Introduction to amphetamine abuse. In "Current Concepts on Amphetamine Abuse" (E. H. Ellinwood and S. Cohen, eds.), pp. 177–184. DHEW Publication No. (HSM)72-9085. U.S. Government Printing Office, Washington, D.C.

Lake, C., and Quirk, R. (1984). Stimulants and look-alike drugs. *Psychiatr. Clin. North Am.* **7**, 689–701.

Lemere, F. (1966). The danger of amphetamine dependency. *Am. J. Psychiatry* **123**(5), 569–572.

Lukas, S. E. (1985). "The Encyclopedia of Psychoactive Drugs. Amphetamines: Danger in the Fast Lane." Chelsea House, New York.

Mausner, J. S., and Kramer, S. (1985). "Mausner and Bahn Epidemiology, An Introductory Text," 2d Ed., Saunders, Philadelphia.

Miller, M. A. (1991). Trends and patterns of methamphetamine smoking in Hawaii. In "Methamphetamine Abuse: Epidemiologic Issues and Implications" (M. A. Miller and N. J. Kozel, eds.), pp. 72–83. National Institute on Drug Abuse Research Monograph Series No. 115. DHHS Publication No. (ADM)91-1836. U.S. Government Printing Office, Washington, D.C.

Miller, M. A., and Tomas, J. M. (1989). Past and current methamphetamine epidemics. In "Epidemiologic Trends in Drug Abuse," Proceedings of the Community Epidemiology Work Group, December, 1989, pp. 58–65. DHHS Publication No. 721-757:20058. U.S. Government Printing Office, Washington, D.C.

Monroe, R. R., and Drell, H. J. (1947). Oral use of stimulants obtained from inhalers. *J. Am. Med. Assoc.* **135**, 909–914.

Morgan, J. P., and Kagan, D. V. (1978). The impact on street amphetamine quality of the 1970 Controlled Substances Act. *J. Psychedelic Drugs* **10**, 303–317.

National Institute on Drug Abuse (1986). "Drug Abuse Trends and Research Issues," Proceedings of the Community Epidemiology Work Group, December, 1986. DHHS Publication No. 181-332:60315. U.S. Government Printing Office, Washington, D.C.

National Institute on Drug Abuse (1989). "Methamphetamine Abuse in the United States." DHHS Publication No. (ADM)89-1608. U.S. Government Printing Office, Washington, D.C.

National Institute on Drug Abuse (1992a). "Statistical Series: Annual Emergency Room Data 1991—Data from the Drug Abuse Warning Network (DAWN)." DHHS Publication No. (ADM)92-1955. U.S. Government Printing Office, Washington, D.C.

National Institute on Drug Abuse (1992b). "Epidemiologic Trends in Drug Abuse," Highlights and Executive Summary: Community Epidemiology Work Group, December, 1992. Compiled for NIDA by Johnson, Bassin, and Shaw under contract #271-90-5300, Washington, D.C.

National Institute on Drug Abuse (1992c). "National Household Survey on Drug Abuse: Population Estimates 1991," (revised November 20, 1992). DHHS Publication No. (ADM)92-1887. U.S. Government Printing Office, Washington, D.C.

National Institute on Drug Abuse (1993a). "National Survey Results on Drug Use from the Monitoring the Future Study, 1975–1992," Vol. II, College Students and Young Adults. DHHS Publication No. (ADM)93-3598. U.S. Government Printing Office, Washington, D.C.

National Institute on Drug Abuse (1993b). "National Survey Results on Drug Use from the Monitoring the Future Study, 1975–1992," Vol. I, Secondary School Students. DHHS Publication No. (ADM)93-3597. U.S. Government Printing Office, Washington, D.C.

National Narcotics Intelligence Consumers Committee (1992). "The NNICC Report 1991." U.S. Department of Justice, Drug Enforcement Administration, Washington, D.C.

Newmeyer, J. A. (1978). The epidemiology of the use of amphetamine and related substances. *J. Psychedelic Drugs* **10**(4), 293–302.

Newmeyer, J. A. (1988). The prevalence of drug use in San Francisco in 1987. *J. Psychoact. Drugs* **20**(2), 185–189.

Perez-Reyes, M., White, R., McDonald, S., Hill, J., Jeffcoat, R. and Cook, C. E. (1991). Pharmacologic effects of methamphetamine vapor inhalation (smoking) in man. *In* "Problems of Drug Dependence 1990," Proceeding of the 52nd Annual Scientific Meeting, The Committee on Problems of Drug Dependence, pp. 575–577. National Institute on Drug Abuse Research Monograph Series No. 105. DHHS Publication No. (ADM)91-1753. U.S. Government Printing Office, Washington, D.C.

Pittel, S. M., and Hofer, R. (1970). The transition to amphetamine abuse. *In* "Current Concepts on Amphetamine Abuse" (E. H. Ellinwood and S. Cohen, eds.). DHEW Publication No. (HSM)72-9085. pp. 105–111. U.S. Government Printing Office, Washington, D.C.

Puder, K. S., Kagan, D. V., and Morgan, J. P. (1988). Illicit methamphetamine: Analysis, synthesis, and availability. *Am. J. Drug Alcohol Abuse* **14**, 463–473.

Sartwell, P. E., and Last, J. M. (1980). Epidemiology. *In* "Maxcy–Rosenau Public Health and Preventive Medicine" (J. M. Last, ed.), 11th Ed., pp. 9–85. Appleton-Century-Crofts, New York.

Smith, R. C. (1970). Compulsive methamphetamine abuse and violence in the Haight–Ashbury district. *In* "Current Concepts on Amphetamine Abuse" (E. H. Ellinwood and S. Cohen, eds.), pp. 205–216. DHEW Publication No. (HSM)72-9085. U.S. Government Printing Office, Washington, D.C.

Substance Abuse and Mental Health Services Administration (1993). "National Household Survey on Drug Abuse: Main Findings 1991." DHHS Publication No. (SMA) 93-1980. U.S. Government Printing Office, Washington, D.C.

U.S. Department of Justice (1991a). "Clandestine Laboratory Seizures in the United States 1990." U.S. Department of Justice, Drug Enforcement Administration, Washington, D.C.

U.S. Department of Justice (1991b). Fact sheet: ICE. *In* "Drugs and Crime Data." Office of Justice Programs, Bureau of Justice Statistics, Washington, D.C.

Susumu Fukui
Kiyoshi Wada
Masaomi Iyo

16
Epidemiology of Amphetamine Abuse in Japan and Its Social Implications

I. INTRODUCTION

The history of drug abuse and dependence in Japan is brief; the watershed came after World War II ended in 1945, when drug abuse became more prevalent worldwide. Methamphetamine (METH) abuse after World War II introduced Japanese society to the attendant social problems of drug abuse; this was the first known incidence of drug abuse in Japan.

Japan is now in its second period of METH abuse. More than two decades of extensive abuse already have caused serious problems in Japanese society and have made METH dependence an important issue in Japanese medical fields.

In contrast to other countries, METH causes the major drug abuse and dependence problem in Japan. However, drug abuse and dependence in general may be one social illness that affects all nations during periods of rapid change or sudden modernization.

In this chapter, the METH abuse situation in Japan is discussed from the viewpoint of the factors contributing to METH abuse:

1. history and social background of METH abuse
2. special demographic features relating to METH abusers
3. supply of METH

The results of a recent survey, that show a shift in the clinical characteristics of psychiatric hospital in-patients and out-patients with METH psychosis, are mentioned also.

The history of METH abuse in Japan can be divided into four periods, including the two epidemics of METH abuse mentioned previously.

II. PERIOD PRIOR TO 1945

Before 1945, drug abuse and dependence were virtually nonexistent in Japan and, therefore, were never thought to be responsible for social problems. However, incidental abuse of opiates and cocaine did occur. Most opiate abusers were from outside Japan, whereas most cocaine abusers consisted of a small number of medical doctors. The estimated total number of abusers was about 400; therefore, the abuse of drugs was neither a social nor a medical problem (Fig. 1; Kuma, 1976).

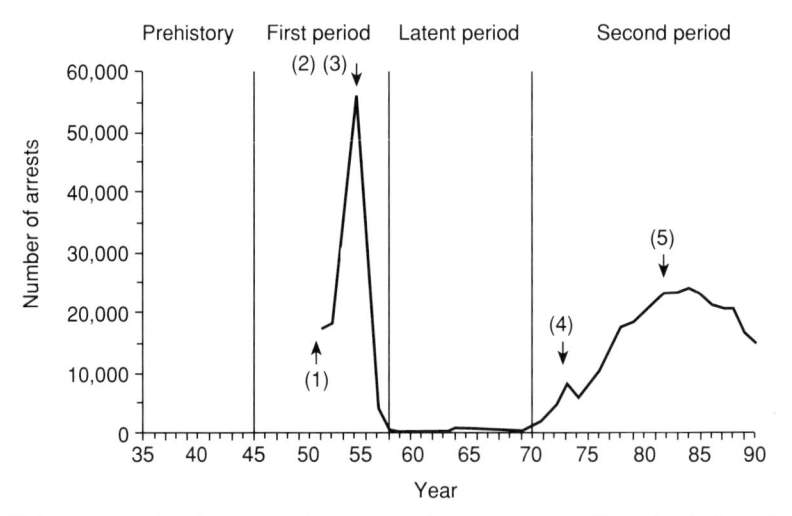

FIGURE 1 History of methamphetamine abuse as change in arrest profile under the Stimulant Control Law in Japan. (1) Enactment of Stimulant Control Law (1951). (2) Kyoko homicide incident (1954). (3) Amendment to Stimulant Control Law (1954). (4) Amendment to Stimulant Control Law (1973). (5) Fukagawa incident (1981).

During World War II, METH was used mainly for Japanese military objectives. At times, soldiers were forced to consume METH to enhance their fighting spirits and to keep them awake while engaged in nocturnal battles. Also, to increase efficiency and productivity in military support industries, workers were coerced into using METH. After 1941, pharmacies sold products containing METH over the counter under commercial names such as Hiropon and Sedorin. However, METH abuse was not a social issue at that time (Tatetsu *et al.*, 1986). Until 1945, Japan was a strong military power with a totalitarian society, in which the Japanese people generally shared the same objectives as the military leaders. In this environment, drug abuse was not recognized by the Japanese. Thus, the social acceptance of the widespread use and availability of METH was the basis of the epidemic of METH abuse that followed the end of World War II.

III. FIRST PERIOD OF METHAMPHETAMINE ABUSE: 1945–1957

The first period of abuse has been divided into three stages: (1) the early stage, during which METH abuse was focused within a group of core abusers; (2) the epidemic stage, during which other groups joined the core abusers; and (3) the ebbing stage, during which the epidemic use of METH was in decline (Fig. 2; Fukui, 1988).

A. Pre-epidemic Stage: 1945–1951

Following the defeat of Japan in World War II, social confusion and an increase in the social problems of the nation stemming from poverty, unemployment, and a scarcity of food and housing resulted in nihilistic and hedonistic attitudes that discouraged and depressed the Japanese people.

Under these circumstances, pharmaceutical manufacturers with large supplies of METH began general over-the-counter sales of the drug using the slogan, "Fight sleepiness and enhance vitality." METH from large army stocks also was becoming available on the black market. Thus, the population was being supplied with METH in abundance but had no awareness of its dangerous effects. The METH abuse epidemic began because the discouraged and depressed Japanese population, especially the younger generation, was eager for the euphoric and mood-elevating effects of the drug.

In the autumn of 1946, the first incidents of METH abuse were seen in the Kansai area (centering on Osaka, Kobe, and Okayama). The number of abusers rapidly increased and spread from the major cities to rural areas throughout the nation (Tatetsu *et al.*, 1986).

The first postwar users of METH were workers in the military support industries and soldiers who had been given the drug during the war. Subse-

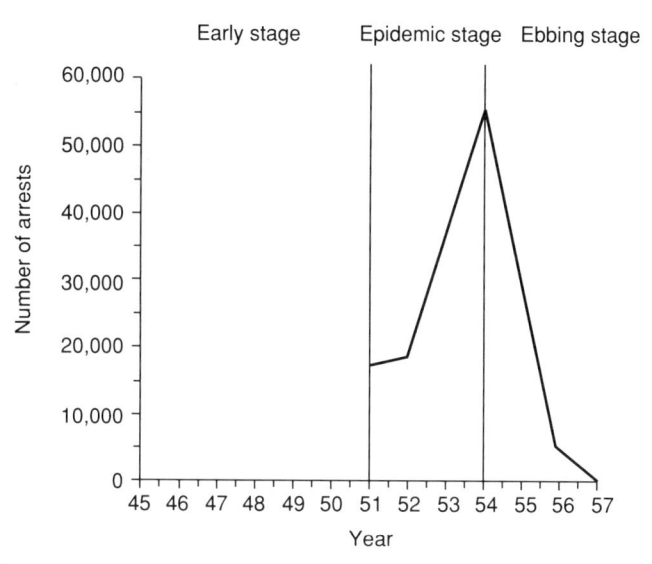

FIGURE 2 Change in arrest profile during the first period of methamphetamine abuse.

quently, writers, journalists, entertainers, artists, students, and factory workers also began to abuse METH to enhance their efficiency while working overnight (Tatetsu et al., 1986). Additionally, juvenile delinquent groups that stayed out all night began to abuse METH as a stimulant and for pleasure. By 1948, 5% of individuals between the ages of 16 and 25 yr were said to be abusing METH (Noda, 1950).

The reasons for the increase in stimulant abuse and dependence in this early stage are threefold. (1) A background of a nihilistic and hedonistic social environment was coupled with an aimless postwar society and a confused population. (2) The supply of METH to the marketplace by pharmaceutical manufacturers was unlimited, without any understanding of the dangerous effects of its abuse and dependence. (3) The younger generation, who had lost their purpose in life, desired relief from depression by the pharmacological effects of METH. These three factors—the social environment, the availability, and a population vulnerable to abuse—combined to bring on an epidemic of METH abuse.

B. Epidemic Stage: 1951–1954

Although the Stimulants Control Law was enacted with strict enforcement in 1951 because of a sudden increase in the number of METH abusers, active illicit manufacturing, smuggling, and sales by the underworld continued to supply a large amount of METH, resulting in an unprecedented epidemic of abuse. During this stage, a shift in the population of abusers occurred to include antisocial and alienated people such as gang members and the unemployed rather than ordinary citizens.

At the same time, injectable METH became available as an alternative to tablets. Taken intravenously, the acute stimulant effect of METH on the central nervous system is enhanced greatly. METH abusers began to abuse the drug intravenously by preference, leading to an increase in the number of homicides and other crimes committed under the pharmacological influence of METH.

METH tablet abuse already had become a public health problem, and METH injection abuse rapidly escalated problems related to societal security (Henmi, 1982; Tatetsu et al., 1986). During the peak abuse period of 1954, the number of METH abusers reached 550,000, including 55,000 METH abusers arrested and 200,000 METH abusers suffering from METH-induced psychosis. Up to 1954, 2,000,000 people were thought to have abused METH in one form or another (Ministry of Health and Welfare, 1958).

C. Ebbing Stage: 1954-1957

In 1954, a youth with a history of METH abuse murdered a 10-year-old girl (Kyoko-chan homicide incident) while under influence of the drug, triggering a civilian awareness movement that pressed for the development and passage of an amendment to the Stimulants Control Law. This amendment led to greater enforcement of the law. After a peak in 1954, METH abuse declined rapidly. Possible reasons for this swift ebbing included (1) severe critical public opinion against METH abuse triggered by the homicide incident; (2) the strict application of punishment under the Stimulants Control Law; (3) the adoption of treatment of chronic addicts under an involuntary admission program based on amendments to the Mental Health Act; (4) economic recovery and improved living standards, thereby reducing the desire for the antidepressant effects of the drug; and (5) blockade of the illicit manufacturing bases (the supply sources of METH) located within Japan by a police force that finally had regained the confidence and trust of the public.

IV. LATENT PERIOD: 1957-1969

Between 1957 and 1969, the rapid industrialization of the country under favorable conditions moved Japanese society into a period of high economic growth. Between 1957 and 1969, the number of arrests under the Stimulants Control Law was less than 1000 annually (see Fig. 1). The problem of METH abuse was no longer an issue in society or in the medical professions.

From about 1957, however, the abuse of other drugs began. For example, heroin abuse began in Osaka, Kobe, and Okayama (the Kansai area) and then appeared in Tokyo and Yokohama, where it prevailed (Kuma, 1976). Although this abuse was recognized as a social problem, it ebbed in 1965. The abuse of tranquilizers such as meprobamate and chlordiazepoxide by middle-aged to elderly individuals and of hypnotics such as methaqualone by teenagers became social problems between 1960 and 1975 (Fukui et al., 1989).

In 1967, volatile solvent abuse began to spread at an alarming rate among young people throughout the nation. The prevalence of this abuse continues to this day. Also during this time period, the abuse of psychotropics was taking root in some groups of mainly young, ordinary citizens (Fukui et al., 1989).

During this same period, METH was being manufactured illegally by organized criminals in the Kansai area, where it was abused during illicit gambling parties and sleepless vigils within Kansai city centers such as Osaka and Hyogo. Japanese organized gangs originated during the 1960s in the Kansai area, and METH had become part of their culture and lifestyle. These factors brought on the second epidemic of METH abuse that followed the latent period (Fukui, 1988).

V. SECOND PERIOD OF METHAMPHETAMINE ABUSE: 1970–PRESENT

After 1970, METH abuse again spread rapidly throughout the nation in an epidemic that is now called the second period of METH abuse.

A. History

Like the first period, the second period of METH abuse can be divided into three stages. (1) the early stage, during which METH abuse spread across the nation but remained within organized crime circles; (2) the epidemic stage, characterized by the swift, epidemic-like spread of METH abuse by ordinary citizens; and (3) the current stage, during which strict control brought about a plateau, followed by a slight decline, in METH abuse (Fukui, 1988).

1. Early Stage: 1970–1974

Between 1970 and 1974, METH abuse swept across Japan from the Kansai area, but was limited to a core group of organized criminals and people within their sphere of influence (Fig. 3). During the latent period, less than 1000 arrests due to violations of the Stimulants Control Law were recorded annually (Fig. 1). This figure increased rapidly from 1970 to a high of 8510 in 1973. Initially only in the Kansai area, the arrests had spread by 1973 throughout the nation (National Police Agency, 1977).

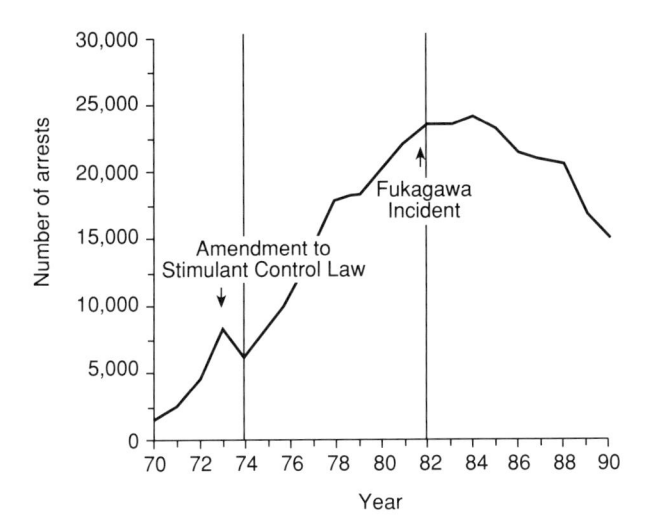

FIGURE 3 Change in arrest profile during second period of methamphetamine abuse.

During this period, 60% of those arrested for violations of the Stimulants Control Law were organized criminals. In addition to this core group, individuals in contact with these criminals (such as the unemployed, workers involved in restaurant and liquor bar industries, bar hostesses and social escorts) became involved in the abuse of METH as the organized crime gangs developed operations across the nation (Fukui, 1988).

As METH abuse spread, an increase in the number of cases of METH-induced psychosis was noticed (Konuma, 1984; Fukui *et al.*, 1989). To cope with this situation, amendments were made to the Stimulants Control Law in 1973 to introduce penal provisions that were equivalent to those of the Narcotics Control Law. Unfortunately, the effectiveness of these amendments in improving the situation was limited.

2. Epidemic Stage: 1975–1981

From 1975 until 1981, a second group of METH abusers appeared, made up of workers with previous criminal convictions and those who were in some way closely associated with organized crime (such as construction workers, truck drivers, and personnel related to trucking; people in the restaurant and liquor bar industries; and workers in the entertainment-related industries). Blue-collar workers, daily-paid laborers, shop assistants, students, salaried workers, housewives, adolescents, and other citizens from the ordinary social stratum who came in close contact with this second group of METH abusers also became involved in METH abuse. A complete epidemic was evident after abuse penetrated the young and elderly populations (Fukui, 1988).

The number of arrests for violations of the Stimulants Control Law exceeded 10,000 in 1976, and continued to increase by 20–30% each year thereafter to reach a figure of 22,000 in 1981 (National Police Agency, 1991). Although the number of latent abusers was unknown, a figure approximately 10 to 20 times that of those arrested was believed to reflect the situation accurately. From a 1981 survey conducted by the Prime Minister's Office, Fujita of the Ministry of Health estimated a total of 300,000 METH abusers in that year, which is comparable to our estimate of 1981 (Fujita, 1982; Fukui et al., 1989).

During the course of this epidemic of METH abuse, many abusers suffered from METH-induced psychosis. The high numbers of cases with aggressive and prominent psychiatric symptoms presented many problems for mental institutions giving treatment (Fukui et al., 1989). Although the exact number of hospitalized patients was unclear, a 1-day survey conducted on June 30, 1981, by the Ministry of Health on the number of hospitalized METH dependents, coupled with our estimates, suggested approximately 3000–4000 hospitalization cases in 1981.

With an increase in the incidence of violent crimes committed by METH abusers under its influence, coupled with the problems caused by its smuggling and illicit sales by organized criminals, METH abuse had developed into a very serious social problem in Japan (National Police Agency, 1991).

3. Current Period: 1982 Onward

The Fukagawa incident in June 1981, when two housewives and two infants were murdered by a METH abuser in Fukagawa, Tokyo, unleashed a flood of strongly critical public opinion on METH abuse. The Japanese government pushed through a number of policies for increased enforcement of the Stimulants Control Law. With markedly strengthened enforcement and movements to eradicate METH, tied in with public campaigns against METH abuse, some changes developed.

The trend of increasing arrests for violations of the Stimulants Control Law declined after the Fukagawa incident. The number of arrests peaked at 24,372 in 1984, and then decreased to 15,267 in 1990, or 63% of the 1984 peak (Ministry of Health and Welfare, Bureau of Pharmaceutical Affairs, 1991). The number of violent crimes committed while under the pharmacological effects of METH increased dramatically between 1975 and 1981, with 553 arrests registered in 1980. However, following the Fukagawa incident, the number rapidly decreased, with only 108 arrests in 1990 (National Police Agency, 1991). The number of individuals hospitalized for METH-induced psychosis has declined also. (Ministry of Health and Welfare, Bureau of Health Service, 1991). The declines in the number of arrests under the Stimulants Control Law and in the number of violent crimes suggest that the measures taken between 1976 and 1981 to overcome the prevalence of METH abuse were effective in suppressing METH abuse.

B. Social Background

The advent of the high economic growth period in the 1960s brought about sudden changes in life-style and social structure in Japan. These sudden changes were likely to have generated the second period of METH abuse. The changes that were relevant to the second period of METH abuse were rapid changes in Japan's industrial structure; rapid increases in the standard of living; rapid urbanization; an increasingly information-oriented society; increasing internationalization; and an increasingly academic achievement-oriented society (Fukui, 1988). Although these rapid changes accompanying the high economic growth brought benefits to the Japanese people, a "distortion" of the social structure was apparent in 1970 (Henmi, 1982; Fukui *et al.*, 1989). Some individuals who could not keep up with or who had turned their backs on these changes resorted to METH abuse.

Through changes in the industrial structure of Japan and the rapid urbanization, hedonistic trends, materialistic values, a drop in social standards, and other social conditions developed, setting a social background conducive to METH and organic solvent abuse. Changes in the industrial structure gave rise to an inability to adapt to new professional challenges and a lack of pleasure in work, and encouraged a society that placed emphasis on higher academic achievements. School dropouts formed antisocial groups, resorting to organic solvent abuse and subsequent METH abuse. The global economic recession following the "Nixon Shock," when Nixon took the United States off the gold standard, and the "Oil Shock" resulted in many newly unemployed. Those bitter with society also resorted to METH abuse (Okae, 1982).

C. Methamphetamine: The Constant Supply to Society

In 1966, Japan was entering a high economic growth phase until the "Nixon Shock" in 1971, followed by the "Oil Shock" in 1973, brought about a global economic slump that affected, in particular, the civil, construction, and service industries in Japan. The organized gangs (Yakuza) felt the financial strain, having relied heavily on these industries as a source of income, and increased their illicit sales and smuggling of METH as an alternative source of income. High profits could be generated from illicit dealings, and METH was suddenly in large supply in Japan. The Yakuza hampered efforts to contain the supply of METH through their well-regulated illicit imports, their control over smuggling routes, and their control of diverse, highly refined, organized illicit dealings over an extensive region. The involvement and methods of the Yakuza are believed to be the greatest factors contributing to the METH abuse that Japan endures even today (Ministry of Health and Welfare, Bureau of Pharmaceutical Affairs, 1991; National Police Agency, 1991).

D. Demographic Features of Methamphetamine Abusers

For a better understanding of the current METH abuse situation in Japan, recent demographic features of METH abusers are presented and the factors causing METH abuse are considered.

The subjects in this study were 233 METH-dependent in-patients and out-patients of 21 psychiatric institutions throughout Japan for the period from May 1, 1988, to October 31, 1988 (Wada and Fukui, 1990). In addition, this report also includes data from police files (Ministry of Health and Welfare, Bureau of Pharmaceutical Affairs, 1991; National Police Agency, 1991) and the results from a nationwide survey through psychiatric institutions (Fukui, 1992).

1. Sex

Of the 233 cases, 196 (84.1%) were males, 34 (14.6%) were females, and 3 (1.3%) were unstated. The number of males greatly exceeded that of females, correlating with the special feature of Bejerot epidemic-type abuse (Bejerot, 1969). As abuse in the general population prevailed, the percentage of female abusers increased as well.

2. Age Distribution

Most dependents were in their thirties, followed by those in their twenties, forties, and, finally, those in their fifties. Individuals in their twenties and thirties make up the majority of METH abusers. Table I shows the 1988 age-group distribution from the survey of mental hospitals throughout the nation.

According to the survey done in 1976, METH dependents in their twenties and thirties composed 93.3% of the patients (Fukui et al., 1989). The surveys of 1981 and 1982 showed that, as METH abuse prevailed in society, the percentage of dependent individuals in their twenties and thirties decreased while the percentage of those in their forties and those over fifty

TABLE I Age of Methamphetamine Users[a]

Age (yr)	N	%
0–19	5	2.1
20–29	67	28.8
30–39	79	33.9
40–49	52	22.3
50–59	22	9.4
60–69	2	0.9
Not stated	6	2.6
Total	233	100

[a]Mean age ± SD = 35.62 ± 9.87.

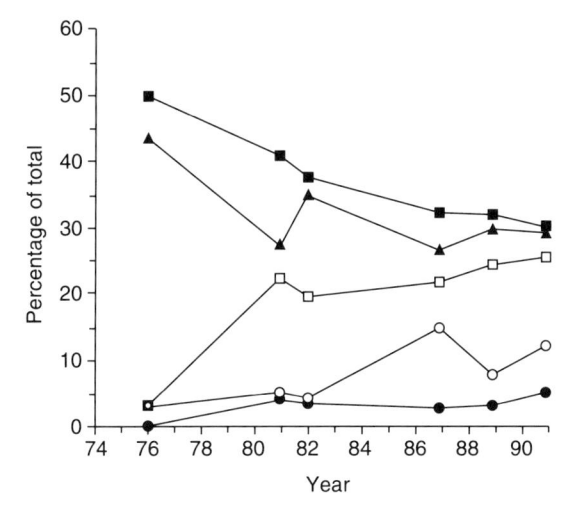

FIGURE 4 Age distribution of methamphetamine abusers visiting hospitals: under 20 (●), 20–29 (▲), 30–39 (■), 40–49 (□), over 50 (○).

increased. In addition, patients aged 19 and under (although a small number) increased. In the surveys of 1987, 1989, and 1991, a decrease in the percentage of METH-dependent individuals in their twenties and thirties was noted, as was an increase in the percentage of those in their forties and those over fifty (Fig. 4; Fukui, 1992).

A National Police Agency report contained an age-group distribution of people arrested on METH abuse and related charges that showed 20- to 39-

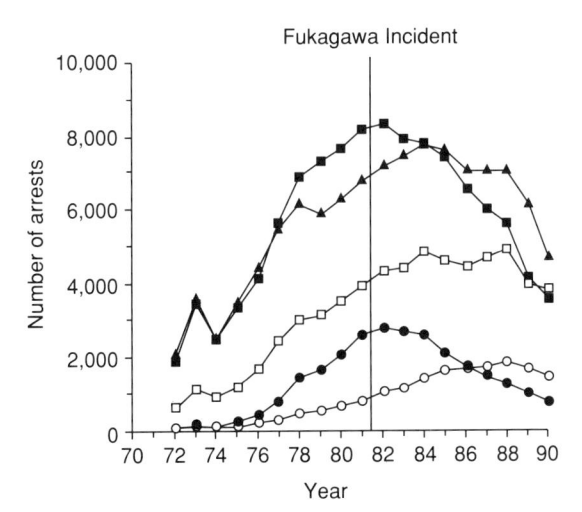

FIGURE 5 Number of arrested methamphetamine abusers given by age: under 20 (●), 20–29 (▲), 30–39 (■), 40–49 (□), over 50 (○).

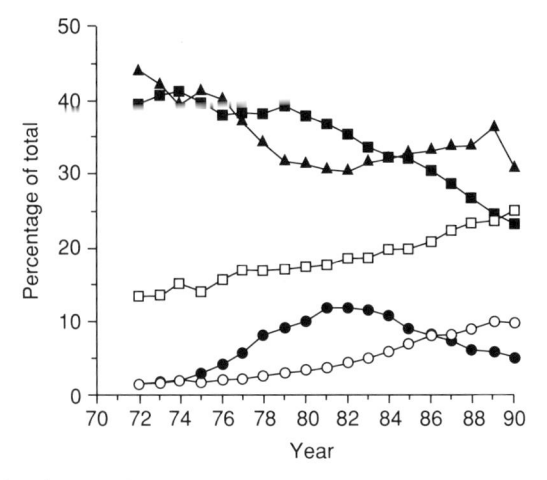

FIGURE 6 Age distribution of arrested methamphetamine abusers: under 20 (●), 20–29 (▲), 30–39 (■), 40–49 (□), over 50 (○).

year-olds to make up the dominant age group abusing METH during the second period of METH abuse, which commenced in 1970 (National Police Agency, 1991). As abuse prevailed in the general population, increases in the number of METH abuse arrests of individuals aged 40–59 and of those aged 19 and under were observed, highlighting the spread of METH abuse across the age groups (Figs. 5,6).

Following the homicides involving a METH abuser in June 1981 (the Fukagawa incident), arrests of minors aged 19 and under for METH abuse crimes slowly declined. However, the numbers for arrests of those aged 40 and over on METH abuse charges continued to rise (Ministry of Health and Welfare, 1991). Thus, the history of arrests suggests that the term of METH abuse for a given abuser has been prolonged.

3. Social Background of Abusers

a. Educational Background Analyzing the educational background of the METH abuse subjects in the survey of hospitals showed that 50% had completed Junior High School, whereas an additional 26% consisted of Senior High School dropouts, indicating that 76% of the METH abusers survey had completed their education only as far as the compulsory level, contrary to popular academic expectations. The complicated family background of abusers and their growing delinquency may have made them unable to complete their high school education.

b. History of Drug Abuse METH abusers with previous drug abuse experience accounted for 119 subjects (51.1%). The largest number, 77

TABLE II Types of Drugs Used Before Methamphetamine[a]

Type of drug	N	%
Organic solvent	77	33.0
Alcohol	34	14.0
Marijuana	10	4.3
Hypnotics	6	2.6
Cough syrups	5	2.1
Anodyne	4	1.7
Minor tranquilizers	1	0.4
Others	10	4.3
Unidentified	3	1.3
None	114	48.9

[a] N = 233; totals include multiple answers by respondents.

(33.0%), had been convicted on organic solvent abuse charges; 34 subjects (14.6%) had been convicted on alcohol abuse charges. The remainder had been convicted on charges involving other substances. Among the many substances available, organic solvents are most likely to be abused by the young in Japan. Almost all the young METH abusers had started with organic solvent abuse, and then changed to METH abuse (see Table II).

c. Criminal Record Of the 233 abusers, 104 (44.6%) were in criminal reform centers for violations of the Stimulants Control Law; an additional 71 subjects (30.5%) were imprisoned. A high percentage of the 233 abusers experienced, in one way or another, problems in their daily lives.

4. First-Time Abuse of METH

a. Age of First METH Use Many addicts began to abuse METH at some time between age 15 and 25. First-time abuse of METH among abusers was predominantly in the late adolescent years or in the early twenties (Table III).

b. Motives for First Use The main motive for first use was temptation (54.5%) or curiosity (42.5%), although most of the METH abuse subjects had been motivated by a combination of both. Usually these subjects had been coaxed by their friends, although some had been mildly curious (Table IV).

c. Key Person Responsible for the First-Time Abuse of METH When asked who had been responsible for the chance to first try METH, the

TABLE III Age at First Methamphetamine Use[a]

Age (yr)	N	%
<15	1	0.4
15–19	61	26.2
20–24	73	31.3
25–29	37	15.9
30–34	27	11.6
35–39	18	7.7
40–44	8	3.4
45–49	1	0.4
50–54	1	0.4
Not stated	6	2.6
Total	233	100

[a] Mean age ± SD = 24.41 ± 7.33.

subjects cited friends and acquaintances more often than illegal dealers. Male abusers, mostly influenced by their male friends (37.8%), started abusing METH with pleasure-seeking objectives or to avoid sleepiness and fatigue when gambling with their friends. However, most female abusers were encouraged by their husbands or boyfriends to abuse METH for a heightened sexual response (Table V).

In response to questions on the profession of the person who had introduced METH abuse, more than half the subjects replied that the person had been connected to gangs in some way. An obvious motive for METH abuse for both males and females was their contact with people connected to gangs (Table VI).

TABLE IV Motives for First Methamphetamine Use[a]

Motive	N	%
Temptation	127	54.5
Curiosity	99	42.5
Seeking stimulation	18	7.7
Voluntarily	18	7.7
Involuntarily	10	4.3
Desperation	4	1.7
Others	8	3.4
Not stated	40	17.2

[a] Totals include multiple answers by respondents.

TABLE V Key Person Who Introduced Methamphetamine[a]

	Male	Female	Not stated	Total
Friends of the same sex	74(37.8)	4(11.8)	2(66.7)	80(34.3)
Acquaintances	33(16.8)	5(14.7)	1(33.3)	39(16.7)
Illegal dealers	21(10.7)	1(2.9)	0(0.0)	22(9.4)
Husband	— —	8(23.5)	0(0.0)	8(3.4)
Friends of the opposite sex	1(0.5)	7(20.6)	0(0.0)	8(3.4)
Cohabitants	0(0.0)	4(11.8)	0(0.0)	4(1.7)
Others	12(6.1)	2(5.9)	0(0.0)	14(6.0)
Not stated	55(28.1)	3(8.8)	0(0.0)	58(24.9)
Total	196(100)	34(100)	3(100)	233(100)

[a]Number of responses given. Percentage in parentheses.

5. Continuous Abuse of METH

a. Motives for Continuous Abuse of METH Many of the subjects resorted to continuous METH abuse because it stimulated the central nervous system in ways such as providing a pleasant feeling (39.1%), relief from sleepiness (28.3%), and relief from fatigue (17.2%).

b. Persons Who Provide METH The percentage of subjects who cited illegal dealers was similar to the percentage who cited friends as the main source of the drug. From the first abuse of METH to continued abuse of METH, contact with METH-abusing friends and acquaintances seemed necessary.

c. Length of METH Use Only 3.4% of the subjects had been involved in METH abuse for less than 1 yr; 15.4% of the subjects were involved in

TABLE VI Key Person Who Introduced Methamphetamine[a]

	Male	Female	Not stated	Total
Persons in contact with gangs	104(53.1)	20(58.8)	3(100)	127(54.5)
Persons without regular occupations	14(7.1)	0(0.0)	0(0.0)	14(6.0)
Persons related to entertaining	8(4.1)	2(5.9)	0(0.0)	10(4.3)
Persons with occupations unrelated to gangs	6(3.1)	2(5.9)	0(0.0)	8(3.4)
Others	4(2.0)	1(2.9)	0(0.0)	5(2.1)
Not stated	60(30.6)	9(26.5)	0(0.0)	69(29.6)
Total	196(100)	34(100)	3(100)	233(100)

[a]Number of responses given. Percentage in parentheses.

TABLE VII Years of Methamphetamine Use[a]

Years (X)	N	%
X < 1	8	3.4
1 ≤ X < 2	6	2.6
2 ≤ X < 3	11	4.7
3 ≤ X < 4	7	3.0
4 ≤ X < 5	4	1.7
5 ≤ X < 10	66	28.3
10 ≤ X < 15	62	26.6
15 ≤ X < 20	27	11.6
20 ≤ X	27	11.6
Not stated	15	6.4
Total	233	100

[a]Mean years ± SD = 11.08 ± 7.94.

METH abuse for less than 5 yr. In contrast, 78.2% had been chronic METH abusers for more than 5 yr, including the 11.6% chronic abusers who had been using METH for more than 20 years (Table VII).

According to previous investigations, the percentage of individuals abusing METH for less than 1 yr was 37% in 1978, decreasing to 28% in 1982. In 1987, the percentage had declined to 4.1% (Fukui et al., 1989). The sharp decline in the percentage of short-term abusers and the increase in abusers with an abuse history of over 5 yr is noticeable. This trend coincides well with increases in the numbers of abusers in the older age groups (Fukui et al., 1989; Fukui, 1992).

As a result of public disapproval and stringent controls that contributed to a gradual decrease in the number of arrests for METH-related crimes, the number of individuals aged 19 and under that are starting METH abuse has declined; decreases in the number of first-time METH abusers have been noted. However, the number of abusers in their forties and over and of long-term abusers continues to increase. The presence of these long-term abusers may explain why the current second period of METH abuse has not seen the number of abusers decline sharply, as at the end of the first period in 1957.

E. Long-Term Abuse and Methamphetamine Psychosis

To learn the current conditions and tendencies of drug dependence in Japan, a survey was performed between August and September 1991 of drug-dependent in-patients and out-patients of mental hospitals (Fukui, 1992). Questionnaires were sent to 1590 psychiatric hospitals throughout Japan; the response rate was 53%. Of the 938 chemically dependent subjects reported by the hospitals, 331 (35.3%) were METH dependent; 382 (40.7%) were organic solvent dependent; 12 (1.3%) were cannabis dependent; 2 (0.2%) were cocaine dependent; 3 (3.4%) were cough syrup dependent; 65 (6.9%) were hypnotics dependent; 25 (2.7%) were anxiolytics dependent; and 61 (6.5%) were analgesic dependent. Other surveys since 1981 have shown similar results; about 80% of the subjects are METH and organic solvent dependents, reflecting the pattern of drug dependence in Japan (Fukui, 1992). In the 1991 survey, however, the percentage of METH dependents was lower than that of organic solvent dependents for the first time, suggesting the start of a decline in METH abuse.

1. Psychosis in Long-Term METH Abusers

Of the METH-related patients visiting psychiatric hospitals, 255 (77.0%) were in a hallucinatory–delusional state. However, 139 cases (42.0%) had not used METH since their last psychotic episode. These individuals visited the psychiatric hospitals because of a prolongation or re-exacerbation of METH-induced psychosis.

During the epidemic stage in the second period of METH abuse in Japan (1975–1981), most METH psychoses were reported as ameliorated within 1 mo of cessation of METH abuse. However, our results support observations of a type of METH psychosis that persists in excess of 1 mo and sometimes several years, or can be reactivated by insomnia, stress, or the intake of a large amount of alcohol (Konuma, 1984; Wade and Fukui, 1990).

The medical doctors reporting the 139 cases were asked to give greater details of their patients' abstinence period. The exact abstinence periods of 117 patients were obtained from prison, hospital, and doctors' records. Determining the abstinence periods of the remaining 22 cases was not possible. Of the 117 cases studied, 64 were cases of prolongation (54.7%) and 53 were cases of re-exacerbation of psychosis (45.3%).

2. Prolonged Psychosis after Cessation of METH Abuse

The period between the first and last abuse of METH in the 64 cases of prolongation had been less than 5 yr for 9 cases (14.1%), between 5 and 10 yr for 27 cases (42.2%), and longer than 10 yr for 28 cases (43.8%). The long-term abusers (over 5 yr) accounted for 86% of the cases of prolongation. These results suggest that the possibility of prolonged psychotic symptoms increases when the METH abuse period exceeds 5 yr. Since the abstinence periods in 32 cases had been between 2 and 5 yr and those in the other 32 cases had been over 5 yr, we may surmise that psychotic symptoms are prolonged extensively after cessation of METH abuse.

3. Re-exacerbated Psychosis after Cessation of METH Abuse

Among the 53 cases with re-exacerbation, 30 cases (56.3%) reported that symptoms of psychosis had been induced by insomnia or stress, 18 cases (34.0%) reported that symptoms had been induced by the intake of a large amount of alcohol, and 5 cases (9.4%) responded that symptoms had been induced by organic solvent abuse.

Of the same 53 cases, the period between first and last abuse of METH had been less than 5 yr for 6 cases (11.3%), between 5 and 10 yr for 23 cases (43.4%), and longer than 10 yr for 24 cases (45.3%). Again, a noticeably higher percentage had been abusers of METH longer than 5 yr. The abstinence period prior to re-exacerbation had ranged from 5 mo to 20 yr, and averaged 60 mo. Re-exacerbation of psychotic symptoms seems more possible when the period of abuse exceeds 5 yr, and may occur even after a long period of abstinence.

The majority of the prolonged and re-exacerbated cases had abused METH longer than 5 yr. Their core symptoms were hallucinations and delusions, coupled with a lack of initiative, all of which are symptoms similar to those in schizophrenia, suggesting a major impact of long-term METH abuse on mental functions.

VI. CURRENT PROBLEMS IN METHAMPHETAMINE ABUSE

The METH abuse problems faced by society and the medical fields during the epidemic stage between 1975 and 1982 now seem to have faded. However, despite severe public criticism, an awareness movement, and measures that include the strict controls resulting from the 1981 Fukagawa incident, the current METH abuse period shows no sign of rapid decline. The circumstances contributing to the rapid decline in the ebbing stage of the first period have changed in the current period.

A. Social Environment

The postwar prevalence of METH abuse was a result of Japan's defeat and the subsequent social confusion. By 1955, however, as Japan's economic recovery and improving living standards brought the postwar confusion to an end, the desire for the effects of METH dropped among the Japanese people.

In the background of the second METH abuse epidemic is the sustained high economic growth that has created distortions in social structure and life-style. Individuals unable to cope with distortions such as nontraditional values, hedonistic trends, materialistic values, and a drop in social standards may resort to METH abuse.

B. Methamphetamine as a Source of Income for Organized Gangs

In the epidemic stage of the first period, unaffiliated organized criminals supplied METH from illicit laboratories within Japan through a unorganized illicit marketing system, which allowed the blockade by law enforcement authorities to be relatively effective.

Current METH smuggling and sales by organized gangs are well organized, making a blockade by law enforcement officials more difficult. The high profit margin of METH sales now has made it the greatest source of income for some gangs. The resulting continuous supply of a large amount of METH is the biggest factor contributing to the present METH abuse epidemic. However, through international cooperation in investigations and heightened enforcement of controls and punishment, current METH abuse is gradually decreasing.

C. Demographics of Methamphetamine Abusers

In the first period, most METH abusers were in their late teens to early twenties. Since these individuals abused METH as a short-term fad, they were able to stop in response to the stricter controls and public pressure because their dependence on the drug was weak.

The majority of the current abusers are more than 30 years old, indicative of a large group of METH abusers whose dependency has been long-term and is well developed. These chronic abusers, who consume large amounts of the drug, are responsible for the current persistence of METH abuse. The particular life-style of METH abusers also contributes to the problem, since interaction with organized gangs and between abusers results in the starting and continuation of METH abuse. One of the recent consequences of continued abuse has been an increase in cases of prolongation and re-exacerbation of psychosis. As a result, the care of long-term METH abusers also has become necessary.

VII. CONCLUSION

METH abuse in Japan, the first incidence worldwide, prevailed against a background of postwar social confusion. For the two decades since 1970, METH has been the main drug of abuse in Japan and has influenced public peace and mental health. Since the supply of METH by organized criminals to society probably will not abate, the abuse of METH is likely to continue.

If a major change in drug abuse in Japan is going to occur, it will be when cocaine replaces METH. In fact, the rapid increase in the quantities of cocaine seized by law enforcement agencies and in hospital admissions of patients with cocaine-related psychoses suggests an imminent cocaine epidemic. Although the features of and measures against drug abuse in Japan have been mentioned, note that closer international collaboration is necessary to prevent drug abuse now and in the future.

REFERENCES

Bejerot, N. (1969). Social medical classification of addiction. *Int. J. Addict* **4**(3), 391–405.
Fujita, D. (1982). Methamphetamine abuse; crime and punishment. *Res. Papers Jpn. Clin. Policy* **19**, 13–20.
Fukui S. (1988). Epidemiology of methamphetamine abuse. *In* "Kakuseizai (Amphetamines)," (N. Kato, ed.). pp. 107–126. Society for the Study of Information on Dependent Drugs. Kyobunsya, Chiba, Japan.
Fukui, S. (1992). Trend and view on drug dependence in Japan. *Clin. Psychiatry* **34**(8), 815–821.
Fukui, S., Wada, K., and Iyo, M. (1989). Drug dependence, recent clinical tendency. *In* "Current Encyclopedia of Psychiatry," (K. Kakeda, Y. Shimazono, T. Okuma, H. Hozaki and R. Takahashi, eds.). pp. 40–65. Nakayama Shoten, Tokyo.
Hemmi, T. (1982). Sociopathology of methamphetamine abuse. *In* "Seishinigaku MOOK No. 3 Methamphetamine and Volataile Solvent Addiction," (Y. Shimazono, H. Hozaki, and N. Kato, eds.). pp. 20–25. Kanehara Syuppan, Tokyo.
Konuma, K. (1984). Multiphasic clinical types of methamphetamine psychosis and its dependence. *Psych. Neurol. Jap.* **86**, 315–339.
Kuma, R. (1976). "Mayaku (Opiates)," pp. 218–229. Hokeneiyosya, Tokyo.
Ministry of Health and Welfare, Bureau of Health Service, Division of Mental Health (1991). "Mental Health in Japan (Handbook of Mental Health)." Ministry of Health and Welfare, Tokyo.
Ministry of Health and Welfare, Bureau of Pharmaceutical Affairs (1958). "Reports on the Evil of Methamphetamine." Ministry of Health and Welfare, Tokyo.
Ministry of Health and Welfare, Bureau of Pharmaceutical Affairs (1991). "General Condition of Administration of Stimulants and Narcotics in 1990." Ministry of Health and Welfare, Tokyo.
National Police Agency, Division of Prevention of Crime, Department of Countermeasures against Drugs (1977). "Statistics of the Violations against Stimulants and Other Drug Related Laws in 1976." National Police Agency, Tokyo.
National Police Agency, Division of Prevention of Crime, Department of Countermeasures against Drugs (1991). "Statistics of the Violations against Stimulants and Other Drug Related Laws in 1990." National Police Agency, Tokyo.
Noda, H. (1950). Research of methamphetamine addiction. *Psych. Neurol. Jap.* **51**, 7.
Okae, A. (1982). Social background of methamphetamine addicts. *Psych. Neurol. Jap.* **84**, 820–825.
Tatetsu, M., Goto, A., and Fujiwara, T. (1986). "In the Methamphetamine Psychosis." Igaku Syoten, Tokyo.
Wada, K., and Fukui, S. (1990). Relationship between years of methamphetamine use and symptoms of methamphetamine psychosis. *Jpn. J. Alcohol Drug Dep.* **25**(3), 143–158.

Charles R. Schuster
Christine R. Hartel

17
Prospects in Amphetamine Research

The purpose of this chapter is not to review the entire history or scope of amphetamine research, which has been done admirably in the preceding sections of this book, but to look to the future. The discussion here is necessarily selective, not comprehensive, and we apologize to individuals who do not find their favorite topics treated here.

When one considers the future of amphetamine research, the old questions remain but are now couched in terms of new approaches and new techniques. What are the mechanisms mediating stimulation of the central nervous system (CNS) by amphetamine? What are the mechanisms underlying tolerance, sensitization, and dependence? What are the mechanisms underlying the neurotoxic actions of amphetamine and its ability to produce psychosis? What are the interactions of this drug with immune and neuroendocrine functions? What is the role of learning and memory in the phenomenon of craving for amphetamine or other drugs? In discovering and defining these mechanisms, what else can be learned about human behavior and biology? With this knowledge, can we develop new treatment and prevention strategies for substance abuse, mental illness, and neurological and behavioral disorders? The purpose of this chapter is to speculate on the important issues that can be addressed with the new tools at hand.

One of the most important lessons learned from the study of amphetamine is that understanding its actions demands an interdisciplinary approach. Amphetamines interact with a wide variety of chemical and neuroanatomical systems in the brain to affect behavior at all levels of integration—from simple spinal reflexes to complex mood changes including, at high doses, a toxic psychosis that is strikingly similar to schizophrenia. The marvels of molecular biology and genetics hold great promise for revealing the mechanisms of action of pharmacological agents; the actions of amphetamine are illuminated by the increasingly molecular view that is possible with current techniques. However, the molecular biological description of events will address mechanism of action only when the focus of study is drug effects that are mediated by one set of homogeneous receptors. A complete understanding of many of the more interesting actions of amphetamine requires a systems approach to the complex set of interactions within and between the molecular, neurochemical, neuroendocrine, and neurobehavioral levels of analysis. This task is extremely challenging and far more complex than any task addressed to date.

I. MULTIPLE ROLES OF AMPHETAMINE IN NEUROPHARMACOLOGY

As prototypical psychomotor stimulants, amphetamines and its analogs have had manifold uses in research. The preceding chapters of this book described the many actions of amphetamine that can serve as tools and models in experimental and clinical neuropharmacology. All these actions serve as keys to the future of amphetamine research.

In unconditioned behavioral paradigms, amphetamine induces increased locomotor activity and intense stereotypic behavior such as grooming and licking, which can be measured reliably by trained human observers or even mechanically. The startle response to an auditory stimulus is an unconditioned behavior that is potentiated by amphetamine. This potentiation serves as a standard of comparison for the effects of other drugs on the same reflex (Swerdlow *et al.*, 1990). In animals, amphetamine has been used to induce a hyperactive (manic) state that serves as one model of psychosis (Iversen, 1986). Enduring changes in brain and behavior produced by long-term amphetamine administration also have provided models of paranoid schizophrenia (Segal and Schuckit, 1983; Robinson and Becker, 1986). These unconditioned effects offer enormous opportunities for the investigation of the molecular actions of the drug that underlie these effects.

The stimulus functions of amphetamine have been studied intensively.

That amphetamine serves as reinforcer for operant behavior has been well established. Given the opportunity, rats, monkeys, and humans all enthusiastically self-administer amphetamine (Johanson and Schuster, 1981). Rats and monkeys will choose amphetamine over placebo and other drugs as well (Johanson, 1978). Given a choice in an operant paradigm, humans reliably choose amphetamine over placebo, but will not necessarily choose other drugs, such as benzodiazepines, over placebo. The reinforcing effects of amphetamine insure that, although its abuse wanes as its adverse effects become generally known, its use will grow again as a new generation, unheeding the past, discovers the powerful reinforcing and mood-altering effects of this drug.

Drug discrimination studies, experiments using punishment, avoidance, and place preference paradigms, and investigations of schedule-controlled behavior and the phenomenon of intracranial self-stimulation all have depended on the consistent and replicable effects of amphetamine as a baseline and a standard of comparison for the effects of other drugs (Goudie, 1991). In the psychostimulant class of drugs, amphetamine serves as the standard by which tolerance, cross-tolerance, sensitization, and dependence are measured and compared (Corfield-Sumner and Stolerman, 1978). Investigations of these characteristics of amphetamine action are cornerstones of behavioral pharmacology and much of modern neuroscience.

The study of the effects of amphetamine in operant paradigms also has given the field of behavioral pharmacology an important general concept. The effects of amphetamine depend largely on the rates and patterns of responding maintained by the operant schedule before drug is given (Dews and DeWeese, 1977). This phenomenon, known as rate dependency, was observed first in amphetamine experiments, but soon became obvious for other drug classes as well. This concept is crucial in correctly interpreting the results of many types of drug-related experiments.

Amphetamine and its analogs have been extremely useful as pharmacological probes for investigating the role of monoamine systems in specific behaviors. For example, the locomotor activation seen in animals after doses of amphetamine is related to the mesocorticolimbic release of dopamine (Koob, 1992). The fact that haloperidol, a dopamine antagonist, blocks amphetamine-induced locomotor activation provides evidence supporting the role of dopamine in mediating this effect of amphetamine (LeMoal and Simon, 1991). Higher doses of amphetamine are neurotoxic, causing irreversible degeneration of dopaminergic neurons; this effect may be the result of the formation of 6-hydroxydopamine, a selective neurotoxin (Ricaurte *et al.*, 1982). These lesions are discrete and specific to various neuroanatomic pathways; as a tool, amphetamine has helped investigators explore these pathways, facilitating identification of the biological mechanisms of stereotypy and locomotor behavior.

Ritz and Kuhar (1989) described another example of the importance of amphetamine as a pharmacological probe in investigating the neurochemical mechanisms underlying the reinforcing effects of drugs. These researchers found that *d*-amphetamine has an affinity for dopamine, norepinephrine, and serotonin uptake transporter binding sites, but that the potency of the reinforcing effects of various phenethylamines was not correlated with their potency as inhibitors of ligand binding to any of the monoamine uptake sites. This result contrasts with results for cocaine and its analogs, which have a high correlation between their potency as reinforcers and their binding to the dopamine uptake transporter, but not their binding with the norepinephrine or serotonin uptake transporters. However, the reinforcing potency of amphetamine and its analogs appears to be related inversely to their potency as serotonin uptake inhibitors, suggesting that activation of serotonin systems decreases the positive reinforcing effects of these drugs.

A series of individual drugs can be imagined, each affecting only one receptor subtype; by adding drug after drug, the effects of amphetamine could be "rebuilt" from a set of "single action" drugs. For example, apomorphine could be used and serotonin agonists selective for various receptor subtypes could be added to "recreate" the serotonergic effects of amphetamine. The results of drug discrimination studies on such a drug series would be fascinating. Then selective agonists from other neurotransmitter systems could be added to study the array of neurochemical interactions that must exist.

II. FUTURE OF AMPHETAMINE RESEARCH

The study of the pharmacology of the amphetamines, which are indirectly acting sympathomimetic drugs, may be conceptualized as the study of the roles of the various neurotransmitter systems with which the drugs interact to produce their effects. In a very real sense, an understanding of the effects of amphetamines is a key to understanding the roles of these neurotransmitter systems in the control of physiology, mood, and behavior. Thus, the study of amphetamines is, in reality, the study of the brain and behavior. In a number of interesting areas in basic neurobehavioral research and clinical psychopharmacology, the study of the actions of amphetamine will be of great importance. The following sections reflect a few of these areas of study.

A. Individual Differences in Response to Amphetamine

Although most individuals of various species reliably self-administer amphetamine, they do so at widely varying base rates. In addition, investigators long have been puzzled by the occasional rat or monkey that cannot be induced to take the drug at all. Differential human drug preferences are also striking. These parametric differences in drug preference seem likely to be the results of complex biological and behavioral characteristics, determined by multiple genes and behavioral histories. Much still can be learned about correlations between underlying biochemical differences and differences in behavior, whether that behavior is drug taking itself or the behavioral effects of the drug taking. Naturally, researchers hope to find that the variance in the effects of amphetamines at the level of physiology or behavior can be correlated with variance in the molecular effects. When a high correlation is found, the effects can be presumed to be related. Perhaps someday investigators will be able to infer causality. Until then, the investigation of individual differences in drug response already is beginning to reveal some important interactions of biology and behavior.

1. Genetics: Baseline Neurochemistry and Molecular Biology

Studies in genetics are providing clues to the phenomena of individual differences and to our understanding of how drugs act in the brain. Immediate-early genes are genes that are activated first in response to an external stimulus, such as amphetamine. Cocaine and amphetamine have been found to activate immediate-early genes in discrete populations of neurons in the CNS (Graybiel *et al.*, 1990; Young *et al.*, 1991). These immediate-early genes in turn regulate the expression of specific target genes, thereby influencing the transcription of proteins from genetic instructions. In the brain, immediate-early genes now are thought to play a key role in mediating the drug-induced long-term changes that are suspected to be related to the development of tolerance and dependence.

Evidence is accumulating that some individuals may have a genetic predisposition to drug use and, further, to move from experimentation to dependence. Data are still insufficient to determine whether such a predisposition has its basis in a genetically determined biological substrate for drug action (e.g., increased numbers of receptors), in genetically determined individual differences in behavior (e.g., increased levels of exploratory behavior), or both. The methodological issues are complex (Svikis and Pickens, 1991), but studies in twins (Pickens *et al.*, 1991) and molecular genetics studies (Smith *et al.*, 1992; Uhl *et al.*, 1992) are beginning to supply the facts needed to answer these questions.

Various strains of rats and mice have demonstrated differential preferences for amphetamine, cocaine, and other drugs of abuse (Elmer *et al.*, 1990; George *et al.*, 1991). These strain variations can be intensified by careful breeding. Thus, observations at the behavioral level suggest that analysis at the molecular genetic level will be useful, but analysis of genetic factors alone to is unlikely to provide a sufficient explanation of the behavioral differences between individuals of the same strain. Correlations of disparities between individuals at different levels of analysis—the genetic, the neurochemical, and the behavioral—will be required to explain these differences.

2. Sex-Based Differences

Only recently have the differences between men and women in their patterns of drug use been demonstrated to have a physical, as well as a social, basis. Beyond questions of sex-specific differences in drug metabolism and disposition, the differential sensitivity of male and female rats to the effects of amphetamine on the catecholaminergic systems are ripe for investigation. Some of these differences seem to be related to drug effects on the endocrine system (Becker and Cha, 1989) and others to effects on the cholinergic and GABA-ergic systems. Some sex-based differences apparently occur because males and females differ in initial levels of hormones and/or neurotransmitters, but others occur even when initial levels are the same. The investigation of sensitization to drug effects as a function of sex and gonadal hormonal status is just beginning (Haney et al., 1992), promising many new insights into sex-based differences in response to stimulant drugs.

3. Behavioral History

Behavioral history modifies the effects of the amphetamines. For example, the usual effects of *d*-amphetamine on behavior suppressed by electric shock (rate decreases) can be reversed after exposure to a condition in which responding postpones shock (Barrett, 1977). Other experiments using punishment and avoidance paradigms have confirmed and extended the finding that amphetamine and other drugs can increase or decrease responding on the same schedule of reinforcement, depending on the previous experience of the animal with other schedules of reinforcement (Barrett and Witkin, 1986).

Environmental factors also play a role in the physical response to drugs. For example, an animal pressing a lever to obtain intravenous injections of cocaine will remain alert and relatively healthy, whereas a yoked control, passively receiving exactly the same amount of drug, will show extreme toxicity and even death (Dworkin et al., 1992).

Behavioral history is also an important factor in determining changes in the effects of drugs after repeated administration, which include both tolerance and sensitization. For example, classical conditioning techniques can be used to evoke drug tolerance (Siegel, 1989). Altering drug effects using classical conditioning may be possible as well. For example, what will happen if the yoked control in the experiment just described is given a signal that a drug infusion is going to occur? Will the drug effects be less debilitating than they are without such a signal? Sensitization to the effects of repeated doses of amphetamine long has been known, but evidence is accumulating that both classical (Drew and Glick, 1988) and operant (Valencia-Flores et al., 1990) conditioning play a role in this phenomenon as well. Thus, the behavioral history of the animal turns out to be a critical variable in a condition once thought to be grounded solely in physiology.

An extraordinary set of studies revealed another layer of complexity in individual response to amphetamine and suggests many new avenues of exploration. Piazza and colleagues (1991a) examined individual variability, not strain differences, in amphetamine response. These researchers found that rats that exhibit high locomotor activity in response to a stress situation (a novel environment) are more likely to learn to self-administer amphetamine than are their lower responding counterparts. The high responder rats exhibited longer duration of corticosterone secretion after exposure to the novel situation; corticosterone also facilitated the acquisition of amphetamine self-administration in low responders. In a second experiment (Piazza *et al.*, 1991b), the high responder rats were found to display a specific neurochemical pattern: a higher ratio of 3,4-dihydroxyphenylacetic acid to dopamine in the nucleus accumbens and striatum and a lower ratio in the prefrontal cortex. These rats also had lower overall serotonin and 5-hydroxyindoleacetic acid levels. The correlations of differences in brain dopaminergic activity and corticosteroid level and the ability to develop amphetamine self-administration are important steps in clarifying individual differences in the stress-related role of the hypothalamic–pituitary–adrenal (HPA) axis in drug self-administration. To complicate the issue further, these interactions of the nervous and endocrine systems also are affected by their interactions with the immune system (see subsequent discussion). The complexity of the situation and the need for further study cannot be overestimated.

B. Effects on Monoamines, Hormones, and the Immune System

Our rapidly increasing knowledge of the nervous, endocrine, and immune systems has demonstrated that these three systems interact in complex and highly integrated ways. Because drug abuse and the AIDS epidemic are tied together so closely, the investigation of the direct effects of drugs of abuse, including amphetamine, on the immune system is already underway. For example, cocaine (Peterson *et al,*, 1992) and opiates (Peterson *et al.*, 1991) have been found to augment the growth of HIV in cultured lymphocytes, although whether this effect has any clinical significancet remains to be determined.

Connections also have been established between amphetamine, amphetamine-induced phenomena, and the HPA axis. As described earlier, Piazza and colleagues (1991a) determined that corticosterone levels correlate with individual vulnerability to amphetamine self-administration. Evidence also suggests that the corticosteroids play a role in amphetamine-induced sensitization. Cole *et al.* (1990) demonstrated that rats pretreated with a corticotropin-releasing factor antiserum showed a significant attenuation of the development of amphetamine-induced behavioral sensitization. Under circumstances of chronic social stress, corticosterone secretion is known to be increased for prolonged periods, leading to adrenal enlargement and thymus involution. The corticosteroids also have immunosuppressive effects, such as decreasing natural killer cell activity and splenocyte reactivity to mitogen. Although Klein et al. (1992) showed that the two phenomena do not always occur together, the consequences of perturbing either system with amphetamine or other drugs that alter the neurochemistry of the brain remain to be determined. Work on the complex interrelationship of the HPA axis, the immune system, and neurochemistry is only beginning. Amphetamine and its analogs undoubtedly will be important tools in examining the dynamic balance among the three systems.

C. Amphetamine and Age-Related Changes in Brain

With age, a decrease occurs in the number of dopamine cells (Hornykiewicz, 1989). Amphetamine and its analogs are still the only drugs of abuse known to cause irreversible neurotoxicity. These facts have grave implications for the many drug users of the late 1960s and 1970s, who have just reached early middle age. If, at an early age, these drug users decreased the numbers of their own striatal dopaminergic neurons by taking amphetamines and, as they age, they lose even more dopaminergic cells, the cumulative cell loss may be tragic. One would expect to find a higher incidence of parkinsonian symptoms in such a population; epidemiological studies are being undertaken to examine this possibility. Success with transplants of dopamine-producing fetal tissue to rats lesioned with 6-hydroxydopamine (Savasta et al., 1992) suggests hope for the treatment of amphetamine neurotoxicity. The aging population of casual drug abusers also may present clinicians with a new type of mental disorder, one rooted in occasional but long-term drug use that has resulted in chronic dopamine or serotonin deprivation with profound effects on mood, aggression, and sleep.

D. Neuroscience and Behavior

1. Mechanisms of Tolerance and Dependence

The development of powerful new tools for the study of brain function is making possible the study of the subcellular, cellular, and behavioral mechanisms responsible for drug tolerance and dependence. In freely moving animals, *in vivo* microdialysis (Kuczenski and Segal, 1990; Tepper et al., 1991) allows simultaneous measurement of neurochemistry and behavior and a closer study of the interactions of the catecholamines and their respective receptors. *In vivo* electrochemistry (Louilot et al., 1989) allows the real time measurement of certain brain neurochemical events in freely moving animals.

In humans, the use of positron emission tomography (PET) and magnetic resonance imaging (MRI) have made possible studies hitherto unimaginable. Researchers can now correlate mood, behavior, and brain activity. Using PET imaging, comparisons already have been made of glucose utilization in the brains of people actually in the process of receiving cocaine and in the brains of addicts who are 6 mo abstinent but are looking at drug paraphernalia. Glucose utilization by both groups is remarkably similar. Many comparative imaging studies must be carried out in the field of drug abuse. Comparisons can be made at various stages of treatment, between people who prefer one drug to another or no drug at all, between those who might be thought to have a genetic vulnerability to drug abuse and those who do not, and so on. Another exciting frontier in imaging is the co-registration of PET and functional MRI images. The enhanced resolution of these images, which blend activity, structure, and function, gives information that enlarges our understanding and stimulates our scientific imaginations.

2. Amphetamines as Anorectic Agents

In a flurry of concern over the exposure of American troops in Somalia to khat, a natural amphetamine analog, the question arose in many minds about the role of that drug in alleviating hunger, a common and useful property in stimulant drugs. We know that food deprivation increases drug self-administration. What is the role of hunger in stimulant abuse? How do the complex hormonal and neurochemical precedents and consequences of amphetamine use and the mechanisms of hunger and satiety interact? Exactly how are the ring-alkylated amphetamines different in their anorectic activity?

The specificity of anorectic effects of analogs such as fenfluramine has not been detailed. For example, although tolerance does not develop to the anorectic effect of fenfluramine, some factor limits the decrease of body weight to about 20%. No reports of fenfluramine abuse have been made, and people do not self-administer the drug under the same experimental conditions that bring about amphetamine self-administration (Johanson and Uhlenhuth, 1982). However, the dependence potential of amphetamine is so high that its role in weight control is necessarily limited. Both amphetamine and fenfluramine may cause sleep disturbances and, in the case of fenfluramine, depression (Hoffman and Lefkowitz, 1990). Since weight control remains a desirable goal for so many people, the role of research on the actions of amphetamine compounds is assured. Clearly, anorectic agents are needed with fewer side effects and with little abuse or dependence potential.

3. Treating Drug Seeking Behavior: Stimulus Effects and Drug Combinations

Our greater understanding of the interactions of behavioral history, stimulus effects, and drug effects will facilitate the development of new approaches to drug abuse treatment and prevention. Amphetamines and other drugs of abuse exert profound stimulus control over behavior; in turn, their usage itself comes under stimulus control. Relapse to drug abuse frequently is associated with this phenomenon, prevention of relapse can occur only when the drug-associated stimuli lose their stimulus properties. The role of learning and memory in this area is still very unclear anatomically, neurochemically, and behaviorally. Some of these phenomena already have been explored in the field of behavioral pharmacology, but much work remains to be done, especially in connection with the brain structures that have become visible *in vivo* only recently with new imaging techniques.

Constrained by limitations in interpreting drug interaction effects, researchers have tried to study drugs of abuse in isolation from each other. As statistical and pharmacological sophistication increases, investigators will be able to study and understand amphetamine abuse as it actually occurs in humans: in combination with alcohol, nicotine, benzodiazepines, and other drugs.

A fuller understanding of the effects of amphetamine also will require an examination of the use of the drug in various social circumstances. Are drug effects under conditions of stress different from those obtained under more relaxed conditions? Foltin and Fischman (1990) showed that the cardiovascular effects of cocaine are potentiated when people do simple performance tests. Thus, even the mildest stress affects the actions of cocaine. The toxicity of amphetamine in rats long has been known to be exacerbated by housing the animals in group cages. In humans, does social use of amphetamine have behavioral and physical effects that differ from those seen when the drug is used in isolation?

III. SUMMARY

Amphetamine first was synthesized in the early 1930s. Since then, researchers have learned a great deal about the drug, its effects in and on the central nervous system, and the mechanisms underlying these actions. The actions of amphetamine occur at all levels—from the behavioral to the physiological to the molecular. At each level, the interaction of amphetamine with the variables controlling behavior, physiology, or molecular activity explains its actions. The complexity of these actions is best described in terms of the operation of complex systems, which are described in ever greater detail as new methods become available. The classical questions about amphetamine remain as investigators search for new answers in new terms. The questions all have been asked before, and some have been partially answered, but as research moves into new realms of study, the old questions become new again and the future of amphetamine reteach becomes the future of neuroscience, behavior, and systems research.

REFERENCES

Barrett, J. E. (1977). Behavioral history as a determinant of the effects of d-amphetamine on punished behavior. *Science* **198**, 67–69.
Barrett, J. E., and Witkin, J. M. (1986). The role of behavioral and pharmacological history in determining the effects of abused drugs. *In* "Behavioral Approaches to Drug Dependence" (S. R. Goldberg and I. P. Stolerman, eds.), pp. 195–223. Academic Press, New York.
Becker, J. B., and Cha, J. H. (1989). Estrous cycle-dependent variation in amphetamine-induced behaviors and striatal dopamine release assessed with microdialysis. *Behav. Brain Res.* **35**, 117–125.
Cole, B. J., Cador, M., Stinus, L., Rivier, C., Rivier, J., Vale, W., Le Moal, M., and Koob, G. F. (1990). Critical role of the hypothalamic pituitary adrenal axis in amphetamine-induced sensitization of behavior. *Life Sci.* **47**, 1715–1720.
Corfield-Sumner, P. K., and Stolerman, I. P. (1978). Behavioural tolerance. *In* "Contemporary Research in Behavioral Pharmacology," (D. E. Blackman and D. J. Sanger, eds.) pp. 391–448. Plenum Press, New York.
Dews, P. B., and DeWeese, J. (1977). Schedules of reinforcement. *In* "Handbook of Psychopharmacology" (L. L. Iversen, S. D. Iversen, and S. H. Snyder, eds.), Vol. 7, pp. 107–150. Plenum Press, New York.
Drew, K. L., and Glick, S. D. (1988). Environment-dependent sensitization to amphetamine-induced circling behavior. *Pharmacol. Biochem. Behav.* **31**, 705–708.
Dworkin, S., Porrino, L. J., and Smith, J. (1992). The importance of behavioral controls in the analysis of on-going events. *NIDA Res. Monogr.* **124**, 173–188.
Elmer, G. I., Meisch, R. A., Goldberg, S. R., and George, F. R. (1990). Ethanol self-administration in long sleep and short sleep mice indicates reinforcement is not inversely related to neurosensitivity. *J. Pharm. Pharmacol.* **254**, 1054.
Foltin, R. W., and Fischman, M. W. (1990). The effects of combinations of intranasal cocaine, smoked marijuana, and task performance on heart rate and blood pressure. *Pharmacol. Biochem. Behav.* **36**, 311–315.
George, F. R., Ritz, M. C., and Elmer, G. I. (1991). Role of genetics in vulnerability to drug dependence. *In* "The Biological Bases of Drug Tolerance and Dependence" (J. A. Pratt, ed.), pp. 265–290. Academic Press, London.
Goudie, A. J. (1991). Animal models of drug abuse and dependence. *In* "Behavioural Models in Psychopharmacology" (P. Willner, ed.), pp. 453–484. Cambridge University Press, Cambridge.
Graybiel, A. M., Moratalla, R., and Robertson, H. A. (1990). Amphetamine and cocaine induce drug-specific activation of the c-fos gene in striosomematrix and limbic subdivisions of the striatum. *Proc. Nat. Acad. Sci. U.S.A.* **87**, 6912–6916.
Haney, M., Castanon, N., Cador, M., Le Moal, M., and Mormede, P. (1992). Cocaine sensitization in Roman high and low avoidance rats is modulated by sex and gonadal hormone status. *Soc. Neurosci. Abstr.* **18**, 543.
Hoffman, B., and Lefkowitz, R. (1990). Catecholamines and sympathomimetic drugs. *In* Goodman and Gilman's "The Pharmacological Basis of Therapeutics" (A. G. Gilman, T. W. Rall, A. S. Nies, and P. Taylor, eds.), 8th Ed., pp. 187–220. Pergamon Press, New York.
Hornykiewicz, O. (1989). Aging and neurotoxins are causative factors in idiopathic Parkinson's disease: A critical analysis of the neurochemical evidence. *Prog. Neuropsychopharmacol. Biol. Psychiatry* **13**, 319–328.
Iversen, S. D. (1986). Animal models of schizophrenia. *In* "The Psychopharmacology and Treatment of Schizophrenia" (P. B. Bradley and S. R. Hirsch, eds.), pp. 71–102. Oxford University Press, Oxford.
Johanson, C. E. (1978). Drugs as reinforcers. *In* "Contemporary Research in Behavioral Pharmacology" (D. E. Blackman and D. J. Sanger, eds.), pp. 325–390. Plenum Press, New York.

Johanson, C. E., and Schuster, C. R. (1981). Animal models of drug self-administration. In "Advances in Substance Abuse: Behavioral and Biological Research" (N. K. Mello, ed.), Vol. 2, pp. 219-297. JAI Press, Greenwich, Connecticut

Johanson, C. E., and Uhlenhuth, E. H. (1982). Drug preferences in humans. Fed. Proc. 41, 228-233.

Klein, F., Lemaire, V., Sandi, C., Vitiello, S., Van der Logt, J., Laurent, P. E., Neveu, P., Le Moal, M., and Mormede, P. (1992). Prolonged increase of corticosterone secretion by chronic social stress does not necessarily impair immune functions. Life Sci. 50, 723-31.

Koob, G. (1992). Drugs of abuse: Anatomy, pharmacology and function of reward pathways. Trends Pharmacol. Sci. 13, 177-184.

Kuczenski, R., and Segal, D. S. (1990). In vivo measures of monoamines during amphetamine-induced behaviors in rats. Prog. Neuropsychopharmacol. Biol. Psychiatry 14, S37-50.

Le Moal, M., and Simon, H. (1991). Mescocorticolimbic dopaminergic network: Functional and regulatory role. Physiol. Rev. 71, 155-234.

Louilot, A., Le Moal, M., and Simon, H. (1989). Opposite influences of dopaminergic pathways to the prefrontal cortex or the septum on the dopaminergic transmission in the nucleus accumbens. An in vivo voltammetric study. Neuroscience 29, 45-56.

Peterson, P. K., Sharp, B. M., Gekker, G., Jackson, B., and Balfour, H. H., Jr. (1991). Opiates, human peripheral blood mononuclear cells, and HIV. Adv. Exp. Med. Biol. 288, 171-178.

Peterson, P. K., Gekker, G., Chao, C. C., Schut, R., Verhoef, J., Edelman, C. K., Erice, A., and Balfour, H. H., Jr. (1992). Cocaine amplifies HIV-1 replication in cytomegalovirus-stimulated peripheral blood mononuclear cell cocultures. J. Immunol. 149, 676-680.

Piazza, P. V., Maccari, S., Deminiere, J. M., Le Moal, M. Mormede, P., and Simon, H. (1991a). Corticosterone levels determine individual vulnerability to amphetamine self-administration. Proc. Nat. Acad. Sci. U.S.A. 88, 2088-2092.

Piazza, P. V., Rouge-Pont, F., Deminiere, J. M., Kharoubi, M., Le Moal, M., and Simon, H. (1991b). Dopaminergic activity is reduced in the prefrontal cortex and increased in the nucleus accumbens of rats predisposed to develop amphetamine self-administration. Brain Res. 567, 169-174.

Pickens, R. W., Svikis, D. S., McGue, M. Lykken, D. T., Heston, L. L., and Clayton, P. J. (1991). Heterogeneity in the inheritance of alcoholism. A study of male and female twins. Arch. Gen. Psychiatry 48, 19-28.

Ricaurte, G. A., Guillery, R. W., Seiden, L. S., Schuster, C. R., and Moore, R. Y. (1982). Dopamine nerve terminal degeneration produced by high doses of methylamphetamine in the rat brain. Brain Res. 235, 93-103.

Ritz, M. C., and Kuhar, M. J. (1989). Relationship between self-administration of amphetamine and monoamine receptors in brain: Comparison with cocaine. J. Pharmacol. Exp. Ther. 248, 1010-1017.

Robinson, T. E., and Becker, J. B. (1986). Enduring changes in brain and behavior produced by chronic amphetamine administration: A review and evaluation of animal models of amphetamine psychosis. Brain Res. 396, 157-198.

Savasta, M., Mennicken, F., Chritin, M., Abrous, D. N., Feuerstein, C., Le Moal, M., and Herman, J. J. (1992). Intrastriatal dopamine-rich implants reverse the changes in dopamine D_2 receptor densities caused by 6-hydroxydopamine lesion of the nigrostriatal pathway in rats. An autoradiographic study. Neurosicence 46, 729-738.

Segal, D. S., and Schuckit, M. A. (1983). Animals models of stimulant-induced psychosis. In "Stimulants: Neurochemical, Behavioral, and Clinical Perspectives" (I. Creese, ed.), pp. 131-167. Raven Press, New York.

Siegel, S. (1989). Classical conditioning and habituation processes in drug tolerance and sensitization. In "Psychoactive Drugs: Tolerance and Sensitization" (A. J. Goudie and M. W. Oglesby, eds.). pp. 115-180. Humana Press, Clifton, New Jersey.

Smith, S. S., O'Hara, B. F., Persico, A. M., Gorelick, D. A., Newlin, D. B., Vlahov, D., Solomon,

L., Pickens, R. W., and Uhl, G. R. (1992). Genetic vulnerability to drug abuse. The D_2 dopamine receptor *Taq* I B1 restriction fragment length polymorphism appears more frequently in polysubstance abusers. *Arch. Gen. Psychiatry* **49**, 723–727.

Svikis, D. S., and Pickens, R. W. (1991). Methodological issues in genetic studies of human substance abuse. *J. Addict. Disorders* **10**, 215–228.

Swerdlow, N. R., Mansbach, R. S., Geyer, M. A., Pulvirenti, L., Koob, G. F., and Braff, D. L. (1990). Amphetamine disruption of prepulse inhibition of acoustic startle is reversed by depletion of mesolimbic dopamine. *Psychopharmacology (Berlin)* **100**, 413–416.

Tepper, J. M., Creese, I., and Schwartz, D. H. (1991). Stimulus-evoked changes in neostriatal dopamine levels in awake and anesthetized rats as measured by microdialysis. *Brain Res.* **559**, 283–292.

Young, S. T., Porrino, L. J., and Iadarola, M. J. (1991). Cocaine induces striatal nonreactive proteins via dopaminergic D_1 receptors. *Proc. Nat. Acad. Sci. U.S.A.* **88**, 1281–1295.

Uhl, G. R., Persico, A. M., and Smith, S. S. (1992). Current excitement with D_2 receptor gene alleles in substance abuse. *Arch. Gen. Psychiatry* **49**, 157–160.

Valencia-Flores, M., Velazquez-Martinez, D. N., and Villarreal, J. E. (1990). Superreactivity to amphetamine toxicity induced by schedule of reinforcement. *Psychopharmacology (Berlin)* **102**, 136–44.

Index

A
γ-Aminobutyric acid
  neurotoxicity role, 287
  release, amphetamine effects, 99
2-Aminoindan, potency, 7–8
2-Aminotetralin, potency, 8
Abuse, amphetamine
  mechanisms, 260–261
  self-administration behavior
    amygdaloid complex role, 248, 252, 259, 260
    dopamine antagonist effect, 245–246
    dopamine role, 244–248, 260
    medial prefrontal cortex role, 248, 252

  microdialysis study, 249–250
  noradrenalin role, 254–255
  nucleus accumbens role, 247–248, 252, 258, 259, 260
  reinforcing effect, 481
  serotonin role, 252–254
Acetylcholine, release, amphetamine effects, 101–102, 137, 138–139
Amphetamines
  aromatic ring modification, 17–21, 24–29
  benzylic carbon oxidation, 6–7
  classification, 198–199
  central nervous system effects, 5, 81, 243
  dealkylation, 46

493

Amphetamines (continued)
  deamination, 51–53
  dopamine axon structure changes with use, 333–335, 344
  glucoronidation, 48
  glutamate release inhibition, 286–287
  halogenation, 18–19, 21, 152
  hydroxylation, 46–47, 50–51
  individual variance of effects, 122–126, 390
  long-term effects, 230–232, 243, 297
  mechanism of action, 9, 21, 33, 82, 86, 103–104, 177
  medical use, 421–422, 441
  metabolism, *in vivo* studies, 49
  methylation, 5–6
  monoamine uptake transporter binding, 482
  N substitution modification, 6, 14–15, 22–24
  pharmacokinetics, 49–50, 69
  pK values, 44
  prevalence of abuse, 396
  psychosis, *see* Psychosis, amphetamine
  Schiff base formation, 48–49
  sensitivity factors
    genetic, 390–391
    sensitization, 392–395
    tolerance, 391–392
  side chain modification, 7–8, 15–17, 29–33
  stereochemistry, 5
  structure, 3, 44
  sulfation, 48
(+)-Amphetamine sulfate, potency, 7–8
Amygdaloid complex, role in self-administration behavior, 248, 252, 259, 260
Anorectic agents, *see* Fenfluramine
Apomorphine, effects, amphetamine inhibition, 131
Ascorbic acid, release, amphetamine effects, 102

B
Behavior, *see also* Abuse
  amphetamine dosage parameter responses
    acute, 117
    chronic, 117–121
  context-dependent conditioning, 122
  dopamine role, 298–299
  effect of duration of amphetamine use, 307
  individual differences in responses to amphetamines, 122–126
  locomotor activity, amphetamine effects, 480
  perseveration, 255–258
  poststimulation
    depression, 121, 140
    withdrawal syndrome, 116, 121, 140
  rewarding behavior of amphetamines, 115
  sensitization to amphetamines, 116, 120, 125, 131, 134, 249–250, 392–395
  serotonin role, 299, 300
  species differences in amphetamine response, 308
  startle habituation, 191–195, 300, 480
  stereotypy rating, 126–127, 255, 480
  stimulus effects, 487–488
  stress reactivity, amphetamine effects, 121, 125
Behavioral Pattern Monitor, in study of effects of amphetamines, 180–185, 190
Benzoic acid, amphetamine catabolic product, 52
Bipolar disorder, amphetamine effects, 396
Blood–brain barrier
  amphetamine destruction, 316
  permeability
    p-chloroamphetamine, 214
    methamphetamine, 425

Brain
  aging effects, 486
  structure, amphetamine effects
    dopaminergic system
      axons, 333–334, 344
      uptake sites, 335, 337
    evaluation, 315–318, 340, 341
    serotonergic system
      axons, 324–329, 330–333, 337–339, 341–344
      cell bodies, 329–330
      uptake sites, 335–337

C
Calcium
  amphetamine neurotoxicity role, 228
  dopamine releasing effect, 94, 96–97
Cathinone
  activity, 67
  metabolism, 48, 67–68
  potency, 6–7
  structure, 67
Chlorimipramine, serotonin uptake inhibitor, 224
p-Chloroamphetamine
  behavioral effects, 192, 193, 194, 195, 210
  blood–brain barrier permeability, 214
  effects on neurotransmitter release
    dopamine, 6, 16, 210, 211, 230
    norepinephrine, 210, 211
    serotonin, 9–10, 14, 16–18, 158, 210–213, 224, 227, 230, 232, 269, 272
  glial cell changes with use, 347
  α-ethyl homolog, 16
  history of use, 209
  long-term effects, 229–232
  metabolism, 62–63, 221–222
  monoamine oxidase inhibition, 210
  neurotoxicity
    behavioral effects, 300–301, 304–305, 380
    calcium role, 228
    dopamine role, 229
    history of study, 269–270
    lesion formation, 213–214
    metabolites of parent compound, 221–222
    ontogeny of susceptibility, 215–216
    phases, 355–356
    prevention
      L-cysteine, 225–226
      serotonin antagonists, 226–227
      serotonin uptake inhibitors, 222–225
    profile, 299–300, 308
    regional specificity, 214–215
    reversibility, 216–217
    serotonergic system effects
      axons, 325, 327–328, 332, 338–339, 341–343, 350–351
      cell bodies, 329, 330
    species susceptibility, 217
    stereoselective effects, 217
  pharmacokinetics, 69, 161
  pK value, 220
  recreational use, 380
  structure, 218
  substitution effects on serotonin depletion
    ring, 218–219
    side-chain, 219–220
p-Chlorophenyl-alanine, tryptophan hydroxylase inhibition, 352
Chlorpromazine, dopamine antagonist, 245
Choline acetyltransferase, amphetamine effects, 273
Clorgyline, monoamine oxidase inhibition, 95, 133
Cocaine
  history, 389–390
  HIV replication response, 485
  individual response variation, 390
  Japanese epidemic, 477
  self-administration behavior, 253, 254, 389
  sensitization, 395
  stress effects on response, 488

Conditioned place preference
  amphetamine induction, 244, 250–252, 255, 259, 260
  effectiveness in studies, 258
Corticosteroids, role in amphetamine sensitization, 485–486
L-Cysteine, protection against amphetamine neurotoxicity, 225–226
Cytochrome P450 system
  isozymes, 45, 52–53, 55
  metabolism of amphetamines, 44–45, 50, 52–55, 60–62, 67, 68
  organization, 44–45

D

Deprenyl
  activity, 65–66
  disposition, 66
  metabolism, 66–67
  monoamine oxidase inhibition, 65–66
Dexfenfluramine, see Fenfluramine
Dihydroxyphenyl acetic acid
  formation from dopamine, 89
  levels, amphetamine effects, 93–94, 133, 159, 169
Dihydroxytryptamine, neurotoxicity, 214–217, 221, 227, 252, 279, 336
2,5-Dimethoxyamphetamine, activity, 153
3,4-Dimethoxyamphetamine, activity, 28
2,5-Dimethoxy-4-bromoamphetamine
  potency, 377–378
  recreational use, 377–378
2,5-Dimethoxy-4-bromophenyliso-propylamine, activity, 151, 152
2,5-Dimethoxy-4-iodophenyliso-propylamine, locomotion effects, 183, 185
2,5-Dimethoxy-4-methylamphetamine, recreational use, 377
2,6-Dimethoxy-4-methylamphetamine, activity, 29, 34
2,5-Dimethoxy-4-methylphenyliso-propylamine, activity, 151, 152

Dopamine
  acute response to amphetamines, 127–131
  axons containing
    morphology, 323
    projection origin, 322–323
    terminal distribution in forebrain, 321–322
  mediation of amphetamine-induced conditioned place preference, 250–252
  metabolites
    dihydroxyphenyl acetic acid, 89
    homovanillic acid, 89
    3-methoxytyramine, 89–90
    neurotoxicity, 279
  mood effects, 298
  motor function effects, 298
  release
    amphetamine effects, 4–5, 9, 16, 83, 86, 88–89, 91–92, 97–99, 103–104, 127–137, 163–164, 249, 306
    exchange–diffusion model, 194–97
    Na^+, K^+-ATPase role, 85
    regional characteristics, 97–99
    response to repeated amphetamine administration, 131–137
  role in amphetamine neurotoxicity, 229, 275–277
  role in amphetamine self-administration, 244–250
  role in Parkinson's disease, 299
  role in schizophrenia, 299
  serotonergic system effects, 277–278
  synthesis, 88, 90
Dopamine β-hydroxylase, 47
Dopamine receptors
  amphetamine interaction sites
    monoamine oxidase, 87–88
    uptake carrier, 83–86, 133–134
    vesicle, 86–87
  autoreceptor sensitivity, 132
  postsynaptic receptor in sensitization, 135, 137
Drug Abuse Warning Network
  description, 446, 447, 449

limitations, 449
target population, 446

E
Ecstasy, see 3,4-Methylenedioxymethamphetamine
Ephedrine
  metabolism, 57–59
  pharmacokinetics, 57, 69
Epidemics, drug abuse
  drug supply
    clandestine manufacture, 444, 445
    diversion of pharmaceutical stores, 443–444
    legal availability, 441–442, 444, 445
    look-alike speed, 444–445
  epidemiological triangle, 439–440
  factors influencing, 440, 454
  Hawaiian ice outbreak, 452–454
  history
    drug development, 441
    population inoculation, 441–442
    post-1960, 443
    World War II, 442–443
  information sources, 445–446
  Japan, 415, 416–417, 422, 434, 439, 459–477
  United States, 441–443, 445–446, 454
Ethylenedioxymethamphetamine, potency, 20
Ethylidenedioxyamphetamine, potency, 19–20
N-Ethyl-3,4-methylenedioxyamphetamine, neurotoxicity, 283
Eve, see 3,4-Methylenedioxy-N-ethylamphetamine

F
Fencamfamine, dopamine uptake blocker, 130
Fenfluramine
  activity, 10, 15, 20–21, 63, 158, 166
  appetite effects, 303, 371, 380, 381, 487
  behavior effects, 190–191, 195, 303–304
  disposition, 65
  dopamine releasing effects, 166, 169
  effect on glial cells, 347
  effect on monoamine uptake sites, 335
  electrophysiological effects, 169–170
  long-term effects, 229–230, 303–304
  mechanism of action, 487
  metabolism, 65
  neurotoxicity, 299–300, 307, 308, 380
  pharmacokinetics
    animals, 64–65, 69, 166
    humans, 63–64, 69
  recreational use, 380
  serotonergic system effects, 325, 327, 329, 330
  serotonin releasing effects, 166–169, 188, 190–191, 380
  side-effects, 304, 306
  stereoselectivity, 166, 167
  structure, 64
  tolerance, 303, 306
  withdrawal effects, 306
Flavin monoxygenase system, amphetamine metabolism, 44, 54–55, 60, 68
Fluoxetine
  combination therapy with 3,4-methylenedioxymethamphetamine, 185, 381
  serotonin uptake inhibition, 185, 194, 223, 344, 374

G
Glial fibrillary acidic protein
  amphetamine-induced effects, 346–348
  function, 346
  synthesis, 346
Glutamate
  decarboxylase, amphetamine effects, 273

Glutamate (*continued*)
  release, 102–103, 139–140, 286–287
  role in amphetamine neurotoxicity, 283–285

H

Hallucinogens
  aromatic ring substitution, 24–29
  conformation effect on serotonin receptor binding, 25–27
  mechanism of action, 33
  nomenclature, 22–23
  N substitution, 23–24
  side chain modification, 29–33
  stereoselectivity, 30–32, 34
Haloperidol
  dopamine antagonist, 481
  methamphetamine psychosis treatment, 432
Homovanillic acid
  accumulation, amphetamine effects, 93–94, 133, 158–159, 168–169
  dopamine metabolite, 89
p-Hydroxyamphetamine, neurotoxicity, 280
6-Hydroxydopamine, neurotoxicity, 481
5-Hydroxyindoleacetic acid, levels, amphetamine effects, 210–212, 269, 274, 376–377
Hyperthermia, effect on amphetamine neurotoxicity, 285, 349

I

Ice, *see* Methamphetamine
Individual variance, amphetamine response
  behavior, 122–126, 482–483
  dopamine release, 128
  factors affecting
    behavioral history, 484–485
    genetics, 483
    sex, 484
  psychosis, 390, 392–395, 410, 411

Isopropylidenedioxyamphetamine, potency, 19–20

J

Japan, methamphetamine use
  cocaine replacement, 477
  demographics of abuser
    age, 468–470, 477
    continuous abuser
      dealers, 473
      duration, 473–474
      motives, 473
    criminal record, 471
    education, 470
    first-time abuser
      age, 471
      introducing person, 471–472
      motives, 471
      prior history of abuse, 470–471
    sex, 468
  dependent use
    compulsive, 424
    cyclic, 423–424, 426–427
    daily high-dose, 424
    introspective abstinence, 424–425
    regular, 423
    single high-dose, 424
    stereotypic behavior, 427
  epidemics, 415, 416–417, 422, 434, 439
  history
    1945 to 1951, 461–462
    1950s, 416–417, 462–463
    1957 to 1969, 463–464
    1970 to 1974, 464–465
    1975 to 1981, 465–466
    current use, 466
    pre-war, 460
    World War II, 416, 459, 461
  social factors, 467, 476
  gang smuggling, 476–477
  medicinal use, 421–422
  mental disorders associated with
    acute syndrome, 425–426
    clinical types, 429
    delusions, 429
    dependence syndrome, 426–428

diagnosis, 428–429, 433
duration, 475–476
hallucinations, 428, 429, 431
interpretation, 433–434
prevalence, 475
psychosis, 415–416, 417–420, 428–430, 433–434, 474–476
relapse phenomenon, 431–433
residual syndrome, 430–431, 432
treatment, 431, 432–433
occasional use, 422–423
production and supply, 420–421, 467, 477

L

Look-alike speed, composition, 444–445
(+)-Lysergic acid diethylamide
classification, 23, 33
locomotion effects, 183, 185
startle habituation effects, 191, 193

M

Magnetic resonance imaging, in study of drug dependence, 487
Mescaline
potency, 24, 177
structure, 23
Methamphetamine
availability, 444, 445
dependent use
compulsive, 424
cyclic, 423–424, 426–427
daily high-dose, 424
introspective abstinence, 424–425
regular, 423
single high-dose, 424
stereotypic behavior, 427
dopamine releasing effects, 270, 324
glutamine releasing effects, 286–287
history of development, 441
ice
characteristics, 453
Hawaii epidemic, 452–454
production, 453–454
routes of administration, 453

Japan
abuse, 417, 422, 434, 439, 459–477
medicinal use, 421–422
military use, 461
production, 420–421
long-term effects, 272–273, 288
mental disorders associated with
acute syndrome, 425–426
clinical types, 429
delusions, 429
dependence syndrome, 426–428
diagnosis, 428–429, 433
hallucinations, 428, 429, 431
psychosis, 394, 415–416, 417–420, 428–430, 433–434
relapse phenomenon, 431–433
residual syndrome, 430–431, 432
treatment, 431, 432–433
worldwide interpretation, 433–434
metabolism, species differences, 53–54
N demethylation, 54–55
neurotoxicity, 227, 270, 273, 299–300, 307–308
γ-aminobutyric acid role, 287, 289
behavior effects, 301, 305
dopamine role, 275–277, 288
glutamate role, 283–285, 286–287, 289
motor function effects, 301–302, 305
serotonin axon loss resulting from, 325
tolerance of individuals, 287–288
occasional use, 422–423
pharmacokinetics, 54, 69
potency, 6
prevalence
emergency room episodes, 449–450
use in the United States, 447, 451–454
ring hydroxylation, 55
serotonin releasing effects, 170, 272
synthesis, 444, 445, 454
tryptophan hydroxylase inhibition, 271–272, 273

p-Methoxyamphetamine
  activity, 152–153
  catecholamine releasing effects, 6, 53
  recreational use, 379
  serotonin releasing effects, 10, 153, 379–380
p-Methoxymethamphetamine
  neurotoxicity, 379
  recreational use, 379
3-Methoxy-4-methylamphetamine, serotonin releasing effects, 17, 21
3-Methoxytyramine
  formation from dopamine, 89–90
  levels, amphetamine effects, 93, 133
N-Methylamphetamine, *see* Methamphetamine
N-Methyl-D-aspartate, tryptophan hydroxylase activity effects, 285–286
N-Methyl-1-(1,3-benzodioxol-5-yl)-2-butanamine
  behavioral effects, 185, 187–188, 192, 193, 194, 195–196
  catecholamine releasing effects, 185
  neurotoxicity, 379
  potency, 178
  recreational use, 379
3,4-Methylenedioxyamphetamine
  behavioral effects, 196
  development, 378
  catecholamine releasing effects, 6, 156–159
  mechanism of action, 11–12
  metabolism, 59–62
  N substitution effects, 14–15
  neurotoxicity, 379
  pharmacokinetics, 59
  potency, 154, 178
  receptor binding, 154–156
  recreational use
    complications, 378
    history, 378–379
  serotonin axon transport changes, 338
  serotonin releasing effects, 12–13, 157–159, 178, 274
  stereoselectivity, 11–12, 14, 154–155, 157–158, 378
3,4-Methylenedioxy-N-ethylamphetamine
  potency, 178, 379
  recreational use, 379
3,4-Methylenedioxymethamphetamine
  α-ethyl homolog, 15–16, 178
  behavioral effects
    antinociception, 195
    Behavioral Pattern Monitor results, 181–185
    drug discrimination, 195–196
    fluoxetine inhibition and, 185, 381
    hyperactivity, 181, 187
    isolation calling rate deficit, 302
    locomotor activity, 180–185
    memory, 306
    mood, 372–373
    morphine analgesia response, 302, 306
    schedule-controlled behavior, 196
    serotonin syndrome, 179–180
    startle habituation, 192, 193, 194–195
    stereotyped, 180
    tolerance, 196–197
  conformation, 154
  demethylenation, 61–62
  development, 12–13
  dopamine releasing effects, 6, 156–158
  effect on glial cells, 346–347
  effect on monoamine uptake sites, 335, 336
  electrophysiological effects, 164–165
  hydroxylation, 62
  long-term effects, 229–230
  mechanism of action, 373–374
  metabolism, 59–62, 279
  neurotoxicity, 161, 299–300, 302–303, 305–306, 376–377, 381
  nitrogen
    alkylation, 154
    demethylation, 60–61
  pharmacokinetics, 59, 69
  potency, 11, 178
  psychotherapy application, 372, 373, 374, 381

receptor binding, 154–156
recreational use
  consequences
    medical complications, 376
    neuropsychiatric difficulties, 375
  patterns, 372, 374–375, 380–381
  serotonergic system effects
    axons, 325, 327, 332, 341–344
    cell bodies, 330
serotonin receptor interaction, 13, 178, 179, 186–187, 197, 198, 200–201
serotonin releasing effects, 13, 16, 19, 157–158, 178–180, 185–188, 373–374
stereoselectivity, 13–14, 154–155, 157–158, 160, 161–162
tryptophan hydroxylase activity effects, 161–162
3,4-Methylenedioxy-5-methoxyamphetamine, serotonin uptake inhibition, 17
1-Methyl-4-phenyltetrahydropyridine
  behavioral effects, 298
  neurotoxicity, 334
p-Methylthioamphetamine, serotonin releasing effects, 18
Microdialysis
  amphetamine self-administration studies, 249–250
  intracerebral techniques, 90–91
MK-801
  effect on body temperature, 285
  glutamate receptor antagonist, 227
  protection against amphetamine neurotoxicity, 227–228, 283–285
Monitoring the Future Survey
  description, 446, 450–451
  limitations, 451
  target population, 446
Monoamine oxidase
  amphetamine inhibition, 44, 65–66, 87, 89, 210, 221, 324
  clorgyline inhibition, 95, 133

N

Na^+, K^+-ATPase, dopamine release role, 85, 96
Narcolepsy, amphetamine therapy, 390, 441
National Household Survey on Drug Abuse
  description, 446
  limitations, 447
  target population, 446
Neurotensin, release, amphetamine effects, 99, 274
Neurotoxicity
  aging effects, 486
  analysis techniques
    anatomic structure, 316, 317–318
    histological, 317, 324, 341
    immunocytochemistry, 317, 324, 340
    staining, 317, 324, 340, 344–346
  irreversibility, 486
  mechanism, 481
  phases, 352–356
  treatment, 486
Noradrenaline, amphetamine self-administration behavior role, 254–255
Norepinephrine, release, amphetamine effects, 4–5, 9, 85–86, 100–101, 137, 138
Norfenfluramine
  catecholamine releasing effects, 166–167
  metabolism, 166, 167
Nucleus accumbens, self-administration behavior role, 247–248, 252, 258, 259, 260

P

Parkinson's disease
  amphetamine therapy, 441
  dopamine role, 299, 307
Perseveration, drug-induced, 255–258
Peyote, mescaline extraction, 378
Phentermine
  hydroxylation, 56–57
  metabolism, 56–57
  pharmacokinetics, 56, 69
Phenylpropanolamine, see Ephedrine

Pimozide
  dopamine antagonist, 245
  effect on amphetamine self-administration, 245
Positron emission tomography, in study of drug dependence, 487
Prefrontal cortex, medial, self-administration behavior role, 248, 252
Pseudephedrine, see Ephedrine
Psychosis, amphetamine
  bipolar disorder and, 396
  case histories, 387–388, 399–401, 404–405
  characteristics
    confusional states, 406–407, 419
    hallucinogenic, 405–406
    nonparanoid presentations, 406–407
    paranoid, 405–406
    schizophrenic, 404–405
    thought disorder, 408–409
  classification, 419
  clinical course, 418–419
  experimental induction, 389
  factors influencing
    dose, 394, 397–398, 407
    duration of use, 398–399
    previous abuse, 399–400
  individual variance in users, 390, 392–395, 410–411
  models, 409–410
  naturally occurring cases, 389
  persistent, 402–404
  prevalence, 387, 388, 396, 397, 475
  schizophrenia and, 395, 397, 401, 402–405, 410, 420
  sensitization, 392–395
  spontaneous, 400–402
  symptoms, 387–388, 393, 407–408, 476, 480

S

Schizophrenia
  amphetamine induction, 395, 420
  delusion characteristics, 404–405
  dopamine role, 299
  prevalence among amphetamine users, 395, 401, 420
Sensitization, amphetamine, 116, 120, 125, 131, 134, 249–250, 392–395, 479, 485–486
Serotonin
  amphetamine effects
    axonal transport, 337–339
    release, 4–5, 9–21, 101, 137, 138, 177–180, 185–188, 190–191, 197–198, 306
    uptake sites, 335–337
  axons containing
    morphology, 319–320
    projection origin, 320–321
    reinnervation, 348–351
    terminal distribution in forebrain, 318–319
    type vulnerability to amphetamines, 357–358
  mood role, 299
  role in amphetamine self-administration, 252–254
  uptake
    carrier substrates, 221
    inhibitor prevention of depletion, 224–225
Serotonin receptors
  drug discrimination role, 196
  electrophysiology, 164–165
  hallucinogen binding, 25–27
  3,4-methylenedioxymethamphetamine interaction, 13, 178, 179, 186–187, 197, 198, 200–201
Serotonin syndrome, characteristics, 179–180
Silver staining, amphetamine neurotoxicity evaluation, 317, 324, 340, 344–346
Speed, look-alike, see Look-alike speed
Startle habituation
  amphetamine effects, 192–195
  defined, 191–192
  serotonin agonist effects, 192–195
  serotonin receptor role, 193–194
STP, see 2,5-Dimethoxy-4-methylamphetamine

T

Tolerance, amphetamine, 196, 197, 287–288, 303, 306, 391–392, 479, 486

2,4,5-Trihydroxyamphetamine, neurotoxicity, 279

2,4,5-Trimethoxyamphetamine, activity, 26, 29

2,4,6-Trimethoxyamphetamine, activity, 28–29

3,4,5-Trimethoxyamphetamine
  origin, 378
  recreational use, 378
  structure, 23

Tryptophan hydroxylase
  amphetamine effects, 159–163, 170–171, 210, 212, 230, 270, 271–272, 273–275, 277, 278, 280–283, 323–324, 329, 330, 343
  cofactor, 280
  inactivation, long-term, 282
  N-methyl-D-aspartate effect, 285–286
  oxidative inactivation, 161–163, 170–171, 280, 282–283

Tyrosine hydroxylase
  activation, 132
  amphetamine effects, 159, 270, 272, 274–277, 280, 324